TRACE AMINES AND NEUROLOGICAL DISORDERS

Companion Web Site:

http://booksite.elsevier.com/9780128036037/

TRACE AMINES AND NEUROLOGICAL DISORDERS
Tahira Farooqui and Akhlaq A. Farooqui

ELSEVIER

ACADEMIC
PRESS

TRACE AMINES AND NEUROLOGICAL DISORDERS

POTENTIAL MECHANISMS AND RISK FACTORS

Edited by

TAHIRA FAROOQUI
Department of Entomology, The Ohio State University,
Columbus, OH, United States

AKHLAQ A. FAROOQUI
Department of Molecular and Cellular Biochemistry, The Ohio State University,
Columbus, OH, United States

AMSTERDAM • BOSTON • HEIDELBERG • LONDON
NEW YORK • OXFORD • PARIS • SAN DIEGO
SAN FRANCISCO • SINGAPORE • SYDNEY • TOKYO
Academic Press is an imprint of Elsevier

British Library Cataloguing-in-Publication Data
A catalogue record for this book is available from the British Library.

Library of Congress Cataloging-in-Publication Data
A catalog record for this book is available from the Library of Congress.

ISBN: 978-0-12-803603-7

For Information on all Academic Press publications
visit our website at https://www.elsevier.com/

Working together
to grow libraries in
developing countries

www.elsevier.com • www.bookaid.org

Publisher: Mara Conner
Acquisition Editor: Melanie Tucker
Editorial Project Manager: Kristi Anderson
Production Project Manager: Chris Wortley
Designer: Victoria Pearson

Typeset by MPS Limited, Chennai, India

THE FUTURE BELONGS TO THOSE WHO BELIEVE IN THE BEAUTY OF THEIR DREAMS.

——Eleanor Roosevelt

Dedication

We dedicate this book to *"all the teachers"* who have made a profound difference in our lives by teaching with passion, making us able to differentiate between right and wrong, and molding our lives in the right direction.

Contents

I

INTRODUCTION AND DESCRIPTION OF TRACE AMINES AND TRACE AMINE-ASSOCIATED RECEPTORS

II

TRACE AMINES AND OLFACTION

13. Trace Amine-Mediated Olfactory Learning and Memory in Mammals and Insects: A Brief Comparative Review

T. FAROOQUI

14. Octopaminergic and Tyraminergic Signaling in the Honeybee (*Apis mellifera*) Brain: Behavioral, Pharmacological, and Molecular Aspects

W. BLENAU AND A. BAUMANN

15. Octopamine and Tyramine Signaling in Locusts: Relevance to Olfactory Decision-Making

Z. MA, X. GUO AND L. KANG

III

TRACE AMINES AND NEUROLOGICAL DISORDERS

IV

PERSPECTIVE

List of Contributors

A. Accorroni Scuola Superiore St. Anna, Pisa, Italy

A. Ahmad Department of Zoology, Maulana Azad National Urdu University, Hyderabad, India

S.H. Al Rokayan Research Chair for Biomedical Applications of Nanomaterials, King Saud University, Riyadh, Saudi Arabia

A.S. Alhomida Department of Biochemistry, King Saud University, Riyadh, Saudi Arabia

G. Andersen German Research Center for Food Chemistry, Leibniz Institute, Freising, Germany

A. Baumann Forschungszentrum Juelich, Institute of Complex Systems – Cellular Biophysics (ICS-4), Juelich, Germany

M.D. Berry Department of Biochemistry, Memorial University of Newfoundland, St. John's, NL, Canada

H. Biebermann Institute of Experimental Pediatric Endocrinology, Charité University Medicine, Berlin, Germany

W. Blenau Institute of Zoology, University of Cologne, Cologne, Germany

L. Carvelli Department of Biomedical Sciences School of Medicine & Health Sciences, University of North Dakota, Grand Forks, ND, United States

S. Espinoza Department of Neuroscience and Brain Technologies, Italian Institute of Technology, Genova, Italy

A.A. Farooqui Department of Molecular and Cellular Biochemistry, The Ohio State University, Columbus, OH, United States

T. Farooqui Department of Entomology, The Ohio State University, Columbus, OH, United States

E.A. Gozal Department of Physiology, Emory University School of Medicine, Atlanta, GA, United States

X. Guo Beijing Institutes of Life Sciences, Chinese Academy of Sciences, Beijing, China; State Key Laboratory of Integrated Management of Pest Insects and Rodents, Institute of Zoology, Chinese Academy of Sciences, Beijing, China

S. Hart Department of Biochemistry, Memorial University of Newfoundland, St. John's, NL, Canada

Y. Hashiguchi Department of Biology, Osaka Medical College, Takatsuki City, Osaka, Japan

S. Hochman Department of Physiology, Emory University School of Medicine, Atlanta, GA, United States

K. Ikemoto Department of Psychiatry, Iwaki Kyoritsu General Hospital, Iwaki, Japan

L. Kang Beijing Institutes of Life Sciences, Chinese Academy of Sciences, Beijing, China; State Key Laboratory of Integrated Management of Pest Insects and Rodents, Institute of Zoology, Chinese Academy of Sciences, Beijing, China

N. Khajavi Institute of Experimental Pediatric Endocrinology, Charité University Medicine, Berlin, Germany

H.A. Khan Department of Biochemistry, King Saud University, Riyadh, Saudi Arabia

G. Kleinau Institute of Experimental Pediatric Endocrinology, Charité University Medicine, Berlin, Germany

J. Köhrle Institute of Experimental Endocrinology, Charité University Medicine, Berlin, Germany

D. Krautwurst German Research Center for Food Chemistry, Leibniz Institute, Freising, Germany

A. Laurino Department of Neuroscience, NEUROFARBA, Section of Pharmacology, University of Florence, Florence, Italy

A. Ledonne Department of Experimental Neurosciences, Santa Lucia Foundation, Rome, Italy

D. Leo Department of Neuroscience and Brain Technologies, Italian Institute of Technology, Genova, Italy

Z. Ma Beijing Institutes of Life Sciences, Chinese Academy of Sciences, Beijing, China; State Key Laboratory of Integrated Management of Pest Insects and Rodents, Institute of Zoology, Chinese Academy of Sciences, Beijing, China

N.B. Mercuri Department of Experimental Neurosciences, Santa Lucia Foundation, Rome, Italy; Department of System Medicine, University of Rome "Tor Vergata", Rome, Italy

A.D. Mosnaim Department of Cellular and Molecular Pharmacology, Neuroimmunopharmacology Laboratory, The Chicago Medical School at Rosalind Franklin University of Medicine and Science, Chicago, IL, United States

C.-U. Pae Department of Psychiatry, The Catholic University of Korea College of Medicine, Seoul, Republic of Korea; Department of Psychiatry and Behavioral Sciences, Duke University Medical Center, Durham, NC, United States; Department of Psychiatry, Bucheon St. Mary's Hospital, Wonmi-Gu, Bucheon, Kyeonggi-Do, Republic of Korea

A. Pryor Department of Biochemistry, Memorial University of Newfoundland, St. John's, NL, Canada

L. Raimondi Pharmacology Unit, Department of Neuroscience, Drug Sciences and Child Health, University of Florence, Florence, Italy

T. Roeder Department of Zoology, Molecular Physiology, Kiel University, Kiel, Germany

S.I. Sherwani Department of Internal Medicine, Division of Pulmonary Medicine, Dorothy M. Davis Heart and Lung Research Institute, The Ohio State University College of Medicine, Columbus, OH, United States

Q. Ullah Department of Chemistry, Maulana Azad National Urdu University, Hyderabad, India

L.J. Wallace Division of Pharmacology, College of Pharmacy, The Ohio State University, Columbus, OH, United States

S.-M. Wang Department of Psychiatry, The Catholic University of Korea College of Medicine, Seoul, Republic of Korea; International Health Care Center, Seoul St. Mary's Hospital, The Catholic University of Korea College of Medicine, Seoul, Republic of Korea; Department of Psychiatry, Seoul St. Mary's Hospital, Seocho-Gu, Seoul, Republic of Korea

M.E. Wolf International Neuropsychiatry Consultants, Highland Park, IL, United States

R. Zucchi Department of Pathology, University of Pisa, Pisa, Italy

Foreword

It is a true honor for me to write this Foreword for Tahira Farooqui and Akhlaq Farooqui's book *Trace Amines and Neurological Disorders: Potential Mechanisms and Risk Factors*, which comes at an opportune moment. Just as long journeys invite the avid traveler to reflect on the wonders and awe-inspiring places discovered along the way, travels through the history of science often lend themselves to contemplation and jubilation. More than a hundred years after Polish biochemist, Marceli Nencki, first isolated phenylethylamine from decomposing gelatine, and following many decades of a tenacious quest for the neurobiological substrate of the mysterious neurophysiological actions of trace amines, a group of naturally occurring and endogenous compounds chemically related to phenylethylamine, research into this most intriguing aminergic system has been burgeoning with ever-increasing momentum. Long regarded as "false transmitters" and believed to exert no more than indirect sympathomimetic actions, this collection of endogenous amines is today regarded as protean neurotransmitters, featuring both transmitter and neuromodulatory actions. Both the topographical distribution of trace amines in the brain and their signaling transduction pathways are currently undergoing extensive mapping and delineation. As new breakthroughs occur in this fast-moving field, new avenues are opened for understanding normal brain function and unraveling the pathogenesis of neurological disease. Today, for us, travelers through the history of neurological science, it is a fitting moment to reflect and rejoice on what we have achieved. This remarkable book is in many ways a celebration of the work of hundreds of physiologists, biochemists, neurologists, and neuroscientists who have delved into the significance of the trace amines, carving out a new understanding of the brain and its functions.

In every journey, the traveler finds historic landmarks, places where history was made and paradigms shifted. If I had to choose one moment in time when a fundamental change of ideas took place, invigorating a renewed interest in the long-known and elusive trace amines, then the turn of the 21st century would be that moment. Albeit another trace amine, octopamine, had previously met the status of bona fide neurotransmitter with the identification of octopamine receptors in *Aplysia* neurons, the historic landmark of this decades-long quest was the serendipitous discovery of the trace aminergic receptor system in vertebrates, owing to the work of Beth Borowsky and her collaborators. In the search for additional subtypes of the serotonin receptor class, a large family of G-protein-coupled receptors was cloned from human and rodent tissue. Such a family of receptors exhibited a high degree of within-species homology but low similarity across orthologs, thereby suggesting that these receptors had been rapidly evolving in the vertebrate brain. Of the newly cloned mammalian receptors, some were shown to be bound and directly activated by trace amines, initiating components of intracellular signal transduction *cascades*. To account for the orphan receptors that were insensitive to trace amines, Lothar Lindemann and collaborators proposed a uniform nomenclature, which has been widely adopted since, with this assortment of receptors being currently referred to as the trace amine-associated receptor (TAAR) family. In addition to the canonical olfactory receptors, evidence showed that many of these receptors played a fundamental role as vertebrate chemosensory receptors. Indeed, the presence of TAARs in the olfactory epithelium of unique subsets of sensory neurons seemed to allow the identification of many odorants, including a number of volatile amines carrying key social information and, possibly, pheromones, thus guiding aversion to, and attraction towards, salient external stimuli. Fittingly, an important part to this book is devoted to the exploration of the prominent role of TAARs in olfactory behavior and olfactory learning.

Most notably, in the course of these ground-breaking investigations it became apparent that one such receptors for trace amines, TAAR1, was exceptionally unique, for only TAAR1 displayed affinity for all of the trace amines and only the *TAAR1* gene was found to be phylogenetically conserved in the brain of all mammalian species studied thus far, including human. Such remarkable discoveries shed new light on this field and brought TAAR1 under the spotlight in the realm of neurology and neuroscience. Although traces amines have long been known to regulate, and be themselves susceptible to regulation by, associated neurotransmitters, evidence of direct interactions between aminergic systems and receptors for trace amines had been lacking. Mapping studies indicated that *Taar1* expression was enriched throughout the limbic and ascending aminergic pathways, encompassing the ventral tegmental area/substantia nigra and dorsal raphe nucleus, regions that contain the dopamine- and serotonin-producing neurons, respectively, and was therefore strategically positioned to regulate the activity of these transmitter systems in the terminal regions of the basal ganglia and limbic brain. Moreover, transgenic mice models revealed that *Taar1* deletion led to elevated discharge rate of dopamine and serotonin neurons, suggesting that TAAR1 activation acted as a brake for aminergic neurotransmission. Soon thereafter, the advent of the first selective agonists and antagonists at TAAR1 contributed to reinforce the notion that TAAR1 might be exquisitely sensitive to fluctuations in aminergic transmitter signaling. Having focused on the dopamine transporter for a considerable time, I was then actively searching for novel pharmacotherapeutic targets for addiction-spectrum disorders and I first learnt about these new compounds through my colleagues at Hoffmann-La Roche Ltd. Indeed, such a discovery was also a turning point in my personal career. Using these newly developed pharmacological tools, the experiments of my colleagues lent further credence to the thesis that TAAR1 activation could indeed modulate aminergic transmission in complex fashion, in fact, producing contrasting effects on dopamine and serotonin neuronal discharge depending on the level of agonist activity at TAAR1. My own experiments showed that TAAR1 activation could reduce a range of drug-related behaviors, including psychomotor stimulant self-administration and relapse to drug-seeking. The overall conclusion of a large body of freshly accrued neurophysiological data was that TAAR1 seemed to be constitutively active and/or tonically activated by endogenous ligands, such that partial agonism would result in antagonistic-like effects. If this held true then it would appear that the use of a partial agonist could be more advantageous in situations where neurochemical imbalances lead to deficient or excessive TAAR1 stimulation, providing a means to "stabilize" TAAR1 activity. Furthermore, through induction of pacemaker activation, partial TAAR1 agonism could potentially promote "normalization" of dopamine and serotonin neuron discharge. What is more, in addition to altering the major aminergic pathways, trace amines have been shown to modulate neuronal signaling mediated by other important neu-rotransmitters, such as glutamate, γ-aminobutyric acid, and acetylcholine, and close interac-tions between TAAR1 and these neurotransmitters systems are beginning to be identified.

Why are these discoveries so relevant for the future of neurology? Progress in developing effective treatments for conditions such as Parkinson's disease, schizophrenia, attention deficit hyperactivity disorder, and addiction has been hindered by the nonspecific side effects asso-ciated with chronic exposure to a wide range of medications based on the pharmacology of classical amines. The prime neuroanatomical location of TAAR1 and its distinctive ability to regulate aminergic neurotransmission both lend support to the hypothesis that this receptor could be a promising candidate for the development of new-generation pharmacotherapies in neurology. Impressive in its reach, a collection of thought-provoking, meticulous and thor-oughly enjoyable reviews deal in this book with the emerging role of trace amines and TAAR1 in the pathophysiology and therapeutics of neurological disease.

The history of the trace amines has been a journey with unexpected detours and sudden accelerations but the field has flourished in ways that few could have predicted only a few years ago. Many aspects of the neurobiology and neuropharmacology of the trace amines and its receptors remain elusive but the advent of new technological advances in areas such as superresolution microscopy, chemogenetics and optogenetics, neuroimaging, and fine-grained behavioral assessment assures us that innovation and further excitement lie ahead, as we push the frontiers of our current understanding. Tahira Farooqui and Akhlaq Farooqui are to be effusively congratulated on their clear vision and sense of mission. I am most confident that this important book will stand as a roadside marker for generations of neurologists and neuroscientists to come and anyone seeking to gain a fresh perspective on the, finally, not-so-mysterious world of the trace amines.

J.J. Canales DPhil
University Reader in Behavioural Neuroscience
University of Leicester

Foreword

It is indeed an honor and privilege for me to accept the Editors' invitation to compose a Foreword for their latest book, *Trace Amines and Neurological Disorders: Potential Mechanisms and Risk Factors*, and thereby, be the first to welcome you to this rapidly growing and exciting area of biomedical research.

With the publication of this volume, the Editors have successfully accomplished what has never been done for this topic before: compiling original contributions from leading research scientists working around the world into a broad, cutting-edge reference text de novo; that is, without the benefit of a prior meeting on the subject.

As far as I can ascertain, four books have had as their focus the so-called "trace amines" and all derive from the proceedings of previous meetings. The first: *Trace Amines and the Brain* (Usdin E and Sandler M, editors, Marcel Dekker, Inc., New York, New York) appeared in 1976 and documents the proceedings of a study group held the previous year during the 14th Annual Meeting of the American College of Neuropsychopharmacology in San Juan, Puerto. The publication of this volume is historic for several reasons, not the least being it occasioned the coining of the exceedingly unfortunate and misleading moniker for which these extremely important biogenic amines are still most widely known, sparking a debate about their nomenclature among those working in the field that continues to this day.

Despite the disagreements that ensued (and continue) over the most appropriate way of referring to these most fundamental of biogenic aryl-alkyl amines: β-phenylethylamine (PEA), *p*-tyramine (TYR), octopamine (OCT), synephrine, and tryptamine the editors of the other three books on the subject all chose to embrace the catchy new term in their titles: *Neurobiology of the Trace Amines − Analytical, Physiological, Pharmacological, Behavioral and Clinical Aspects* (Boulton AA, Baker GB, Dewhurst WG, Sandler M, editors, Human Press, Clinton New York, 1984; based on the proceedings of the 9th International Society for Neurochemistry meeting held in Vancouver, BC Canada in 1983); *Neuropsychopharmacology of the Trace Amines − Experimental & Clinical Aspects* (Boulton AA, Maitre L, Biecke PR, Riederer P, editors, Human Press, Clinton New Jersey, 1985; based on the proceedings of the 2nd Trace Amines Symposium held at Weitenberg Castle Tübingen, Germany in 1985); and *Trace Amines − Comparative & Clinical Neurology*, (Boulton AA, Juorio AV, Downer RGH, editors, 1988. Humana Press, Clifton, New Jersey; based on the proceedings of a satellite meeting of the 11th International Society for Neurochemistry held on Isla Margarita, Venezuela in 1987).

Reading the preambles accompanying each of these four earlier compendia one cannot help but detect the editors' growing excitement about the imminent transformative impact they expected these essential amines were sure to have on neuroscience, in general and psychiatry, in particular. Unfortunately, with the benefit of hindsight, we now know that instead of ushering in a new period of promise and prominence, the volume published in 1988 marked the beginning of the end of an era that had its modern inception in 1876 Bern, Switzerland.

Two years after receiving his PhD in chemistry from the prestigious Prussian University of Berlin in 1870, Polish-born physician and chemist Wilhelm Marceli Nencki accepted a position at the University of Bern, Switzerland, where he pursued the identification of microbial fermentation products. In the course of his research Nencki discovered a novel ptomaine in the liquors generated from microbial decomposition of both gelatin and egg white in the presence of pancreatic extract. Nencki determined the chemical composition of his new compound to be $C_8H_{11}N$ and initially suggested it was a novel collidine. However, when Schulze and Barbieri (1881) demonstrated phenylalanine can be decarboxylated by microbial action Nencki (1882) conceded PEA is the most likely structure of his compound; a conclusion confirmed by Spiro in 1901.

Soon thereafter, in a collaborative effort, the London-based Wellcome Research team composed of chemists George Barger and George S. Walpole, physician and future Nobel laureate Henry H. Dale, and Cambridge pharmacologist Walter E. Dixon conducted the first in vivo physiologic experiments using pure PEA and TYR. As the pressor action produced by these primary amines was adrenaline-like the authors coined the term "sympathomimetic" to describe them (1910). During the 75 years that followed, interest in these compounds continued to grow, attracting the attention of many leading neuroscientists of the time including Nobel laureate Julius Axelrod, Alan A. Boulton, William R. Martin, Menek Goldstein, and James A. Nathanson.

However, by the late 1980s and early 1990s all of the vertebrate catecholamine and serotonin receptor-coding genes had been discovered and despite the cloning of invertebrate receptors for TYR and OCT, genes coding for vertebrate trace amine receptors remained elusive. Unfortunately, this lacuna was interpreted by most grant reviewers, journal editors, and scientific policymakers of the day as validating the opinion of a few influential individuals who advocated the notion these amines were "false" transmitters and nothing more than trivial (ie, "trace") metabolites of bona fide neurotransmitters. As a consequence, funding dried up and the field came to a halt—until 2001.

In 2001 two publications (Boroswsky et al.; Bunzow et al.) chronicled the discovery and characterization of vertebrate G-protein-coupled receptors that are activated by trace amines. The significance of this development was not lost on the journals' editors who commissioned commentaries to accompany each report. In their remarks the authors of these perspectives emphasized that the existence of receptors for the trace amines elevates these compounds to neuromodulatory/neurotransmitter status and consequently deserving of vigorous and immediate investigation.

However, inertia is hard to overcome and substantive public granting agency funding of proposals in the United States and Canada have yet to materialize. Fortunately, despite a lack of financial support, a few North American research groups have been able to continue their studies of these fundamental amines and their receptors in brain and spinal cord, their metabolism and involvement in maintaining metabolic homeostasis as well as appetitive, compulsive, and psychotic behaviors. Some of this exciting research can be found among the chapters comprising this book. But by far the field has received the most interest and support among European scientists, funding agencies, and pharmaceutical companies.

Recently the German government invested a considerable sum of money to underwrite the formation of an integrated international consortium of collaborators interested in emergent areas of thyroid hormone research with two of three funded areas being the behavior, molecular biology, pharmacology, and physiology of trace amine receptors and their ligands (especially 3-iodothyronamine) the recently discovered endogenous thyroid hormone-like TAAR1 agonist. Some of this groundbreaking research is included in this volume. Others working in Germany and Italy have contributed chapters to this book describing their exploration of TAAR1-mediated signaling with respect to dopamine neuron signaling in a variety of contexts including normal and abnormal appetitive behavior, decision-making, movement control, and mental illness such as psychosis and depression. In addition, evidence continues to rapidly accumulate in support of the interpretation that these receptors serve fundamental alerting functions in cells of the immune and cardiovascular systems. International interest in the field is also growing, made evident by the several chapters contributed by scientists from China, India, Japan, Korea, and Saudi Arabia on such diverse topics, such as methods of trace amine analysis and synthesis, receptor evolution, neuropsychopharmacology, and invertebrate olfaction.

By assembling a tome of such depth and breadth of scope the editors have generated a unique and scholarly work on the topic that will be widely referred to and cited for years to come. Perhaps more importantly however, this book serves as a stentorian announcing to the world that these ultimate amines and their receptors have finally "arrived" and that the time to research their roles in health and disease is at hand.

In honor of their unwavering dedication and fundamental contributions to the field it is with humility and the utmost respect that I dedicate this Foreword to Drs. Alan A. Boulton and Alexander T. "Sasha" Shulgin.

D.K. Grandy MSc, PhD
Portland, Ore, March 2016

Preface

In addition to well-known classical monoamine neurotransmitters (such as dopamine, norepinephrine, and serotonin), a group of endogenous trace amines, including m- and p-tyramine, tryptamine, m- and p-octopamine, β-phenylethylamine, and synephrine, are heterogeneously distributed throughout the mammalian peripheral and central nervous systems. Unlike the classical monoamine neurotransmitters, trace amines are present in very low concentrations. They can be synthesized within parent monoamine neurotransmitter systems. Therefore they are similar to classical monoamine neurotransmitters with regard to their structure, metabolism, and tissue. The exact role of these trace amines in humans is not known. However, it is becoming increasingly evident that trace amines exert their effect by activating Trace Amine-Associated Receptors (TAARs). Trace amines play significant roles in the coordination of biogenic monoamine-based synaptic physiology. At high concentrations, they exert well-characterized presynaptic "amphetamine-like" effects on monoamine metabolism. However, at lower concentrations, they possess postsynaptic modulatory effects, potentiating the activity of other neurotransmitters such as dopamine and serotonin. At present, TAAR1 is the only human receptor that binds with endogenous trace amines. It plays a significant role in brain by regulating dopamine, norepinephrine, and serotonin neurotransmission, as well as also influences immune system. TAAR1 is a negative regulator of dopamine transmission, thus making it a target for neuropsychiatric disorders, such as schizophrenia. TAAR1 also plays a role in the modulation of N-methyl-D-aspartate (NMDA) receptor in the prefrontal cortex (a region critically involved in high-order cognitive processes), implicating the potential for developing TAAR1-based drugs to treat disorders related to aberrant cortical functions. Like dopamine, norepinephrine, and serotonin, trace amines have been implicated in the pathophysiology of many neurological disorders, including phenylketonuria, migraine, attention deficit hyperactivity disorder (ADHD), depression, and schizophrenia. Therefore, understanding of the molecular mechanisms and developing selective agonists and antagonists for TAARs can become a good approach for treating above-mentioned diseases.

This book "Trace Amines and Neurological Disorders: Potential Mechanisms and Risk Factors" focuses on molecular mechanisms underlying trace amine-mediated neurological diseases. It will help in developing new drugs to treat these chronic diseases. Information presented in this edited book would be useful for pharmacology, neuroscience and biology graduate students, teachers and basic science researchers, and drug company workers.

1. This book has 25 chapters, which are organized into four sections: (I) Introduction and Description of Trace Amines and TAARS (see chapters: Trace Amines: An Overview; Methods of Trace Amine Analysis in Mammalian Brain; Synthesis and Neurochemistry of Trace Amines; The Origin and Evolution of the Trace Amine-Associated Receptor

Family in Vertebrates; Differential Modulation of Adrenergic Receptor Signaling by Octopamine, Tyramine, Phenylethylamine, and 3-Iodothyronamine; Effects of Trace Amines on the Dopaminergic Mesencephalic System; Trace Amine-Associated Receptors in the Cellular Immune System; Trace Amines and Their Receptors in the Control of Cellular Homeostasis; Trace Amine-Associated Receptor 1 Modulation of Dopamine System; Trace Amines as Intrinsic Monoaminergic Modulators of Spinal Cord Functional Systems; Trace Amine-Associated Receptors: Ligands and Putative Role in the Central Nervous System; β-Phenylethylamine Requires the Dopamine Transporter to Release Dopamine in *Caenorhabditis elegans*), (II) Trace Amines and Olfaction (see chapters: Trace Amine-Mediated Olfactory Learning and Memory in Mammals and Insects: A Brief Comparative Review; Octopaminergic and Tyraminergic Signaling in the Honeybee (*Apis mellifera*) Brain: Behavioral, Pharmacological, and Molecular Aspects; Octopamine and Tyramine Signaling in Locusts: Relevance to Olfactory Decision-Making), (III) Trace Amines and Neurological Disorders (see chapters: Neurochemical Aspects of Neurological Disorders; Trace Amines and Their Relevance to Neurological Disorders: A Commentary; Trace Amines in Neuropsychiatric Disorders; β-Phenylethylamine-Class Trace Amines in Neuropsychiatric Disorders: A Brief Historical Perspective; Involvement of So-Called D-Neuron (Trace Amine Neuron) in the Pathogenesis of Schizophrenia: D-Cell Hypothesis; 3-Iodothyronamine, a New Chapter in Thyroid Story: Implications in Learning Processes; Trace Amine Receptors and Mood Disorders: Focusing on Depression; Trace Amine-Associated Receptor 1: Implications for Treating Stimulant Drug Addiction; Trace Amines and Their Potential Role in Primary Headaches: An Overview), and (IV) Perspective (see chapter: Perspective and Directions for Future Research on Trace Amines and Neurological Disorders). In Section I, the chapter "Trace Amines: An Overview" elegantly reviews the role of trace amines in vertebrates and invertebrates. Chapter "Methods of Trace Amine Analysis in Mammalian Brain" updates various methods for the estimation of trace amines in the mammalian brain. Chapter "Synthesis and Neurochemistry of Trace Amines" deals with the synthesis and neurochemistry of trace amines. Chapter "The Origin and Evolution of the Trace Amine-Associated Receptor Family in Vertebrates" addresses the origin and evolutionary aspects of trace amine-associated receptor gene family. Chapters "Differential Modulation of Adrenergic Receptor Signaling by Octopamine, Tyramine, Phenylethylamine, and 3-Iodothyronamine," "Effects of Trace Amines on the Dopaminergic Mesencephalic System," "Trace Amine-Associated Receptors in the Cellular Immune System," "Trace Amines and Their Receptors in the Control of Cellular Homeostasis," "Trace Amine-Associated Receptor 1 Modulation of Dopamine System," "Trace Amines as Intrinsic Monoaminergic Modulators of Spinal Cord Functional Systems," "Trace Amine-Associated Receptors: Ligands and Putative Role in the Central Nervous System" offer functional aspects of trace amines affecting different physiological systems, such as differential modulation of adrenergic receptor signaling by multitarget trace amine ligands, effects of trace amines on dopaminergic mesencephalic system, TAARs in the cellular immune system, and trace amines and their receptors in the control of cellular homeostasis, TAAR1 modulation of dopamine system, trace amines as intrinsic monoaminergic modulators of spinal cord functional systems, and TAARS ligands and their putative role in the central nervous system. Chapter "β-Phenylethylamine Requires the Dopamine Transporter to Release Dopamine in *Caenorhabditis elegans*" demonstrates β-phenylethylamine increases synaptic dopamine by interacting with dopamine transporter in *Caenorhabditis elegans*. In Section II, chapter "Trace Amine-Mediated Olfactory Learning and Memory in Mammals and Insects: A Brief Comparative Review" focuses on comparing trace amines-mediated olfactory learning and memory in mammals and insects. Chapters "Octopaminergic and Tyraminergic Signaling in the Honeybee (*Apis mellifera*) Brain: Behavioral, Pharmacological, and Molecular Aspects" and "Octopamine and Tyramine Signaling in Locusts: Relevance to Olfactory Decision-Making" deal with insect model systems, demonstrating octopaminergic and tyraminergic signaling and olfactory behavior in the Honeybee (*Apis mellifera*) and Locusts (*Locusta migratoria*), respectively. In section III, chapter "Neurochemical Aspects of Neurological Disorders" provides neurochemical aspects of neurological disorders to understand these diseases. Chapter "Trace Amines and Their Relevance to Neurological Disorders: A Commentary" then comments on the trace amines and their relevance to neurological disorders, showing depression as a characteristic in most of these diseases. Chapter "Trace Amines in Neuropsychiatric Disorders" reviews the role of trace amines in neuropsychiatric disorders. Chapter "β-Phenylethylamine-Class Trace Amines in Neuropsychiatric Disorders: A Brief Historical Perspective" provides a historical perspective and an update on phenylethylamines in neuropsychiatric disorders. Chapter "Involvement of So-Called D-Neuron (Trace Amine Neuron) in the Pathogenesis of Schizophrenia: D-Cell Hypothesis" explains the D-neuron (trace amine-producing neuron) involvement in the pathogenesis of schizophrenia. Chapter "3-Iodothyronamine, a New Chapter in Thyroid Story: Implications in Learning Processes" provides readers information on the involvement of 3-iodothyronamine (T1AM) that is endowed with prolearning and antiamnesic features. Chapter "Trace Amine Receptors and Mood Disorders: Focusing on Depression" briefly reviews the potential role of trace amine receptors in the development and treatment for mood disorders. Chapter "Trace Amine-Associated Receptor 1: Implications for Treating Stimulant Drug Addiction" focuses on TAAR1 and its potential role in treating stimulant drug addiction. Chapter "Trace Amines and Their Potential Role in Primary Headaches: An Overview" overviews the pathophysiology and genetic factors associated with the cluster and migraine headaches and suggests a potential molecular mechanism underlying these headaches. In Section IV, chapter "Perspective and Directions for Future Research on Trace Amines and Neurological Disorders" provides readers with an in-depth perspective and directions for future research on trace amines and neurological disorders.

It is hoped that "Trace Amines and Neurological Disorders: Potential Mechanisms and Risk Factors" will further enhance young and senior scientists to perform more research on involvement and roles of trace amines in neurological diseases.

T. Farooqui and A.A. Farooqui

Acknowledgments

We thank all the authors of this book who shared their expertise by contributing chapters of a high standard, thus making our editorial task much easier. We are also grateful for the cooperation and patience of Kristi Anderson, Senior Editorial Project Manager; Melanie Tucker, Senior Acquisitions Editor; and Chris Wortley, Book Production Project Manager at Elsevier/Academic Press, for their full cooperation and professional handling of the manuscript.

INTRODUCTION AND DESCRIPTION OF TRACE AMINES AND TRACE AMINE-ASSOCIATED RECEPTORS

Trace Amines: An Overview

T. Roeder

Department of Zoology, Molecular Physiology, Kiel University, Kiel, Germany

INTRODUCTION

Biogenic amines such as epinephrine, norepinephrine, dopamine, serotonin, or histamine act as classical neurotransmitters or neuroactive compounds in the mammalian central nervous system. They are involved in the regulation of countless behaviors and take central roles for the proper functioning of the nervous system in general. Consequently, deregulation of aminergic signaling has been associated with a great number of neurological diseases such as Parkinson's disease, schizophrenia, depression, and many others. Moreover, biogenic amine metabolism is central to our reward system, thus being centrally involved in almost all types of drug action and the development of addiction. Taking this importance into account, synthesis, storage, but also removal of these biogenic amines is tightly regulated in order to keep their concentrations in physiologically relevant ranges. Termination of aminergic signaling is usually achieved by the combined action of specific reuptake systems and their enzymatic degradation. A complex web of enzymatic activities is required to regulate synthesis of these bioactive amines, but it also produces intermediate products and so-called trace amines (TAs) (Fig. 1.1).[1–3]

In order to elicit physiological responses in target cells, sets of highly specific receptor molecules tailored to transmit only information carried by the specific biogenic amines are utilized. This set of receptors for biogenic amines has been used extensively as targets for countless therapeutic interventions using specific receptor agonists and/or antagonists. In

Trace Amines and Neurological Disorders
DOI: http://dx.doi.org/10.1016/B978-0-12-803603-7.00001-X

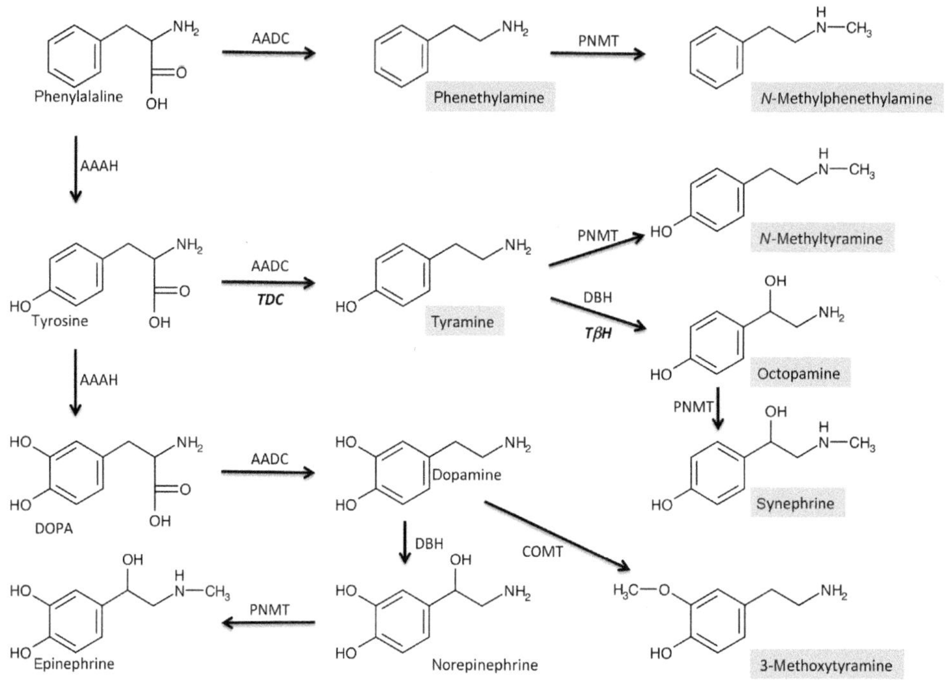

FIGURE 1.1 Biosynthesis pathways that give rise to TAs. Starting with the amino acids phenylalanine and tyrosine, two enzymes, the aromatic L-amino acid decarboxylase (AADC) and the biopterin-dependent aromatic amino acid hydroxylases (AAAH) produce phenethylamine, tyramine (AADC), and L-3,4-diidroxyphenylalanine (L-DOPA, AAAH), respectively. In invertebrates the tyrosine decarboxylase (TDC) produces tyramine from tyrosine. Moreover, the AADC also produces dopamine from L-DOPA. The phenylethanolamine N-methyltransferase (PNMT) catalyzes a number of different reactions including the production of N-methylphenethylamine from phenethylamine. Octopamine is, on the other hand, produced by the dopamine β-monooxygenase (DBH). In invertebrates, this is achieved by the tyramine β-hydroxylase (TβH). DBH is also responsible for the synthesis of norepinephrine from dopamine. Finally, the catechol-O-methyltransferase (COMT) converts dopamine into 3-methoxytyramine. TAs are marked by gray boxes and the enzymes that are required to produce tyramine and octopamine in invertebrates are shown in bold and italics.[1–3]

addition to this handful of biogenic amines that are present at high levels in the mammalian central nervous system, other amines have been described that were found only in very low concentration. Consequently, these amines have been named TAs. As already pointed out, TAs are connected to the classical aminergic transmitter substances as they are usually produced within the same network of enzymes giving rise to these compounds. The very low concentration of TAs and the lack of information about specific receptor molecules led for several decades to the assumption that TAs have no particular transmitter or hormone function in the mammalian nervous system. This situation was changed only recently with the identification of a particular class of G-protein-coupled receptors named *trace amine activated receptors* (TAARs).[4] The original definition of the term TAs was very vague; usually, they are defined as monoamines that fulfill a small set of criteria: (1) they do not act as classical neurotransmitters in vertebrates; (2) they are present in only minute amounts (mostly in the nanomolar range) in the mammalian brain; and (3) they usually have a sympathomimetic activity. Compounds such as phenylethanolamine, octopamine, tyramine, tryptamine, and synephrine are typical TAs (Fig. 1.1). Despite the fact that their particular physiological role was mostly neglected, deregulation of their concentration have been linked to numerous psychotic disorders such as depression, schizophrenia, migraine, or attention deficit hyperactivity disorder. The synthesis pathways that give rise to the different biogenic amines, as well as to the TAs listed above, are shown in Fig. 1.1.[1,2] Interestingly, both, conventional biogenic amines with an unequivocal transmitter function as well as the TAs share the same synthesis network comprising of proteins with the corresponding enzymatic activities. One early explanation for the presence of TAs was that they are unavoidable byproducts of these complex synthesis pathways.

More than 10 years ago, deorphanization of bioamine receptors revealed that a number of different receptors in the mammalian brain react at nanomolar concentrations to TAs. These receptors called *trace amine-associated receptors* (TAARs) revitalized the interest in studying the biology of TAs dramatically.[4] A more detailed characterization of TAARs is found below.

The second highly interesting area of research where TAs have been focused on is their physiological role in invertebrates. Two of these TAs, octopamine, and tyramine, act as highly potent and highly important transmitter compounds in almost all invertebrates. In these animals they take the role of the classical aminergic neuroactive compounds epinephrine and norepinephrine. As for the TAARs, the role of TAs in invertebrates will be introduced below.

Trace Amine-Associated Receptors

In 2001, deorphanization of G-protein-coupled receptors revealed that TAs specifically activate some receptor molecules.[5,6] The first member of this family name trace amine-associated receptor 1 (TAAR1), was shown to be specifically activated by a set of TAs including phenylethanolamine (PEA) and tryptamine in the micro- to nanomolar range. TAAR1 was the founding member of a new family of GPCRs that shares significant similarities with members of the conventional biogenic amine receptor families. Apparently, the occurrence of TAARs is tightly associated with vertebrate evolution; thus this family of GPCRs is absent in invertebrates.[7,8] In humans only six TAARs were found, while rodents have more than twice the number of TAAR genes in their genomes (17 in rats and 15 in mice). In fish genomes, a highly variable number of genes can be assigned to the TAAR family. While some fish species, such as some cichlids (from 12 to 44) or the pufferfish *Fugu* (13) have gene number very similar to those seen in rodents, others, such as the zebrafish contain more than 100 different TAAR genes.[9] Despite the structural similarities shared by all known members of the TAAR family, more detailed analyses revealed that this group could be further divided into two or three subclasses based on their primary structures. Despite naming the entire family of proteins TAARs, most of them are insensitive towards confrontation with classical TAs. Only TAAR1 has unequivocally been shown to be sensitive to TA confrontation, responding with specific activation and a suitable physiological response.[10] Research in recent years was dichotomous with regard to the TAAR family. While a modulatory role in the nervous system that controls the activities of biogenic amine receptors and transporters has been assigned to some of them (eg, TAAR1), others are obviously olfactory receptors specifically tuned to detect volatile amines.[11]

Nanomolar concentrations of some TAs could specifically activate TAAR1. Tyramine and PEA are the most effective, meaning that they have the potential to induce the full agonistic response at the lowest concentrations. Binding of the agonists activates the Gαs G-protein, thus leading to an increase in cyclic AMP (cAMP) levels. The major function of TAAR1 appears to be restricted to monoaminergic systems. Different aspects in these monoaminergic systems, more precisely in dopaminergic and serotonergic systems, are regulated by neurotransmission involving TAARs. Especially dopaminergic and serotonergic systems are involved, presumably via regulation of the corresponding uptake systems. These neuronal systems are tightly associated with a set of different neuronal disorders. Newly synthesized TAAR1 agonists gained substantial interest as they might have potent anxiolytic and antipsychotic effects.[12,13]

As already pointed out, the second group of TAARs that are not involved in regular aminergic signaling have been identified as olfactory receptors. They are tailored to recognize volatile amines, thus complementing the comprehensive set of conventional olfactory receptors in order to generate the complete olfactory input.[14] These receptors are expressed within the olfactory system utilizing its signal transduction machinery.[15] The different TAARs that are part of the olfactory system recognize different amines, thus covering a complex set of volatile compounds that are relevant for the organism's survival.[16,17] The volatile amines are often components of urine or feces, thus, they give information about conspecifics or potential predators. Functionally these volatile amines with high information content often serve as phero- or kairomones.[18,19] Consequently, some of these volatile amines induce attraction, while others lead to aversive reactions. It is still a matter of debate, whether the primary ligands for the founding members of this family were indeed TAs or volatile amines. Thus, it is still not clear whether the olfactory system hijacked this part of the aminergic system or if aminergic signaling used this primarily olfactory system.

TAs in Invertebrates

Only two TAs, the monoamines octopamine and tyramine, yielded substantial interest of neurobiologists and pharmacologists focusing on invertebrates as they take very particular roles in invertebrates. In contrast to the situation found in the mammalian brain, both amines are present at very high concentrations throughout the nervous systems of almost all invertebrates studied so far. They act as regular transmitters in these animals, where they take the roles of epinephrine and norepinephrine that are mostly absent or present in only minor amounts in invertebrates, thus representing invertebrate TAs. Since its discovery more than 60 years ago,[20] octopamine became a pet compound of invertebrate neurobiologists and pharmacologists. The source from which octopamine was first isolated gave it its name, the salivary glands of *Octopus*, where it could be found in millimolar concentrations.[20] Tyramine, on the other hand, functions as the biological precursor of octopamine, but has its own physiological role, independent of octopamine.[3,21,22] Nevertheless, the problems associated with disentangling the roles of octopamine and tyramine led most researchers to focus on octopamine rather than on tyramine.[23] Despite the fact that both monoamines are TAs in the vertebrate nervous system, their naming as TAs in the context of invertebrate neurobiology is not really justified.

Mostly, vertebrates and invertebrates use the same biogenic amines in similar functional contexts. Only the "adrenergic" signaling compounds make an exception to this generally applicable rule. While epinephrine and norepinephrine are the adrenergic compounds operative in vertebrates, their role is taken by octopamine and tyramine in invertebrates. Structurally, these compounds are highly related, with the major difference being that epinephrine and norepinephrine are catecholamines, while octopamine and tyramine carry only a single hydroxyl group at their phenolic ring. Despite the fact that the structures of the signaling compounds differ, adrenergic systems of vertebrates and invertebrates share a surprisingly high degree of functional and architectural similarities. Compounds are released in very similar conditions and they control or modulate very similar physiological aspects of life, which implies that the last common ancestor of vertebrates and invertebrates already had a well-defined "adrenergic" system that was in charge of controlling very similar aspects of life.[24] Apparently, either the progenitors of vertebrates or those of invertebrates switched from using epinephrine/norepinephrine to octopamine/tyramine or vice versa.

Beside the multifarious role of octopamine and tyramine for the control of different behaviors, they are also responsible for modulation of a great variety of different peripheral organs in order to adapt their physiology to the needs of the organism in a particular situation. Octopamine is centrally involved in the control of learning and memory, but also part of the neurochemical circuitry that controls aggression.[25,26] Moreover, this neuroactive compound is able to modulate the input of different sensory organs at different levels and it appears to take a decisive role for controlling various aspects of movement activity.[27–29] Receptors that transmit the physiological effects of octopamine and tyramine are all G-protein-coupled, but they share no homologies with the TAAR receptor families.[30] In contrast, they are classical biogenic amine receptors, clustering together with vertebrate α- or β-adrenergic receptors.[31,32] As octopamine and tyramine receptors are the only invertebrate amine receptors that have no direct counterparts in vertebrates, they attracted the interest of invertebrate neurobiologists. Octopamine receptors were specially chosen as targets for highly specific agonists.[33] Compounds with affinities in the subnanomolar range have been developed and some of them, including amitraz and chlordimeform, have been extensively used as insecticides.[34,35]

CONCLUSION

Research within the field of TAs has developed from an exotic subject into different flourishing research areas. In the invertebrate field, octopamine and tyramine, as typical TAs, developed into the most interesting amines, attracting a larger number of researchers to this field, as these monoamines function as regular transmitters, involved in regulating a plethora of behaviors and physiological processes. The situation in the vertebrate field completely changed after identification of the first TAAR in 2001. Identification of this new family of G-protein receptors has led to the emergence of two completely new fields of research: (1) the role of TAs in regulation of monoaminergic systems and (2) the activity of TAARs as highly specific odorant receptors tailored to detect volatile amines, thus representing receptors for various phero- and kairomones. These developments will be presented in depth in the different chapters of this book.

References

1. Pendleton RG, Gessner G, Sawyer J. Studies on lung N-methyltransferases, a pharmacological approach. *Naunyn Schmiedebergs Arch Pharmacol* 1980;**313**(3):263–8.
2. Broadley KJ. The vascular effects of trace amines and amphetamines. *Pharmacol Ther* 2010;**125**(3):363–75.
3. Roeder T. Octopamine in invertebrates. *Prog Neurobiol* 1999;**59**(5):533–61.
4. Grandy DK. Trace amine-associated receptor 1-family archetype or iconoclast? *Pharmacol Ther* 2007;**116**(3):355–90.
5. Borowsky B, Adham N, Jones KA, et al. Trace amines: identification of a family of mammalian G protein-coupled receptors. *Proc Natl Acad Sci USA* 2001;**98**(16):8966–71.
6. Bunzow JR, Sonders MS, Arttamangkul S, et al. Amphetamine, 3,4-methylenedioxymethamphetamine, lysergic acid diethylamide, and metabolites of the catecholamine neurotransmitters are agonists of a rat trace amine receptor. *Mol Pharmacol* 2001;**60**(6):1181–8.
7. Lindemann L, Ebeling M, Kratochwil NA, Bunzow JR, Grandy DK, Hoener MC. Trace amine-associated receptors form structurally and functionally distinct subfamilies of novel G protein-coupled receptors. *Genomics* 2005;**85**(3):372–85.
8. Lindemann L, Hoener MC. A renaissance in trace amines inspired by a novel GPCR family. *Trends Pharmacol Sci* 2005;**26**(5):274–81.
9. Azzouzi N, Barloy-Hubler F, Galibert F. Identification and characterization of cichlid TAAR genes and comparison with other teleost TAAR repertoires. *BMC Genom* 2015;**16**:335.
10. Ferrero DM, Wacker D, Roque MA, Baldwin MW, Stevens RC, Liberles SD. Agonists for 13 trace amine-associated receptors provide insight into the molecular basis of odor selectivity. *ACS Chem Biol* 2012;**7**(7):1184–9.
11. Xie Z, Miller GM. Trace amine-associated receptor 1 as a monoaminergic modulator in brain. *Biochem Pharmacol* 2009;**78**(9):1095–104.
12. Revel FG, Moreau JL, Gainetdinov RR, et al. TAAR1 activation modulates monoaminergic neurotransmission, preventing hyperdopaminergic and hypoglutamatergic activity. *Proc Natl Acad Sci USA* 2011;**108**(20):8485–90.
13. Miller GM. Avenues for the development of therapeutics that target trace amine associated receptor 1 (TAAR1). *J Med Chem* 2012;**55**(5):1809–14.
14. Pacifico R, Dewan A, Cawley D, Guo C, Bozza T. An olfactory subsystem that mediates high-sensitivity detection of volatile amines. *Cell Rep* 2012;**2**(1):76–88.
15. Yoon KH, Ragoczy T, Lu Z, et al. Olfactory receptor genes expressed in distinct lineages are sequestered in different nuclear compartments. *Proc Natl Acad Sci USA* 2015;**112**(18):E2403–9.
16. Johnson MA, Tsai L, Roy DS, et al. Neurons expressing trace amine-associated receptors project to discrete glomeruli and constitute an olfactory subsystem. *Proc Natl Acad Sci USA* 2012;**109**(33):13410–5.
17. Zhang J, Pacifico R, Cawley D, Feinstein P, Bozza T. Ultrasensitive detection of amines by a trace amine-associated receptor. *J Neurosci* 2013;**33**(7):3228–39.
18. Hussain A, Saraiva LR, Ferrero DM, et al. High-affinity olfactory receptor for the death-associated odor cadaverine. *Proc Natl Acad Sci USA* 2013;**110**(48):19579–84.
19. Enjin A, Suh GS. Neural mechanisms of alarm pheromone signaling. *Mol Cells* 2013;**35**(3):177–81.
20. Erspamer V, Boretti G. Identification and characterization, by paper chromatography, of enteramine, octopamine, tyramine, histamine and allied substances in extracts of posterior salivary glands of octopoda and in other tissue extracts of vertebrates and invertebrates. *Arch Int Pharmacodyn Ther* 1951;**88**(3):296–332.
21. Roeder T. Tyramine and octopamine: ruling behavior and metabolism. *Annu Rev Entomol* 2005;**50**:447–77.
22. Roeder T. Biochemistry and molecular biology of receptors for biogenic amines in locusts. *Microsc Res Tech* 2002;**56**(3):237–47.
23. Lange AB. Tyramine: from octopamine precursor to neuroactive chemical in insects. *Gen Comp Endocrinol* 2009;**162**(1):18–26.
24. Adamo SA, Baker JL. Conserved features of chronic stress across phyla: the effects of long-term stress on behavior and the concentration of the neurohormone octopamine in the cricket, *Gryllus texensis*. *Horm Behav* 2011;**60**(5):478–83.
25. Hoyer SC, Eckart A, Herrel A, et al. Octopamine in male aggression of *Drosophila*. *Curr Biol* 2008;**18**(3):159–67.
26. Schwaerzel M, Monastirioti M, Scholz H, Friggi-Grelin F, Birman S, Heisenberg M. Dopamine and octopamine differentiate between aversive and appetitive olfactory memories in *Drosophila*. *J Neurosci* 2003;**23**(33):10495–502.

27. Wong R, Lange AB. Octopamine modulates a central pattern generator associated with egg-laying in the locust, *Locusta migratoria*. *J Insect Physiol* 2014;**63**:1–8.
28. Selcho M, Pauls D, El Jundi B, Stocker RF, Thum AS. The role of octopamine and tyramine in <u>Drosophila</u> larval locomotion. *J Comp Neurol* 2012;**520**(16):3764–85.
29. Brembs B, Christiansen F, Pfluger HJ, Duch C. Flight initiation and maintenance deficits in flies with genetically altered biogenic amine levels. *J Neurosci* 2007;**27**(41):11122–31.
30. Balfanz S, Strunker T, Frings S, Baumann A. A family of octopamine [corrected] receptors that specifically induce cyclic AMP production or Ca2+ release in *Drosophila* melanogaster. *J Neurochem* 2005;**93**(2):440–51.
31. Evans PD, Maqueira B. Insect octopamine receptors: a new classification scheme based on studies of cloned *Drosophila* G-protein coupled receptors. *Invert Neurosci* 2005;**5**(3–4):111–8.
32. El-Kholy S, Stephano F, Li Y, Bhandari A, Fink C, Roeder T. Expression analysis of octopamine and tyramine receptors in *Drosophila*. *Cell Tissue Res* 2015;**361**(3):669–84.
33. Roeder T. Pharmacology of the octopamine receptor from locust central nervous tissue (OAR3). *Br J Pharmacol* 1995;**114**(1):210–6.
34. Roeder T, Degen J, Dyczkowski C, Gewecke M. Pharmacology and molecular biology of octopamine receptors from different insect species. *Prog Brain Res* 1995;**106**:249–58.
35. Roeder T, Seifert M, Kahler C, Gewecke M. Tyramine and octopamine: antagonistic modulators of behavior and metabolism. *Arch Insect Biochem Physiol* 2003;**54**(1):1–13.

CHAPTER

2

Methods of Trace Amine Analysis in Mammalian Brain

H.A. Khan[1], Q. Ullah[2], A. Ahmad[3], A.S. Alhomida[1] and S.H. Al Rokayan[4]

[1]Department of Biochemistry, King Saud University, Riyadh, Saudi Arabia
[2]Department of Chemistry, Maulana Azad National Urdu University, Hyderabad, India [3]Department of Zoology, Maulana Azad National Urdu University, Hyderabad, India [4]Research Chair for Biomedical Applications of Nanomaterials, King Saud University, Riyadh, Saudi Arabia

OUTLINE

INTRODUCTION

Trace amines include β-phenylethylamine (PEA), tryptamine (T), phenylethanolamine (PEOHA), *m*- and *p*-tyramine (*m*- and *p*-TA), *m*- and *p*-octopamine (*m*- and *p*-OA), and synephrine (SYN). Trace amines are typically found throughout the central nervous system. These amines are structurally similar, containing a benzene ring in common except the T that contains an indole ring (Fig. 2.1). The term "trace amines" used for these amines is due to the fact that they are present at very low concentrations (usually 0.1–100 ng/g) in the central nervous system with a heterogeneous distribution.[1] Trace amines were initially considered as metabolic byproducts of no physiologic or pathologic relevance. However, the discovery of a family of G-protein-coupled receptors which appear to be selectively activated by trace amines has prompted the interest of investigators to evaluate the potential physiological relevance of these compounds.[1–3] The trace amine-associated receptors (TAARs) are phylogenetically and functionally distinct from other G-protein-coupled receptor families and from OA and TA receptors in invertebrates.[4] In humans, TAARs are located in various areas in the brain and their dysfunction has been associated with impaired neurological function.[5–7]

Several studies have shown that trace amines can act as neuromodulators for monoamine neurotransmitters.[8–10] The trace amines can alter the release and uptake of norepinephrine (NE), dopamine (DA), and 5-hydroxytryptamine (5-HT) not only by regulating the active transport of these neurotransmitters across the plasma membrane but also by involving mechanisms of action targeting the neurotransmitter vesicles themselves.[11] Because of their modulatory role on classical neurotransmitters, trace amines have been implicated in a number of psychiatric and neurological disorders, including depression, schizophrenia, phenylketonuria, Reye's syndrome, Parkinson's disease, attention deficit hyperactivity disorder, Tourette's syndrome, epilepsy, and migraine headaches.[9,12–17] The neuromodulatory effects of low levels of trace amines as well as the existence of trace amine receptor

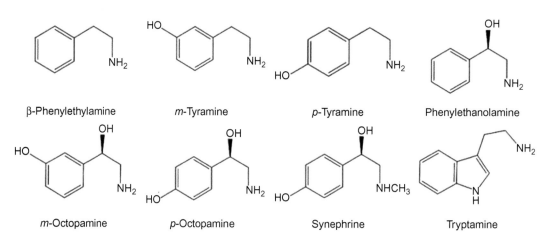

FIGURE 2.1   Chemical structures of various trace amines.

TABLE 2.1 A Comparative View of Trace Amines and Monoamine Neurotransmitter Levels in Rat Brain

Trace Amines (ng/g)		Monoamine Neurotransmitters (ng/g)	
β-Phenylethylamine	1.80[19]	Dopamine	600[26]
	1.50[20]		1150[27]
p-Tyramine	1.10[21]	Norepinephrine	490[26]
	1.00[22]		834[27]
m-Tyramine	0.32[23]	Serotonin	716[27]
	0.22[22]		462[28]
Tryptamine	0.24[24]		
	0.15[22]		
p-Octopamine	0.90[25]		
	0.77[22]		
m-Octopamine	0.20[25]		

polymorphisms and mutations to distinct clinical conditions would make them ideal candidates for the development of novel therapeutics for a wide range of neurological disorders.[10,18]

The endogenous levels of trace amines are several-hundred-fold below those of the classical neurotransmitters such as DA, NE, and 5-HT as shown in Table 2.1. The levels of trace amines in brain are affected by several drugs used to treat neuropsychiatric disorders.[29] Administration of monoamine oxidase inhibitor antidepressants, such as phenelzine and tranylcypromine, results in a greater increase in brain levels of trace amines as compared to classical neurotransmitter amines.[30] Although their rate of synthesis is equivalent to that of DA and NE, trace amines have an exceedingly rapid turnover rate, with an endogenous pool half-life of approximately 30 s.[28,31] Determination of trace amines in brain is crucially important to better understand their role in neuropsychiatric disorders. Owing to the low absolute concentrations and high turnover of trace amines, it is difficult to accurately measure their concentrations in nervous tissue. However, the recent advents in chromatographic science and mass spectrometric detection have made it possible to precisely measure the levels of trace amine discrete regions of the brain. This chapter provides a review of various methods for the estimation of trace amines in mammalian brain.

METHODS OF TRACE AMINE ANALYSIS

Several analytical methods including integrated ion-current mass spectrometry,[21] gas chromatography (GC),[32–34] GC coupled with mass spectrometry,[35,36] high-performance liquid chromatography (HPLC),[37,38] and radioenzymatic methods[39,40] have been used for the analysis of trace amines in brain. Hess and Udenfriend[41] described a fluorometric method sufficiently sensitive to determine 0.1–5 μg of T in body fluids and tissues. The method involves an extraction with benzene, cyclization with formaldehyde to produce tetrahydronorharman followed by a dehydrogenation to form the highly fluorescent product norharman. Cooper and Venton[42] developed an electrochemical method using fast-scan cyclic voltammetry at carbon-fiber microelectrodes to detect fast changes in TA and OA. The technique has limits of detection of 18 nM for TA and 30 nM for OA, much lower than expected levels in insects and lower than basal levels in some brain regions of mammals. Various chromatographic and radioenzymatic methods used for the analysis of trace amines are described in the following text.

Gas Chromatography

Baker et al.[32] developed a rapid, specific, and sensitive procedure for the routine analysis of endogenous p-TA in rat brain, using capillary columns with electron capture gas–liquid chromatography. The detection limit of the method was less than 1 ng/g of brain whereas 14 structurally related amines did not interfere with the assay. A control p-TA level in whole rat brain was found to be 2.4 ± 0.3 ng/g.[32] A gas chromatographic procedure with electron capture detection (GC-ECD) has been described for the analysis of PEA in tissues and body fluids.[33] The method involves the use of pentafluoro-benzenesulfonyl chloride for extraction and derivatization of PEA followed by separation and analysis of the derivatized amine on a gas chromatograph equipped with a fused-silica capillary column and an electron capture detector. The procedure is rapid, provides a stable and sensitive derivative, and has been applied to analysis of PEA in brain, heart, kidney, liver, lung, spleen, and blood from the rat and urine from human subjects. The level of PEA in control rat brain was found to be 1.88 ± 0.36 ng/g.[33] This method is more rapid than a previously reported GC-ECD method which provided for simultaneous analysis of the PEA, m- and p-TA, normetanephrine, and 3-metanephrine.[34]

Gas Chromatography-Mass Spectrometry

Dourish et al.[36] used a gas chromatography-mass spectrometry (GC-MS) method for the determination of PEA, m-TA, and p-TA, and their respective major acid metabolites, phenylacetic acid, meta-hydroxyphenylacetic acid, and para-hydroxyphenylacetic acid in brain, plasma, and urine samples of isolated aggressive mice and in group housed controls. The brain levels of PEA, m-TA, and p-TA in control mice were 1.5 ± 0.5, 2.0 ± 0.3, and 5.4 ± 0.6 ng/g. The plasma concentrations of PEA, m-TA, and p-TA were below the limits of detection; whereas the urinary levels of these trace amines were comparatively higher than their respective brain levels.[36] Trace amines can be absorbed on a SEP-PAK Cl8 cartridge from biological samples such as urine, plasma, and tissue homogenates and subsequently eluted as isothiocyanate derivatives by ethyl acetate containing carbon disulfide, thus combining elution and derivatization in a single step, with high recovery rates.[43] An ultrasensitive negative chemical ion (GC-MS) method was used for determination of PEA, m-TA, and p-TA in different regions of rat brain.[44] Using this procedure, the concentrations of PEA in different regions of rat brain were: caudate ($2.71 \pm 0.73\,ng/g$), hypothalamus ($0.45 \pm 0.15\,ng/g$), cerebellum ($0.09 \pm 0.02\,ng/g$), olfactory bulb ($0.35 \pm 0.11\,ng/g$), brain stem ($0.13 \pm 0.03\,ng/g$), hippocampus ($0.20 \pm 0.11\,ng/g$), and cortex ($0.69 \pm 0.13\,ng/g$). The levels of m-TA were as follows: caudate ($2.69 \pm 0.19\,ng/g$), hypothalamus ($0.32 \pm 0.16\,ng/g$), cerebellum ($0.07 \pm 0.04\,ng/g$), olfactory bulb ($0.09 \pm 0.04\,ng/g$), brain stem ($0.04 \pm 0.01\,ng/g$), hippocampus ($0.07 \pm 0.02\,ng/g$), and cortex ($0.18 \pm 0.15\,ng/g$). The same procedure showed the brain levels of p-TA as follows: caudate ($8.99 \pm 1.60\,ng/g$), hypothalamus ($0.93 \pm 0.13\,ng/g$), cerebellum ($0.78 \pm 0.27\,ng/g$), olfactory bulb ($0.70 \pm 0.13\,ng/g$), brain stem ($0.90 \pm 0.36\,ng/g$), hippocampus ($0.40 \pm 0.06\,ng/g$), and cortex ($1.78 \pm 0.28\,ng/g$).[44]

High-Performance Liquid Chromatography

A reversed-phase HPLC method with fluorometric detection was reported for the analysis of PEA using two columns, containing $200\,\mu L$ of Dowex 50-X8 and Amberlite CG-50, whereas p-methoxyphenylethylamine was used as an internal standard.[38] The recoveries of PEA and p-methoxyphenylethylamine were $53.9 \pm 9.4\%$ and $68.1 \pm 12.4\%$, respectively. Regional distributions of PEA in rat and mouse brains showed its highest concentrations in hypothalamus and hippocampus in both these animals.[38] Huebert et al.[45] developed a sensitive and specific HPLC method for the measurement of PEA in rat brain and human plasma. The method involves solvent extraction of PEA with hexane in the presence of amphetamine or phenylpropylamine (PPA) as internal standards. Automated precolumn derivatization with o-phthalaldehyde and 2-mercaptoethanol followed by reverse-phase HPLC separated PEA and PPA from endogenous interferences. Detection and quantification were carried out by amperometric detection at $+0.75\,V$ relative to a Ag/AgCl reference electrode or by coulometric detection with cell potentials set at $+0.29$ and $+0.50\,V$. The limit of detection for PEA was $10\,pg$ and the limit of quantification in plasma was $60\,pg/mL$. By using this procedure, the basal whole brain and striatal levels of PEA were found to be 0.584 ± 0.243 and $2.89 \pm 1.03\,ng/g$, respectively.[45]

Larson and Dalo[37] extracted the endogenously formed T from the rat brain tissue using ethyl acetate, purified on a weak cation-exchange resin and determined by HPLC with fluorometric detection. A μBondapak C18 reversed-phase column was used under isocratic conditions. Using this method, the concentration of T in the whole brain of normal rats was found to be $0.60 \pm 0.06\,ng/g$ of tissue. Downer and Martin[46] developed a HPLC method with coulometric electrochemical detection for p-OA, N-acetyl-p-OA, p-SYN, TA, norepinephrine, epinephrine, 5-hydroxytryptophan, normetanephrine, DA, metanephrine, 3,4-dihydroxyphenylacetic acid, N-acetyldopamine, tryptophan, 5-hydroxyindoleacetic acid, 5-HT, N-acetyl-5-hydroxytryptamine, homovanillic acid, and tyrosine. The procedure was applied to study monoamine degradation in the insect brain. Leung and Tsao[47] optimized an HPLC mobile phase for the simultaneous determination of 24 neurochemicals including trace amines (SYN, TA, OA), catecholamine, serotonin, their precursors, and metabolites in mouse brain. This mobile phase contained sodium acetate ($0.04\,M$), citric acid ($0.01\,M$), sodium chloride ($0.0126\,M$), sodium octyl sulfate ($91\,mg/L$), tetrasodium EDTA ($50\,mg/L$), and 10% (v/v) methanol.

Thin-Layer Chromatography

A thin-layer chromatography (TLC) coupled with high-resolution mass spectrometry method was developed for the determination of SYN.[48] Using this procedure p- and m-SYN in the mouse hypothalamus were found to be below the limit of sensitivity of the method (<3.8 ng/g). A fluorometric test based on reaction of PEA with ninhydrin in the presence of L-leucyl-L-alanine to yield a highly fluorescent compound has been described.[49] This test is highly specific for PEA as only a few other compounds (phenylalanine and 5-HT) produced little or no fluorescence. The amount of PEA in rat brain could be determined by this reaction after its separation from phenylalanine and 5-HT by n-heptane extraction. Thus, TLC was coupled with this fluorometric detection method in order to separate PEA from other amines that interfere with this test. Moreover, brains from 10 to 12 rats were pooled to determine PEA levels (5 ng/g) because this method is not as sensitive as GC or HPLC.

Radioenzymatic Methods

Danielson et al.[40] described a specific and sensitive radioenzymatic method for simultaneous determination of p-OA, m-OA, and PEOH, based on the N-methylation of p-hydroxylated phenylethylamines by the enzyme PEOHA N-methyl transferase using [3H] methyl-S-adenosyl-L-methionine as a donor. The sensitivity of this method was increased 10-fold with the use of radioactive dansyl chloride and radioactive S-adenosyl-methionine.[50] David and Delacour[51] also used a radioenzymatic method for determination of brain PEOH, m-OA, and p-OA in three different strains of rats. The range of these trace amines in the hypothalamus was as follows: PEOH (0.69–1.33 ng/g), m-OA (0.62–1.33 ng/g), and p-OA (4.85–12.15 ng/g). In brain stem, these values were 0.47–0.85, 0.45–0.70, and 1.80–4.10 ng/g, respectively.[51] Tallman et al.[39] developed an enzymatic-isotopic assay for the measurement of tyramine with a sensitivity of 1.0 ng. They used this method for measuring the endogenous content of tyramine in various tissues from adult rats. The highest tyramine content was found in rat heart atria, followed by salivary gland, kidney, and brain. Within the brain the distribution of tyramine is heterogeneous and the highest tyramine content was localized in the striatum.[39] It is important to note that the radioenzymatic procedure does not differentiate p-TA from the structural isomer, m-TA, which is also present in brain. Philips[52] reviewed the radioenzymatic techniques for assaying the trace amines OA, PEOHA, PEA, TA, and T and the classical neurotransmitter amines including norepinephrine, epinephrine, DA, and serotonin.

PROTOCOLS FOR TRACE AMINE ANALYSIS IN BRAIN

Some useful protocols for the analysis of trace amines in brain samples are described below. More details of the methodology including the sources of chemicals, reagents, and apparatus can be obtained from the respective citations.

Determination of p-TA in Rat Brain

A rapid, specific and sensitive procedure has been developed for the routine analysis of endogenous p-TA in rat brain, using GC with electron capture detection.[32] After removing the rat brains, immerse them immediately in isopentane on dry-ice, and then transfer to vials kept on dry-ice; store them at −80°C until used. Homogenize the whole brain in five volumes of ice-cold perchloric acid solution (0.1 mol/L) containing 10 mg % (w/v) disodium EDTA. The homogenates are centrifuged at 10,000g for 20 min and the resulting supernatants are collected for subsequent extraction.

An aliquot (4 mL) of brain supernatant is mixed with the internal standards including 500 ng of benzylamine, 3-phenylpropylamine, and 2-(4-chlorophenyl)ethylamine. The pH is adjusted to 7.8 with potassium bicarbonate followed by addition of 400 µL of sodium phosphate buffer (0.25 mol/L; pH 7.8) and centrifugation to remove the potassium perchlorate formed during neutralization. The supernatant is removed and agitated for 1.0 min with 5.0 mL of a 2.5% (v/v) solution of di(2-ethylhexyl)-phosphoric acid in chloroform. After separation of the phases by centrifugation, the aqueous layer is aspirated and discarded, whereas the chloroform layer is transferred to a tube containing 2.5 mL of HCI (0.5 mol/L) and the mixture is agitated for 1.0 min. After a brief centrifugation, the aqueous layer is retained and basified by adding a small amount of solid sodium bicarbonate. Three hundred microliters of acetic anhydride are added with shaking. Additional small quantities of sodium bicarbonate are added intermittently with agitation until all effervescence had ceased. Ethyl acetate (3.0 mL) is then added to each aqueous phase and the mixture is shaken for 2.0 min. To the separated ethyl acetate phase, 400 µL of 10 mol/L ammonium hydroxide are added and the mixture is vigorously agitated for 40 min. Then 300 µL of 8 mol/L HCl are added and the mixture is shaken (15 s) and centrifuged (1 min). The organic layer is removed and evaporated to dryness under a stream of nitrogen followed by the addition of ethyl acetate (25 µL) and trifluoroacetic anhydride (75 µL) and the reaction was allowed to proceed at room temperature for 30 min. The 300 µL of cyclohexane and 3.0 mL of saturated aqueous sodium tetraborate are added to each tube followed by agitation for 15 s. The organic layer is isolated after a brief centrifugation and 1.0 µL is used for GC analysis.[32]

GC was performed using a AA WCOT SP2100 glass capillary column (10 m) (Supelco, Belleforte, PA, United States) and 15 mCi 63 Ni source linear electron capture detector (Hewlett Packard, United States). The carrier gas was helium maintained at 7 psi, with an auxiliary flow of argon:methane (90:10) to the detector at a flow rate of 35 mL/min. The oven temperature was programed as follows: the initial column temperature of 80°C was maintained for 0.6 min and increased to 120°C at a rate of 30°C/min. The temperature of both the injection port and detector was 250°C. Retention times of derivatized p-TA and the N-acetyl-N-trifluoroacetyl derivatives of benzylamine, 3-phenylpropylamine, and 2-(4-chlorophenyl)-ethylamine were 8.8, 3.3, 6.7, and 9.4 min, respectively.[32]

For construction of calibration curves, varying amounts of p-TA (1–50 ng) and internal standards (500 ng of each) were added to 4.0 mL of 0.1 mol/L perchloric acid containing 10 mg% EDTA and extracted as mentioned above. The recovery of this procedure is about 65% and the detection limit for p-TA is less than 1 ng/g of brain tissue. The representative chromatograms are shown in Fig. 2.2. Using this method, the mean ± SEM value for p-TA in control rat brain was found to be 2.4 ± 0.3 ng/g, which is in good agreement with levels obtained using a mass spectrometric-integrated ion-current technique[21] and a GC-MS method.[35]

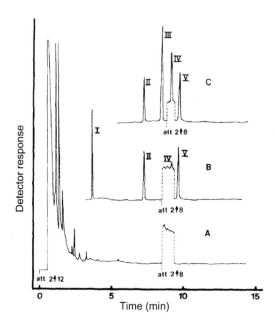

FIGURE 2.2　Gas chromatograms obtained in the assay of *p*-TA (peak IV) in brain tissue, with the internal standards benzylamine (peak I), 3-phenylpropylamine (peak II), and 2-(4-chlorophenyl)ethylamine (peak V). Attenuation changes programmed into the run are marked. (A) perchloric acid blank, (B) control rat brain, and (C) brain from a rat treated with the monoamine oxidase inhibitor tranylcypromine (peak III). *Source: Reprinted from Baker GB, LeGatt DF, Coutts RT. A gas chromatographic procedure for quantification of para-tyramine in rat brain. J Neurosci Method 1982;5:181–8 with permission from Elsevier.*

Determination of PEA, *m*-TA, and *p*-TA in Rat Brain

An ultrasensitive negative chemical ion (GC-MS) method for the analysis of PEA, *m*-TA, and *p*-TA in different regions of rat brain has been reported by Durden and Davis.[44] Rat brains are quickly removed after stunned killing and decapitation. The isolated brains are placed in ice-chilled saline and the brain regions are dissected out on an ice-chilled petri dish. Brain tissues are weighed (up to 200 mg) and homogenized in a 1.5 mL disposable tube containing 500 μL distilled water, 25 μL perchloric acid, 100 μL 0.1% Triton X-100, and 100 μL of a mixture of deuterated internal standards including 1700 pg of phenylethylamine-1,1,2,2-d_4 (PEA-d_4), 1300 pg of *m*-tyramine-1,1,2,2-d_4 (*m*-TA-d_4), and 1100 pg of *p*-tyramine-1,1,2,2-d_4 (*p*-TA-d_4). Blank and standard solutions contained distilled water or nondeuterated trace amines, respectively. The tubes are centrifuged at 6000g for 15 min, the supernatants are transferred to new tubes and mixed with 400 μL of 0.5 M Na_2CO_3, and 25 μL of acetic anhydride, followed by addition of 125 μL 23% NaOH and another 25 μL acetic anhydride. The products are extracted twice with 3 mL ethyl acetate:hexane (1:1, v/v) and the organic phase evaporated to dryness. The residue is allowed to react with 3 μL of pentafluorobenzoyl chloride and 1.7 μL of triethylamine in 200 μL hexane at 50°C for 15 min. The contents are cooled and washed with 100 μL of phosphate buffer (pH 8.2) followed by washing with 100 μL of distilled water and then kept in a freezer ($-17°C$) for 4 h to precipitate pentafluorobenzoic anhydride. The organic phase is used for sample injection. Standard curves are prepared by using distilled water solutions (1.0 mL) containing various amounts of the PEA, m-TA, and p-TA (0–400 pg) with constant amounts of PEA-d_4, m-TA-d_4, and p-TA-d_4.

The samples are analyzed by GC-MS using a capillary column (Supelco DB5, 30 m × 0.32 mm id, 0.25 μm bonded film) connected to a 1.0 m length of 0.53 mm id fused-silica retention gap by a Supelco GlasSeal connector to exit directly into the mass spectrometer ion source. The sample injection volume is 1.0 μL and the carrier gas is helium. The oven temperature is programmed as follows: 140°C for 2 min, 10°C/min to 290°C, and then isothermal for 4 min. A negative ion electron capture chemical ionization mode is used for mass spectrometry. For selected ion monitoring analysis of N-acetyl-N-pentafluorobenzoyl phenylethylamine the mass reference is 330.9792 perfluorokerosene (PFK) and the sample ions are 357.0810 (PEA) and 361.1061 (PEA-d_4). For N,O-diacetyl-N-pentafluorobenzoyl derivatives of m-TA and p-TA the detected ions are 415.0843 and 419.1094.[44]

The mass spectra of N,O-diacetyl-N-pentafluorobenzoyl-m-tyramine and N,O-diacetyl-N-pentafluorobenzoyl-p-tyramine obtained from derivatized standards of a mixture of the protio and deutero amine are shown in Fig. 2.3. The spectrum of N-acetyl-N-pentafluorobenzoyl phenylethylamine has been reported earlier.[53] The spectra are characterized by intense molecular ions with relatively little fragmentation, which enhances the sensitivity. The low-resolution mass spectra of the PEA and p-TA from a rat caudate sample are shown in Fig. 2.4. Using this procedure, PEA can be detected at 75 fg and m-TA and p-TA at approximately 200 fg using pure solutions. The sensitivity of the tyramine analyses when working at the picogram levels in the tissues is 6–8 times lower than for PEA because of losses during extraction and derivatization steps.[44]

Determination of T in Rat Brain

Larson and Dalo[37] described an HPLC method for determination of extracted endogenously formed T from rat brain tissue using ethyl acetate, purified on a weak cation-exchange resin and determined by HPLC with fluorometric detection. The rats are killed by decapitation and the whole brains are removed and immediately frozen in liquid nitrogen until being analyzed. Preweighed brains are homogenized (1:2, w/v) in 0.75 M tribasic phosphate buffer (pH 12.5), containing 10 g/L sodium metabisulfite and then extracted with 2 mL of ethyl acetate per 1 mL of homogenate. Tubes are shaken for 15 min and then centrifuged at 3000 rpm for 5 min. The organic layer is removed and the extraction is repeated once again, extractants are pooled and evaporated using a vacuum centrifuge. The residue is first suspended in 0.2 mL of 0.2 M phosphoric acid and then mixed with 9.8 mL of 0.2 M sodium potassium phosphate buffer (pH 6.1).

A precolumn cleanup is performed as follows. The solution obtained from the extraction of T is filtered through a small amount of glass wool to get rid of gross residual particles and then placed directly onto Bio-Rex 70 columns, washed with 10 mL of 0.02 M phosphate buffer (pH 6.1) and then with 10 mL of distilled water. T is eluted from the columns with 3 mL of 2.0 M formic acid. The eluent is dried in a vacuum evaporator and the residue resuspended in 100 μL of mobile phase for injection into the reversed-phase column.[37]

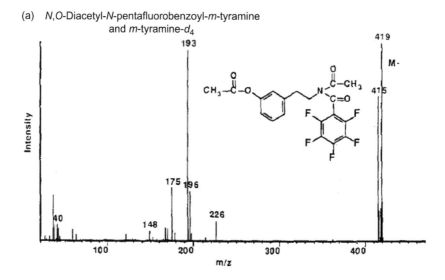

(a) N,O-Diacetyl-N-pentafluorobenzoyl-m-tyramine and m-tyramine-d4

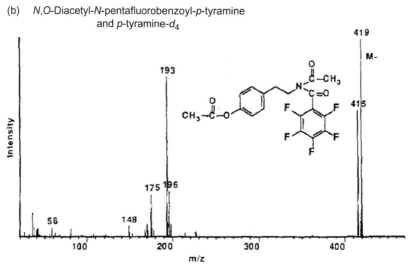

(b) N,O-Diacetyl-N-pentafluorobenzoyl-p-tyramine and p-tyramine-d4

FIGURE 2.3 Negative ion mass spectra of the N,O-diacetyl-N-pentafluorobenzoyl derivatives of (A) m-tyramine and 1,1,2,2-tetradeutero-m-tyramine and (B) p-TA and 1,1,2,2-tetradeutero-p-TA. *Source: Reprinted from Durden DA, Davis BA. Determination of regional distributions of phenylethylamine and meta- and para-tyramine in rat brain regions and presence in human and dog plasma by an ultra-sensitive negative chemical ion gas chromatography-mass spectrometric (NCI-GC-MS) method. Neurochem Res 1993;18:995–1002 with permission from Springer.*

An isocratic HPLC system connected to a 300 × 3.9 mm μBondapak Cl8 column (Waters, Milford, MA, United States) and a variable-wavelength fluorometer is used for separation and determination of T. The mobile phase is composed of 0.01 M sodium acetate buffer (pH 4.6) containing 20% methanol. Before use, the mobile phase is filtered through a 0.3-μm filter (Millipore, Bedford, MA, United States) and then degassed. The flow rate of the mobile

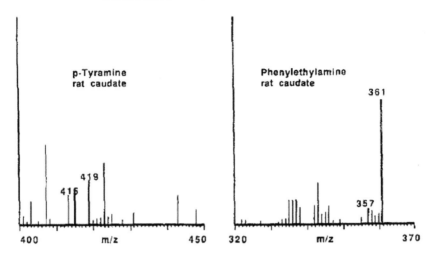

FIGURE 2.4 Partial mass spectra of the molecular ion regions of the compounds recorded at the retention times of *p*-TA and phenylethylamine from a derivatized extract obtained from a rat caudate. *Source: Reprinted from Durden DA, Davis BA. Determination of regional distributions of phenylethylamine and meta- and para-tyramine in rat brain regions and presence in human and dog plasma by an ultra-sensitive negative chemical ion gas chromatography-mass spectrometric (NCI-GC-MS) method. Neurochem Res 1993;18:995–1002 with permission from Springer.*

phase is set at 2.5 mL/min. It was observed that methanol concentrations less than 20% provided better separation but at the cost of decreasing the peak height. Out of the 24 different indoles tested using this method, the only compound which overlapped with T (retention time, 5.51 min) was *N*-acetyl-5-hydroxytryptamine (retention time, 5.88 min). Although this overlap was only partial, it became more pronounced as the column aged.[37]

Compounds eluted with a strong acid from the weak cation-exchange resin in addition to T include 5-HT, 5-methoxytryptamine, 5-methyltryptamine, *N*-methyltryptamine, and the tryptolines (Fig. 2.5; Table 2.2). After testing various acids, 2.0 M formic acid was found to be the most effective eluent of T from the resin, resulting in a total recovery of 97.9% of the T from the precolumn. Using the procedure described above, the average concentration of T in the whole rat brain was found to be 0.60 ± 0.22 ng/g. A typical chromatogram obtained from such an extraction is shown in Fig. 2.6.

Determination of PEOH, *m*-OA, and *p*-OA in Rat Brain

David and Delacour[51] modified the radioenzymatic methods described by Molinoff and Axelrod[54] and Saavedra and Axelrod[55] for the analysis of PEOH, *m*-OA, and *p*-OA in rat brain. In the modified method, the tissues are homogenized in five volumes of ice-cold 0.05 M Tris–HCl buffer at pH 8.6 containing 1 mM pargyline. The brain tissue homogenates are heated in a boiling water bath for 3 min. After centrifugation at 1000g for 5 min, aliquots of supernatants (150 μL) are incubated with 60 μL of 0.05 M Tris–HCl (pH 8.6) containing 40 μL of PEOHA *N*-methyltransferase and 0.04 nmol of [3H] *S*-adenosyl-methionine. The reaction is stopped after 45 min by the addition of 200 μL of 0.5 M borate buffer (pH 11.0)

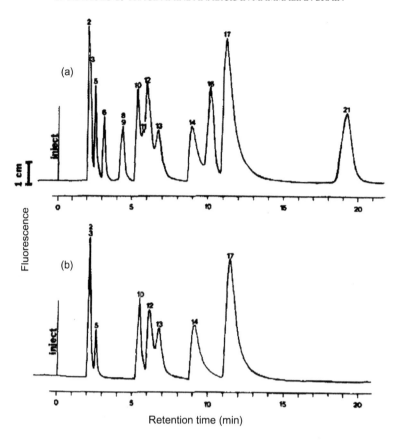

FIGURE 2.5   (A) Chromatogram resulting from the injection of 150 ng of each indole compound indicated by number (please refer to Table 2.2). (B) Chromatogram of the formic acid eluent of weak cation-exchange precolumns after application of the same mixture of indole compounds as indicated in (A). Peak number 10 is for T. *Source: Reprinted from Larson AA, Dalo NL. Quantification of tryptamine in brain using high-performance liquid chromatography. J Chromatogr 1986;375:37–47 with permission from Elsevier.*

saturated with sodium chloride and containing 2 μg each of *p*-synephrine, phenylephrine, and *N*-methylphenylethanolamine. The trace amines are extracted with ethyl acetate and the solvent is evaporated under nitrogen. The residues are dissolved in 1 mL of saturated sodium carbonate and 0.5 mL of acetone containing 8 mg/mL of dansyl chloride and the contents are allowed to react in the dark.

The dansylated amines are separated by TLC according to Danielson et al.[40] using three successive and different mobile phases: chloroform:*n*-butyl acetate (5:2, v/v), toluene:triethylamine:methanol (50:5:1), and cyclohexane:ethyl acetate (25:35). The respective spots are then scratched from the stationary coating and the radioactivity is measured. The standard solutions are prepared by dissolving known amounts of PEOH, *m*-OA, and *p*-OA in 0.1 M Tris–HCl (pH 8.6) followed by serial dilutions.[50] Using this method, the range of trace amines in hypothalamus of three different strains of rats were as follows: PEOH (0.69–1.33 ng/g), *m*-OA (0.62–1.33 ng/g), and *p*-OA (4.85–12.15 ng/g).[51]

TABLE 2.2   Retention Times of Various Trace Amines Analyzed by HPLC (Please Refer to Fig. 2.5 for Chromatograms)

Peak No.	Compound	Retention Time (min)
2	5-Hydroxytryptoline	2.07
3	5-Hydroxytryptamine	2.15
5	5-Hydroxymethyltryptoline	2.57
6	L-Tryptophan	3.02
8	5-Hydroxytryptophol	4.14
9	DL-Methoxytryptophan	4.39
10	Tryptamine	5.51
11	N-Acetyl-5-hydroxytryptamine	5.88
12	N-Methyltryptamine	6.20
13	5-Methoxytryptamine	6.75
14	Tryptoline	8.70
15	5-Methoxy-N,N-dimethyltryptamine	9.10
17	Methyltryptoline	11.32
21	5-Methoxytryptophol	19.50

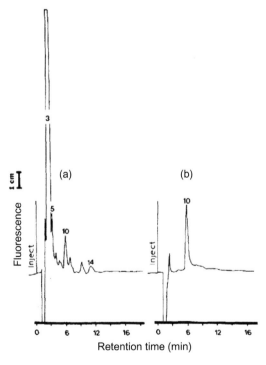

FIGURE 2.6   (A) Chromatogram of a typical final product of an extraction of T from rat brain. Quantitation of the T (peak 10) indicates a value of 0.62 ng/g. (B) Chromatogram of a smaller and more dilute sample of extract from the brain of a rat pretreated with DL-p-chlorophenylalanine, tranylcypromine, and L-tryptophan prior to sacrifice. Quantitation of T in drug-treated rats indicates a concentration of 95 ng/g tissue. *Source: Reprinted from Larson AA, Dalo NL. Quantification of tryptamine in brain using high-performance liquid chromatography. J Chromatogr 1986;375:37–47 with permission from Elsevier.*

CONCLUSION

Trace amines are present in the nervous system at very minute concentrations as compared to the classic neurotransmitter amines. These amines are heterogeneously distributed throughout the central and peripheral nervous systems and play important roles in neurotransmission and neuromodulation. Determination of trace amines in brain is crucially important to better understand their role in neuropsychiatric disorders. Owing to the low absolute concentrations and high turnover of trace amines, it is important to accurately measure their concentrations in nervous tissue. Various chromatographic techniques such as TLC, GC, and HPLC have been reported for the determination of trace amines in brain tissue. However, coupling the chromatography with MS significantly improves the sensitivity and specificity of the analytical method. Radioenzymatic methods are also sensitive enough to determine the levels of trace amines but they are largely replaced by chromatographic methods to avoid the use of radioactive materials. Not only the analytical technique but also proper animal handling, sample collection, and storage play important roles in accuracy and reproducibility of results. Moreover, an optimized extraction method significantly improves the recovery of trace amines from the tissues and minimizes the interference from other related compounds during the analysis.

Acknowledgments

This work was supported by the Research Chair for Biomedical Applications of Nanomaterials, Deanship of Scientific Research, King Saud University, Riyadh, Saudi Arabia.

References

1. Berry MD. Mammalian central nervous system trace amines. Pharmacologic amphetamines, physiologic neuromodulators. *J Neurochem* 2004;**90**:257–71.
2. Borowsky B, Adham N, Jones KA, et al. Trace amines: identification of a family of mammalian G protein-coupled receptors. *Proc Natl Acad Sci USA* 2001;**98**:8966–71.
3. Bunzow JR, Sonders MS, Arttamangkul S, et al. Amphetamine, 3,4-methylenedioxy-amphetamine, lysergic acid diethylamide, and metabolites of the catecholamine neurotransmitters are agonists of a rat trace amine receptor. *Mol Pharmacol* 2001;**60**:1181–8.
4. Lindemann L, Hoener MC. A renaissance in trace amines inspired by a novel GPCR family. *Trends Pharmacol Sci* 2005;**26**:274–81.
5. Wolinsky TD, Swanson CJ, Smith KE, Zhong H, Borowsky B, Seeman P, et al. The trace amine 1 receptor knockout mouse: an animal model with relevance to schizophrenia. *Genes Brain Behav* 2007;**6**:628–39.
6. Pae CU, Drago A, Mandelli L, De Ronchi D, Serretti A. TAAR 6 and HSP-70 variations associated with bipolar disorder. *Neurosci Lett* 2009;**465**:257–61.
7. Ludewick HP, Schwab SG, Albus M, Lerer B, Maier W, Trixler M, et al. No support for an association with TAAR6 and schizophrenia in a linked population of European ancestry. *Psychiatry Genet* 2008;**18**:208–10.
8. Lindemann L, Meyer CA, Jeanneau K, Bradaia A, Ozmen L, Bluethmann H, et al. Trace amine-associated receptor 1 modulates dopaminergic activity. *J Pharmacol Exp Ther* 2008;**324**:948–56.
9. Boulton AA, Bieck PR, Maitre L, Riederer P, editors. *Neuropsychopharmacology of the trace amines: experimental and clinical aspects*. Clifton, NJ: Humana Press; 1985.
10. Berry MD. The potential of trace amines and their receptors for treating neurological and psychiatric diseases. *Rev Recent Clin Trials* 2007;**2**:3–19.
11. Raiteri M, Del Carmine R, Bertollini A, Levi G. Effect of sympathomimetic amines on the synaptosomal transport of noradrenaline, dopamine and 5-hydroxytryptamine. *Eur J Pharmacol* 1977;**41**:133–43.

12. Boulton AA. Trace amines and mental disorders. *Can J Neurol Sci* 1980;**7**:261–3.
13. Baker GB, Bornstein RA, Yeragani VK. Trace amines and Tourette's syndrome. *Neurochem Res* 1993;**18**:951–6.
14. D'Andrea G, Terrazzino S, Leon A, Fortin D, Perini F, Granella F, et al. Elevated levels of circulating trace amines in primary headaches. *Neurology* 2004;**62**:1701–5.
15. D'Andrea G, Nordera G, Pizzolato G, Bolner A, Colavito D, Flaibani R, et al. Trace amine metabolism in Parkinson's disease: low circulating levels of octopamine in early disease stages. *Neurosci Lett* 2010;**469**:348–51.
16. Tomlinson S, Baker GB. Trace amines. In: Stolerman IP, editor. *Encyclopedia of psychopharmacology*. Berlin: Springer; 2010.
17. Narang D, Tomlinson S, Holt A, Mousseau DD, Baker GB. Trace amines and their relevance to psychiatry and neurology: a brief overview. *Bull Clin Psychopharmacol* 2011;**21**:73–9.
18. Branchek TA, Blackburn TP. Trace amine receptors as targets for novel therapeutics: legend, myth and fact. *Curr Opin Pharmacol* 2003;**3**:90–7.
19. Juorio AV, Greenshaw AJ, Zhu MY, Paterson IA. The effects of some neuroleptics and *d*-amphetamine on striatal 2-phenylethylamine in the mouse. *Gen Pharmacol* 1991;**22**:407–13.
20. Baker GB. Chronic administration of monoamine oxidase inhibitors: implications for interactions between trace amines and catecholamines. In: Dahlstrom A, Belmaker RH, Sandler M, editors. *Progress in catecholamine research. Part A: Basic aspects and peripheral mechanisms*. New York: Alan R. Liss; 1988.
21. Durden DA, Philips SR. Kinetic measurements of the turnover rates of phenylethylamine and tryptamine in vivo in the rat brain. *J Neurochem* 1980;**34**:1725–32.
22. Paterson IA, Juorio AV, Boulton AA. 2-Phenylethylamine: a modulator of catecholamine transmission in the mammalian central nervous system? *J Neurochem* 1990;**55**:1827–37.
23. Durden DA, Philips SR, Boulton AA. Identification and distribution of β-phenylethylamine in the rat. *Can J Biochem* 1973;**51**:995–1002.
24. Bertler A, Rosengren E. Occurrence and distribution of catecholamines in the brain. *Acta Physiol Scand* 1959;**47**:350–61.
25. Saavedra JM. Enzymatic isotopic assay for and presence of β-phenylethylamine in brain. *J Neurochem* 1974;**22**:211–6.
26. Jackson DM, Smythe DB. The distribution of β-phenylethylamine in discrete regions of the rat brain and its effect on brain noradrenaline, dopamine and 5-hydroxytryptamine levels. *Neuropharmacology* 1973;**12**:663–8.
27. Philips SR, Durden DA, Boulton AA. Identification and distribution of *p*-tyramine in the rat. *Can J Biochem* 1974;**2**:366–73.
28. Philips SR, Rozdilsky B, Boulton AA. Evidence for the presence of *m*-tyramine, *p*-tyramine, tryptamine, and phenylethylamine in the rat brain and several areas of the human brain. *Biol Psychiatry* 1978;**13**:51–7.
29. Philips SR, Davis BA, Durden DA, Boulton AA. Identification and distribution of *m*-tyramine in the rat. *Can J Biochem* 1975;**53**:65–9.
30. Philips SR, Durden DA, Boulton AA. Identification and distribution of tryptamine in the rat. *Can J Biochem* 1974;**52**:447–51.
31. Juorio AV, Sloley BD. Determination of trace amines and related compounds by high performance liquid chromatography with electrochemical or fluorometric detection. In: Holman RB, Cross AJ, Joseph MH, editors. *High performance liquid chromatography in neuroscience research*. Hoboken, NJ: John Wiley & Sons; 1993, p. 217–42.
32. Baker GB, LeGatt DF, Coutts RT. A gas chromatographic procedure for quantification of para-tyramine in rat brain. *J Neurosci Method* 1982;**5**:181–8.
33. Baker GB, Rao TS, Coutts RT. Electron-capture gas chromatographic analysis of beta-phenylethylamine in tissues and body fluids using pentafluorobenzenesulfonyl chloride for derivatization. *J Chromatogr* 1986;**381**:211–7.
34. Coutts RT, Baker GB, LeGatt DF, McIntosh GJ, Hopkinson G, Dewhurst WG. Screening for amines of psychiatric interest in urine using gas chromatography with electron-capture detection. *Progr Neuropsychopharmacol* 1981;**5**:565–8.
35. Karoum F, Nasrallah H, Potkin S, Chuang L, Moyer-Schwing J, Phillips I, et al. Mass fragmentography of phenylethylamine, *m*- and *p*-tyramine and related amines in plasma, cerebrospinal fluid, urine, and brain. *J Neurochem* 1979;**33**:201–12.
36. Dourish CT, Davis BA, Dyck LE, Jones RSG, Boulton AA. Alterations in trace amine and trace acid concentrations in isolated aggressive mice. *Pharmacol Biochem Behav* 1982;**17**:1291–4.
37. Larson AA, Dalo NL. Quantification of tryptamine in brain using high-performance liquid chromatography. *J Chromatogr* 1986;**375**:37–47.

38. Taga C, Tsuji M, Nakajima T. Rapid and sensitive determination of beta-phenylethylamine in animal brains by high performance liquid chromatography with fluorometric detection. *Biomed Chromatogr* 1989;**3**:118–20.
39. Tallman JF, Saavedra JM, Axelrod J. A sensitive enzymatic-isotopic method for the analysis of tyramine in brain and other tissues. *J Neurochem* 1976;**27**:465–9.
40. Danielson TJ, Boulton AA, Robertson HA. *m*-Octopamine, *p*-octopamine and phenylethanolamine in mammalian brain: a sensitive, specific assay and effects of drugs. *J Neurochem* 1977;**29**:1131–5.
41. Hess SM, Udenfriend S. A fluorometric procedure for the measurement of tyrptamine in tissues. *J Pharmacol Exp Ther* 1959;**127**:175–7.
42. Cooper SE, Venton BJ. Fast-scan cyclic voltammetry for the detection of tyramine and octopamine. *Anal Bioanal Chem* 2009;**394**:329–36.
43. Mumtaz M, Narasimhachari N, Friedel RO. Assay of trace amines by gas chromatography with nitrogen detector in biological samples using Cl8 SEP-PAK cartridges for sample cleanup. *Anal Biochem* 1982;**126**:365–73.
44. Durden DA, Davis BA. Determination of regional distributions of phenylethylamine and meta- and para-tyramine in rat brain regions and presence in human and dog plasma by an ultra-sensitive negative chemical ion gas chromatography-mass spectrometric (NCI-GC-MS) method. *Neurochem Res* 1993;**18**:995–1002.
45. Huebert ND, Schwach V, Richter G, Zreika M, Hinze C, Haegele KD. The measurement of β-phenylethylamine in human plasma and rat brain. *Anal Biochem* 1994;**221**:42–7.
46. Downer RG, Martin RJ. Analysis of monoamines and their metabolites by high performance liquid chromatography with coulometric electrochemical detection. *Life Sci* 1987;**41**:833–6.
47. Leung PY, Tsao CS. Preparation of an optimum mobile phase for the simultaneous determination of neurochemicals in mouse brain tissues by high-performance liquid chromatography with electrochemical detection. *J Chromatogr* 1992;**576**:245–54.
48. Durden DA, Juorio AV, Davis BA. Thin-layer chromatographic and high resolution mass spectrometric determination of beta-hydroxyphenylethylamines in tissues as dansyl-acetyl derivatives. *Anal Chem* 1980;**52**:1815–20.
49. Suzuki O, Yagi K. A fluorometric assay for beta-phenylethylamine in rat brain. *Anal Biochem* 1976;**75**:192–200.
50. David JC. An improved radioenzymatic and dansylation method for the determination of moctopamine, *p*-octopamine and phenylethanolamine in biological samples. *Ann Biol Anim Bioch Biophys* 1977;**17**:1101–5.
51. David JC, Delacour J. Brain contents of phenylethanolamine, *m*-octopamine and *p*-octopamine in the roman strains of rats. *Brain Res* 1980;**195**:231–5.
52. Philips SR. Radioenzymatic analysis of neurotransmitters. *Life Sci* 1987;**41**:877–80.
53. Durden DA, Davis BA, Boulton AA. Quantification of plasma phenylethylamine by electron capture negative ion gas chromatography-mass spectrometry of the *N*-acetyl-*N*-pentafluorobenzoyl derivative. *Biol Mass Spectrom* 1991;**20**:375–81.
54. Molinoff P, Axelrod J. Octopamine: normal occurrence in sympathetic nerves of rats. *Science* 1969;**164**:428–9.
55. Saavedra JM, Axelrod J. Demonstration and distribution of phenylethanolamine in brain and other tissues. *Proc Natl Acad Sci USA* 1973;**70**:769.

3

Synthesis and Neurochemistry of Trace Amines[*]

A. Pryor, S. Hart and M.D. Berry

Department of Biochemistry, Memorial University of Newfoundland,
St. John's, NL, Canada

OUTLINE

INTRODUCTION

Trace amines are endogenous compounds found in the nervous system of all species examined.[1] They are chemically related to the classic monoamine neurotransmitters dopamine, norepinephrine, and serotonin (Fig. 3.1), and consist primarily of 2-phenylethylamine, *p*-tyramine, *p*-octopamine, and tryptamine. In invertebrates, trace amines serve roles similar to the vertebrate adrenergic system, acting as neurotransmitters, neuromodulators, and neurohormones, with neuronal release suggested to occur from synaptic vesicles in a depolarization-dependent manner.[2,3] In contrast, in vertebrates, trace amines do not appear to

[*]All authors contributed equally to this work.

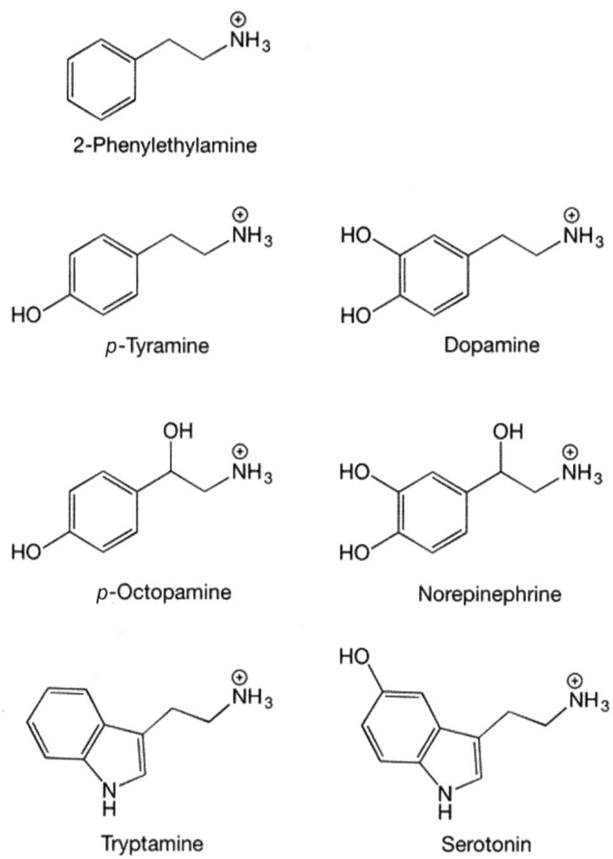

FIGURE 3.1   The structure of trace amines and the corresponding monoamine neurotransmitters. Trace amines show strong structural similarity to the monoamine neurotransmitters, differing only in the number of ring hydroxyl groups present.

act as neurotransmitters but instead modulate the activity of coexisting monoamine neuro-transmitters. This modulation is the subject of other chapters in this volume and the reader is referred to the chapters by Leone et al., Leo and Espinoza, and Berry for further details. Consistent with this divergence of function, vertebrate trace amine receptors are evolu-tionarily distinct from their invertebrate counterparts.[4,5] A family of trace amine-associated receptors (TAARs) is present in vertebrates, some of which are selectively activated by mul-tiple trace amines.[6,7] In invertebrate species, however, dedicated receptors for p-tyramine and p-octopamine are present.[2] In vertebrates, endogenous levels of trace amines are several hundred-fold lower than monoamine neurotransmitters,[1] and form the basis for the fam-ily name. In both invertebrates and vertebrates the primary synthetic route for trace amines is the enzymatic decarboxylation of L-amino acids, although the enzymes responsible vary. This chapter will review the current state of knowledge with respect to the control of the synthesis, storage, and release of trace amines in vertebrates and invertebrates, and will highlight the differences between the two in the trace amine receptor populations present.

VERTEBRATE TRACE AMINES

The synthesis of the major vertebrate trace amines is shown in Fig. 3.2. L-Tyrosine, L-phenylalanine, and L-tryptophan are all decarboxylated by aromatic L-amino acid decarboxylase (AADC; EC 4.1.1.28) giving rise to *p*-tyramine, 2-phenylethylamine, and tryptamine, respectively.[8-12] *Meta-* and *ortho-* isomers of tyramine have also been identified,[13,14] but are present endogenously in even lower levels than the *para*-isoforms and will not be considered further here. *p*-Octopamine can subsequently be formed from *p*-tyramine by the action of dopamine-β-hydroxylase (EC 1.14.17.1).[8,9] These two enzymes are common to the synthetic pathways of monoamine neurotransmitters, where they are regarded as being present in large excess and not rate-limiting. One possible exception to this situation is during L-3,4-dihydroxyphenylalanine (L-DOPA) therapy for Parkinson's disease where the exogenous administration of L-DOPA, in combination with marked loss of dopaminergic neurons, may result in AADC becoming rate-limiting.[15] By virtue of it being the

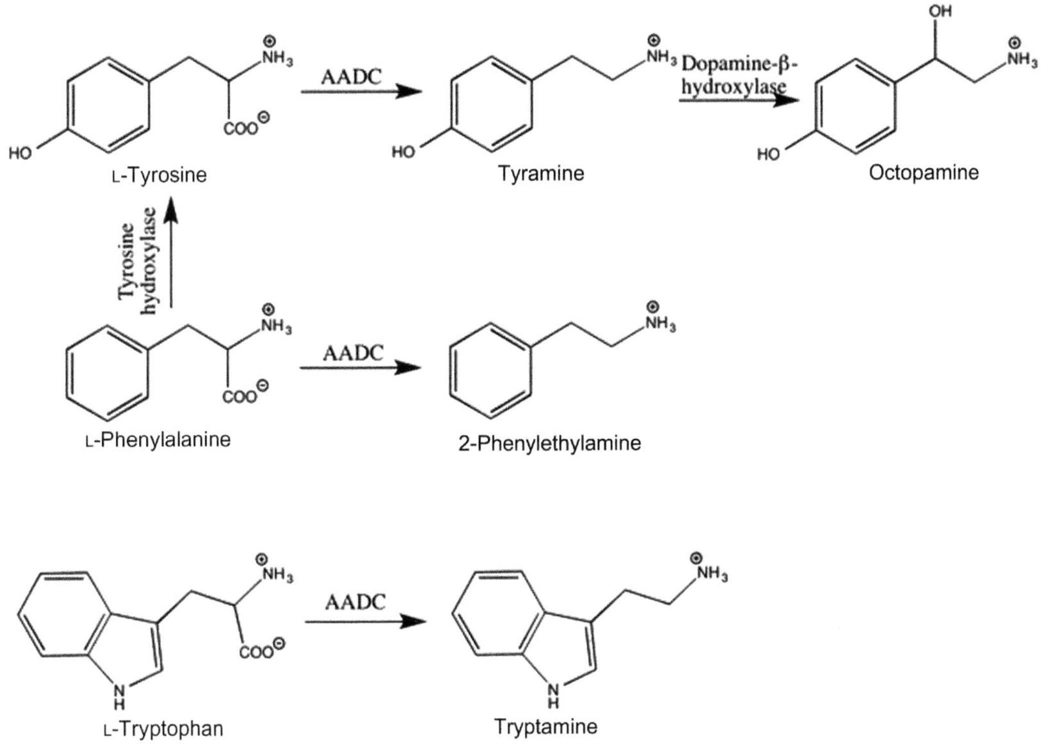

FIGURE 3.2 The primary synthetic route for the major vertebrate trace amines. Vertebrate trace amines are synthesized using the same enzymes as those involved in catecholamine and indoleamine neurotransmitter synthesis, with the exception that the initial ring hydroxylation step(s) are omitted. For simplicity L-phenylalanine conversion to L-tyrosine has been shown to occur via tyrosine hydroxylase. In some tissues and species, dedicated phenylalanine hydroxylase enzymes may also be present. AADC, aromatic L-amino acid decarboxylase.

only enzyme present, AADC is, however, rate-limiting for the synthesis of *p*-tyramine, 2-phenylethylamine, and tryptamine,[1,16] and is now known to be regulated.

Trace Amine Synthesis

The enzyme now known as AADC was first described in 1938,[17] as a component of guinea pig kidney responsible for catalyzing the decarboxylation of L-DOPA. In 1942, the findings of L-DOPA decarboxylation in kidney and liver of various other species was confirmed, and the reaction was also shown to be stereospecific, D-DOPA not being a substrate.[18] Decarboxylation of 5-hydroxytryptophan to serotonin by kidney preparations was subsequently described,[19] and initially ascribed to a novel decarboxylating activity. It was not until 1962 that decarboxylation by a common enzyme was suggested.[20] While L-5-hydroxytryptophan and L-DOPA were shown to be the preferred substrates, the trace amine precursors L-tryptophan, L-phenylalanine, and L-tyrosine were also shown to be decarboxylated by the same enzyme.[20–22] As a result of the broad substrate selectivity the enzyme was named AADC.

Active AADC is homodimeric with one pyridoxal 5′-phosphate cofactor bound per dimer.[23,24] The AADC gene consists of 15 exons spanning more than 85 kb on human chromosome 7p12.1-12.3.[25] This gives rise to monomer subunits 53.9 kDa in size and consisting of 480 amino acids. The AADC gene undergoes alternative splicing within its 5′ untranslated region allowing distinct control of neuronal and nonneuronal expression of the protein.[26–29] In addition to this, alternative splicing within the coding region is also possible, generating distinct isoforms of AADC.[30–33] One such splice variant lacks exon 3 from the full-length transcript,[30,32] giving a truncated monomer consisting of 442 amino acids ($AADC_{442}$). This novel splice variant is primarily found in neuronal cells, predominating over the full-length $AADC_{480}$ in most peripheral and central nervous system tissues. In contrast $AADC_{480}$ predominates in only a few tissues, including the pituitary gland, retina, olfactory bulb, hypothalamus, and mammillary body.[30] Of particular note $AADC_{442}$ is reported to not exhibit catalytic activity towards L-DOPA or L-5-hydroxytryptophan.[30] This raises an interesting question of why an apparently nonfunctional variant would be predominantly expressed in many neurons.

An even shorter isoform of AADC has also been described, consisting of only 338 amino acids.[33] The mRNA coding for this form of the protein is lacking exons 11–15 from the full-length transcript and also includes an alternative form of exon 10, which includes 23 amino acids that do not appear in other variants. This isoform is not found in the brain, but is abundantly expressed in the kidney, where AADC is thought to be involved in the regulation of sodium transport.[34]

The functional significance of such alternative transcripts is poorly defined. It is worth noting in this respect that the trace amine precursors were originally described to be very poor substrates for AADC with K_m values close to the limits of solubility.[21] Whether one, or more, of these shorter AADC variants shows increased activity toward trace amine synthesis is an interesting possibility that requires systematic investigation.

AADC Tissue Distribution

As alluded to above, AADC is found in both neuronal and nonneuronal tissues in vertebrates. Within nervous tissue AADC distribution largely mirrors that of central and peripheral monoaminergic neurons, being found within the primary catecholaminergic and serotonergic nuclei and projection areas, as would be expected for a monoamine synthetic enzyme. In addition, however, AADC is also found in a group of neurons that contain neither tyrosine hydroxylase nor serotonin.[35–37] Fifteen distinct groups of these so-called D-neurons have been identified. Phenotypically these cells appear ideally positioned to provide a trace aminergic system. Of particular note in this regard many of these groups interconnect with catecholaminergic neurons,[36] which would be consistent with the regulation of catecholaminergic neurotransmission reported for TAAR1 (see other chapters in this issue for a more detailed discussion of the effects of TAAR1 activation). In addition to neurons, AADC has also been reported to be expressed in glial cells,[38,39] although the putative role of glial AADC remains unknown.

Outside of the nervous system AADC is also found in various mammalian cell types, including in humans.[40–42] In the kidney, AADC is thought to be involved in the conversion of L-DOPA to dopamine during the regulation of sodium transport in the renal tubules.[34] In addition, various groups of endocrine cells, previously classified as the Amine Precursor Uptake and Decarboxylation (APUD) system, contain abundant AADC.[41] These cells are found in the pancreas, gastrointestinal tract, and thyroid, where they are involved in the secretion of various small polypeptides.[43] Such cells could synthesize trace amines, and indeed TAARs have been identified in many of these cell types.[44–46] Of particular note, TAAR1 was recently reported to be involved in the control of the endocrine function of gastric D-cells,[47] an effect that was proposed to occur following the uptake of dietary amino acids.

AADC Regulation

AADC is regulated in a biphasic manner, with an early-onset, short-term change in enzyme activity due to changes in phosphorylation status, followed by delayed-onset, longer-lasting changes in de novo protein synthesis.[48–51]

Regulation of AADC was first reported in the retina where increased activity was seen in response to light,[52] effects that were subsequently shown to be mimicked by both α_2-adrenoceptor and dopamine receptor antagonists.[53,54] In contrast, dopamine receptor agonists caused the opposite effect, both decreasing AADC activity and preventing the increase seen in response to light.[54] The same effects of dopamine receptor agonists and antagonists were subsequently seen in rat striatum,[55,56] rat striatal synaptosomes,[57] and various mouse brain regions.[49,58–60] While the approximate 30% change in AADC activity observed is insufficient to change endogenous dopamine levels,[59,61] such treatments have been shown to cause parallel changes in trace amine levels[62]; dopamine receptor antagonism increasing 2-phenylethylamine levels, while dopamine receptor agonists decrease them. Such effects are consistent with the suggestion that AADC activity is only rate-limiting with respect to trace amine synthesis.[1] Similar regulatory effects of neurotransmitter receptors on AADC activity have also been reported at 5-HT$_{1A}$ and 5-HT$_{2A}$ receptors,[60] as well as NMDA receptors.[63,64] As such there is good evidence that changes in neuronal AADC activity result in altered trace amine synthesis capacity and that this occurs in response to the activation status of neurotransmitter receptors.

This early, transient, regulation of AADC has now been established to be due to protein kinase activation. Consensus sequences for multiple protein kinases are present in the AADC protein and protein kinases A,[65,66] C,[57,67] and G[68] have all been shown to phosphorylate AADC with resultant increases in enzyme activity. The increased AADC activity appears to be due to an increase in V_{max} with no change in K_m for either L-DOPA or the pyridoxal phosphate cofactor.[55,67,68] It is worth noting that all of the above-described regulatory effects on AADC activity have been measured almost exclusively using L-DOPA as a substrate. Whether similar activity changes are seen with other substrates is largely unknown. Given that the regulation of AADC does not change endogenous dopamine levels, but does change trace amine levels, it would appear worthwhile reexamining these effects with substrates other than L-DOPA, in particular L-phenylalanine and L-tyrosine. In this respect it is important to note that there have been previous reports of differential regulation of L-DOPA and L-5-hydroxytryptophan decarboxylating activity by AADC.[23,69] Further, a preferential stimulation of trace amine synthesis via AADC in response to a variety of organic solvents has also previously been reported.[70,71]

Recently AADC expression/activity has been reported to be selectively increased in response to spinal cord transection.[72–74] The significance of the changes in response to spinal cord injury are poorly understood at present, although a role for spinal trace amines in controlling locomotor activity has also recently been proposed.[75] The cells in which the AADC increase following transection occurs is a matter of some conjecture, with evidence for increases in both neuronal and nonneuronal cells presented.[72] While most studies have focused on the regulation of neuronal AADC, a number of stimuli have also been reported to regulate AADC activity in various nonneuronal cell populations.[76,77]

Trace Amine Degradation

Trace amines, much like the closely related monoamine neurotransmitters, are metabolized primarily via monoamine oxidase (MAO). 2-Phenylethylamine remains the only known endogenous compound that shows a high selectivity for MAO-B.[78] The other trace amines are mixed substrates, being metabolized by both MAO-A and MAO-B.[79,80] Not being catechols (Fig. 3.1), trace amines do not appear to be substrates for catechol-O-methyl transferase, and as such O-methoxy derivatives have not been observed. In addition to oxidative deamination, the primary amine is also subject to methylation, and endogenous N-methyl and N,N-dimethyl derivatives of trace amines are possible.[81] The enzymes responsible for this conversion are unclear, but likely involve indoleamine-N-methyl transferase(s) (tryptamine) and phenylethanolamine-N-methyl transferase (other trace amines). Metabolism by cytochrome P450 isozymes is also possible,[82] but this is thought to be a very minor route of metabolism under normal conditions.[83]

Trace amines show an incredibly rapid turn-over, with the half-lives for the endogenous trace amine pool estimated at 16 s for tryptamine and 24 s for 2-phenylethylamine.[80] Such a rapid turn-over suggests that, unlike monoamine neurotransmitters, trace amines are not stored, and this indeed appears to be the case.

Trace Amine Storage and Release

Trace amines readily diffuse across synthetic lipid bilayers,[84] and their rate of release from synaptosomal preparations does not increase in response to K^+-induced depolarization,[85,86] and may even decrease.[84] This suggests that trace amines are not stored in synaptic vesicles, and simply diffuse across biological membranes, an interpretation consistent with observations that their rate of release is solely dependent on tissue levels.[86] Extracellular levels of trace amines also are not increased by reserpine treatment, further indicating they are not stored in synaptic vesicles,[87,88] or at least not in reserpine-sensitive vesicles. Such observations are fully consistent with the observed short half-life of the endogenous trace amine pool.[80] In such a system the extracellular levels of trace amines would be expected to be in a steady state, determined by the relative rates of synthesis and degradation. It is worth noting, however, that while K^+-induced depolarization does not increase release, it has been reported that veratridine-induced depolarization does increase the release of p-tyramine.[86] Although this requires confirmation, it raises the possibility of a subset of differentially regulated synaptic vesicles that contain trace amines.

Trace amine passage across nonneuronal membranes generally shows the same characteristics as above. Mosnaim et al.[89] recently showed that 2-phenylethylamine, and several of its derivatives, readily crossed the blood–brain barrier in a manner consistent with passive diffusion. Likewise, p-tyramine passage across intestinal epithelial membranes appears to be a diffusion-mediated process,[90] although entry into red blood cells has been suggested to involve a transporter.[91] 2-Phenylethylamine transport across intestinal epithelial cells has also been reported to be mediated by an unknown transporter.[92] On the basis of an apparent depolarization-induced decrease in trace amine release from synaptosomes, we have also suggested the presence of a trace amine transporter in neuronal membranes.[84] A number of known transporters do include trace amines in their substrate profile. At high concentrations trace amines can be transported by the dopamine,[93] norepinephrine,[94] and serotonin transporters,[95] and they have also been reported to be substrates for a number of the so-called "uptake 2" transporters,[96,97] including a nanomolar affinity at organic cation transporter 1 (SLC22A1).[98,99]

In summary, the available evidence indicates that trace amines readily diffuse across biological membranes and are not actively stored. There is some evidence that this diffusion can be modified by the presence of membrane transporters, although transporters selective for trace amines at physiological concentrations have not yet been identified.

Vertebrate Trace Amine Receptors

In 2001 two groups independently identified a new class of G-protein-coupled receptors that exhibited selectivity for trace amines.[6,7] Subsequently named TAARs, the majority of this family remain orphan receptors, with endogenous ligands awaiting identification. TAAR1 and TAAR4 are, however, activated by endogenous trace amines,[100] and provide a molecular target for trace amine-mediated effects. Both TAAR1 and TAAR4 appear to be coupled to the G_s signal transduction cascade,[100] with G-protein-independent coupling of TAAR1 to β-arrestin 2 also recently reported.[101]

Although trace amines have been identified in every species examined, both vertebrate and invertebrate, TAARs are only found in vertebrate species.[4,5] As described below, evolutionarily distinct trace amine receptors are present in invertebrates, indicating that the ability to detect (and therefore utilize) trace amines has arisen at least twice during evolution. Twenty-eight different subfamilies of TAAR have been identified, with multiple isoforms of many of these present.[102,103] With the exception of TAAR1, all TAAR subtypes are expressed in the olfactory epithelium,[104–106] where they appear to participate in both conspecific and heterospecific olfactory signaling.[107–110] TAAR10–28 are only found in aquatic vertebrates,[103,111] and this is thought to be a function of the different physicochemical characteristics required for between individual signaling molecules in aquatic versus terrestrial environments. Although involved in olfaction, a number of TAAR isoforms are also expressed throughout the body, including neuronal,[6,112,113] immune,[114,115] cardiovascular,[116,117] skeletal muscle,[118] and gastrointestinal systems.[45,119]

A thorough description of the pharmacology of TAAR is beyond the scope of this chapter and the reader is referred to the chapter by Berry in this volume for a detailed review of known TAAR-mediated effects. TAAR1 is, however, garnering considerable attention as a novel target for the pharmacotherapy of various psychiatric disorders. It is interesting to note in this respect that loss of the putative trace aminergic D-neurons was previously reported in schizophrenia,[120] and a combined trace amine/D-neuron hypothesis of schizophrenia was recently suggested.[121]

INVERTEBRATE TRACE AMINES

Trace Amines as Invertebrate Neurotransmitters

In invertebrate species the trace amines p-tyramine and p-octopamine are well-established neurotransmitters, thought to serve the role of invertebrate versions of norepinephrine and epinephrine.[2,122] Consistent with this, and as described briefly below, invertebrate receptors for p-tyramine and p-octopamine show close homology to vertebrate adrenergic receptors, but are only distantly related to TAAR. Synthesis of p-tyramine and p-octopamine in invertebrates is analogous to vertebrate synthesis (cf Figs. 3.2 and 3.3) but dedicated tyrosine decarboxylase enzymes exist for the synthesis of tyramine.[123–125] For example, there have been two functional gene transcripts identified for tyrosine decarboxylase in Drosophila,[126] which appear to serve the function of neuronal and nonneuronal isoforms. These are in addition to AADC-like enzymes, some of which (at least in Drosophila) do not contribute significantly to p-tyramine and p-octopamine synthesis.[127] The p-tyramine formed can subsequently be converted to p-octopamine by the action of a dedicated

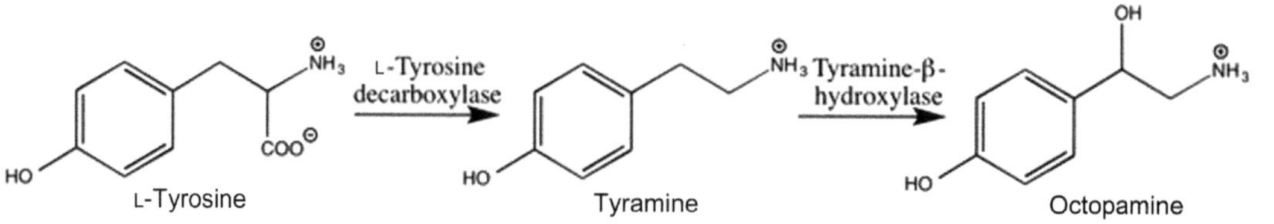

FIGURE 3.3 Synthetic route for p-tyramine and p-octopamine in invertebrates. Invertebrate synthesis of trace amines occurs by comparable processes to those in invertebrates, but involves unique enzymes that have evolved to show high selectivity for L-tyrosine and p-tyramine as substrates. This contrasts with vertebrate synthesis where the polysubstrate AADC is used for p-tyramine synthesis and dopamine-β-hydroxylase for subsequent conversion to p-octopamine. Although homologs of these two vertebrate enzymes exist in at least some invertebrate species, they do not appear to utilize L-tyrosine or p-tyramine as substrates.

tyramine-β-hydroxylase enzyme (Fig. 3.3),[128–130] with multiple isoforms of this enzyme also present in some species.[130] This contrasts with the use of dopamine-β-hydroxylase for the synthesis of *p*-octopamine by vertebrates.

Invertebrate Trace Amine Receptors

Much like the case with TAAR, expression of *p*-tyramine and *p*-octopamine receptors is not limited to the nervous system of invertebrates but is widely distributed throughout the body.[2,131] This widespread distribution is consistent with the proposed physiological effects of *p*-tyramine and *p*-octopamine, spanning learning, and memory,[132] cardiac function,[133] locomotion,[134] energy metabolism,[2] feeding behavior,[135] olfaction-mediated behaviors,[136,137] and reproduction.[138,139] *p*-Tyramine and *p*-octopamine signaling in invertebrates is the subject of dedicated chapters in this volume by Blenau and Baumann, and Ma et al., and the reader is referred to these for a more detailed discussion. Below we give a very brief overview of the receptors identified in various invertebrate species, by way of comparison to the vertebrate TAAR previously described. It should be noted that although *p*-octopamine- and *p*-tyramine-mediated effects have been widely studied in various species, the receptors through which such effects are mediated are much less studied, and it is often unclear whether described receptors are truly novel isoforms or simply orthologs of receptors previously described in other species.

p-Octopamine Receptors

At least four distinct octopamine receptors, OAMB, Oct1βR, Oct2βR, and Oct3βR,[131,132] have been identified, the latter three of which bear strong structural homology to the vertebrate β1–3 adrenergic receptors.[140] The first octopamine receptor described was cloned from the mushroom bodies of *Drosophila melanogaster* brain,[141] and therefore named OAMB. This (and its orthologs) have subsequently been proposed to be termed OctαR on the basis of their homology to α-adrenergic receptors,[132,142] although the OAMB terminology remains widely used in the literature. Subsequent studies have identified OAMB orthologs in *Apis mellifera*[143] and *Anopheles gambiae*,[144] amongst others. Historically OAMB has been the most studied octopamine receptor with activation reported to be coupled to accumulation of calcium.[132,141] Oct1βR, Oct2βR, and Oct3βR, in contrast, appear to be coupled to accumulation of cAMP following stimulation of adenylate cyclase.[140,145] In the most recent phylogenetic analysis, OAMB was reported to be evolutionarily closer related to the *p*-tyramine receptors (see below) than to the OctβR isoforms.[131]

p-Tyramine Receptors

At least three G-protein-coupled *p*-tyramine receptors, TyrRI, TyrRII, and TyrRIII have been described, having first been suggested in 1990.[146] Characterization of the original *p*-tyramine receptor in *Drosophila* led to the suggestion that it was a mixed octopamine/tyramine receptor,[147,148] although subsequent studies have shown a higher affinity for *p*-tyramine in most species. As a consequence the receptor has become known as the tyramine 1 (TyrRI) receptor.[132,142] At least in some species TyrRI may exhibit agonist-specific coupling[147]; *p*-tyramine binding couples to adenylate cyclase inhibition, whereas *p*-octopamine binding results in an increase of cytosolic calcium levels.[132,147]

In *Drosophila*, both TyrRII and TyrRIII show much greater selectivity for *p*-tyramine, with no crossreactivity to *p*-octopamine.[142,149] Orthologs of TyrRII have been reported in various other invertebrate species,[142,150] although TyrRIII orthologs have not yet been identified. TyrRII is coupled to a release of intracellular calcium stores.[149,150] In contrast, TyrRIII appears to couple to an inhibition of adenylate cyclase,[142] and is responsive to a wide variety of amines in addition to *p*-tyramine.

At least one ionotropic *p*-tyramine activated receptor (LGC-55) has also been reported in *Caenorhabditis elegans*.[151] This *p*-tyramine-gated chloride channel is involved in locomotor coordination through hyperpolarization of the neck muscles. Orthologs of LGC-55 were also identified in other species including *Caenorhabditis briggsae*, *Caenorhabditis remanei*, *Pristionchus pacificus*, and *Brugia malayi*.[151]

Trace Amine Synthesis by Prokaryotes

In recent years there has been a growing appreciation of the role of the gastrointestinal microbiome in health and disease, with links to various neurologic and psychiatric disorders suggested.[152] Production of trace amines by various prokaryotes, particularly in the gastrointestinal tract, is well established,[153] and provides a possible molecular mechanism for such relationships. A vast array of decarboxylase enzymes are present in prokaryotes, and many of these include L-amino acids in their substrate profile.[154,155] A review of prokaryote enzymes of potential relevance to trace amine metabolism is far beyond the scope of this chapter. Below we provide some brief examples of prokaryotes known to be commensal to the human gastrointestinal system that have been shown to synthesize trace amines.

Enterococcus faecium contains decarboxylase enzymes that use both L-phenylalanine and L-tyrosine as substrates, producing 2-phenylethylamine and *p*-tyramine, respectively.[156] *Clostridium sporogenes* and *Ruminococcus gnavus* also both contain decarboxylase activities capable of producing trace amines.[157] Thought originally to be selective for L-tyrosine, and previously known as a tyrosine decarboxylase, following purification the enzyme CLOSO_02083 was found to preferentially decarboxylate L-tryptophan to tryptamine.[157] Similarly in *R. gnavus*, the RUMGNA_01526 enzyme showed over 1000-fold selectivity toward L-tryptophan as a substrate in comparison to L-tyrosine and L-phenylalanine. In both cases the tryptamine produced was subsequently secreted into the extracellular fluid.[157] In addition to the production of trace amines, the precursor amino acids L-phenylalanine and L-tryptophan, which are essential amino acids in humans, can be produced and released from commensal gut microbes such as *Bifidobacterium* sp., and this has been linked to clinical conditions.[158] Further, antibiotic treatment has been reported to modify host markers of L-tyrosine, L-phenylalanine, and L-tryptophan metabolism as a result of changes in the gut microbiota.[154] As such, production of both trace amines and their precursors by commensal microbes is possible, and this may lead to alterations in endogenous trace amine levels, and hence altered host physiology following activation of TAAR by the produced trace amines.

Given the increasing evidence for a role of TAAR1 (and presumably other TAARs) in central nervous system function, immune responses, and possibly nutrient metabolism, and the growing appreciation of the role of commensal microbes in human health and disease, it is anticipated that a role for trace amine production in such effects will receive growing attention in future years. Such a relationship may not be limited to the gastrointestinal microbiome, but could also include the genitourinary system. For example, biogenic amine-producing bacterial taxa have been reported to be associated with bacterial vaginosis, with cadaverine, putrescine, and *p*-tyramine particularly prevalent amines.[155] Notably, both cadaverine and *p*-tyramine are known TAAR agonists.

CONCLUDING STATEMENTS

Although trace amines are found in both vertebrate and invertebrate species, it is now well established that the ability to utilize these compounds in cellular signaling arose independently. Invertebrates and vertebrates have evolved different synthetic enzymes for the production of trace amines, which, although analogous, are evolutionarily unrelated. Similarly, the invertebrate and vertebrate families of trace amine receptors are not homologous. Consistent with the evolutionary distance between vertebrates and invertebrates for trace amine synthetic and detection capabilities, trace amine functioning is also distinct between the two. In invertebrates *p*-tyramine and *p*-octopamine serve the role of an adrenergic system, acting as bona fide neurotransmitters, stored in synaptic vesicles and released in an activity-dependent manner. In contrast, in vertebrates, trace amines appear to regulate the efficacy of coexisting transmitters, and are neither stored in synaptic vesicles, nor released in an activity-dependent, exocytotic manner. As such, although invertebrate species are often used as a model system for vertebrate neuronal function, the available evidence suggests that they may not be suitable model systems for studying responses involving trace amines.

References

1. Berry MD. Mammalian central nervous system trace amines. Pharmacologic amphetamines, physiologic neuromodulators. *J Neurochem* 2004;**90**(2):257–71.
2. Roeder T. Tyramine and octopamine: ruling behavior and metabolism. *Annu Rev Entomol* 2005;**50**:447–77.
3. Lange AB. Tyramine: from octopamine precursor to neuroactive chemical in insects. *Gen Comp Endocrinol* 2009;**162**(1):18–26.
4. Gloriam DE, Bjarnadottir TK, Schioth HB, Fredriksson R. High species variation within the repertoire of trace amine receptors. *Ann NY Acad Sci* 2005;**1040**:323–7.
5. Lindemann L, Ebeling M, Kratochwil NA, Bunzow JR, Grandy DK, Hoener MC. Trace amine-associated receptors form structurally and functionally distinct subfamilies of novel G protein-coupled receptors. *Genomics* 2005;**85**(3):372–85.
6. Borowsky B, Adham N, Jones KA, et al. Trace amines: identification of a family of mammalian G protein-coupled receptors. *Proc Natl Acad Sci USA* 2001;**98**(16):8966–71.
7. Bunzow JR, Sonders MS, Arttamangkul S, et al. Amphetamine, 3,4-methylenedioxymethamphetamine, lysergic acid diethylamide, and metabolites of the catecholamine neurotransmitters are agonists of a rat trace amine receptor. *Mol Pharmacol* 2001;**60**(6):1181–8.
8. Boulton AA, Wu PH. Biosynthesis of cerebral phenolic amines. I. In vivo formation of *p*-tyramine, octopamine, and synephrine. *Can J Biochem* 1972;**50**(3):261–7.
9. Boulton AA, Wu PH. Biosynthesis of cerebral phenolic amines. II. In vivo regional formation of *p*-tyramine and octopamine from tyrosine and dopamine. *Can J Biochem* 1973;**51**(4):428–35.
10. Snodgrass SR, Iversen LL. Formation and release of 3H-tryptamine from 3H-tryptophan in rat spinal cord slices. *Adv Biochem Psychopharmacol* 1974;**10**:141–50.

11. Silkaitis RP, Mosnaim AD. Pathways linking L-phenylalanine and 2-phenylethylamine with *p*-tyramine in rabbit brain. *Brain Res* 1976;**114**(1):105–15.

12. Dyck LE, Yang CR, Boulton AA. The biosynthesis of *p*-tyramine, *m*-tyramine, and β-phenylethylamine by rat striatal slices. *J Neurosci Res* 1983;**10**(2):211–20.

13. Boulton AA. Identification, distribution, metabolism, and function of meta and para tyramine, phenylethylamine and tryptamine in brain. *Adv Biochem Psychopharmacol* 1976;**15**:57–67.

14. Davis BA. Biogenic amines and their metabolites in body fluids of normal, psychiatric and neurologic subjects. *J Chromatogr* 1989;**466**:89–218.

15. Berry MD, Juorio AV, Li XM, Boulton AA. Aromatic L-amino acid decarboxylase: a neglected and misunderstood enzyme. *Neurochem Res* 1996;**21**(9):1075–87.

16. Paterson IA, Juorio AV, Boulton AA. 2-Phenylethylamine: a modulator of catecholamine transmission in the mammalian central nervous system? *J Neurochem* 1990;**55**(6):1827–37.

17. Holtz P, Heise R, Ludtke K. Fermentativer abbau von l-dioxyphenylalanin (dopa) durch niere. *Naunyn Schmiedebergs Arch Pharmacol* 1938;**191**(1):87–118.

18. Blaschko H. The activity of l(–)-dopa decarboxylase. *J Physiol* 1942;**101**(3):337–49.

19. Udenfriend S, Clark CT, Titus E. 5-Hydroxytryptophan decarboxylase—a new route of metabolism of tryptophan. *J Am Chem Soc* 1953;**75**(2):501–2.

20. Lovenburg W, Weissbach H, Udenfriend S. Aromatic L-amino acid decarboxylase. *J Biol Chem* 1962;**237**(1):89–93.

21. Christenson JG, Dairman W, Udenfriend S. Preparation and properties of a homogeneous aromatic L-amino acid decarboxylase from hog kidney. *Arch Biochem Biophys* 1970;**141**(1):356–67.

22. Saavedra JM, Axelrod J. Effect of drugs on the tryptamine content of rat tissues. *J Pharmacol Exp Ther* 1973;**185**(3):523–9.

23. Siow YL, Dakshinamurti K. Effect of pyridoxine deficiency on aromatic L-amino acid decarboxylase in adult rat brain. *Exp Brain Res* 1985;**59**(3):575–81.

24. Burkhard P, Dominici P, Borri-Voltattorni C, Jansonius JN, Malashkevich VN. Structural insight into Parkinson's disease treatment from drug-inhibited DOPA decarboxylase. *Nat Struct Biol* 2001;**8**(11):963–7.

25. Sumi-Ichinose C, Ichinose H, Takahashi E, Hori T, Nagatsu T. Molecular cloning of genomic DNA and chromosomal assignment of the gene for human aromatic L-amino acid decarboxylase, the enzyme for catecholamine and serotonin biosynthesis. *Biochemistry* 1992;**31**(8):2229–38.

26. Albert VR, Lee MR, Bolden AH, Wurzburger RJ, Aguanno A. Distinct promoters direct neuronal and nonneuronal expression of rat aromatic L-amino acid decarboxylase. *Proc Natl Acad Sci USA* 1992;**89**(24):12053–7.

27. Ichinose H, Sumi-Ichinose C, Ohye T, Hagino Y, Fujita K, Nagatsu T. Tissue-specific alternative splicing of the first exon generates two types of mRNAs in human aromatic L-amino acid decarboxylase. *Biochemistry* 1992;**31**(46):11546–50.

28. Hahn SL, Hahn M, Kang UJ, Joh TH. Structure of the rat aromatic L-amino acid decarboxylase gene: evidence for an alternative promoter usage. *J Neurochem* 1993;**60**(3):1058–64.

29. Sumi-Ichinose C, Hasegawa S, Ichinose H, et al. Analysis of the alternative promoters that regulate tissue-specific expression of human aromatic-L amino acid decarboxylase. *J Neurochem* 1995;**64**(2):514–24.

30. O'Malley KL, Harmon S, Moffat M, Uhland-Smith A, Wong S. The human aromatic L-amino acid decarboxylase gene can be alternatively spliced to generate unique protein isoforms. *J Neurochem* 1995;**65**:2409–16.

31. Rorsman F, Husebye ES, Winqvist O, Bjork E, Karlsson FA, Kampe O. Aromatic-L-amino acid decarboxylase, a pyridoxal phosphate-dependent enzyme, is a b-cell autoantigen. *Proc Natl Acad Sci USA* 1995;**92**:8626–9.

32. Chang YT, Mues G, Hyland K. Alternative splicing in the coding region of human aromatic L-amino acid decarboxylase mRNA. *Neurosci Lett* 1996;**202**(3):157–60.

33. Vassilacopoulou D, Sideris DC, Vassiliou AG, Fragoulis EG. Identification and characterization of a novel form of the human L-dopa decarboxylase mRNA. *Neurochem Res* 2004;**29**(10):1817–23.

34. Aperia A, Hokfelt T, Meister B, et al. The significance of L-amino acid decarboxylase and DARPP-32 in the kidney. *Am J Hypertens* 1990;**3**(6 Pt 2):11S–3S.

35. Jaeger CB, Teitelman G, Joh TH, Albert VR, Park DH, Reis DJ. Some neurons of the rat central nervous system contain aromatic-L-amino acid decarboxylase but not monoamines. *Science* 1983;**219**(4589):1233–5.

36. Jaeger CB, Ruggiero DA, Albert VR, Park DH, Joh TH, Reis DJ. Aromatic L-amino acid decarboxylase in the rat brain: immunohistochemical localization in neurons of the brain stem. *Neuroscience* 1984;**11**(3):691–713.

37. Kitahama K, Ikemoto K, Jouvet A, et al. Aromatic L-amino acid decarboxylase-immunoreactive structures in human midbrain, pons, and medulla. *J Chem Neuroanat* 2009;**38**(2):130–40.

38. Li XM, Juorio AV, Paterson IA, Walz W, Zhu MY, Boulton AA. Gene expression of aromatic L-amino acid decarboxylase in cultured rat glial cells. *J Neurochem* 1992;**59**(3):1172–5.

39. Juorio AV, Li XM, Walz W, Paterson IA. Decarboxylation of L-dopa by cultured mouse astrocytes. *Brain Res* 1993;**626**(1–2):306–9.

40. Lindstrom P, Sehlin J. Mechanisms underlying the effects of 5-hydroxytryptamine and 5-hydroxytrptophan in pancreatic islets. A proposed role for L-aromatic amino acid decarboxylase. *Endocrinology* 1983;**112**:1524–9.

41. Lauweryns JM, Van Ranst L. Immunocytochemical localization of aromatic L-amino acid decarboxylase in human, rat, and mouse bronchopulmonary and gastrointestinal endocrine cells. *J Histochem Cytochem* 1988;**36**(9):1181–6.

42. Takayanagi M, Watanabe T. Immunocytochemical colocalization of insulin, aromatic L-amino acid decarboxylase, dopamine-beta-hydroxylase, S-100 protein and chromogranin A in B-cells of the chicken endocrine pancreas. *Tissue Cell* 1996;**28**(1):17–24.

43. Pearse AG. The diffuse neuroendocrine system: peptides, amines, placodes and the APUD theory. *Prog Brain Res* 1986;**68**:25–31.

44. Regard JB, Kataoka H, Cano DA, et al. Probing cell type-specific functions of Gi in vivo identifies GPCR regulators of insulin secretion. *J Clin Invest* 2007;**117**(12):4034–43.

45. Ito J, Ito M, Nambu H, et al. Anatomical and histological profiling of orphan G-protein-coupled receptor expression in gastrointestinal tract of C57BL/6J mice. *Cell Tissue Res* 2009;**338**(2):257–69.

46. Revel FG, Moreau JL, Pouzet B, et al. A new perspective for schizophrenia: TAAR1 agonists reveal antipsychotic- and antidepressant-like activity, improve cognition and control body weight. *Mol Psychiatry* 2013;**18**(5):543–56.

47. Adriaenssens A, Lam BY, Billing L, et al. A transcriptome-led exploration of molecular mechanisms regulating somatostatin-producing D-cells in the gastric epithelium. *Endocrinology* 2015;**156**:3924–36.

48. Buckland PR, O'Donovan MC, McGuffin P. Changes in dopa decarboxylase mRNA but not tyrosine hydroxyase mRNA levels in rat brain following antipsychotic treatment. *Psychopharmacology (Berl)* 1992;**108**(1–2):98–102.

49. Hadjiconstantinou M, Wemlinger TA, Sylvia CP, Hubble JP, Neff NH. Aromatic L-amino-acid decarboxylase activity of mouse striatum is modulated via dopamine-receptors. *J Neurochem* 1993;**60**(6):2175–80.

50. Buckland PR, Spurlock G, McGuffin P. Amphetamine and vigabatrin down regulate aromatic L-amino acid decarboxylase mRNA levels. *Brain Res Mol Brain Res* 1996;**35**(1–2):69–76.

51. Buckland PR, Marshall R, Watkins P, McGuffin P. Does phenylethylamine have a role in schizophrenia?: LSD and PCP up-regulate aromatic L-amino acid decarboxylase mRNA levels. *Mol Brain Res* 1997;**49**(1–2):266–70.

52. Hadjiconstantinou M, Rossetti Z, Silvia C, Krajnc D, Neff NH. Aromatic L-amino acid decarboxylase activity of the retina is modulated in vivo by environmental light. *J Neurochem* 1988;**51**:1560–4.

53. Rossetti Z, Krajnc D, Neff NH, Hadjiconstantinou M. Modulation of retinal aromatic L-amino acid decarboxylase via α_2 adrenoceptors. *J Neurochem* 1989;**52**:647–52.

54. Rossetti ZL, Silvia CP, Krajnc D, Neff NH, Hadjiconstantinou M. Aromatic L-amino-acid decarboxylase is modulated by D1 dopamine-receptors in rat retina. *J Neurochem* 1990;**54**(3):787–91.

55. Zhu MY, Juorio AV, Paterson IA, Boulton AA. Regulation of aromatic L-amino acid decarboxylase by dopamine receptors in the rat brain. *J Neurochem* 1992;**58**(2):636–41.

56. Zhu MY, Juorio AV, Paterson IA, Boulton AA. Regulation of striatal aromatic L-amino acid decarboxylase: effects of blockade or activation of dopamine receptors. *Eur J Pharmacol* 1993;**238**(2–3):157–64.

57. Zhu M-Y, Juorio AV, Paterson IA, Boulton AA. Regulation of aromatic L-amino acid decarboxylase in rat striatal synaptosomes: effects of dopamine receptor agonists and antagonists. *Br J Pharmacol* 1994;**112**:23–30.

58. Cho S, Neff NH, Hadjiconstantinou M. Regulation of tyrosine hydroxylase and aromatic L-amino acid decarboxylase by dopaminergic drugs. *Eur J Pharmacol* 1997;**323**(2–3):149–57.

59. Cho S, Duchemin AM, Neff NH, Hadjiconstantinou M. Tyrosine hydroxylase, aromatic L-amino acid decarboxylase and dopamine metabolism after chronic treatment with dopaminergic drugs. *Brain Res* 1999;**830**(2):237–45.

60. Neff NH, Wemlinger TA, Duchemin AM, Hadjiconstantinou M. Clozapine modulates aromatic L-amino acid decarboxylase activity in mouse striatum. *J Pharmacol Exp Ther* 2006;**317**(2):480–7.

61. Berry MD, Scarr E, Zhu M-Y, Paterson IA, Juorio AV. The effects of administration of monoamine ocidase inhibitors on rat striatal neurone responses to dopamine. *Br J Pharmacol* 1994;**113**:1159–66.

62. Juorio AV, Greenshaw AJ, Zhu MY, Paterson IA. The effects of some neuroleptics and D-amphetamine on striatal 2-phenylethylamine in the mouse. *Gen Pharmacol* 1991;**22**(2):407–13.

63. Hadjiconstantinou M, Rossetti Z, Wemlinger TA, Neff NH. Dizocilpine enhances striatal tyrosine hydroxylase and aromatic L-amino acid decarboxylase activity. *Eur J Pharmacol* 1995;**289**(1):97–101.

64. Fisher A, Biggs CS, Starr MS. Differential effects of NMDA and non-NMDA antagonists on the aromatic L-amino acid decarboxylase activity in the nigrostriatal pathway of the rat. *Brain Res* 1998;**792**:126–32.

65. Young EA, Neff NH, Hadjiconstantinou M. Evidence for cyclic AMP-mediated increase of aromatic L-amino acid decarboxylase activity in the striatum and midbrain. *J Neurochem* 1993;**60**(6):2331–3.

66. Duchemin AM, Berry MD, Neff NH, Hadjiconstantinou M. Phosphorylation and activation of brain aromatic L-amino acid decarboxylase by cyclic AMP-dependent protein kinase. *J Neurochem* 2000;**75**(2):725–31.

67. Young EA, Neff NH, Hadjiconstantinou M. Phorbol ester administration transiently increases aromatic L-amino acid decarboxylase activity of the mouse striatum and midbrain. *J Neurochem* 1994;**63**:694–7.

68. Duchemin AM, Neff NH, Hadjiconstantinou M. Aromatic L-amino acid decarboxylase phosphorylation and activation by PKGIalpha in vitro. *J Neurochem* 2010;**114**(2):542–52.

69. Rahman MK, Nagatsu T, Kato T. Aromatic L-amino-acid decarboxylase activity in central and peripheral-tissues and serum of rats with L-Dopa and L-5-hydroxytryptophan as substrates. *Biochem Pharmacol* 1981;**30**(6):645–9.

70. Juorio AV, Yu PH. Effects of benzene and other organic solvents on the decarboxylation of some brain aromatic-L-amino acids. *Biochem Pharmacol* 1985;**34**(9):1381–7.

71. Juorio AV, Yu PH. Effects of benzene and pyridine on the concentration of mouse striatal tryptamine and 5-hydroxytryptamine. *Biochem Pharmacol* 1985;**34**(20):3774–6.

72. Li Y, Li L, Stephens MJ, et al. Synthesis, transport, and metabolism of serotonin formed from exogenously applied 5-HTP after spinal cord injury in rats. *J Neurophysiol* 2014;**111**(1):145–63.

73. Wienecke J, Ren LQ, Hultborn H, et al. Spinal cord injury enables aromatic L-amino acid decarboxylase cells to synthesize monoamines. *J Neurosci* 2014;**34**(36):11984–2000.

74. Azam B, Wienecke J, Jensen DB, Azam A, Zhang M. Spinal cord hemisection facilitates aromatic L-amino acid decarboxylase cells to produce serotonin in the subchronic but not the chronic phase. *Neural Plast* 2015;**2015**:549671.

75. Gozal EA, O'Neill BE, Sawchuk MA, et al. Anatomical and functional evidence for trace amines as unique modulators of locomotor function in the mammalian spinal cord. *Front Neural Circuits* 2014;**8**:134.

76. Hayashi M, Yamaji Y, Kitajima W, Saruta T. Aromatic L-amino acid decarboxylase activity along the rat nephron. *Am J Physiol* 1990;**258**(1 Pt 2):F28–33.

77. Wessel TC, Joh TH. Parallel upregulation of catecholamine-synthesizing enzymes in rat brain and adrenal gland: effects of reserpine and correlation with immediate early gene expression. *Brain Res Mol Brain Res* 1992;**15**(3–4):349–60.

78. Yang H-YT, Neff NH. -Phenylethylamine: a specific substrate for type B monoamine oxidase. *J Pharmacol Exp Ther* 1973;**187**:365–71.

79. Philips SR, Boulton AA. The effect of monoamine oxidase inhibitors on some arylalkylamines in rat striatum. *J Neurochem* 1979;**33**:159–67.

80. Durden DA, Philips SR. Kinetic measurements of the turnover rates of phenylethylamine and tryptamine in vivo in the rat brain. *J Neurochem* 1980;**34**(6):1725–32.

81. Lindemann L, Hoener MC. A renaissance in trace amines inspired by a novel GPCR family. *Trends Pharmacol Sci* 2005;**26**(5):274–81.

82. Niwa T, Murayama N, Umeyama H, Shimizu M, Yamazaki H. Human liver enzymes responsible for metabolic elimination of tyramine; a vasopressor agent from daily food. *Drug Metab Lett* 2011;**5**(3):216–9.

83. Yu AM, Granvil CP, Haining RL, et al. The relative contribution of monoamine oxidase and cytochrome p450 isozymes to the metabolic deamination of the trace amine tryptamine. *J Pharmacol Exp Ther* 2003;**304**(2):539–46.

84. Berry MD, Shitut MR, Almousa A, Alcorn J, Tomberli B. Membrane permeability of trace amines: evidence for a regulated, activity-dependent, nonexocytotic, synaptic release. *Synapse* 2013;**67**(10):656–67.

85. Henry DP, Russell WL, Clemens JA, Plebus LA. Phenylethylamine and *p*-tyramine in the extracellular space of the rat brain: quantification using a new radioenzymatic assay and in situ microdialysis. In: Boulton AA, Juorio AV, Downer RGH, editors. *Trace amines; comparative and clinical neurobiology*. Totowa, NJ: Humana Press; 1988. p. 239–50.

86. Dyck LE. Release of some endogenous trace amines from rat striatal slices in the presence and absence of a monoamine-oxidase inhibitor. *Life Sci* 1989;**44**(17):1149–56.

87. Boulton AA, Juorio AV, Philips SR, Wu PH. The effects of reserpine and 6-hydroxydopamine on the concentrations of some arylalkylamines in rat brain. *Br J Pharmacol* 1977;**59**:209–14.

88. Juorio AV, Greenshaw AJ, Wishart TB. Reciprocal changes in striatal dopamine and β-phenylethylamine induced by reserpine in the presence of monoamine oxidase inhibitors. *Naunyn Schmiedebergs Arch Pharmacol* 1988;**338**:644–8.

89. Mosnaim AD, Callaghan OH, Hudzik T, Wolf ME. Rat brain-uptake index for phenylethylamine and various monomethylated derivatives. *Neurochem Res* 2013;**38**(4):842–6.

90. Tchercansky DM, Acevedo C, Rubio MC. Studies of tyramine transfer and metabolism using an in-vitro intestinal preparation. *J Pharm Sci* 1994;**83**(4):549–52.

91. Blakeley AG, Nicol CJ. Accumulation of amines by rabbit erythrocytes in vitro. *J Physiol* 1978;**277**:77–90.

92. Fischer W, Neubert RHH, Brandsch M. Transport of phenylethylamine at intestinal epithelial (Caco-2) cells: mechanism and substrate specificity. *Eur J Pharm Biopharm* 2010;**74**(2):281–9.

93. Liang YJ, Zhen J, Chen N, Reith ME. Interaction of catechol and non-catechol substrates with externally or internally facing dopamine transporters. *J Neurochem* 2009;**109**(4):981–94.

94. Danek Burgess KS, Justice Jr. JB. Effects of serine mutations in transmembrane domain 7 of the human norepinephrine transporter on substrate binding and transport. *J Neurochem* 1999;**73**(2):656–64.

95. Raiteri M, Del Carmine R, Bertollini A, Levi G. Effect of sympathomimetic amines on the synaptosomal transport of noradrenaline, dopamine and 5-hydroxytryptamine. *Eur J Pharmacol* 1977;**41**:133–43.

96. Engel K, Wang J. Interaction of organic cations with a newly identified plasma membrane monoamine transporter. *Mol Pharmacol* 2005;**68**(5):1397–407.

97. Schomig E, Lazar A, Grundemann D. Extraneuronal monoamine transporter and organic cation transporters 1 and 2; a review of transport efficiency. *Handb Exp Pharmacol* 2006;**175**(175):151–80.

98. Iseki K, Sugawara M, Saitoh N, Miyazaki K. The transport mechanisms of organic cations and their zwitterionic derivatives across rat intestinal brush-border membrane. II. Comparison of the membrane potential effect on the uptake by membrane vesicles. *Biochim Biophys Acta* 1993;**1152**(1):9–14.

99. Breidert T, Spitzenberger F, Grundemann D, Schomig E. Catecholamine transport by the organic cation transporter type 1 (OCT1). *Br J Pharmacol* 1998;**125**(1):218–24.

100. Alexander SP, Benson HE, Faccenda E, et al. The concise guide to pharmacology 2013/14: G protein-coupled receptors. *Br J Pharmacol* 2013;**170**(8):1459–581.

101. Harmeier A, Obermueller S, Meyer CA, et al. Trace amine-associated receptor 1 activation silences GSK3beta signaling of TAAR1 and D2R heteromers. *Eur Neuropsychopharmacol* 2015;**25**:2049–61.

102. Hashiguchi Y, Nishida M. Evolution of trace amine associated receptor (TAAR) gene family in vertebrates: lineage-specific expansions and degradations of a second class of vertebrate chemosensory receptors expressed in the olfactory epithelium. *Mol Biol Evol* 2007;**24**(9):2099–107.

103. Hussain A, Saraiva LR, Korsching SI. Positive Darwinian selection and the birth of an olfactory receptor clade in teleosts. *Proc Natl Acad Sci USA* 2009;**106**(11):4313–8.

104. Liberles SD, Buck LB. A second class of chemosensory receptors in the olfactory epithelium. *Nature* 2006;**442**(7103):645–50.

105. Horowitz LF, Saraiva LR, Kuang D, Yoon KH, Buck LB. Olfactory receptor patterning in a higher primate. *J Neurosci* 2014;**34**(37):12241–52.

106. Kanageswaran N, Demond M, Nagel M, et al. Deep sequencing of the murine olfactory receptor neuron transcriptome. *PLoS One* 2015;**10**(1):e0113170.

107. Ferrero DM, Lemon JK, Fluegge D, et al. Detection and avoidance of a carnivore odor by prey. *Proc Natl Acad Sci USA* 2011;**108**(27):11235–40.

108. Dewan A, Pacifico R, Zhan R, Rinberg D, Bozza T. Non-redundant coding of aversive odours in the main olfactory pathway. *Nature* 2013;**497**(7450):486–9.

109. Zhang J, Pacifico R, Cawley D, Feinstein P, Bozza T. Ultrasensitive detection of amines by a trace amine-associated receptor. *J Neurosci* 2013;**33**(7):3228–39.

110. Li Q, Liberles SD. Aversion and attraction through olfaction. *Curr Biol* 2015;**25**(3):R120–129.

111. Hashiguchi Y, Furuta Y, Nishida M. Evolutionary patterns and selective pressures of odorant/pheromone receptor gene families in teleost fishes. *PLoS One* 2008;**3**(12):e4083.

112. Xie Z, Vallender EJ, Yu N, et al. Cloning, expression, and functional analysis of rhesus monkey trace amine-associated receptor 6: evidence for lack of monoaminergic association. *J Neurosci Res* 2008;**86**(15):3435–46.
113. Dinter J, Muhlhaus J, Wienchol CL, et al. Inverse agonistic action of 3-iodothyronamine at the human trace amine-associated receptor 5. *PLoS One* 2015;**10**(2):e0117774.
114. Nelson DA, Tolbert MD, Singh SJ, Bost KL. Expression of neuronal trace amine-associated receptor (TAAR) mRNAs in leukocytes. *J Neuroimmunol* 2007;**192**(1–2):21–30.
115. Babusyte A, Kotthoff M, Fiedler J, Krautwurst D. Biogenic amines activate blood leukocytes via trace amine-associated receptors TAAR1 and TAAR2. *J Leukoc Biol* 2013;**93**(3):387–94.
116. Chiellini G, Frascarelli S, Ghelardoni S, et al. Cardiac effects of 3-iodothyronamine: a new aminergic system modulating cardiac function. *FASEB J* 2007;**21**(7):1597–608.
117. Fehler M, Broadley KJ, Ford WR, Kidd EJ. Identification of trace-amine-associated receptors (TAAR) in the rat aorta and their role in vasoconstriction by beta-phenylethylamine. *Naunyn Schmiedebergs Arch Pharmacol* 2010;**382**(4):385–98.
118. Vanti WB, Muglia P, Nguyen T, et al. Discovery of a null mutation in a human trace amine receptor gene. *Genomics* 2003;**82**(5):531–6.
119. Kubo H, Shibato J, Saito T, Ogawa T, Rakwal R, Shioda S. Unraveling the rat intestine, spleen and liver genome-wide transcriptome after the oral administration of lavender oil by a two-color dye-swap DNA microarray approach. *PLoS One* 2015;**10**(7):e0129951.
120. Ikemoto K, Nishimura A, Oda T, Nagatsu I, Nishi K. Number of striatal D-neurons is reduced in autopsy brains of schizophrenics. *Legal Med* 2003;**5**:S221–4.
121. Ikemoto K. D-neuron in schizophrenia research. *Adv Biosens Bioelectron* 2013;**2**(3):35–8.
122. Homberg U, Seyfarth J, Binkle U, Monastirioti M, Alkema MJ. Identification of distinct tyraminergic and octo-paminergic neurons innervating the central complex of the desert locust, *Schistocerca gregaria*. *J Comp Neurol* 2013;**521**(9):2025–41.
123. Ishida Y, Ozaki M. Aversive odorant causing appetite decrease downregulates tyrosine decarboxylase gene expression in the olfactory receptor neuron of the blowfly, *Phormia regina*. *Naturwissenschaften* 2012;**99**(1):71–5.
124. McCoole MD, Atkinson NJ, Graham DI, et al. Genomic analyses of aminergic signaling systems (dopamine, octopamine, and serotonin) in *Daphnia pulex*. *Comp Biochem Physiol Part D Genomics Proteomics* 2012;**7**(1):35–58.
125. Christie AE, Fontanilla TM, Roncalli V, Cieslak MC, Lenz PH. Identification and developmental expression of the enzymes responsible for dopamine, histamine, octopamine, and serotonin biosynthesis in the copepod crustacean *Calanus finmarchicus*. *Gen Comp Endocrinol* 2014;**195**:28–39.
126. Cole SH, Carney GE, McClung CA, Willard SS, Taylor BJ, Hirsh J. Two functional but noncomplementing *Drosophila* tyrosine decarboxylase genes: distinct roles for neural tyramine and octopamine in female fertility. *J Biol Chem* 2005;**280**(15):14948–55.
127. Han Q, Ding H, Robinson H, Christensen BM, Li J. Crystal structure and substrate specificity of *Drosophila* 3,4-dihydroxyphenylalanine decarboxylase. *PLoS One* 2010;**5**(1):e8826.
128. Alkema MJ, Hunter-Ensor M, Ringstad N, Horvitz HR. Tyramine functions independently of octopamine in the *Caenorhabditis elegans* nervous system. *Neuron* 2005;**46**(2):247–60.
129. Nishimura K, Kitamura Y, Inoue T, et al. Characterization of tyramine beta-hydroxylase in planarian *Dugesia japonica*: cloning and expression. *Neurochem Int* 2008;**53**(6–8):184–92.
130. Chatel A, Murillo L, Bourdin CM, Quinchard S, Picard D, Legros C. Characterization of tyramine beta-hydroxylase, an enzyme upregulated by stress in *Periplaneta americana*. *J Mol Endocrinol* 2013;**50**(1):91–102.
131. El-Kholy S, Stephano F, Li Y, Bhandari A, Fink C, Roeder T. Expression analysis of octopamine and tyramine receptors in *Drosophila*. *Cell Tissue Res* 2015;**361**(3):669–84.
132. Evans PD, Maqueira B. Insect octopamine receptors: a new classification scheme based on studies of cloned *Drosophila* G-protein coupled receptors. *Invert Neurosci* 2005;**5**(3–4):111–8.
133. Zornik E, Paisley K, Nichols R. Neural transmitters and a peptide modulate *Drosophila* heart rate. *Peptides* 1999;**20**(1):45–51.
134. Saraswati S, Fox LE, Soll DR, Wu CF. Tyramine and octopamine have opposite effects on the locomotion of *Drosophila* larvae. *J Neurobiol* 2004;**58**(4):425–41.
135. Yang Z, Yu Y, Zhang V, Tian Y, Qi W, Wang L. Octopamine mediates starvation-induced hyperactivity in adult *Drosophila*. *Proc Natl Acad Sci USA* 2015;**112**(16):5219–24.
136. Kutsukake M, Komatsu A, Yamamoto D, Ishiwa-Chigusa S. A tyramine receptor gene mutation causes a defective olfactory behavior in *Drosophila melanogaster*. *Gene* 2000;**245**(1):31–42.

137. Ma Z, Guo X, Lei H, Li T, Hao S, Kang L. Octopamine and tyramine respectively regulate attractive and repulsive behavior in locust phase changes. *Sci Rep* 2015;**5**:8036.
138. Lee H-G, Seong C-S, Kim Y-C, Davis RL, Han K-A. Octopamine receptor OAMB is required for ovulation in *Drosophila melanogaster*. *Dev Biol* 2003;**264**:179–90.
139. Lim J, Sabandal PR, Fernandez A, et al. The octopamine receptor Octbeta2R regulates ovulation in *Drosophila melanogaster*. *PLoS One* 2014;**9**(8):e104441.
140. Maqueira B, Chatwin H, Evans PD. Identification and characterization of a novel family of *Drosophila* beta-adrenergic-like octopamine G-protein coupled receptors. *J Neurochem* 2005;**94**(2):547–60.
141. Han KA, Millar NS, Davis RL. A novel octopamine receptor with preferential expression in *Drosophila* mushroom bodies. *J Neurosci* 1998;**18**(10):3650–8.
142. Bayliss A, Roselli G, Evans PD. A comparison of the signalling properties of two tyramine receptors from *Drosophila*. *J Neurochem* 2013;**125**(1):37–48.
143. Blenau W, Baumann A. Molecular and pharmacological properties of insect biogenic amine receptors: lessons from *Drosophila melanogaster* and *Apis mellifera*. *Arch Insect Biochem Physiol* 2001;**48**(1):13–38.
144. Kastner KW, Shoue DA, Estiu GL, et al. Characterization of the *Anopheles gambiae* octopamine receptor and discovery of potential agonists and antagonists using a combined computational-experimental approach. *Malar J* 2014;**13**:434–47.
145. Chang DJ, Li XC, Lee YS, et al. Activation of a heterologously expressed octopamine receptor coupled only to adenylyl cyclase produces all the features of presynaptic facilitation in *Aplysia* sensory neurons. *Proc Natl Acad Sci USA* 2000;**97**(4):1829–34.
146. Saudou F, Amlaiky N, Plassat JL, Borrelli E, Hen R. Cloning and characterization of a *Drosophila* tyramine receptor. *EMBO J* 1990;**9**(11):3611–7.
147. Robb S, Cheek TR, Hannan FL, Hall LM, Midgley JM, Evans PD. Agonist-specific coupling of a cloned *Drosophila* octopamine/tyramine receptor to multiple second messenger systems. *EMBO J* 1994;**13**(6):1325–30.
148. Reale V, Hannan F, Midgley JM, Evans PD. The expression of a cloned *Drosophila* octopamine/tyramine receptor in *Xenopus* oocytes. *Brain Res* 1997;**769**(2):309–20.
149. Cazzamali G, Klaerke DA, Grimmelikhuijzen CJ. A new family of insect tyramine receptors. *Biochem Biophys Res Commun* 2005;**338**(2):1189–96.
150. Huang J, Ohta H, Inoue N, et al. Molecular cloning and pharmacological characterization of a *Bombyx mori* tyramine receptor selectively coupled to intracellular calcium mobilization. *Insect Biochem Mol Biol* 2009;**39**(11):842–9.
151. Pirri JK, McPherson AD, Donnelly JL, Francis MM, Alkema MJ. A tyramine-gated chloride channel coordinates distinct motor programs of a *Caenorhabditis elegans* escape response. *Neuron* 2009;**62**(4):526–38.
152. Dinan TG, Cryan JF. The impact of gut microbiota on brain and behaviour: implications for psychiatry. *Curr Opin Clin Nutr Metab Care* 2015;**18**(6):552–8.
153. Yang YX, Mu CL, Luo Z, Zhu WY. Bromochloromethane, a methane analogue, affects the gut microbiota and metabolic profiles of the rat gastrointestinal tract. *Appl Environ Microbiol* 2016;**82**:778–87.
154. Zheng X, Xie G, Zhao A, et al. The footprints of gut microbial-mammalian co-metabolism. *J Proteome Res* 2011;**10**(12):5512–22.
155. Nelson TM, Borgogna JL, Brotman RM, Ravel J, Walk ST, Yeoman CJ. Vaginal biogenic amines: biomarkers of bacterial vaginosis or precursors to vaginal dysbiosis? *Front Physiol* 2015;**6**:253.
156. Marcobal A, de las Rivas B, Munoz R. First genetic characterization of a bacterial beta-phenylethylamine biosynthetic enzyme in Enterococcus faecium RM58. *FEMS Microbiol Lett* 2006;**258**(1):144–9.
157. Williams BB, Van Benschoten AH, Cimermancic P, et al. Discovery and characterization of gut microbiota decarboxylases that can produce the neurotransmitter tryptamine. *Cell Host Microbe* 2014;**16**(4):495–503.
158. Aarts E, Ederveen T, Naaijen J, et al. Gut microbiome in ADHD and its relation to brain function. *Soc Neurosci Abstr* 2015 683.610.

The Origin and Evolution of the Trace Amine-Associated Receptor Family in Vertebrates

Y. Hashiguchi

Department of Biology, Osaka Medical College, Takatsuki City, Osaka, Japan

INTRODUCTION

Trace amine-associated receptors (TAARs) are a family of G-protein-coupled receptors (GPCRs) that are considered to detect trace amines (TAs) in the nervous system of vertebrates. The discovery of the first TAAR (TAAR1) was made by two research groups independently.[1,2] Subsequently, multiple copies of TAARs have been found in mammals, and genes coding for TAARs are located in a specific chromosomal region as a gene cluster.[3] A characteristic peptide fingerprint motif that largely overlaps with the seventh transmembrane domain is observed in all TAARs.[3,4]

Initially, studies of the biological function of TAARs focused on TAAR1 that is expressed in the brain and various other tissues.[1] TAAR1 responds to TAs p-tyramine and β-phenylethylamine, but is not sensitive to classical biogenic amines such as catecholamines.[2] TAAR1 also interacts with psychotropic agents such as amphetamine.[2,5] In humans, genetic linkage studies consistently suggest that there is a significant association between the TAAR locus (ie, TAAR gene cluster in 6q23.2) and susceptibility to schizophrenia and/or bipolar disorder.[5] It seems noteworthy that TAAR1 is one of the GPCR involved in higher-order brain functions.

However, subsequent work revealed a functional dichotomy in the TAAR family. All TAAR subtypes except TAAR1 function as chemosensory receptors in the mouse olfactory epithelium (OE), and detect volatile amines derived from organisms.[6,7] Recent studies demonstrated that amines in the urine of the same or different species that are recognized by the "olfactory-type" TAARs frequently elicit innate behavioral responses such as attraction or aversion.[8,9] Thus, TAARs are also considered to be an important focus for further understanding the evolution of adaptive behaviors in mammals.

Studies based on the evolutionary genomics have clarified the origin and evolution of TAAR families.[3,10–12] TAARs likely originated in the common ancestor of jawed and jawless vertebrates,[11] although another proposed hypothesis suggests that TAARs arose specifically in the jawed vertebrates.[12] In jawed vertebrates, the TAAR gene family can be classified into five subfamilies that diverged before the separation of teleost fishes and tetrapods.[11] Genomics and database studies also showed that the repertoires of TAAR gene families are highly variable among vertebrate taxa. The number of TAAR gene copies is highly expanded in some teleost fishes.[10–13] In contrast, amphibians, reptiles, and birds possess only a few (3–6) TAAR genes.[11,13–15] In mammals, copy numbers of TAAR genes are variable among species[3,11,12,16] and some TAAR genes have been lost repeatedly in the primate lineage.[17]

This chapter summarizes recent research on the origin and evolution of the TAAR family and consists of three main sections as follows:

1. In the first section, the evolutionary origin of TAARs is discussed. As mentioned above, there are two conflicting hypotheses about the origin of TAARs. First, these two hypotheses are briefly described and their validity is assessed. Second, the timing of the origin of TAARs and the most closely related GPCR is considered.
2. In the second section, the evolution of the TAAR gene family, based on the phylogenetic tree and the chromosomal positions of TAAR genes in representative vertebrate species, is discussed. Evolutionary patterns of each of five TAAR subfamilies are compared.

The evolutionary dynamics of the TAAR family in teleost fishes, amphibians, reptiles, birds, and mammals are discussed separately.

3. In the third section, adaptive evolution of TAARs is considered with a focus on the frequent duplications and positive selection in some teleost fish TAAR genes, functional differentiations of mammalian TAARs, and repeated loss (ie, pseudogenization) of TAAR genes in canine and primate lineages.

THE EVOLUTIONARY ORIGIN OF TAARs

TAAR Genes in Various Vertebrate Species

TAAR genes have been found in all vertebrate taxa, but copy numbers are highly variable among taxa and even among species within a taxonomic group (Table 4.1). Genomics studies have revealed that the TAAR gene family in teleost fishes is much larger and variable than that in tetrapods,[10–12] whereas only a few TAAR genes have been identified in the frog, anole lizard, and chicken.[11–13] More recently, TAAR or TAAR-like GPCR genes have been identified from ancestral vertebrate species. In the elephant shark, only two intact TAAR genes were found.[12] In contrast, jawless vertebrates possess a relatively large number of TAAR genes. Twenty-one intact TAAR-like genes and 17 pseudogenes were identified from the draft genome of sea lamprey *Petromyzon marinus*[11] and at least seven genes were expressed in the olfactory organ.[18]

TAARs Originated From the Common Ancestor of Jawless and Jawed Vertebrates

Phylogenetic and evolutionary analyses have clarified the origin of TAARs. Gloriam et al.[10] conducted phylogenetic analysis of mammalian and teleost fish TAARs and suggested that TAARs consist of several related subgroups that diverged before the separation of mammalian and teleost fish lineages. Vertebrate TAARs do not show a close evolutionary relationship to the invertebrate TA-binding receptors,[10] indicating that TAARs originated within the vertebrate lineage. Using TAAR genes from the genomes of three mammals (human, mouse, and opossum), chicken, frog, four teleost fishes (zebrafish, medaka, stickleback, and fugu), and sea lamprey, a more detailed phylogenetic analysis showed that all lamprey TAAR-like genes formed a monophyletic clade and the lamprey TAAR clade was a sister group of TAAR genes in jawed vertebrates.[11] A newly reconstructed phylogenetic tree that includes shark and coelacanth TAAR genes also robustly reproduced this evolutionary relationship (Fig. 4.1). This provides strong support that TAARs originated before the divergence of jawless and jawed vertebrates. Several lamprey TAAR genes are expressed in the olfactory organ,[18,20] indicating that lamprey TAAR genes are likely used as chemosensory receptors. Lamprey-specific repeated duplications and pseudogenizations of TAAR genes might imply the adaptive response of TAARs to rapidly changing environmental odor chemicals. The evolutionary pattern of lamprey TAARs seems to correspond with the birth-and-death evolution as frequently observed in vertebrate olfactory and pheromone receptor genes.[21]

TABLE 4.1   Copy Numbers of Putatively Functional TAAR Genes and Nonfunctional Pseudogenes in Representative Vertebrate Species

Species	TAAR Genes	TAAR Pseudogenes	Total	References
Sea lamprey	21	17	38	[11]
Elephant shark	3	2	5	Y. Hashiguchi (unpublished)
Zebrafish	109	10	119	[11]
Medaka	25	7	32	[11]
Stickleback	49	15	64	[11]
Fugu	13	6	19	[11]
Coelacanth	20	5	25	Y. Hashiguchi (unpublished)
Frog	6	1	7	[11]
Anole lizard	3	0	3	Y. Hashiguchi (unpublished)
Chicken	3	2	5	[11]
Opossum	22	3	25	[11]
Mouse	15	1	16	[3]
Human	5	3	8	[3]

However, some researchers have claimed that the origin of TAARs is much younger than that of olfactory receptors (ORs) which originated before the divergence between jawed and jawless vertebrates.[12,13] Hussain et al.[12] concluded that TAARs originated after the divergence of lamprey and other vertebrate lineages based on a phylogenetic analysis using TAARs and related GPCRs, aminergic receptors and ORs, in vertebrates. Their phylogenetic analysis suggested that the lamprey TAAR-like GPCRs (called lamprey AmRs) were more closely related to the vertebrate aminergic receptors, not to TAARs of jawed vertebrates. Hussain et al.[12] argued that the analysis of Hashiguchi and Nishida[11] led to spurious results because of inclusion of just a few aminergic receptors in their phylogenetic analysis. However, the phylogenetic tree shown in Fig. 4.1 was reconstructed using all aminergic receptors contained in the analysis of Hussain et al.[12] except ORs, and supports the monophyletic relationship of the lamprey TAARs (lamprey AmRs) and the jawed vertebrate TAARs with a high (99%) bootstrap probability. Hussain et al.[12] may reflect bias with the inclusion of ORs, which show very low amino acid homology to TAARs, in the dataset of their phylogenetic analysis. To identify the origin of TAARs, adequate outgroups that diverged before the separation of jawed and jawless vertebrates should be selected. Tessarolo et al.[13] also argued that the lamprey genes are not TAARs but likely some other aminergic receptors based on a phylogenetic tree consisting of 246 putatively functional

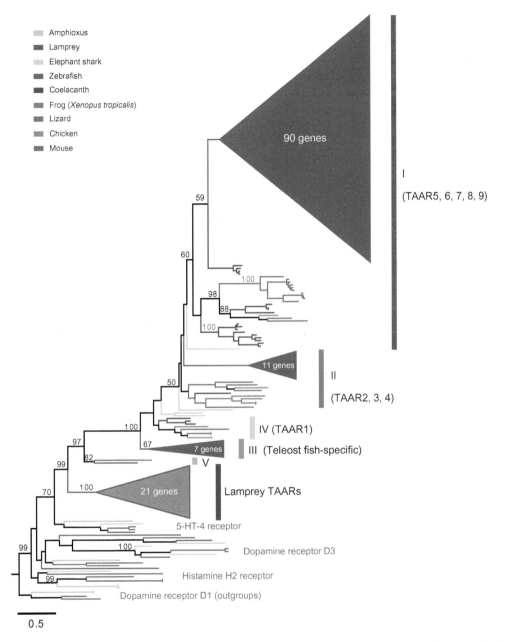

FIGURE 4.1 Phylogenetic tree of the nucleotide sequences of TAAR and related aminergic GPCR genes in chordates. The tree was newly reconstructed by the author, using the maximum likelihood (ML) method implemented in the RAxML program[19] with the rapid-bootstrap analysis option (100 bootstrap replicates). Numbers on each tree node indicate bootstrap values (showing major clades only). The color of each tree branch indicates the species. The majority of zebrafish (blue) and sea lamprey (brown) TAAR genes were collapsed in the phylogenetic tree. The dataset of this phylogenetic tree consists of TAARs, 5-HT-4 receptors, 5-HT-2 receptors, dopamine receptor D1, dopamine receptor D3, histamine H2 receptors, and adrenergic receptors.

TAARs from 15 vertebrate species. However, their phylogenetic tree appears to support the close relationship of lamprey TAARs and jawed vertebrate TAARs, similar to the phylogenetic trees by Hashiguchi and Nishida[11] and Fig. 4.1 in this chapter.

5-HT-4 Receptors Are the Most Closely Related GPCRs to TAARs

Serotonin (5-HT)-4 receptors in mice, humans, and zebrafish show relatively high (~40%) amino acid identity to TAARs.[1,3,11] Phylogenetic analysis also indicates that 5-HT-4 receptors cluster with TAARs and this relationship is supported with a high (70%) bootstrap probability (Fig. 4.1). Thus, it seems likely that 5-HT-4 receptors are the most closely related GPCRs to TAARs. Intriguingly, an ortholog of 5-HT-4 receptor is found in the amphioxus genome (see Fig. 4.1). This implies that ancestral genes of 5-HT-4 receptors and TAARs diverged before the separation of cephalochordate and vertebrate lineages. However, orthologs of TAAR genes have not been identified from the amphioxus genome (Y. Hashiguchi, unpublished data) and thus, TAARs were likely to be lost within the evolutionary process of the amphioxus lineage.

EVOLUTION OF TAAR GENE FAMILY

Subfamilies of Vertebrate TAARs

Phylogenetic analysis of TAAR genes in 10 vertebrate species indicated that the TAAR family can be subdivided into at least five subfamilies (subfamilies I–V) that diverged before the separation of teleosts and tetrapods, and one lamprey-specific subfamily.[11] Five TAAR subfamilies in jawed vertebrates and one lamprey-specific subfamily are indicated in Fig. 4.1. It should be noted that Hussain et al.[12] classified vertebrate TAAR genes into three classes (ie, classes I–III) based on the clades of their phylogenetic tree. However, their classification seems not to reflect the evolutionary history of TAARs because the classes are defined by an incorrectly rooted phylogenetic tree. To avoid confusion, in this chapter, only the evolution-based classification (ie, subfamilies I–V) is used.

Chromosomal positions of the TAAR genes and pseudogenes in seven vertebrate species are shown in Fig. 4.2. TAAR genes in human, mouse, chicken, and anole lizard cluster in a single chromosome region. In these species, TAAR genes in subfamilies I, II, and IV are located within the same chromosomal region as the gene cluster (Fig. 4.2). In the frog, subfamilies II and IV genes are located within a 100-kb region of scaffold 172 and in zebrafish, several subfamily I genes and all subfamilies II, III, and IV genes are located within a 1.7-Mb region of chromosome 20 (Fig. 4.2). Conserved arrangements of subfamilies I–IV TAAR genes in vertebrates strongly suggest the close evolutionary relationship among TAARs in these subfamilies. On the other hand, subfamily V TAAR genes in the frog, zebrafish, and stickleback are found in a different chromosomal region (Fig. 4.2), indicating that subfamily V includes the most ancestral TAARs in jawed vertebrates. This is also supported by the phylogenetic analysis of TAAR genes (Fig. 4.1).

Phylogenetic analysis shows that genes in each TAAR subfamily display very different evolutionary patterns, suggesting functional differentiation of TAARs among different subfamilies. Evolutionary dynamics of each TAAR subfamily are described below.

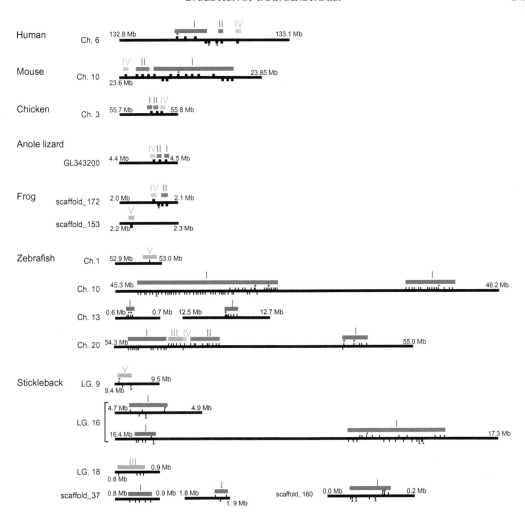

FIGURE 4.2 Chromosomal locations of TAAR genes in the human, mouse, chicken, anole lizard, frog, zebrafish, and stickleback. Each subfamily is indicated in the same color as in Fig. 4.1. *Asterisks* indicate pseudogenes.

Subfamily I: This is the largest subfamily in the TAAR family and contains TAARs from both teleost fishes and tetrapods except *Xenopus tropicalis*. TAARs 5–9 in mammals, defined by Lindemann et al.[3], are included in this subfamily. In mice, most of the subfamily I TAARs are expressed in the olfactory organs and one TAAR (mTAAR5) was shown to function as a chemosensory receptor.[6] Intriguingly, expansion of teleost fish TAAR genes occurred mainly in this subfamily. Substantially large subfamily I TAAR clades were recognized in the phylogenetic tree (Fig. 4.1), particularly in zebrafish. Species-specific expansions of the subfamily I TAAR genes were also observed in medaka, stickleback, fugu, European eel, and spotted gar (Fig. 4.3). In zebrafish, stickleback, and European eel, several subfamily I TAAR genes

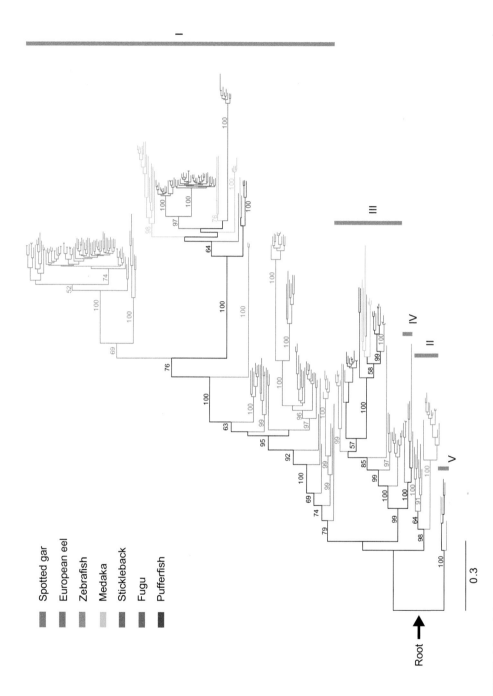

FIGURE 4.3 Phylogenetic relationships of the TAAR genes in seven teleost fishes. This tree is reconstructed by the ML method implemented in the RAxML program[19] with the rapid-bootstrap analysis option (1000 bootstrap replicates). Numbers on each node indicate bootstrap values (showing major clades only). The colors of the tree branches indicate species. The *arrow* indicates the root of the phylogenetic tree.

are expressed specifically in the olfactory organ,[6,11,22] suggesting that teleost subfamily I TAARs are actually chemosensory receptors. It is interesting to note that teleost subfamily I TAARs lost the aminergic ligand-binding motif, which was highly conserved in other TAAR subfamilies.[12]

Subfamily II: This subfamily consists of several tetrapod and zebrafish TAAR genes (Fig. 4.1). In mammals, subfamily II consists of TAARs 2, 3, and 4,[3] all of which are expressed in the olfactory organ. Mouse TAAR3 and TAAR4 are chemosensory receptors that respond to several primary amines.[6] In teleost fishes, genes in this subfamily were found in zebrafish (Figs. 4.1 and 4.3), European eel, and spotted gar (Fig. 4.3) but not in other model fishes. Monophyly of subfamily II TAARs was weakly supported in the phylogenetic analysis of Hashiguchi and Nishida[11] but not supported in the ML tree of Fig. 4.1 in this chapter. Further studies are needed to clarify the evolutionary relationship of mammalian and teleost subfamily II TAAR genes.

Subfamily III: This subfamily is teleost fish-specific. TAAR genes in this subfamily were found in all teleost species examined (Fig. 4.3). Biological functions of subfamily III TAARs are unknown, although expression of some zebrafish and stickleback genes was detected in the olfactory organs.[11]

Subfamily IV: This subfamily corresponds to TAAR1 that was identified in mammals.[1–3,6] Subfamily IV TAAR genes were also found in the chicken, anole lizard, frog, and coelacanth genomes.[11–13] Phylogenetic analysis suggests that one zebrafish TAAR gene was also included in this subfamily[11,12] although this relationship was not well supported statistically. In mammals, subfamily IV TAAR (ie, TAAR1) is expressed in the brain and various tissues and responds to TAs, thyroid hormone derivatives, and some psychoactive chemicals.[1,5,23] The subfamily IV TAAR presumably retains ancestral functions.

Subfamily V: This group of TAAR genes was identified only in frogs (*X. tropicalis*) and teleost fishes.[11] The subfamily V TAAR gene is located in a different chromosomal region from subfamilies I–IV TAAR genes (Fig. 4.2). Phylogenetic analysis indicates that subfamily V TAAR genes first diverged from the TAAR genes in other subfamilies (Fig. 4.1). This strongly suggests that subfamily V TAAR genes are the most ancestral group of TAAR genes in jawed vertebrates.[11] At present, biological functions of subfamily V TAARs are completely unknown. In teleost fishes, expression of subfamily V TAAR genes was not detected in the nose, brain, or other tissues.[11]

Lamprey TAARs: Phylogenetic analysis has shown that sea lamprey TAARs form a monophyletic clade (Fig. 4.1). This indicates that frequent gene duplications and losses have occurred in the lamprey lineage. As mentioned above, some lamprey TAAR genes are expressed specifically in the nose,[18,20] suggesting that the lamprey TAARs are really used as ORs.

Expansion of Subfamily I TAAR Genes in Teleost Fishes

Analysis of genome databases revealed that the copy numbers of TAAR genes in teleost fishes, particularly in zebrafish and stickleback, are much larger than those in terrestrial vertebrates (Table 4.1). This is mainly attributed to the expansions of subfamily I TAAR genes in teleost fishes (Fig. 4.1). Phylogenetic trees of the four evolutionary distinct chemosensory receptor families (OR, TAAR, V1R, and V2R) in teleost fishes indicates that species-specific

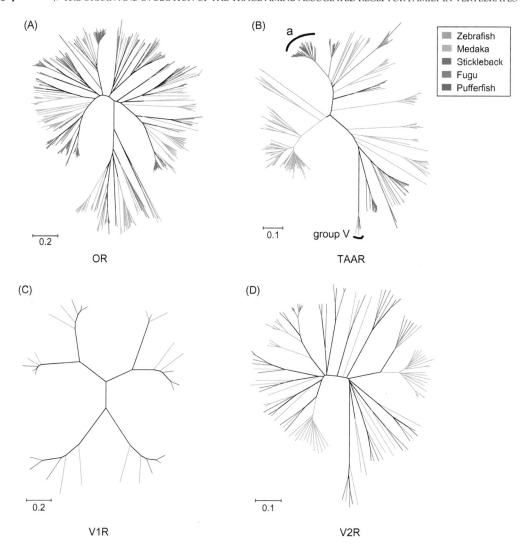

FIGURE 4.4   Phylogenetic trees of the four evolutionary distinct classes of chemosensory receptors in five teleost fishes.[24] The trees were reconstructed using the neighbor-joining method with Poisson-corrected protein distances. The colors of the tree branches indicate species.

gene duplications are observed more frequently in TAARs than in other chemosensory receptor families (Fig. 4.4).[24] In teleost species, the proportion of "species-specific" genes (ie, genes that originated from species-specific gene duplications) in TAARs was significantly higher than in ORs and V2Rs.[24] Also, teleost TAAR genes show relatively high nonsynonymous substitution rates.[12,24,25] These observations imply that teleost TAARs have evolved under the selective pressure of functional divergence. Intriguingly, similar evolutionary patterns are also observed in mammalian vomeronasal receptors that detect pheromones.[26]

Teleost TAARs might also be used for detecting some species-specific odor chemicals, such as sex pheromones. A recent study shows that one zebrafish subfamily I TAAR, designated as TAAR13c, responds to cadaverine, putrescine, and related diamines,[27] suggesting that the majority of teleost TAARs also recognize amines, similar to mammalian TAARs. Amines are major olfactory stimuli in fishes, playing crucial roles in chemical communication within and among species.[28–30] Cadaverine and putrescine are chemicals emitted from decaying flesh, and zebrafish exhibit powerful and innate avoidance behavior to these diamines.[27] Although zebrafish TAAR13c is a receptor for aversive chemicals and not a pheromone receptor, some teleost TAARs may be used for detecting amines for chemical communication within and/or among species.

Degeneration of TAAR Genes in Amphibian, Reptiles, and Birds

In contrast to teleost fishes, only a few TAAR genes have been found in amphibians, reptiles, and birds (Table 4.1). Also, species-specific gene gains and losses were rarely identified in the TAAR genes of these vertebrate taxa.[26] All of these species possess one copy of subfamily IV (ie, TAAR1) genes and 1–4 copies of subfamily II (ie, TAARs 2, 3, and 4) genes (Fig. 4.1). One subfamily I TAAR gene was identified in anole lizard and chicken, but not in X. tropicalis (Fig. 4.1). On the other hand, one subfamily V TAAR gene was found only in X. tropicalis (Fig. 4.1).

In situ hybridization analysis of the TAAR genes in the OE of X. tropicalis larva has shown that two subfamily II TAAR genes, called TAAR4a and TAAR4b, were sparsely expressed in the OE.[15,31] These TAARs are considered to function as chemosensory receptors in the OE of X. tropicalis. Expression zones of both of these TAARs highly overlap with the amine-response region of the X. tropicalis OE,[15] suggesting that these TAARs detect amines. It is interesting to note that detection of amines by TAARs is an evolutionary conserved feature among teleost fishes, amphibians, and mammals, despite very different evolutionary dynamics of TAAR genes between teleost fishes and tetrapods. In contrast to TAAR4a and TAAR4b, one subfamily IV TAAR (ie, TAAR1) of X. tropicalis was not expressed in the OE.[31] It is known that mammalian TAAR1 is also not expressed in the OE, but is instead expressed in the brain and is a receptor for neurotransmitters.[5] Thus, TAAR1 may have a nonolfactory role in amphibians, as has been reported for mammalian orthologs.[6] Unlike X. tropicalis, expression patterns and functions of reptile and avian TAARs are completely unknown. Further studies are needed to understand the functional evolution of TAARs in these vertebrate groups.

A Slight Increase of Mammalian TAAR Genes

The number of TAAR genes in mammals is substantially larger than those in other tetrapod lineages (Table 4.1). Mammalian TAARs consist of nine subtypes (ie, TAARs 1–9) defined by Lindemann et al.[3] In mammals, lineage-specific gene duplications and losses have occurred in a few TAAR subtypes. For example, five TAAR7 and three TAAR8 genes were identified in the mouse genome[3,7] and three TAAR4 and seven TAAR9 paraloges were found in the opossum genome.[11] On the other hand, in the primate lineage, some TAAR subtypes are known to be frequently lost.[17] In humans, TAAR3, TAAR4, and TAAR7 genes are nonfunctional pseudogenes.[3] Detailed evolutionary dynamics of the primate TAAR genes is considered in the next section.

Summary: The Evolutionary History of TAAR Gene Family

The long-term evolutionary history of the vertebrate TAAR gene family is summarized in Fig. 4.5A. Phylogenetic analysis of TAARs and related GPCRs (Fig. 4.1) shows that TAARs possibly originated in the common ancestor of chordates by a gene duplication of ancestral GPCR shared with 5-HT-4 receptors. In the amphioxus lineage, however, orthologs of the ancestral TAARs seem to be lost. TAARs are considered to have already acquired the chemosensory receptor function in the most recent common ancestor (MRCA) of jawless and jawed vertebrates, because TAAR genes are expressed in the olfactory organs in both lineages. In lamprey, the number of TAARs has increased within the lineage by repeated gene duplications and losses. In contrast, only a few (2–5 copies) TAAR genes have been found in the shark genome[12] (Y. Hashiguchi, unpublished data). Phylogenetic analysis shows that all shark TAARs are included in the subfamilies of jawed vertebrate TAARs and do not form a lineage-specific clade (Fig. 4.1). This indicates that the five TAAR subfamilies already existed in the MRCA of cartilaginous and bony fishes. After the divergence of cartilaginous fishes, copy numbers of TAAR genes increased in teleost fishes by frequent gene duplications, resulting in the highly species-specific repertoires of teleost TAARs. Expansion of TAAR genes also occurred in the mammals, but some TAAR genes were pseudogenized in the primate lineage.

Fig. 4.5B shows a schematic model of the evolutionary changes in copy numbers and chromosomal locations of the five TAAR subfamilies in jawed vertebrates.[11] The MRCA of teleost fishes and tetrapods had at least five ancestral TAARs corresponding to the five subfamilies. In the tetrapod lineages, the overall structure of the ancestral TAAR gene cluster has been retained (Fig. 4.2). On the other hand, teleost TAAR genes are located separately in multiple chromosomal regions (Fig. 4.2). The complicated genomic distributions of teleost TAAR genes could be explained by considering the teleost fish-specific whole genome duplication (3R-WGD) and subsequent lineage-specific gene duplications and losses.

ADAPTIVE EVOLUTION OF TAARs

As previously mentioned, TAARs can be divided into olfactory and nonolfactory types. Interestingly, both types of TAARs appear to be directly involved in animal behaviors. A "nonolfactory type" TAAR, TAAR1, is known to interact with chemicals that can elicit innate behavioral responses, such as thyroid hormone derivatives and psychotropic agents such as amphetamine and 3,4-methylenedioxymetamphetamine (MDMA).[2,5,7,23] On the other hand, "olfactory-type" TAARs are used to recognize amines that induce attractive or aversive behaviors.[6–8,27] Mutations in coding and/or regulatory regions of both types of TAAR genes can alter the behavioral responses to specific chemical substances. Thus, in vertebrates, TAARs are likely to evolve under strong selective constraints for adaptive behavioral responses. In this section, the focus is on the adaptive evolution of TAAR genes by considering studies on the natural selection operating on TAAR genes in teleost fishes and mammals.

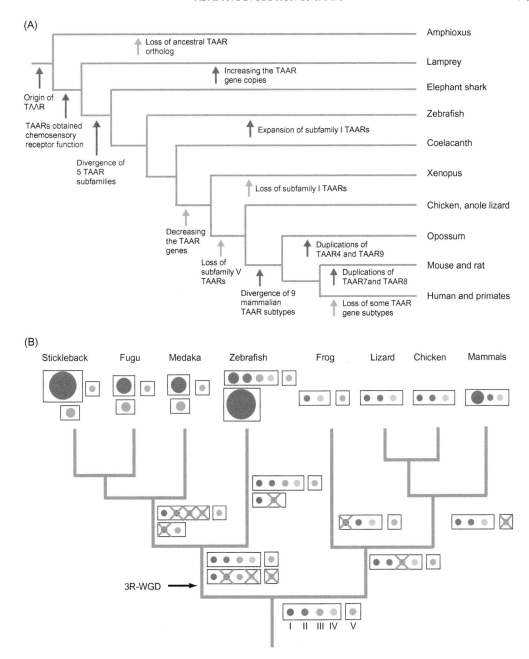

FIGURE 4.5 Schematic illustrations of the evolutionary trajectory of the TAAR gene family in vertebrates. (A) Evolutionary events identified by phylogenetic analysis of vertebrate TAAR genes. *Magenta* and *blue arrows* indicate the evolutionary events caused by gene duplications and losses, respectively. (B) Evolution of copy numbers and chromosomal locations of TAAR genes in jawed vertebrates.[11] *Squares* surrounding *colored circles* indicate chromosomes. *Colored circles* indicate the five subfamilies identified in the jawed vertebrates. The size of each circle roughly reflects the number of genes belonging to each subfamily. The *arrow* indicates the timing of the fish-specific whole genome duplication (ie, 3R-WGD).

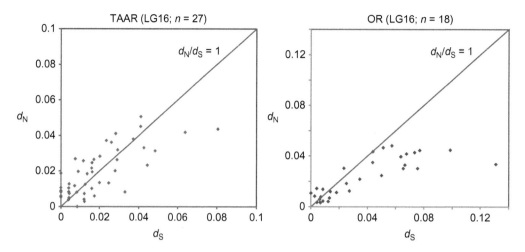

FIGURE 4.6   Synonymous (d_S) and nonsynonymous (d_N) nucleotide substitution rates of TAAR and OR genes located in the stickleback linkage group 16 (LG16). The stickleback TAAR genes in LG16 correspond to "clade *a*" of Fig. 4.4. Values of d_S and d_N were estimated by "free-ratio model" of branch-specific estimation of ω ($=d_N/d_S$) implemented in PAML4.7a program package.[32] *Diagonal lines* indicate $d_N/d_S = 1$.

Teleost Fishes

Molecular evolutionary analysis has shown that the synonymous to nonsynonymous nucleotide substitution rate ratios ($\omega = d_N/d_S$) of TAAR genes in teleost fishes are much higher than those in tetrapods.[12] Also, in teleost fishes, ω values of TAAR genes tend to be higher than those of other chemosensory receptor (OR and V2R) genes.[24] In particular, nonsynonymous substitution rates in the "clade *a*" of stickleback TAAR genes (see Fig. 4.4) frequently exceed synonymous substitution rates (Fig. 4.6), suggesting that positive selection has operated on these genes.[24,25] In stickleback, population genomics studies have detected footprints of directional positive selection on TAAR genes. A genome scan study using 157 microsatellite markers linked to physiologically important genes identified a footprint of directional selection on a subfamily III TAAR gene in European stickleback populations.[33] A genome-wide survey of parallel marine–freshwater divergence in stickleback based on whole genome sequences of 20 wild individuals also detected the directional selection on the regulatory region of a subfamily III TAAR gene.[34] These studies imply that TAARs played crucial roles in repeated adaptations to freshwater environments that occurred in marine (anadromous) sticklebacks.

Mammals

In mice, rats, and humans, high-throughput chemical screens and medicinal chemistry approaches have revealed ligands for several TAARs.[1,2,6,23,35] These studies have also shown that TAAR orthologs from different species generally recognize similar ligands.[7] Exceptionally, in TAAR7s that evolve rapidly in mice and rats, recent mutations have changed ligand recognition properties among paralogs.[35] Identification of the TAAR ligands

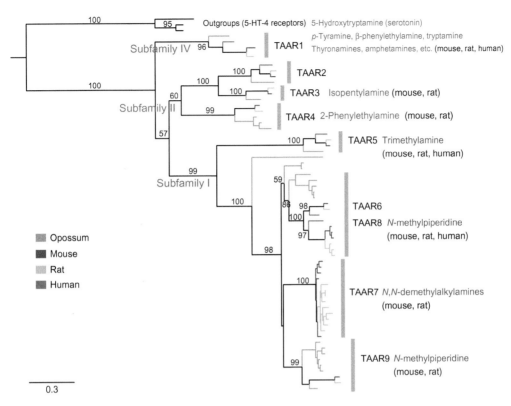

FIGURE 4.7 Phylogenetic tree of the nucleotide sequences of mammalian TAAR genes. Ligands of each TAAR subtype are mapped in the tree. Primary and tertiary amines are shown by magenta and dark blue, respectively. The colors of branches indicate species. The tree is reconstructed by the ML method using RAxML program[19] with the rapid-bootstrap analysis option (1000 bootstrap replicates). *Numbers* on each node indicate bootstrap values (showing major clades only).

enabled reconstruction of the evolution of TAAR functions in mammals. Liberles[7] pointed out that TAARs detecting primary or tertiary amines form different clades on the phylogenetic tree. Mapping the ligands of each TAAR subtype to the correctly rooted phylogenetic tree shows that TAARs recognizing tertiary amines are derived from ancestral TAARs that prefer primary amines (Fig. 4.7). TAARs recognizing tertiary amines correspond to the subfamily I TAARs (ie, TAARs 5–9).

As mentioned above, ligands of mammalian TAARs are biogenic amines that frequently induce innate behavioral responses such as attraction and avoidance. In mice and rats, TAAR4 detects 2-phenylethylamine, contained in the urine of bobcats.[8] A screening of 38 mammalian species indicated that 2-phenylethylamine is present at a higher concentration in the urine of carnivores[8] and elicits an innate avoidance response in rodents.[8] These results show that the TAAR4 is a receptor for kairomone, which is used for avoiding predators, and may provide a strong evolutionary advantage for mice and rats. In contrast to TAAR4, TAAR5 is an example of a chemosensory receptor for detecting chemicals that induce species-specific behavioral responses. TAAR5 detects trimethylamine in male mouse urine,[6] which elicits different behavioral responses among species.[9] Behavioral analysis showed that trimethylamine was an aversive chemical to rats and humans, although it was attractive to mice.[9] It is interesting to note that mice normally release >1000-fold more urinary trimethylamine in comparison with humans, and they do so in a sex-specific manner. Male-specific release of trimethylamine in mice is caused by male-specific repression of the flavin-containing monooxygenase 3 that oxidizes trimethylamine.[9]

In mammals, secondary losses of specific subtypes of TAAR genes have also been reported. In dogs, TAAR1 is pseudogenized by several frameshift mutations.[16] A pseudogenization event of TAAR1 seems to have occurred prior to the divergence of the tribe Canini but after the divergence from the feliforms.[16] Assessing the effect of the TAAR1 agonist on [³H] uptake in canine striatal synaptosomes and comparing the degree and pattern of uptake inhibition in other mammals suggests that a TAAR1 pseudogenization event resulted in an uncompressed loss of function.[16] It is not known why TAAR1 was specifically lost in dogs and their relatives despite the evolutionary conservation of TAAR1 in all tetrapod lineages. Comprehensive identification and phylogenetic analysis of TAAR3, TAAR4, and TAAR5 genes in various mammalian species showed that these TAAR genes were repeatedly pseudogenized in different primate species.[17] For example, pseudogenization events of TAAR3 have occurred at least three times in the evolutionary history of primates. This study suggests that functional significance of TAARs 3–5 is decreased in the primate lineage.

CONCLUSION AND PERSPECTIVES

Recent genomics studies have expanded our knowledge on the origin and evolution of chemosensory GPCRs in vertebrates. Phylogenetic analysis revealed that the main ORs originated in the common ancestor of chordates before the separation of the amphioxus lineage[36] and that TAARs also originated before the divergence of jawed and jawless vertebrates, presumably in the common ancestor of chordates (Fig. 4.5A). Subsequently, the TAAR family was expanded in the lineage of teleost fishes,[10,11] in contrast to the OR family that was highly expanded in the terrestrial vertebrates (ie, tetrapods) possibly to detect diverse volatile odor chemicals within the air.[21,37] In both fishes and mammals, known ligands of olfactory-type TAARs are amines.[6,27] Amphibian TAARs are also thought to detect amines.[15] Limited repertoires of TAARs in tetrapods might be caused by functional specialization of TAARs to detect amines, which occurred in the early stage of vertebrate evolution. Of course, this hypothesis remains largely speculative because biological functions (ie, endogenous ligands) of most teleost TAARs are still unknown. It is reported that three-fourths of teleost fish TAARs lack the aminergic ligand-binding motif,[12] suggesting that these TAARs do not detect amines. Further studies are needed to understand the evolutionary mechanisms of diversification and degeneration in vertebrate TAARs.

Advances in genome-sequencing technology enable us to access whole genome information in various nonmodel organisms. However, at the present time, complete repertoires of the TAAR family have only been identified in very limited model organisms. To reconstruct a more detailed evolutionary history of TAARs, it will be useful to identify the

TAAR repertoires in many more vertebrate species. In particular, to study organisms that use amines for some lineage-specific purposes may be of interest. For instance, in anemone fish genus *Amphiprion*, it is suggested that amines are involved in the symbiosis between anemone fish and sea anemones.[28] In this case, chemicals secreted by the sea anemone elicit symbiotic behavior of the anemone fish. More interestingly, different amines are used in different symbiotic pairs of sea anemone–anemone fish species.[28] Evolutionary analysis of TAARs in these anemone fishes may provide significant insights into the evolution of the species-specific symbiosis between anemone fish and sea anemone.

Recent studies conducted in mammals and fishes suggest that amines recognized by TAARs frequently elicit innate behavioral responses such as attraction and aversion.[8,9,27] Thus, TAARs are considered interesting targets to understand the evolution of adaptive behaviors in vertebrates. In the future, studies of TAARs will provide valuable insights into various adaptive behaviors, such as reproduction, avoidance of predators, and chemical communications within and between species.

References

1. Borowsky B, Adham N, Jones KA, Raddatz R, Artymyshyn R, Ogozalek KL, et al. Trace amines: identification of a family of mammalian G protein-coupled receptors. *Proc Natl Acad Sci USA* 2001;**98**:8966–71.
2. Bunzow JR, Sonders MS, Arttamangkul S, Harrison LM, Zhang G, Quigley DI, et al. Amphetamine, 3,4-methylenedioxymethamphetamine, lysergic acid diethylamide, and metabolites of the catecholamine neurotransmitters are agonists of a rat trace amine receptor. *Mol Pharmacol* 2001;**60**:1181–8.
3. Lindemann L, Ebeling M, Kratochwil NA, Bunzow JR, Grandy DK, Hoener MC. Trace amine-associated receptors form structurally and functionally distinct subfamilies of novel G protein-coupled receptors. *Genomics* 2005;**85**:372–85.
4. Lindemann L, Hoener MC. A renaissance in trace amines inspired by a novel GPCR family. *Trends Pharmacol Sci* 2005;**26**:274–81.
5. Zucchi R, Chiellini G, Scanlan TS, Grandy DK. Trace amine-associated receptors and their ligands. *Br J Pharmacol* 2006;**149**:967–78.
6. Liberles S, Buck LB. A second class of chemosensory receptors in the olfactory epithelium. *Nature* 2006;**442**:645–50.
7. Liberles SD. Trace amine-associated receptors: ligands, neural circuits, and behaviors. *Curr Opin Neurobiol* 2015;**34**:1–7.
8. Ferrero DM, Lemon JK, Fluegge D, Pashkovski SL, Korzan WJ, Datta SR, et al. Detection and avoidance of a carnivore odor by prey. *Proc Natl Acad Sci USA* 2011;**108**:11235–40.
9. Li Q, Korzan WJ, Ferrero DM, Chang RB, Roy DS, Buchi M, et al. Synchronous evolution of an odor biosynthesis pathway and behavioral response. *Curr Biol* 2013;**23**:11–20.
10. Gloriam DEI, Bjarnadóttir TK, Yan Y-L, Postlethwait JH, Schiöth HB, Fredriksson R. The repertoire of trace amine G-protein-coupled receptors: large expansion in zebrafish. *Mol Phylogenet Evol* 2005;**35**:470–82.
11. Hashiguchi Y, Nishida M. Evolution of trace amine-associated receptor (TAAR) gene family in vertebrates: lineage-specific expansions and degradations of a second class of vertebrate chemosensory receptors expressed in the olfactory epithelium. *Mol Biol Evol* 2007;**24**:2099–107.
12. Hussain A, Saraiva LR, Korsching SI. Positive Darwinian selection and the birth of an olfactory receptor clade in teleosts. *Proc Natl Acad Sci USA* 2009;**106**:4313–8.
13. Tessarolo JA, Tabesh MJ, Nesbitt M, Davidson WS. Genomic organization and evolution of the trace amine-associated receptor (TAAR) repertoire in Atlantic salmon. *G3* 2014;**4**:1135–41.
14. Lagerström MC, Hellström AR, Gloriam DE, Larsson TP, Schiöth HB, Fredriksson R. The G protein-coupled receptor subset of the chicken genome. *PLoS Comput Biol* 2006;**2**:e54.
15. Syed AS, Sansone A, Röner S, Nia SB, Manzini I, Korsching SI. Different expression domains for two closely related amphibian TAARs generate a bimodal distribution similar to neuronal responses to amine odors. *Sci Rep* 2014;**5**:13935.

16. Vallender EJ, Xie Z, Westmoreland SV, Miller GM. Functional evolution of the trace amine associated receptors in mammals and the loss of TAAR1 in dogs. *BMC Evol Biol* 2010;**10**:51.

17. Stäubert C, Böselt I, Bohnekamp J, Römpler H, Enard W, Schöneberg T. Structural and functional evolution of the trace amine-associated receptors TAAR3, TAAR4 and TAAR5 in primates. *PLoS One* 2010;**5**:e11133.

18. Libants S, Carr K, Wu H, Teeter JH, Chung-Davidson Y-W, Zhang Z, et al. The sea lamprey *Pertomyzon marinus* genome reveals the early origin of several chemosensory receptor families in the vertebrate lineage. *BMC Evol Biol* 2009;**9**:180.

19. Stamatakis A. RAxML Version 8: a tool for phylogenetic analysis and post-analysis of large phylogenies. *Bioinformatics* 2014;**30**:1312–3.

20. Berghard A, Dryer L. A novel family of ancient vertebrate odorant receptors. *J Neurobiol* 1998;**37**:383–92.

21. Nei M, Niimura Y, Nozawa M. The evolution of animal chemosensory receptor gene repertoires: roles of chance and necessity. *Nat Rev Genet* 2008;**9**:951–63.

22. Churcher AM, Hubbard PC, Marques JP, Canário AVM, Huertas M. Deep sequencing of the olfactory epithelium reveals specific chemosensory receptors are expressed at sexual maturity in the European eel *Anguilla anguilla*. *Mol Ecol* 2015;**24**:822–34.

23. Scanlan TS, Suchland KL, Hart ME, Chiellini G, Huang Y, Kruzich PJ, et al. 3-Iodothyronamine is an endogenous and rapid-acting derivative of thyroid hormone. *Nat Med* 2004;**10**:638–42.

24. Hashiguchi Y, Furuta Y, Nishida M. Evolutionary patterns and selective pressures of odorant/pheromone receptor gene families in teleost fishes. *PLoS One* 2008;**3**:e4083.

25. Hashiguchi Y, Nishida M. Adaptive evolution of odorant receptor gene families in Japanese threespine stickleback populations. *Kaiyo Monthly* 2010;**42**:372–8. [In Japanese].

26. Grus WE, Zhang J. Distinct evolutionary patterns between chemoreceptors of 2 vertebrate olfactory systems and the differential tuning hypothesis. *Mol Biol Evol* 2008;**25**:1593–601.

27. Hussain A, Saraiva LR, Ferrero DM, Ahuja G, Krishna VS, Liberles SD, et al. High-affinity olfactory receptor for the death-associated odor cadaverine. *Proc Natl Acad Sci USA* 2013;**110**:19579–84.

28. Murata M, Miyagawa-Koshima K, Nakanishi K, Naya Y. Characterization of compounds that induce symbiosis between sea anemone and anemone fish. *Science* 1986;**234**:585–7.

29. Hubbard PC, Barata EN, Canário AVM. Olfactory sensitivity to catecholamines and their metabolites in the goldfish. *Chem Senses* 2003;**28**:207–18.

30. Rolen SH, Sorensen PW, Mattson D, Caprio J. Polyamines as olfactory stimuli in the goldfish *Carassius auratus*. *J Exp Biol* 2003;**206**:1683–96.

31. Gliem S, Syed AS, Sansone A, Kludt E, Tantalaki E, Hassenklöver T, et al. Bimodal processing of olfactory information in an amphibian nose: odor responses segregate into a medial and lateral stream. *Cell Mol Life Sci* 2013;**70**:1965–84.

32. Yang Z. PAML4: phylogenetic analysis by maximum likelihood. *Mol Biol Evol* 2007;**24**:1586–91.

33. Shimada Y, Shikano T, Merliä J. A high incidence of selection on physiologically important genes in the three-spined stickleback, *Gasterosteus aculeatus*. *Mol Biol Evol* 2011;**28**:181–93.

34. Jones FC, Grabherr MG, Chan YF, Russell P, Mauceli E, Johnson J, et al. The genomic basis of adaptive evolution in threespine sticklebacks. *Nature* 2012;**484**:55–61.

35. Ferrero DM, Wacker D, Roque MA, Baldwin MW, Stevens RC, Liberles SD. Agonists for 13 trace amine-associated receptors provide insight into the molecular basis of odor selectivity. *ACS Chem Biol* 2012;**7**:1184–9.

36. Niimura Y. On the origin and evolution of vertebrate olfactory receptor genes: comparative genome analysis among 23 chordate species. *Genome Biol Evol* 2009;**1**:34–44.

37. Niimura Y, Nei M. Evolutionary dynamics of olfactory receptor genes in fishes and tetrapods. *Proc Natl Acad Sci USA* 2005;**102**:6039–44.

Differential Modulation of Adrenergic Receptor Signaling by Octopamine, Tyramine, Phenylethylamine, and 3-Iodothyronamine

G. Kleinau[1], N. Khajavi[1],
J. Köhrle[2] and H. Biebermann[1]

[1]Institute of Experimental Pediatric Endocrinology, Charité University Medicine,
Berlin, Germany [2]Institute of Experimental Endocrinology, Charité University
Medicine, Berlin, Germany

OUTLINE

Trace Amines and Neurological Disorders
DOI: http://dx.doi.org/10.1016/B978-0-12-803603-7.00005-7

INTRODUCTION

Aminergic Receptors with Partially Overlapping Ligand-Binding Profiles

Several tyramine (TYR) receptors (TAR) and octopamine (OA) receptors (OAR) have been identified in insects[1–4] and mollusks.[5] Their corresponding ligands are trace amines (for a detailed description, see Section I). Trace amines act as neurotransmitters in invertebrates.[3] They are involved in the regulation of metabolism, reproduction, and behavioral functions, which is in agreement with the expression of their receptors in different organs and tissues.[6] The TYR/OA system in invertebrates is a counterpart to the mammalian adrenergic system (see Section I).[3,7] Trace amines in mammalians act as modulators of classical monoamine neurotransmitters,[8,9] and have been implicated in a diverse array of human diseases ranging from schizophrenia to affective disorders (reviewed in Sections I and III).[10]

The TARs/OARs and adrenergic receptors (α- and β-subtypes) belong to the rhodopsin-like family of G-protein-coupled receptors (GPCRs). A third aminergic receptor group, the trace amine-associated receptors (TAARs), were first described in 2001 (see Section I).[11–13] Of note, most TAAR subtypes were assumed to be odorant receptors (see Section II).[14–20] Specific olfactory neurons expressing TAARs constitute an olfactory subsystem.[21,22] Furthermore, several TAARs have been postulated to be associated with neurological disorders such as bipolar disease,[23,24] schizophrenia,[25,26] depression, and Parkinson's disease.[27,28]

TAAR1 is activated by trace amines,[29] such as TYR, β-phenylethylamine (PEA), OA,[30,31] and has been reported to induce signaling via the Gs protein/adenylyl cyclase system. Ligands of the dopamine, serotonin, histamine, or adrenergic receptors also activate TAAR1-mediated signaling.[13,30–32] This high ligand promiscuity of TAAR1 is most likely based on the conservation of specific amino acid motifs in the ligand-binding region, conserved throughout particular aminergic receptors.[16,30,33–36] Finally, TAAR1 was assigned as a receptor target of 3-iodothyronamine (3-T1AM).[37]

The concomitance of the aforementioned ligands in mammals raises the question of cross-interrelations either on the level of their respective receptors, for example, in heterodimerization,[38] or on the level of ligand–receptor interactions. Hypothetically, in case of overlapping receptor cell localization, specific aminergic receptors should interact with a diverse set of shared ligands. Such interrelations may, therefore, have a multiple impact on orchestrated signaling events and physiological (neurological) responses. For example, the α2A-adrenergic receptor (ADRA2A), the histamine receptor 1, and TAAR1 are all expressed in the hypothalamus, whereby TAAR1 and ADRA2A are also expressed in pancreatic β-cells (*IUPHAR database*,[39] *Expression Atlas*, EMBL-EBI[40]). These receptors interact with shared ligands, respectively, which raise the question of an ordered regulation in information transfer (signaling).

From a general perspective, the occurrence of identical or similar receptors at different organs and their diverse activity regulation by shared ligands should be interpreted as an option for fine-tuned synchronization of physiological reactions including resulting behavior. This implies that detailed in vitro and in vivo studies are required to estimate and decipher the full set of potential, combinatorial scenarios for actions via these receptor/ligand systems. This becomes even more complex with respect to the modulatory function of trace amines on classical monoamine ligands. Recent studies have unraveled potential signal induction by OA, TYR, PEA, and 3-T1AM at the β-adrenergic receptors 1 and 2 (ADRB1 and ADRB2),[41,42] and of 3-T1AM at the ADRA2A, as well as human and mouse TAAR5.[43,44] These findings are summarized in this chapter and also include previously published findings.

INDICATIONS FOR BINDING AND ACTION OF THE TRACE AMINES OA, PEA, AND TYR AT ADRENERGIC RECEPTORS

What do we know about the direct binding and action of trace amines at adrenergic receptors? Several examples of publications over recent decades have provided direct and indirect indications and will be described in the following section, although due to the extensive amount of prior publications,[45] this is unfortunately incomplete.

OA functions as a neuromodulator, neurotransmitter, and neurohormone in insect nervous systems and thereby is involved in regulation of many physiological processes such as learning and memory, rhythmic behaviors, or mobilization of lipids and carbohydrates.[46] In vertebrates, the (invertebrate) receptors for OA are not present, but OA may exert roles in biogenic amine-based neurosynaptic physiology and may function as a neuromodulator (see Section II).[29,47] Moreover, it has been suggested that OA is an agonist at the β3-adrenergic receptor (ADRB3) in mammalian fat cells.[48] Furthermore, OA activates the ADRA2A that is coupled to the Gi/o signaling system.[49,50] α-Adrenoreceptors are involved in numerous physiological processes such as the regulation of blood pressure, heart rate, lipolysis, or insulin release.[51,52]

A pathophysiological role for OA has been discussed in depression, Parkinson's disease,[27] and other neurological disorders (see Section III). An increased synthesis of OA and TYR occurs in migraine,[53] and an increase in the OA concentration in patients with severe renal disease has been observed.[54] Interestingly, studies on the mechanisms of migraine have revealed that the blocking of ADRB (coupled mainly to the Gs/adenylylcyclase system) reverted the inhibitory effect of OA on the release of nitric oxide that may be of importance in astroglial cells related to the development of cluster headache.[55]

The complex role and potency of OA to potentially interact with a diverse set of aminergic receptors has also been suggested in the study performed by Broadley et al.[56] By using agonists and blockers of α- and β-adrenergic receptors, they concluded that there are different binding sites (receptors) of OA, while the physiological effect related to vasoconstriction was hypothesized to be potentially mediated by TAAR1.

In humans, PEA may exert effects as a neuromodulator of aminergic synapses.[57] There is little information available on the agonistic and/or antagonistic activities of this compound on adrenoreceptor subtypes. PEA most likely acts as an α2A/2C-adrenergic receptor antagonist, which has been confirmed by in vitro studies in HEK293 and CHO cells.[58] Furthermore, PEA has been found to modulate vascular effects and most likely interacts with both the α1- and α2-adrenergic receptors.[59] Based on studies of the rat cerebral cortex, PEA has also been reported as a partial α1-receptor agonist.[60]

TYR-induced mobilization of norepinephrine stores have been shown to produce elevated ventricular contractility consistent with enhanced sympathetic neuroeffector properties, and this effect may be related to ADRB1.[61] Both PEA and TYR constrict rat aortic rings via a mechanism that does not involve α1-receptors,[56,62,63] but potentially TAAR1.

These findings collectively implicate direct interactions between PEA, TYR, OA, and specific adrenergic receptors. Consequently, obtaining further detailed insight into the molecular interactions of these amines at aminergic receptors is of importance to understand the complex system of signaling regulation by trace amines at GPCRs.

Signaling Modification at ADRB1 and ADRB2 by OA, TYR, and PEA

A recent study has investigated the effect of the TAAR1 agonists TYR, PEA, and OA on the signaling properties of ADRB1 and ADRB2 (Fig. 5.1A).[42] The experimental design

(A)

Receptor	Antagonist	Agonist	Modulator
hADRA2A	3-T1AM decreases NorEpi induced MAPK activation	3-T1AM stimulates Gi/o signaling	
hADRB1	TYR, PEA inhibiting ISOP induced Gs activation	OA stimulates Gs signaling; 3-T1AM activates related Ca²⁺ channel	
hADRB2	TYR, PEA, OA inhibiting ISOP induced Gs activation	3-T1AM activates related Ca²⁺ channels and influx	3-T1AM enhances ISOP induced Gs signaling
hTAAR1		3-T1AM, OA, TA, PEA stimulating Gs signaling	
hTAAR5	3-T1AM inverse agonist on basal Gq signaling and MAPK		
Co-expressed hTAAR1/hADRA2A (hetero-oligomer)	3-T1AM abolished NorEpi induced Gi/o signaling		

(B)

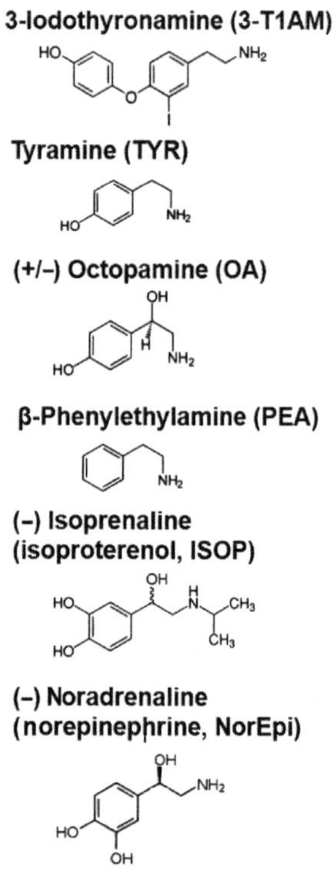

3-Iodothyronamine (3-T1AM)

Tyramine (TYR)

(+/−) Octopamine (OA)

β-Phenylethylamine (PEA)

(−) Isoprenaline (isoproterenol, ISOP)

(−) Noradrenaline (norepinephrine, NorEpi)

FIGURE 5.1   Effects of TYR, OA, 3-T1AM, and PEA on the signaling properties of specific human aminergic receptors.[41-44] This schematic summary reflects the spectrum of signaling modifications by specific trace amines and 3-T1AM that recently have been investigated by in vitro studies at diverse aminergic receptors. They may act as agonists, antagonists, or modulators. Interestingly, antagonistic action at heterodimeric receptors has also been observed.

included single and coapplication of ligands in vitro (Fig. 5.1B) to analyze potential synergistic or antagonistic effects in a combinatorial approach. In the aforementioned study, TYR and PEA were found to act as antagonists for the endogenous agonist isoprenaline (ISOP) at both ADRB, whereas OA was observed to be a weak ADRB2-antagonist and ADRB1-agonist (Fig. 5.1A).

This information characterizes particular trace amines as ligands for ADRB1 and ADRB2. Despite the high amino acid sequence, similarities between TAAR1 and ADRBs (Fig. 5.2), inhibition of ISOP-induced signaling at ADRB1 and ADRB2 by TYR and PEA point to differences in ligand interaction and causative effects compared with TAAR1 and other aminergic receptors such as OARs. Of note, the observed partial antagonistic effects (decrease in maximum response in cAMP accumulation) in costimulation experiments (no shift of the EC_{50} in the concentration–response curve) lead to the conclusion of allosteric (noncompetitive) TYR and PEA action on ISOP effects. Namely, the binding sites or pockets inside the binding region of ADRB1 and ADRB2 may differ for diverse ligands, either inducing activation or inhibiting signaling.

It remains speculative how these ligands are bound to the receptors and how the antagonism may be explained.[42] Generally, the discrepancies between PEA/TYR- and OA-induced effects may be explained by the detailed intermolecular interaction pattern including differences in ligand properties (Fig. 5.1B) and interrelated amino acid side chain interactions, for example, at the transmembrane helices 6 and 7. A scenario that may explain the in vitro

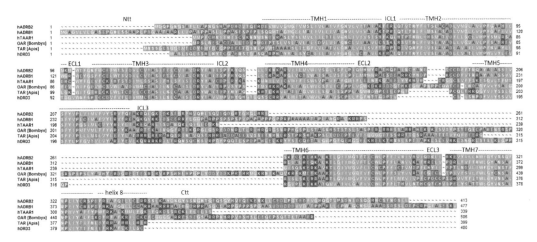

FIGURE 5.2 Amino acid sequence comparison between hTAAR1, hADRB1, hADRB2, bombyxOAR, apisTAR, and hDRD3. The alignment compares amino acids of different aminergic receptors: ADRB1, β1 adrenergic receptor; ADRB2, β2 adrenergic receptor; TAAR1, trace amine-associated receptor 1; ADRA2A, α2A adrenergic receptor; OAR, octopamine receptor; D3R, dopamine-3 receptor; and TAR, tyramine receptor. Particular background colors indicate specific biophysical properties of the amino acid side chains: black, proline; blue, positively charged; cyan/green, aromatic and hydrophobic; green, hydrophobic; red, negatively charged; gray, hydrophilic; dark red, cysteines; and magenta, histidine. The putative helix dimensions and loop regions are assigned according to observable features in the crystal structure of the ADRB2 (pdb entry code 2RH1). Sequence similarities are (Blossum 62 Matrix, entire sequence, examples): TAAR1/ADRB1, 34%; TAAR1/ADRB2, 39%; TAAR1/OAR, 30%; TAAR1/TAR, 36%; TAAR1/DRD3, 33%; OAR/TAR, 36%; ADRB1/ADRB2, 50%; ADRB1/TAR, 33%; and ADRB1/OAR, 32%.

observed allosteric antagonistic effects is the capacity for oligomerization (here oligomerization also comprises of the possibility of dimerization as the smallest possible oligomer constitution) between GPCRs that may alter ligand-binding properties such as ligand affinity and efficiency.[64–70] In a hetero-oligomeric constitution of pairs between different receptor-protomers, lateral allosteric mutual effects of the receptors on each other may lead to improved or inhibited ligand-binding. Furthermore, transactivation has been observed in oligomeric GPCRs, whereby one ligand activates two GPCR molecules or binds to one protomer and activates the second protomer.[65] Such interactions between aminergic receptors, particularly TAARs and adrenergic receptors, have not yet been investigated, with one exception. Co-expressed hTAAR1 and hADRA2A, which are both expressed in, for example, pancreatic β-cells, undergo dimerization in vitro and the signaling capacities of ADRA2A were reported to be altered (uncoupling from Gi signaling pathway) via this receptor arrangement.[43] In conclusion, hetero-oligomerization between amine-sensitive GPCRs that are expressed in the same cell type and tissue may also have an impact on trace amine/catecholamine binding and mediated effects, respectively. This signal modulation by receptor pairs is, so far, underestimated in the field of trace amine research and may be stressed to a greater extent in future studies given that hundreds of different GPCRs may be expressed in one cell type,[71] depending on several factors (eg, age, gender, diseases), and may undergo intermediate receptor pairs.[72] Finally, this also has been, is, and will be an issue for advanced pharmacological studies and drug discovery,[65] as targeting GPCR dimers should increase ligand selectivity and avoid undesired side effects.

To date, it remains unknown how these findings concerning the effects of OA, TYR, and PEA at ADRBs can be observed from a concrete physiological or pathophysiological perspective. According to several studies available in different databases (eg, http://kidbdev.med.unc.edu/databases/ShaunCell/home.php,[71] http://www.nextprot.org, or EMBL-EBI https://www.ebi.ac.uk/gxa/home,[40]) ADRB1 is expressed in various tissues such as the lung, brain, and heart, and also in the retina and testes. ADRB2 has been reported to be located in the lung, skeletal muscle, bone, eye, and skin. Although overlapping expression profiles of ADRBs exist, this does not necessarily imply that both receptors are expressed in the same cellular compartment as has been shown for the distribution of ADRB1 and ADRB2 on cardiomyocytes.[73] Therefore, compartmentation of expressed GPCRs in one cell type should be taken into consideration.

In conclusion, modulatory effects on ADRB1 and ADRB2 signaling by trace amines may be feasible in particular cell systems and may be likely relevant under specific conditions. Accordingly, studies on the background mechanisms of migraines have demonstrated that OA, which is increased in patients suffering from migraine or cluster headache, is capable of modulating astroglial activation in response to inflammatory stimuli via ADRB1 and ADRB2 (see Section III).[55] This example also points to the important fact that the effects of trace amines are concentration-dependent and would, therefore, require a comprehensive evaluation of their physiological concentrations under particular conditions. Generally, as long as systemic, cellular, or local concentrations of trace amines, including those of endogenous thyronamines, have not been determined under both physiological and pathophysiological conditions, data on receptor binding and signaling effects generated from pharmacological and in vitro experiments should be interpreted with caution and not audaciously extrapolated as evidence for the relevant in vivo function of trace amines.

3-T1AM IS A MULTI-GPCR TARGET LIGAND

Thyronamines—A Peculiar Subclass of Thyroid Hormone-Derived Trace Amines

The thyroid hormone (TH)-derived biogenic amine 3-T1AM differs from classical trace amines in several aspects. Though 3-T1AM binds to TAAR1 and other receptors located in the plasma and intracellular membranes (see below), its physicochemical properties, biosynthesis pathway(s), transport in blood, distribution to target tissues, cellular action, metabolism, and elimination are untypical compared with other classical trace amines.

In contrast to these and biogenic amines that originate from their amino acid precursors by decarboxylation and may typically be stored in and released from secretory vesicles, a biosynthesis pathway of the main thyronamine, 3-T1AM, from TH precursors such as T4, T3, and 3,5-T2 by a combination of deiodinase and decarboxylase reactions has been suggested.[74] All three deiodinase enzymes accept iodothyronamines as substrates,[75] eventually generating iodine-free thyronamine (T0AM) that have been identified in tissues and in in vitro experiments.[76,77]

While aromatic amino acid decarboxylase, initially proposed to generate iodothyronines from TH iodothyronines,[78] apparently do not accept THs as ligands,[79] ornithine decarboxylase efficiently generates iodothyronamines from THs in mouse gastrointestinal tissue and in vitro using human recombinant enzyme.[74] Whether other amino acid decarboxylases also generate iodothyronamines, including 3-T1AM, and which tissues, apart from gastrointestinal mucosa, contribute to 3-T1AM formation in vivo remains to be elucidated. Currently, there is limited evidence suggesting that 3-T1AM originates in the thyroid gland itself,[80] and no evidence has been presented for de novo biosynthesis for 3-T1AM by first coupling two iodine-free tyrosine residues and subsequently iodinating the thyronamine specifically in the 3-position independent of thyroglobulin polypeptide chain-based TH biosynthesis.

A further relevant difference between 3-T1AM and the classical trace amines and biogenic amines is its very long half-life in human blood due to its high affinity binding to apolipoprotein B100 that may prevent its rapid inactivation by monoamine oxidases.[81] (Intra-)cellular metabolism of 3-T1AM by inhibitor-sensitive monoamine oxidases, yielding thyroacetic acid metabolites with a rather short half-life, has been demonstrated in vivo and in vitro for several tissues and cell lines.

The thyroacetic acid metabolites are poor ligands for TAAR but may interact with the classical T3 receptors as shown for 3,3′,5-triiodo-thyroacetic acid (Triac),[82,83] and with other yet to be identified receptors in 3-iodo-thyroacetic acid.[84]

To date, only 3-T1AM and iodine-free-thyronamine T0AM have been unequivocally identified in vivo in (sub-)nanomolar concentration range via mass spectrometry of immunoassays.[37,85,86]

3-T1AM has been reported to induce numerous physiological responses and is detectable in peripheral blood and various peripheral organs such as the liver or heart.[37,77,86–88] Application of 3-T1AM in rodents mediates a switch from glucose metabolism to lipid oxidation, thus influencing energy consumption,[85] and shows dose-dependent effects on feeding behavior or body weight (see also Section III).[89–91] 3-T1AM has been demonstrated as a lipolytic agent.[92]

Since the assignment of TAAR1 as the receptor target of 3-T1AM,[37] numerous studies have further concentrated on mechanisms of 3-T1AM.[86,87,93–100] Surprisingly, the previously reported hypothermic 3-T1AM effects on thermoregulation were observed to persist in *mTAAR1* knockout mice.[37,96] This unanticipated observation suggests potential further receptor target(s) for 3-T1AM in vivo. What do we know about the action of thyronamines at GPCRs? An antagonistic effect on β-adrenergic signaling has already been reported for T3AM, 3,5-T2AM, and T0AM. Thyronamines are likely to interfere with ligand binding to adrenergic receptors expressed at the plasma membrane of turkey erythrocytes, and have been reported to inhibit the activation of cAMP production.[101,102] In addition, a physiological role for 3-T1AM as an endogenous adrenergic-blocking neuromodulator in the central noradrenergic system has been recently suggested.[94] Agonistic effects of T1AM and T0AM have not been detected at ADRB2 or the dopamine receptor 1.[37] Moreover, 3-T1AM has been reported to bind to the α2A-adrenergic receptor, a receptor that influences glucose homeostasis.[103] Besides pancreatic β-cells, the ADRA2A is expressed in tissues such as the intestine, several regions of the brain (eg, hypothalamus), the eye, and blood vessels. As these collective results strongly suggest ADRA2A to be an important mediator of 3-T1AM action, detailed studies on signaling regulation by 3-T1AM at ADRA2A have been conducted.[43]

3-T1AM Mediates Selective Signaling at ADRA2A and TAAR1/ADRA2A Hetero-Oligomerization Modifies ADRA2A Signaling

The signal transduction of ADRA2A and coexpression of TAAR1/ADRA2A in response to 3-T1AM, respectively, provided evidence that 3-T1AM acts directly on ADRA2A and induces Gi/o signaling,[43] as known for the classical adrenergic receptor agonist NorEpi (Fig. 5.1). This observed agonistic effect of 3-T1AM is in accordance with previously reported physiological and postulated functional effects at ADRA2A.[103] Moreover, a synergistic effect in Gi/o activation by simultaneous treatment with the ligands NorEpi and 3-T1AM was not observed. Therefore, it can be postulated that 3-T1AM and NorEpi both bind at ADRA2A (Fig. 5.3), although differences in the receptor–ligand interaction pattern exist that may result in two specific findings: (1) the observed lack of MAPK induction by 3-T1AM, which is in contrast to the NorEpi-induced signaling profile and (2) the antagonistic effect of 3-T1AM on NorEpi-induced MAPK (Fig. 5.1). In conclusion, NorEpi and 3-T1AM share Gi/o activation at ADRA2A, but differ in respect to MAPK activation. The latter is particularly interesting, considering that MAPK activation in pancreatic β-cells has been described to have effects on apoptosis, cell growth, and differentiation.[111]

Moreover, the potential formation of hetero-oligomers between TAAR1 and ADRA2A has been confirmed,[43] as both receptors are expressed in vivo in β-cells. Of note, cotransfection of ADRA2A with TAAR1 abolished NorEpi-induced signaling via Gi/o, pointing towards the uncoupling of ADRA2A from the primary signaling cascade. This is in accordance with the previously reported mechanism of the uncoupling of the ADRA2A signaling pathway due to hetero-oligomerization with the μ-opioid receptor,[69] and has also been reported for ADRB2 interaction with the prostaglandin receptor.[112,113] The decreased Gi/o-mediated signaling in ADRA2A hetero-oligomers may have two molecular explanations. A so-called

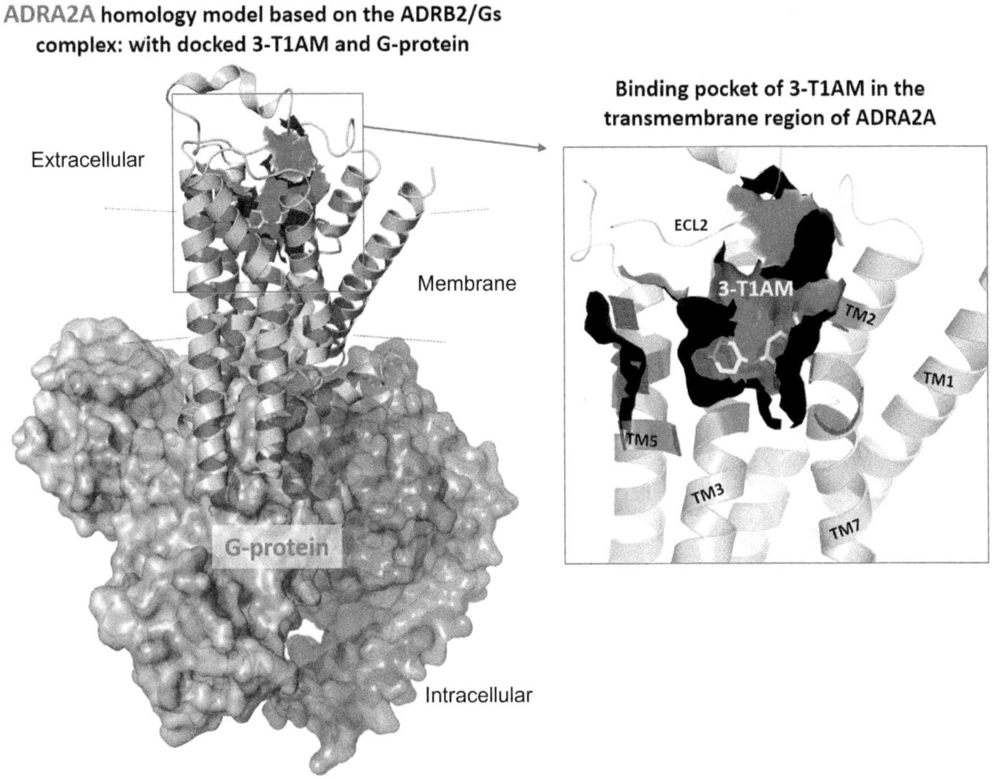

ADRA2A homology model based on the ADRB2/Gs complex: with docked 3-T1AM and G-protein

Binding pocket of 3-T1AM in the transmembrane region of ADRA2A

FIGURE 5.3 Homology model of a complex comprising of the receptor ADRA2A, agonistic ligand 3-T1AM, and the heterotrimeric G-protein Gs. This homology model is based on the determined crystal structure of the ADRB2/Gs complex published in 2011 by Rasmussen et al.[104] Modeling procedures were used according to an already-described modeling protocol.[42] Additionally, the Gs protein from the template structure complex was included in the active receptor complex model. The ligand in the original structure was deleted (agonistic ligand BI-167107). Instead, 3-T1AM (designed manually by using implemented Sybyl ligand-building tools), was located manually in the general transmembrane loop ligand-binding region for family A GPCRs (described in Ref.[105]). For justification of the ligand-binding mode, the highly conserved aspartate in TMH3 of aminergic receptors[106] was chosen as a structural-functional constraint, as its importance has been previously described.[107–109] Furthermore, molecular dynamics simulations (4ns) with flexible side chains (ligand and receptor) were used to estimate a potential detailed ligand justification and intermolecular interactions. The result (entire complex model) presented here shows a binding pocket (partial inner surface) of 3-T1AM in the transmembrane region (clipped presentation, insert window) embedded between the transmembrane helices and the extracellular loop 2 (ECL2). Of specific note, it has been recently reported[110] that some odorant trace amine receptor subtypes of teleost's have a highly conserved aspartate at helix 3, but also, or alternatively, a second aspartate in the same spatial level in helix 5 that strongly impacts the capacities of these receptors for amine ligand-binding and selectivity parameters. The models were subjected to energy minimization until converging at a termination gradient of 0.05 kcal/molÅ by fixing the backbone atoms. The Amber 7 force field FF99 (modeling software Sybyl-X2.0, Certara, NJ, United States) was used. For model representation PyMOL was used (Molecular Graphics System, version 1.3, Schrödinger, LLC).

"lateral off-target allosterism" between the receptors in a hetero-oligomeric constellation leads to suppression of the Gi activation capacity of ADRA2A. This may be due to direct interactions with TAAR1 and may result in structural/dynamic constraints for signaling at ADRA2A. On the other hand, sterical exclusion of G-protein binding may occur via direct protomer–protomer interaction (oligomerization). This signifies that in specific protomer–protomer arrangements, such as potentially by protomer contacts at TMH5, the receptor/G-protein molecule ratio may not be at a 2:2 ratio, but rather a 2:1 ratio should be assumed. Consequently, the signaling capacity of a dimer, which is dependent on the amount of G-protein activation, would be decreased. Altogether, the hetero-oligomer constellation between ADRA2A and TAAR1 results in a modified ADRA2A signaling profile and is a further example for a phenomenon known for other GPCRs.[66]

3-T1AM Interacts with β-Adrenergic Receptor 2

Based on the aforementioned findings,[43] a consequent question concerns the potential interplay between ADRB1 or ADRB2 and the thyronamine 3-T1AM. The aim of a recent study was to ascertain whether 3-T1AM-induced cellular signaling is linked with ADRB1 and ADRB2.[41] This study revealed that 3-T1AM does not activate Gs or MAPK signaling via the analyzed ADRB in vitro, but enhances Gs-mediated signaling of ISOP-stimulated ADRB2 as a positive modulator at specific concentrations (between 10 and 100 nM). This effect of 3-T1AM in costimulation experiments was only marginally detected with the second endogenous ligand NorEpi. ISOP and NorEpi differ in their affinities for ADRB2, whereby ISOP has a higher affinity. The positive modulatory effect observed for the combination between 3-T1AM and ISOP evokes the question: How can such an increase in signaling capacity be explained? In this respect, two differing scenarios related to allosterism can be discussed, whereby potential dimerization again may help to unravel the molecular mechanisms and effects of 3-T1AM.

1. In the case of so-called "off-target" effects, dimers constituting two interacting receptor-protomers would mutually influence one another as has been described above,[65] and binding of 3-T1AM at one protomer should increase the capacity for ISOP-induced signaling at the second protomer, an effect known as positive allosterism. Several examples of signaling (or binding) modification in GPCR dimers have already been reported and ADRB2 is known to comprise of dimeric arrangements.[72,114]
2. In the second potential scenario of "on-target" allosterism, 3-T1AM would bind at an allosteric receptor-binding site and improve, by spatial rearrangements of the monomeric receptor conformation, the signaling capacity for ISOP bound in the orthosteric binding site. Such a principal scenario of two different ligand-binding sites at a GPCR monomer has been previously described in detail for the lutropin receptor.[115]

Interestingly, costimulation experiments with 3-T1AM and ISOP suggested that higher 3-T1AM concentrations (>100 nM and fixed 1 μM concentrations of ISOP), result in a diminished enhancing effect on ISOP-induced signaling. This observed effect at higher 3-T1AM concentrations compared with lower concentrations would also be compatible with suggested mechanisms generally known for GPCR/ligand complexes: (1) A two-binding site model for 3-T1AM, whereby one site would have a stimulating effect and the second an inhibitory influence. This would also conform to a combination of the two aforementioned scenarios concerning the positive allosteric effect of 3-T1AM. Such general mechanisms have been previously described for the M3-muscarinic acetylcholine receptor or the ADRB1.[116,117] (2) Hypothetically, the effect of a decreased Gs-mediated cAMP accumulation at increasing 3-T1AM concentrations in costimulation experiments with a constant ISOP concentration may be related to the activation of Gi. This would inhibit Gs-mediated signaling. β-Adrenergic receptors are known to be promiscuous for different G-protein subtypes and their differentiated activation may be dependent on the ligand variant or concentration.[118]

In conclusion, a detailed signaling pathway induced directly by 3-T1AM at ADRB2 still awaits characterization. A few indications point to the potential involvement of the Gi-mediated pathway, such as a decrease in ISOP/3-T1AM-induced Gs signaling by higher 3-T1AM concentrations, but also 3-T1AM related activation of the *transient receptor potential cation channel subfamily M member 8* (TRPM8) via ADRB, which will be described in detail below. Moreover, an enhancing effect on endogenous ligand-induced Gs-activation at ADRB2 has been found in costimulation experiments with 3-T1AM. In principle, the involvement of receptor dimer constellations on the regulatory capacity of 3-T1AM, but also TYR, PEA, and OA has been suggested in recent findings. Homo- and hetero-oligomerization between aminergic receptors may support regulatory fine-tuning of investigated receptors influenced by 3-T1AM, whereby positive or negative allosteric effects may be relevant. Of note, while TAAR1 oligomerization and oligomerization of adrenergic receptors is well-known,[43,119] such events have not yet been reported for OA or TYR receptors.

3-T1AM Is an Inverse Agonist at Human TAAR5

A further member of the TAAR group, TAAR5, has been investigated for its interaction with 3-T1AM.[44] Both TAAR1 and TAAR5 are expressed in primates.[36] TAAR5 is a highly conserved TAAR subtype among characterized mammalian species investigated so far. Volatile amines such as di- and trimethylethylamine have been reported to be bound by mTAAR5.[19,120]

Overlapping localization of mTAAR1 and mTAAR5 has been confirmed in the amygdala and ventromedial hypothalamus.[44]

To unravel the potential spectrum of the signaling capacity of a TAAR5/3-T1AM complex, G protein-dependent and -independent pathways of mouse and human TAAR5 (hTAAR5) under ligand-independent conditions and following application of 3-T1AM have been examined.[44] Murine and human TAAR5 orthologs both display a slight basal activity in Gq/11 activation, but demonstrate significant differences in the basal activity in Gs and MAP kinase signaling. Moreover, in contrast to mTAAR5, 3-T1AM application at hTAAR5 resulted in a significant reduction of basal IP3 formation and MAP kinase signaling. However, no agonistic activity of 3-T1AM was detected. These studies have excluded TAAR5 as a mediator of observed 3-T1AM-induced pharmacological actions but suggest a potential affinity of TAAR5 for thyronamines, or designed TAAR-acting molecules, whereby potential differences between mouse and human must be considered.[35]

An Axis Between Thyronamines, TRP Channel-Mediated Ca^{2+} Flux, and β-Adrenergic Receptors

A previous study reported the identification of 3-T1AM as a molecule that increases Ca^{2+} influx and whole cell current via *transient receptor potential melastatin 8* (TRPM8) channel activation.[121] Moreover, 3-T1AM decreases interleukin-6 release induced by capsaicin-mediated activation of *transient receptor potential vanilloid 1* (TRPV1); however, the underlying mechanism has not yet been thoroughly deciphered in both cases. A recent study provided deeper insights into 3-T1AM mechanisms in conjunction with TRP channels.[41] Eye cells are known to express adrenergic receptors and an association between voltage-dependent Ca^{2+} channels and adrenergic receptors has been described.[71,122–124] TRPM8 and TRPV1 are not voltage-dependent,[125] but are thermosensitive and respond to many extracellular or intracellular stimuli such as G-protein subunits.[126] Therefore, to initially ascertain whether adrenergic receptors respond to 3-T1AM, human conjunctival epithelial cells (IOBA-NHC) were investigated. NorEpi and 3-T1AM were both found to increase intracellular Ca^{2+} levels in IOBA-NHC cells. This effect was blocked by the nonselective adrenergic receptor blocker timolol.[41] This prompted the suggestion that 3-T1AM effects may be attributable to an axis comprised of 3-T1AM/adrenergic receptor/TRP channel. 3-T1AM-induced effects may be directly mediated via protein interactions between the GPCR and the particular TRP channel. Such direct protein–protein interactions have been reported for GPCRs and other channels such as KIR channels that form a macromolecular ion channel–GPCR signaling complex.[127] A second scenario may be via an indirect signal transmission. For example, $G\alpha q$ subunit preforms a complex with TRPM8 and inhibits TRPM8,[128] or the β-γ subunits of Gi/o may be activated by the ADRB and consequently induce TRPM8 activation on the intracellular site.[118] Further evidence of the direct interaction between 3-T1AM and aminergic receptors expressed in particular tissues is required in complementation with G-protein-mediated signaling that may eventually result in the modulation of certain ion channels. Finally, these insights also concern the afore-described action of trace amines such as OA, which activates the ADRB2.[42]

Pathophysiological aspects related to the modification of ion channels by, for example, 3-T1AM are inflammation, pain development, and manifestation. In a recent review on the pathogenesis of migraine,[53] thus far unrecognized genes such as *TRPM8* that is also expressed in trigeminal and dorsal root ganglia were concluded to possibly play an important role in the initiation of headache attacks.[129] Moreover, characterization of the ion channel TRPV1 has demonstrated it to be a polymodal nocisensor.[130] Furthermore, the role of TRPV1 in the enhancement of proinflammatory cytokine release has raised enormous interest in TRPV1 as a prime transducer of pathological pain and inflammation.[131] Since 3-T1AM-induced TRPM8 activation leads to a decline in TRPV1 activity, this crosstalk effect may be of therapeutic benefit in a clinical setting.

CONCLUSION

Previously and recently in vitro identified interrelations of trace amines and 3-T1AM at adrenergic receptors implicate that trace amines and 3-T1AM have a multitude of GPCR

Targets of 3-T1AM

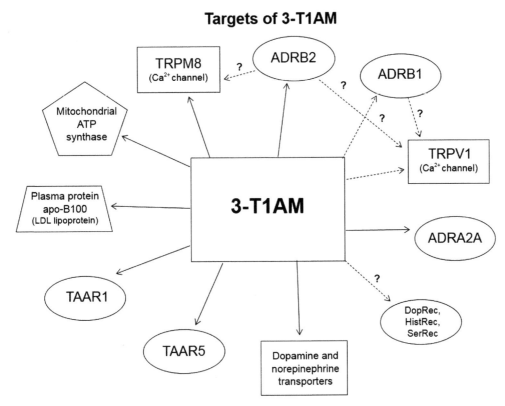

FIGURE 5.4 Identified targets of 3-T1AM. Several additional family A GPCRs, aside from TAAR1, interrelate with 3-T1AM as shown in recent and prior molecular in vitro studies. There is initial evidence indicating that TRP channels and Ca^{2+} release may be regulated by 3-T1AM-induced signaling cascades, although the exact mechanism has not yet been clarified. Finally, 3-T1AM has a functional impact on a multitude of proteins.[77] Hypothetically, 3-T1AM may additionally target further aminergic receptors such as members of the dopamine (DopRec), histamine (HistRec), or serotonine (SerRec) receptor groups.

targets (Fig. 5.4). These ligand/receptor complexes may also have an impact in vivo as physiological modulators.[103]

Moreover, their signaling effects are diverse, including antagonistic, modulatory, and agonistic effects. Of specific note, the molecular mechanisms that explain the observed signaling modulation probably include allosteric effects and oligomerization of the GPCR targets. In regard to the assumption that trace amines modulate physiological reactions, the diverse spectrum of signaling modulation would conform to a fine-tuning of signaling under certain conditions. Finally, new components in regulation cascades, such as TRP channels, are suggested in relation to 3-T1AM/adrenergic receptor systems. Based on the relatively scarce systematic information particularly concerning in vivo concentrations, GPCR targets, their oligomerization, and missing structural insights, studies focusing on these issues are recommended to elucidate and characterize physiological and pathophysiological conditions.

Acknowledgments

This work was supported by the Deutsche Forschungsgemeinschaft: Priority program SPP1629 Thyroid Trans Act BI 893/5-2 and KO 922/16-1/2.

References

1. Arakawa S, Gocayne JD, McCombie WR, et al. Cloning, localization, and permanent expression of a *Drosophila* octopamine receptor. *Neuron* 1990;**4**(3):343–54.
2. Cazzamali G, Klaerke DA, Grimmelikhuijzen CJ. A new family of insect tyramine receptors. *Biochem Biophys Res Commun* 2005;**338**(2):1189–96.
3. Roeder T. Tyramine and octopamine: ruling behavior and metabolism. *Annu Rev Entomol* 2005;**50**:447–77.
4. Saudou F, Amlaiky N, Plassat JL, Borrelli E, Hen R. Cloning and characterization of a *Drosophila* tyramine receptor. *EMBO J* 1990;**9**(11):3611–7.
5. Gerhardt CC, Bakker RA, Piek GJ, et al. Molecular cloning and pharmacological characterization of a molluscan octopamine receptor. *Mol Pharmacol* 1997;**51**(2):293–300.
6. El-Kholy S, Stephano F, Li Y, Bhandari A, Fink C, Roeder T. Expression analysis of octopamine and tyramine receptors in *Drosophila*. *Cell Tissue Res* 2015;**361**(3):669–84.
7. David JC, Coulon JF. Octopamine in invertebrates and vertebrates. A review. *Prog Neurobiol* 1985;**24**(2):141–85.
8. Berry MD. Mammalian central nervous system trace amines. Pharmacologic amphetamines, physiologic neuromodulators. *J Neurochem* 2004;**90**(2):257–71.
9. Zucchi R, Chiellini G, Scanlan TS, Grandy DK. Trace amine-associated receptors and their ligands. *Br J Pharmacol* 2006;**149**(8):967–78.
10. Berry MD. The potential of trace amines and their receptors for treating neurological and psychiatric diseases. *Rev Recent Clin Trials* 2007;**2**(1):3–19.
11. Liberles SD. Trace amine-associated receptors: ligands, neural circuits, and behaviors. *Curr Opin Neurobiol* 2015;**34**:1–7.
12. Maguire JJ, Parker WAE, Foord SM, Bonner TI, Neubig RR, Davenport AP. International Union of Pharmacology. LXXII. Recommendations for trace amine receptor nomenclature. *Pharmacol Rev* 2009;**61**(1):1–8.
13. Lindemann L, Ebeling M, Kratochwil NA, Bunzow JR, Grandy DK, Hoener MC. Trace amine-associated receptors form structurally and functionally distinct subfamilies of novel G protein-coupled receptors. *Genomics* 2005;**85**(3):372–85.
14. Dewan A, Pacifico R, Zhan R, Rinberg D, Bozza T. Non-redundant coding of aversive odours in the main olfactory pathway. *Nature* 2013;**497**(7450):486–9.
15. Ferrero DM, Wacker D, Roque MA, Baldwin MW, Stevens RC, Liberles SD. Agonists for 13 trace amine-associated receptors provide insight into the molecular basis of odor selectivity. *ACS Chem Biol* 2012;**7**(7):1184–9.
16. Hussain A, Saraiva LR, Korsching SI. Positive Darwinian selection and the birth of an olfactory receptor clade in teleosts. *Proc Natl Acad Sci USA* 2009;**106**(11):4313–8.
17. Krautwurst D. Human olfactory receptor families and their odorants. *Chem Biodivers* 2008;**5**(6):842–52.
18. Liberles SD. Trace amine-associated receptors are olfactory receptors in vertebrates. *Ann NY Acad Sci* 2009;**1170**(1):168–72.
19. Liberles SD, Buck LB. A second class of chemosensory receptors in the olfactory epithelium. *Nature* 2006;**442**(7103):645–50.
20. Yoon KH, Ragoczy T, Lu Z, et al. Olfactory receptor genes expressed in distinct lineages are sequestered in different nuclear compartments. *Proc Natl Acad Sci USA* 2015;**112**(18):E2403–2409.
21. Johnson MA, Tsai L, Roy DS, et al. Neurons expressing trace amine-associated receptors project to discrete glomeruli and constitute an olfactory subsystem. *Proc Natl Acad Sci USA* 2012;**109**(33):13410–5.
22. Pacifico R, Dewan A, Cawley D, Guo C, Bozza T. An olfactory subsystem that mediates high-sensitivity detection of volatile amines. *Cell Rep* 2012;**2**(1):76–88.
23. Pae CU, Drago A, Mandelli L, De Ronchi D, Serretti A. TAAR 6 and HSP-70 variations associated with bipolar disorder. *Neurosci Lett* 2009;**465**(3):257–61.
24. Vanti WB, Nguyen T, Cheng R, Lynch KR, George SR, O'Dowd BF. Novel human G-protein-coupled receptors. *Biochem Biophys Res Commun* 2003;**305**(1):67–71.

25. Duan J, Martinez M, Sanders AR, et al. Polymorphisms in the trace amine receptor 4 (TRAR4) gene on chromosome 6q23.2 are associated with susceptibility to schizophrenia. *Am J Human Genet* 2004;**75**(4):624–38.

26. Bly M. Examination of the trace amine-associated receptor 2 (TAAR2). *Schizophr Res* 2005;**80**(2–3):367–8.

27. D'Andrea G, Nordera G, Pizzolato G, et al. Trace amine metabolism in Parkinson's disease: low circulating levels of octopamine in early disease stages. *Neurosci Lett* 2010;**469**(3):348–51.

28. Pae CU, Drago A, Kim JJ, et al. TAAR6 variations possibly associated with antidepressant response and suicidal behavior. *Psychiatry Res* 2010;**180**(1):20–4.

29. Burchett SA, Hicks TP. The mysterious trace amines: protean neuromodulators of synaptic transmission in mammalian brain. *Prog Neurobiol* 2006;**79**(5–6):223–46.

30. Borowsky B, Adham N, Jones KA, et al. Trace amines: identification of a family of mammalian G protein-coupled receptors. *Proc Natl Acad Sci USA* 2001;**98**(16):8966–71.

31. Bunzow JR, Sonders MS, Arttamangkul S, et al. Amphetamine, 3,4-methylenedioxymethamphetamine, lysergic acid diethylamide, and metabolites of the catecholamine neurotransmitters are agonists of a rat trace amine receptor. *Mol Pharmacol* 2001;**60**(6):1181–8.

32. Wolinsky TD, Swanson CJ, Smith KE, et al. The trace amine 1 receptor knockout mouse: an animal model with relevance to schizophrenia. *Genes Brain Behav* 2007;**6**(7):628–39.

33. Gloriam DEI, Bjarnadóttir TK, Yan Y-L, Postlethwait JH, Schiöth HB, Fredriksson R. The repertoire of trace amine G-protein-coupled receptors: large expansion in zebrafish. *Mol Phylogenet Evol* 2005;**35**(2):470–82.

34. Grandy DK. Trace amine-associated receptor 1—family archetype or iconoclast? *Pharmacol Ther* 2007;**116**(3):355–90.

35. Stäubert C, Böselt I, Bohnekamp J, Römpler H, Enard W, Schöneberg T. Structural and functional evolution of the trace amine-associated receptors TAAR3, TAAR4 and TAAR5 in primates. *PLoS One* 2010;**5**(6):e11133.

36. Vallender EJ, Xie Z, Westmoreland SV, Miller GM. Functional evolution of the trace amine associated receptors in mammals and the loss of TAAR1 in dogs. *BMC Evol Biol* 2010;**10**:51.

37. Scanlan TS, Suchland KL, Hart ME, et al. 3-Iodothyronamine is an endogenous and rapid-acting derivative of thyroid hormone. *Nat Med* 2004;**10**(6):638–42.

38. Rozenfeld R, Devi LA. Exploring a role for heteromerization in GPCR signalling specificity. *Biochem J* 2011;**433**(1):11–18.

39. Alexander SP, Benson HE, Faccenda E, et al. The concise guide to pharmacology 2013/14: G protein-coupled receptors. *Br J Pharmacol* 2013;**170**(8):1459–581.

40. Petryszak R, Burdett T, Fiorelli B, et al. Expression Atlas update—a database of gene and transcript expression from microarray- and sequencing-based functional genomics experiments. *Nucleic Acids Res* 2014;**42**(Database issue):D926–32.

41. Dinter J, Khajavi N, Muhlhaus J, et al. The multitarget ligand 3-iodothyronamine modulates beta-adrenergic receptor 2 signaling. *Eur Thyroid J* 2015;**4**(Suppl 1):21–9.

42. Kleinau G, Pratzka J, Nurnberg D, et al. Differential modulation of beta-adrenergic receptor signaling by trace amine-associated receptor 1 agonists. *PLoS One* 2011;**6**(10):e27073.

43. Dinter J, Muhlhaus J, Jacobi SF, et al. 3-Iodothyronamine differentially modulates alpha-2A-adrenergic receptor-mediated signaling. *J Mol Endocrinol* 2015;**54**(3):205–16.

44. Dinter J, Muhlhaus J, Wienchol CL, et al. Inverse agonistic action of 3-iodothyronamine at the human trace amine-associated receptor 5. *PLoS One* 2015;**10**(2):e0117774.

45. Stohs SJ. Physiological functions and pharmacological and toxicological effects of p-octopamine. *Drug Chem Toxicol* 2015;**38**(1):106–12.

46. Farooqui T. Octopamine-mediated neuromodulation of insect senses. *Neurochem Res* 2007;**32**(9):1511–29.

47. David JC, Coulon JF, Cavoy A, Delacour J. Effects of aging on p- and m-octopamine, catecholamines, and their metabolizing enzymes in the rat. *J Neurochem* 1989;**53**(1):149–54.

48. Carpene C, Galitzky J, Fontana E, Atgie C, Lafontan M, Berlan M. Selective activation of beta3-adrenoceptors by octopamine: comparative studies in mammalian fat cells. *Naunyn Schmiedebergs Arch Pharmacol* 1999;**359**(4):310–21.

49. Airriess CN, Rudling JE, Midgley JM, Evans PD. Selective inhibition of adenylyl cyclase by octopamine via a human cloned alpha 2A-adrenoceptor. *Br J Pharmacol* 1997;**122**(2):191–8.

50. Brown CM, McGrath JC, Midgley JM, et al. Activities of octopamine and synephrine stereoisomers on alpha-adrenoceptors. *Br J Pharmacol* 1988;**93**(2):417–29.

51. Fagerholm V, Haaparanta M, Scheinin M. Alpha2-adrenoceptor regulation of blood glucose homeostasis. *Basic Clin Pharmacol Toxicol* 2011;**108**(6):365–70.

52. Giovannitti Jr. JA, Thoms SM, Crawford JJ. Alpha-2 adrenergic receptor agonists: a review of current clinical applications. *Anesth Prog* 2015;**62**(1):31–9.

53. D'Andrea G, D'Arrigo A, Dalle Carbonare M, Leon A. Pathogenesis of migraine: role of neuromodulators. *Headache* 2012;**52**(7):1155–63.

54. Kinniburgh DW, Boyd ND. Determination of plasma octopamine and its level in renal disease. *Clin Biochem* 1979;**12**(1):27–32.

55. D'Andrea G, D'Arrigo A, Facchinetti F, et al. Octopamine, unlike other trace amines, inhibits responses of astroglia-enriched cultures to lipopolysaccharide via a beta-adrenoreceptor-mediated mechanism. *Neurosci Lett* 2012;**517**(1):36–40.

56. Broadley KJ, Fehler M, Ford WR, Kidd EJ. Functional evaluation of the receptors mediating vasoconstriction of rat aorta by trace amines and amphetamines. *Eur J Pharmacol* 2013;**715**(1–3):370–80.

57. Sabelli HC, Javaid JI. Phenylethylamine modulation of affect: therapeutic and diagnostic implications. *J Neuropsychiatry Clin Neurosci* 1995;**7**(1):6–14.

58. Ma G, Bavadekar SA, Schaneberg BT, Khan IA, Feller DR. Effects of synephrine and beta-phenethylamine on human alpha-adrenoceptor subtypes. *Planta Med* 2010;**76**(10):981–6.

59. Narang D, Kerr PM, Lunn SE, et al. Modulation of resistance artery tone by the trace amine beta-phenylethylamine: dual indirect sympathomimetic and alpha1-adrenoceptor blocking actions. *J Pharmacol Exp Ther* 2014;**351**(1):164–71.

60. Dyck LE, Boulton AA. Effects of beta-phenylethylamine on polyphosphoinositide turnover in rat cerebral cortex. *Neurochem Res* 1989;**14**(1):63–7.

61. Clarke GL, Bhattacherjee A, Tague SE, Hasan W, Smith PG. ß-Adrenoceptor blockers increase cardiac sympathetic innervation by inhibiting autoreceptor suppression of axon growth. *J Neurosci* 2010;**30**(37):12446–54.

62. Broadley KJ. The vascular effects of trace amines and amphetamines. *Pharmacol Ther* 2010;**125**(3):363–75.

63. Broadley KJ, Akhtar Anwar M, Herbert AA, et al. Effects of dietary amines on the gut and its vasculature. *Br J Nutr* 2009;**101**(11):1645–52.

64. Fung JJ, Deupi X, Pardo L, et al. Ligand-regulated oligomerization of beta(2)-adrenoceptors in a model lipid bilayer. *EMBO J* 2009;**28**(21):3315–28.

65. George SR, O'Dowd BF, Lee SP. G-protein-coupled receptor oligomerization and its potential for drug discovery. *Nat Rev Drug Discov* 2002;**1**(10):808–20.

66. Lambert NA. GPCR dimers fall apart. *Sci Signal* 2010;**3**(115):pe12.

67. Lohse MJ. Dimerization in GPCR mobility and signaling. *Curr Opin Pharmacol* 2010;**10**(1):53–8.

68. Mancia F, Assur Z, Herman AG, Siegel R, Hendrickson WA. Ligand sensitivity in dimeric associations of the serotonin 5HT2c receptor. *EMBO Rep* 2008;**9**(4):363–9.

69. Vilardaga JP, Nikolaev VO, Lorenz K, Ferrandon S, Zhuang Z, Lohse MJ. Conformational cross-talk between alpha2A-adrenergic and mu-opioid receptors controls cell signaling. *Nat Chem Biol* 2008;**4**(2):126–31.

70. White JF, Grodnitzky J, Louis JM, et al. Dimerization of the class A G protein-coupled neurotensin receptor NTS1 alters G protein interaction. *Proc Natl Acad Sci USA* 2007;**104**(29):12199–204.

71. Regard JB, Sato IT, Coughlin SR. Anatomical profiling of G protein-coupled receptor expression. *Cell* 2008;**135**(3):561–71.

72. Smith NJ, Milligan G. Allostery at G protein-coupled receptor homo- and heteromers: uncharted pharmacological landscapes. *Pharmacol Rev* 2010;**62**(4):701–25.

73. Nikolaev VO, Moshkov A, Lyon AR, et al. Beta2-adrenergic receptor redistribution in heart failure changes cAMP compartmentation. *Science* 2010;**327**(5973):1653–7.

74. Hoefig CS, Wuensch T, Rijntjes E, et al. Biosynthesis of 3-iodothyronamine from T4 in murine intestinal tissue. *Endocrinology* 2015;**156**(11):4356–64.

75. Piehl S, Heberer T, Baliz G, et al. Thyronamines are isozyme-specific substrates of deiodinases. *Endocrinology* 2008;**149**(6):3037–45.

76. Agretti P, De Marco G, Russo L, et al. 3-Iodothyronamine metabolism and functional effects in FRTL5 thyroid cells. *J Mol Endocrinol* 2011;**47**(1):23–32.

77. Zucchi R, Accorroni A, Chiellini G. Update on 3-iodothyronamine and its neurological and metabolic actions. *Front Physiol* 2014;**5**:402.

78. Dratman MB. On the mechanism of action of thyroxin, an amino acid analog of tyrosine. *J Theor Biol* 1974;**46**(1):255–70.

79. Hoefig CS, Renko K, Piehl S, et al. Does the aromatic L-amino acid decarboxylase contribute to thyronamine biosynthesis? *Mol Cell Endocrinol* 2012;**349**(2):195–201.

80. Hackenmueller SA, Marchini M, Saba A, Zucchi R, Scanlan TS. Biosynthesis of 3-iodothyronamine (T1AM) is dependent on the sodium-iodide symporter and thyroperoxidase but does not involve extrathyroidal metabolism of T4. *Endocrinology* 2012;**153**(11):5659–67.

81. Roy G, Placzek E, Scanlan TS. ApoB-100-containing lipoproteins are major carriers of 3-iodothyronamine in circulation. *J Biol Chem* 2012;**287**(3):1790–800.

82. Paris M, Escriva H, Schubert M, et al. Amphioxus postembryonic development reveals the homology of chordate metamorphosis. *Curr Biol* 2008;**18**(11):825–30.

83. Wagner RL, Huber BR, Shiau AK, et al. Hormone selectivity in thyroid hormone receptors. *Molec Endocrinol* 2001;**15**(3):398–410.

84. Laurino A, De Siena G, Saba A, et al. In the brain of mice, 3-iodothyronamine (T1AM) is converted into 3-iodothyroacetic acid (TA1) and it is included within the signaling network connecting thyroid hormone metabolites with histamine. *Eur J Pharmacol* 2015;**761**:130–4.

85. Braulke LJ, Klingenspor M, DeBarber A, et al. 3-Iodothyronamine: a novel hormone controlling the balance between glucose and lipid utilisation. *J Comp Physiol B* 2008;**178**(2):167–77.

86. Saba A, Chiellini G, Frascarelli S, et al. Tissue distribution and cardiac metabolism of 3-iodothyronamine. *Endocrinology* 2010;**151**(10):5063–73.

87. Piehl S, Hoefig CS, Scanlan TS, Kohrle J. Thyronamines—past, present, and future. *Endocr Rev* 2011;**32**(1):64–80.

88. Hoefig CS, Kohrle J, Brabant G, et al. Evidence for extrathyroidal formation of 3-iodothyronamine in humans as provided by a novel monoclonal antibody-based chemiluminescent serum immunoassay. *J Clin Endocrinol Metab* 2011;**96**(6):1864–72.

89. Dhillo WS, Bewick GA, White NE, et al. The thyroid hormone derivative 3-iodothyronamine increases food intake in rodents. *Diabetes Obes Metab* 2009;**11**(3):251–60.

90. Manni ME, De Siena G, Saba A, et al. 3-Iodothyronamine: a modulator of the hypothalamus-pancreas-thyroid axes in mice. *Br J Pharmacol* 2012;**166**(2):650–8.

91. Haviland JA, Reiland H, Butz DE, et al. NMR-based metabolomics and breath studies show lipid and protein catabolism during low dose chronic T(1)AM treatment. *Obesity (Silver Spring)* 2013;**21**(12):2538–44.

92. Mariotti V, Melissari E, Iofrida C, et al. Modulation of gene expression by 3-iodothyronamine: genetic evidence for a lipolytic pattern. *PLoS One* 2014;**9**(11):e106923.

93. Ghelardoni S, Suffredini S, Frascarelli S, et al. Modulation of cardiac ionic homeostasis by 3-iodothyronamine. *J Cell Mol Med* 2009;**13**(9b):3082–90.

94. Gompf HS, Greenberg JH, Aston-Jones G, Ianculescu AG, Scanlan TS, Dratman MB. 3-Monoiodothyronamine: the rationale for its action as an endogenous adrenergic-blocking neuromodulator. *Brain Res* 2010;**1351**:130–40.

95. Ianculescu AG, Scanlan TS. 3-Iodothyronamine (T(1)AM): a new chapter of thyroid hormone endocrinology? *Mol Biosyst* 2010;**6**(8):1338–44.

96. Panas HN, Lynch LJ, Vallender EJ, et al. Normal thermoregulatory responses to 3-iodothyronamine, trace amines and amphetamine-like psychostimulants in trace amine associated receptor 1 knockout mice. *J Neurosci Res* 2010;**88**(9):1962–9.

97. Piehl S, Heberer T, Balizs G, Scanlan TS, Kohrle J. Development of a validated liquid chromatography/tandem mass spectrometry method for the distinction of thyronine and thyronamine constitutional isomers and for the identification of new deiodinase substrates. *Rapid Commun Mass Spectrom* 2008;**22**(20):3286–96.

98. Pietsch CA, Scanlan TS, Anderson RJ. Thyronamines are substrates for human liver sulfotransferases. *Endocrinology* 2007;**148**(4):1921–7.

99. Scanlan TS. Minireview: 3-iodothyronamine (T1AM): a new player on the thyroid endocrine team? *Endocrinology* 2009;**150**(3):1108–11.

100. Zucchi R, Ghelardoni S, Chiellini G. Cardiac effects of thyronamines. *Heart Fail Rev* 2010;**15**(2):171–6.

101. Cody V, Meyer T, Dohler KD, Hesch RD, Rokos H, Marko M. Molecular structure and biochemical activity of 3,5,3'-triiodothyronamine. *Endocr Res* 1984;**10**(2):91–9.

102. Meyer T, Hesch RD. Triiodothyronamine—a beta-adrenergic metabolite of triiodothyronine? *Horm Metab Res* 1983;**15**(12):602–6.
103. Regard JB, Kataoka H, Cano DA, et al. Probing cell type-specific functions of Gi in vivo identifies GPCR regulators of insulin secretion. *J Clin Invest* 2007;**117**(12):4034–43.
104. Rasmussen SG, DeVree BT, Zou Y, et al. Crystal structure of the beta2 adrenergic receptor-Gs protein complex. *Nature* 2011;**477**(7366):549–55.
105. Kleinau G, Jaeschke H, Worth CL, et al. Principles and determinants of G-protein coupling by the rhodopsin-like thyrotropin receptor. *PLoS One* 2010;**5**(3):e9745.
106. Huang ES. Construction of a sequence motif characteristic of aminergic G protein-coupled receptors. *Protein Sci* 2003;**12**(7):1360–7.
107. Tan ES, Groban ES, Jacobson MP, Scanlan TS. Toward deciphering the code to aminergic G protein-coupled receptor drug design. *Chem Biol* 2008;**15**(4):343–53.
108. Tan ES, Miyakawa M, Bunzow JR, Grandy DK, Scanlan TS. Exploring the structure–activity relationship of the ethylamine portion of 3-iodothyronamine for rat and mouse trace amine-associated receptor 1. *J Med Chem* 2007;**50**(12):2787–98.
109. Tan ES, Naylor JC, Groban ES, et al. The molecular basis of species-specific ligand activation of trace amine-associated receptor 1 (TAAR1). *ACS Chem Biol* 2009;**4**(3):209–20.
110. Li Q, Tachie-Baffour Y, Liu Z, Baldwin MW, Kruse AC, Liberles SD. Non-classical amine recognition evolved in a large clade of olfactory receptors. *Elife* 2015;**4**:e10441.
111. Verga Falzacappa C, Panacchia L, Bucci B, et al. 3,5,3′-Triiodothyronine (T3) is a survival factor for pancreatic β-cells undergoing apoptosis. *J Cell Physiol* 2006;**206**(2):309–21.
112. McGraw DW, Mihlbachler KA, Schwarb MR, et al. Airway smooth muscle prostaglandin-EP1 receptors directly modulate beta2-adrenergic receptors within a unique heterodimeric complex. *J Clin Invest* 2006;**116**(5):1400–9.
113. Barnes PJ. Receptor heterodimerization: a new level of cross-talk. *J Clin Invest* 2006;**116**(5):1210–2.
114. Sartania N, Appelbe S, Pediani JD, Milligan G. Agonist occupancy of a single monomeric element is sufficient to cause internalization of the dimeric beta2-adrenoceptor. *Cell Signal* 2007;**19**(9):1928–38.
115. Heitman LH, Kleinau G, Brussee J, Krause G, Ijzerman AP. Determination of different putative allosteric binding pockets at the lutropin receptor by using diverse drug-like low molecular weight ligands. *Mol Cell Endocrinol* 2012;**351**(2):326–36.
116. Thor D, Schulz A, Hermsdorf T, Schoneberg T. Generation of an agonistic binding site for blockers of the M(3) muscarinic acetylcholine receptor. *Biochem J* 2008;**412**(1):103–12.
117. Baker JG, Hill SJ. Multiple GPCR conformations and signalling pathways: implications for antagonist affinity estimates. *Trends Pharmacol Sci* 2007;**28**(8):374–81.
118. Wenzel-Seifert K, Seifert R. Molecular analysis of beta(2)-adrenoceptor coupling to G(s)-, G(i)-, and G(q)-proteins. *Mol Pharmacol* 2000;**58**(5):954–66.
119. Mercier JF, Salahpour A, Angers S, Breit A, Bouvier M. Quantitative assessment of beta 1- and beta 2-adrenergic receptor homo- and heterodimerization by bioluminescence resonance energy transfer. *J Biol Chem* 2002;**277**(47):44925–31.
120. Wallrabenstein I, Kuklan J, Weber L, et al. Human trace amine-associated receptor TAAR5 can be activated by trimethylamine. *PLoS One* 2013;**8**(2):e54950.
121. Khajavi N, Reinach PS, Slavi N, et al. Thyronamine induces TRPM8 channel activation in human conjunctival epithelial cells. *Cell Signal* 2015;**27**(2):315–25.
122. Enriquez de Salamanca A, Siemasko KF, Diebold Y, et al. Expression of muscarinic and adrenergic receptors in normal human conjunctival epithelium. *Invest Ophthalmol Vis Sci* 2005;**46**(2):504–13.
123. Messina Baas O, Pacheco Cuellar G, Toral-Lopez J, et al. ADRB1 and ADBR2 gene polymorphisms and the ocular hypotensive response to topical betaxolol in healthy mexican subjects. *Curr Eye Res* 2014;**39**(11):1076–80.
124. Bavencoffe A, Gkika D, Kondratskyi A, et al. The transient receptor potential channel TRPM8 is inhibited via the alpha 2A adrenoreceptor signaling pathway. *J Biol Chem* 2010;**285**(13):9410–9.
125. Minke B, Cook B. TRP channel proteins and signal transduction. *Physiol Rev* 2002;**82**(2):429–72.

126. Veldhuis NA, Poole DP, Grace M, McIntyre P, Bunnett NW. The G protein-coupled receptor-transient receptor potential channel axis: molecular insights for targeting disorders of sensation and inflammation. *Pharmacol Rev* 2015;**67**(1):36–73.
127. Doupnik CA. GPCR-Kir channel signaling complexes: defining rules of engagement. *J Recept Signal Transduct Res* 2008;**28**(1–2):83–91.
128. Zhang X, Mak S, Li L, et al. Direct inhibition of the cold-activated TRPM8 ion channel by Galphaq. *Nat Cell Biol* 2012;**14**(8):851–8.
129. Su L, Wang C, Yu YH, Ren YY, Xie KL, Wang GL. Role of TRPM8 in dorsal root ganglion in nerve injury-induced chronic pain. *BMC Neurosci* 2011;**12**:120.
130. Hanack C, Moroni M, Lima WC, et al. GABA blocks pathological but not acute TRPV1 pain signals. *Cell* 2015;**160**(4):759–70.
131. Bley KR. Recent developments in transient receptor potential vanilloid receptor 1 agonist-based therapies. *Expert Opin Investig Drugs* 2004;**13**(11):1445–56.

Effects of Trace Amines on the Dopaminergic Mesencephalic System

A. Ledonne[1] and N.B. Mercuri[1,2]

[1]Department of Experimental Neurosciences, Santa Lucia Foundation, Rome,
Italy [2]Department of System Medicine, University of Rome "Tor Vergata",
Rome, Italy

OUTLINE

Trace Amines and Neurological Disorders
DOI: http://dx.doi.org/10.1016/B978-0-12-803603-7.00006-9

INTRODUCTION

Trace amines (TAs) are a class of endogenous compounds, which are heterogeneously distributed throughout the mammalian brain and peripheral nervous tissues, despite being at relatively low levels. The group includes β-phenylethylamine (β-PEA), tyramine (TYR), octopamine (OCT), synephrine (SYN), and tryptamine (TRP). TAs have been identified in several prokaryotic and eukaryotic organisms and in all vertebrate and invertebrate species studied so far, including humans.[1] It is now largely accepted that TAs are not inactive byproducts of amino acid metabolism but indeed they are considered important neuromodulators involved in key physiological brain functions.

Here we present an overview of the close interplay between TAs and dopamine (DA). We discuss growing evidence demonstrating that TAs critically modulate the mesencephalic dopaminergic (DAergic) system, by means of several cellular mechanisms. Since DAergic neurotransmission is crucial for essential physiological functions such as locomotion, attention and cognition, and its dysfunction is associated with different neuropsychiatric disorders, the TA-induced modulation of the DAergic system could have relevant physiopathological implications.

THE DAergic MESENCEPHALIC SYSTEM

The DAergic system is principally constituted by the soma and dendrites of DA neurons located in the ventral mesencephalon and their axonal projections and synaptic boutons to the terminal fields. In mammals, most DA-containing neurons cluster within three major mesencephalic groups: the retrorubral field (A8), the substantia nigra pars compacta (SNpc) (A9), and the ventral tegmental area (VTA) (A10).[2]

Based on the somatic localization and projection areas, the DAergic mesencephalic system is subdivided into the nigrostriatal and mesolimbic systems; the latter includes the mesoaccumbal and the mesocortical pathways. The nigrostriatal system originates in the SNpc and projects to the dorsal striatum. The retrorubral area DAergic neurons also project to the dorsal striatum, being considered a SNpc caudal extension. The mesoaccumbal and mesocortical pathways originate in the VTA and project to limbic areas, nucleus accumbens (NAcc), amygdala, olfactory tubercle, and prefrontal, cingulate, and entorhinal cortex.

The DAergic system is critically involved in the control of key physiological functions, such as voluntary movement and posture, motivated behaviors, and different cognitive functions. According to a traditional opinion, different DAergic pathways mediate specific physiological functions. In particular it is thought that the nigrostriatal pathway is involved in the control of motor functions,[3] while the mesoaccumbal and mesocortical pathways are mainly implicated in reward, will, and cognitive functions.[4] This functional segregation has been also corroborated by the evidence that Parkinson's disease (PD), a neurological disorder mainly characterized by motor inabilities, is primarily due to the selective degeneration of SNpc DAergic neurons, whereas other neuropsychiatric disorders, such as schizophrenia, attention deficit hyperactivity disorder (ADHD), and addiction are principally associated with mesolimbic/mesocortical DAergic pathway dysregulation.

Nowadays, however, this functional/physiopathological subdivision of the DAergic system is no longer sustainable. Indeed, the SNpc and the striatum also represent essential stations for behavioral aspects of rewards, craving, and aversion, as well as for the regulation of cognitive functions.[5,6] Moreover, although a major dysfunction of mesolimbic/mesocortical pathways is traditionally related to schizophrenia, there are increasing data recognizing a role of the nigrostriatal DAergic system in this disease.[7,8]

DAᴇʀɢɪᴄ TRANSMISSION

DA is produced in DAergic neurons starting from the ʟ-amino acids phenylalanine and tyrosine. Tyrosine (diet derived or produced from ʟ-phenylalanine) is converted into ʟ-3,4-diidroxyphenylalanine (ʟ-DOPA) by tyrosine hydroxylase. DA derives from direct decarboxylation of ʟ-DOPA, by ʟ-aromatic amino acid decarboxylase (ʟ-AADC).

Once synthesized, DA is stored in vesicles by the vesicular monoamine transporter 2 and is released following neuronal depolarization. Extracellular DA levels are mainly regulated by the DA membrane transporter (DAT), which mostly reuptakes released DA into the cytosol, but can also mediate DA reversed transport, thus leading to its non-exocytotic release.[9] DAT is the biological target of several psychostimulant drugs of abuse, such as amphetamine, cocaine, and structurally related compounds.

DA is degraded by two enzymatic systems: monoaminoxidase (MAO) and catechol-O-methyltransferase (COMT). Intracellularly DA is catabolized by mitochondrial MAOs with the production of dihydroxyphenylacetic acid, whereas extracellular DA is converted in homovanillic acid and 3-methoxytyramine, by a combined action of extracellular COMT and glial MAOs.

The physiological effects of DA are mediated by five G-protein-coupled receptors (GPCRs) that are divided into two major subclasses: the D1-like and D2-like receptor families.[10]

D1-like family receptors (D1 and D5) are coupled to $G_{s/olf}$ protein and induce stimulation of adenylate cyclase (AC), cAMP production, and activation of cAMP-dependent signaling pathways. They are localized prevalently postsynaptically in DA projection areas.

D2-like family receptors (D2, D3, and D4) are coupled to $G_{i/o}$ proteins, thus inducing inhibition of AC, activation of G-protein gated inwardly rectifying K^+ channels (GIRK) and closure of voltage-activated Ca^{2+} channels. D2-like receptors are localized both postsynaptically and presynaptically on DAergic neurons, where they act as inhibitory autoreceptors.[11]

Although traditionally DA receptors have been depicted as monomeric entities, they can form several oligomeric complexes,[12] resulting from the association of different DA receptor subtypes either alone or with other GPCRs and ligand-gated channels. Oligomeric complex formation greatly enhances the complexity of DA-activated signaling pathways. Several oligomeric complexes have been described containing DA receptors associated with the adenosine A1 and A2, serotoninergic 5-HT_{2A}, glutamatergic mGluR5 and NMDAR, and histaminergic H3, as well as different homomeric dimers, like D1–D2, D2–D4, D1–D3, D2–D3, and D2–D5.[12]

A peculiar characteristic of the DAergic mesencephalic system is represented by the capability to release DA, not only from synaptic terminals, in the projection areas but also at the somatodendritic level, in the mesencephalon itself. The DA release from both soma-dendrites and terminals largely obeys the same rules, since it is evoked by neuronal activity and is Ca^{2+}-dependent.[13]

The firing activity of DAergic neurons plays a crucial role in the control of DA release. Midbrain DA neurons in vivo display a tonic background firing in a narrow frequency range (1–8 Hz). This is often interrupted either by transient (phasic, <500 ms) sequences of high-frequency firing (>15 Hz), so-called "bursts," or by transient pauses in electrical activity, where DA neurons remain silent.[14] The pacemaker activity of DAergic neurons is regulated by the integration of several excitatory/inhibitory synaptic inputs that dictate the changing firing pattern (bursts/pause). The DA somatodendritic release participates, throughout the D2 activation, with the negative-feedback mechanisms involved in pausing/decreasing the DAergic neuronal activity. Indeed D2 activation causes GIRK channel opening and consequently a reduction in firing rate.[11]

Other inhibitory mechanisms are represented by GABAergic inputs, arising from adjacent SN pars reticulata and from the pallidum and the striatum.

Excitatory glutamatergic inputs, mainly originating in the cortex and in the subthalamic nucleus, sustain bursting behavior of DAergic neurons by activating NMDARs and mGluR1.[15] Therefore, the interplay between the GABA/DA-mediated inhibition and the glutamate-mediated excitation causes the burst-like firing of the DAergic neurons, which is very efficient at releasing DA in the extracellular space.

TRACE AMINES

TAs are structurally similar to DA. The simplest chemical structure of TAs is represented by an aromatic ring linked to a short aliphatic chain (β-PEA structure), differing from DA structure only by the absence of two hydroxylic groups. DA and TAs share the same biosynthetic and catabolic enzymatic pathways. All are derived from phenylalanine and tyrosine and L-AADC, the key common biosynthetic enzyme, directly produce β-PEA and TYR, from phenylalanine and tyrosine respectively, and DA from L-DOPA. Both DA and TAs catabolism occurs mainly via MAO, with the production of phenylacetic acid, hydroxyphenyl acetic acid, and dihydroxyphenylacetic acid from β-PEA, TYR, and DA, respectively.

TAs cerebral levels are two orders of magnitude lower than monoamines. Nonetheless, TAs rate of synthesis is comparable to DA thus also suggesting that TAs could play key roles in modulating neuronal functions.

Although some evidence on the synaptosomal localization of TAs has been reported,[16,17] the proof of specific mechanisms for their vesicular storage and exocytotic release is still lacking. Indeed, TAs are not released following depolarization[18] and the emptying of monoamine vesicle content, with reserpine, does not deplete their tissue levels.[19] In consideration of their highly lipophilic nature, particularly of β-PEA, TAs might be mainly *released* from neuronal terminals by diffusion across the cell membrane.[20] Recently TAs transmembrane diffusion has been demonstrated together with the evidence that their synaptic levels are regulated by different transporters.[21]

Therefore, TAs cerebral levels, without a specific vesicle storage, are strictly dependent on the equilibrium between synthesis and catabolism. Thus, despite the high TAs rate of synthesis, their low levels are dependent on their extremely short half-life (30 s).

TRACE AMINES-ASSOCIATED RECEPTORS

The importance of endogenous TAs in the modulation of physiological brain functions has been definitively validated by the discovery of a class of GPCRs activated by TAs.[22,23] This class of the so-called trace amine-associated receptors (TAARs) is composed of a variable number of subtypes among different species.[1] However only two members, TAAR1 and TAAR4, also called TA$_1$ and TA$_2$, respectively, according to recent IUPHAR nomenclature,[24] are sensitive to TAs.

TAAR1, the best-characterized member of the TAAR family, is a Gs-protein-coupled receptor that increases cAMP levels, consequent to AC activation. In rodent and primate brains, TAAR1 mRNA and proteins were found in several brain areas, including the VTA, the SNpc, the limbic system, the locus coeruleus, the dorsal raphe, and the basal ganglia.[23,25–27] The TAAR1 pharmacological profile is rather complex, since it is activated, besides by all TAs, by monoaminergic neurotransmitters and different psychostimulants, such as amphetamine and MDMA, as well as by iodothyronamines, which are aminergic metabolites of thyroid hormones.[1]

TAAR1 is considered the prototype receptor mediating TAs' cerebral effects, with the other family components mainly regulating olfactory functions. However, other TAAR receptors, besides TAAR1, might be involved in TAs' neuronal effects.

TAAR4, like TAAR1, is a Gs-coupled receptor and signals via the activation of the AC pathway. TAAR4 is expressed in several human brain regions, including cerebellum, spinal cord, basal ganglia, frontal cortex, SN, amygdala, and hippocampus.[28] Although the TAAR4 cerebral role is still largely elusive, the observation that TAAR4 is a susceptibility gene for schizophrenia[28,29] supports its potential involvement in the modulation of the DAergic system, functionally compromised in this disorder.

ROLE OF TAs IN THE MODULATION OF DAergic MESENCEPHALIC SYSTEM

Several evidence indicate that TAs play key regulatory roles in the modulation of the DAergic system. TAs, β-PEA and TYR, are expressed at relatively high levels in the mesencephalic DAergic pathways,[17] consistent with the notion that there are synthesized in the DAergic neurons.[30] Earlier studies on TA binding sites demonstrate high-affinity binding sites for TYR[31] and β-PEA[32] in the striatum and other DA-rich areas.

More recently, TAAR expression in the mesencephalic DAergic system has been reported. TAAR1 mRNA and proteins were found in both SNpc and VTA.[23,25–27] Interestingly a colocalization between TAAR1 and DAT has been reported in DAergic neurons in mouse and primate SNpc,[25] suggesting an important interplay between TAAR1 and DAT in the presynaptic modulation of DAergic transmission. Moreover, a reciprocal functional interaction between TAAR1 and D2 receptors has also been demonstrated.[25,32,33] TAAR1 is considered an important modulator of the DAergic system as well as the target of addictive drugs.[27,32–34] Furthermore, TAAR4 is also expressed in mesencephalic DAergic nuclei,[28] but its role in the DAergic system modulation has not yet been investigated.

Besides the TAAR1-mediated effects, TAs can modulate the DAergic system by activating different presynaptic and/or postsynaptic mechanisms, resulting in a direct modulation of DAergic, glutammatergic, and GABAergic receptors function.[35–39]

TAs-INDUCED DA RELEASE

Direct Interaction of TAs with DAT

The first characterized role of TAs in the regulation of DAergic mesencephalic system is represented by their ability to induce extracellular DA release, consequent to its displacement from synaptic vesicles and reversal of DAT. For this amphetamine-like effect TAs were considered mere "false neurotransmitters," synthesized as byproducts of amino acid metabolism and also called "endogenous amphetamines."[20] The TA-induced DA release has been demonstrated in vitro using synaptosomal[40] and striatal brain slice preparations,[41] as well as, in vivo by microdialysis techniques, showing that both β-PEA and TYR induce DA outflow in the striatum[42] and NAcc[43] of freely moving rats, causing ipsilateral rotation in unilateral 6-hydroxydopamine-lesioned animals.[44]

In the mesencephalon, the more evident functional effect produced by TAs, as a direct consequence of increased DA release, is an inhibition of the firing activity of nigral DAergic neurons, dependent on somatodendritic D2 autoreceptor activation. The inhibitory effect of TAs, TYR and OCT, on the firing of SNpc DAergic neurons was first reported using extracellular recordings in midbrain rat slices.[45] A similar inhibitory role was later reported for β-PEA in vivo, since its intravenous application reduced firing frequency and burst occurrence.[46]

The mechanism underlying the TAs-induced inhibition of SNpc DAergic neuron activity has been further investigated by our group. In particular, by performing intracellular electrophysiological in vitro recordings, we demonstrated that the TAs-induced reduction of firing rate of nigral DAergic neurons is dependent on an indirect activation of D2 receptors, consequent to increased DA release.[47] In relation to the specific mechanisms by which TAs produce DA release, earlier studies demonstrating that TAs are substrates of DAT led to the largely accepted idea that TAs mainly affect the DAergic system indirectly, by interfering with DA uptake.

Interestingly, by using multiple approaches to both deplete DA vesicular storage or interfere with DAT function, we demonstrated that TA-induced DA release is both DAT-dependent and -independent. Thus, TAs-induced DA release is mainly caused by the displacement from a cytoplasmic pool of newly synthesized DA not yet stored in synaptic vesicles.[47]

TAAR1-Mediated DA Release

More recently several evidences have demonstrated that TAs modulate DAergic transmission by means of other mechanisms, besides a direct interaction with DAT.[48]

Since TAAR1 shows overlapping distribution with mesolimbic DA pathways, it is considered to be a key modulator of DA release. A physical and functional interaction between TAAR1 and DAT has been reported in DAergic neurons of mouse and rhesus monkey

SNpc.[45,49] Indeed, it has been demonstrated that the TAs-induced DA release is dependent on a direct action of TAs on TAAR1, resulting in a consequent modulation of DAT function.[25,26]

Since TAAR1 also mediates DA release induced by different amphetamine-like drugs, it has been suggested that TAAR1, rather than DAT, is the direct target of these psychostimulants.

Notably TAAR1–DAT functional interaction seems to be reciprocal.[25,50] In consideration of the preferential intracellular localization of TAAR1, DAT might mediate a preferential route of entry of TAs in the intracellular compartment, thus allowing more efficient TA-activated TAAR1 signaling.[25] Indeed, DAT coexpression with TAAR1 in heterologous expression systems greatly enhances TAAR1 signaling in a DAT-dependent manner.[25,26]

TAAR1-INDUCED MODULATION OF DAᴇʀɢɪᴄ NEURON FIRING ACTIVITY

Remarkably, TAAR1s modulate the DAergic mesencephalic system by directly regulating neuronal firing activity, in addition to controlling DA release. The recent introduction of different selective TAAR1 modulators, as well as the availability of transgenic mice lines with TAAR1 deletion, allowed better investigation of the TAAR1 physiological role in the modulation of DAergic mesencephalic system.

Interestingly, some reports demonstrate that TAAR1 is a tonic inhibitory modulator of spontaneous firing activity of DAergic neurons. In fact, the TAAR1 full agonist, RO5166017, inhibits the firing rate of VTA DAergic neurons in midbrain mice slices.[51] Regarding the mechanisms underlying TAAR1-dependent regulation of DAergic neuron firing activity, it has been reported that TYR inhibits VTA DAergic neurons by inducing the GIRK channel opening, through a direct activation of TAAR1.[52] Therefore, the specific TAAR1 activation induces an inhibition of spontaneous firing activity of VTA DAergic neurons.

Moreover the spontaneous firing frequency of VTA DAergic neurons is higher in TAAR1 knockout (TAAR1 KO) animals with respect to wild-type (WT) littermates.[27,51] This evidence supports the contention that TAs tonically inhibit neuronal activity through TAAR1 receptors.

Accordingly, the pharmacological inhibition of TAAR1, with the selective TAAR1 antagonist, N-(3-ethoxy-phenyl)-4-pyrrolidin-1-yl-3-trifluoromethyl-benzamide, increased the spontaneous firing frequency of VTA DAergic neurons in WT, but not in TAAR1 KO mice.[52]

Regarding the functional relevance of TAAR1 signaling in the modulation of the mesencephalic DAergic system, it has been reported that TAAR1 tonic activation effectively regulates the amount of DA released in the projection areas. In particular the TAAR1 pharmacological inhibition increases DA outflow in the NAcc, but not in the striatum, of WT mice, whereas the TAAR1 activation produces a specular reduction of DA release.[34] Moreover TAAR1 KO mice display increased extracellular basal DA levels in NAcc with respect to WT littermates.[34]

Several reports indicate that TAAR1 deletion induces hypersensitivity of the DAergic mesencephalic system in vivo. Indeed both neurochemical and behavioral effects of several amphetamine-like drugs are increased in TAAR1 KO mice with respect to WT, thus producing a stronger locomotor activation and DA release.[27,49]

Moreover, specific activation of TAAR1 in vivo counteracts the behavioral manifestations dependent on hyperactivation of the DAergic mesencephalic system, like the one induced by interfering with DA uptake mechanisms.[51] Indeed both pharmacological and genetic manipulations of DAT, obtained with cocaine or in DAT KO mice, produce a locomotor hyperactivation which is reduced by in vivo administration of the TAAR1 agonist, RO5166017.[51]

TAs-INDUCED MODULATION OF D2 RECEPTORS

TAs can also modulate the DAergic mesencephalic system by regulating the function of D2 receptors.

By using electrophysiological recordings of SNpc DAergic neurons in midbrain rodent slices, we have reported an inhibitory effect of TAs, β-PEA and TYR, on D2 receptor function in nigral DAergic neurons.[37] In our conditions, TA-induced modulation of D2 activity was TAAR1-independent, since TAs-induced depression of D2 function was still present in TAAR1 KO mice and the pharmacological activation of TAAR1 did not mimic the TA-induced inhibition of D2 function.[37]

Other evidence in VTA neurons strongly supports an involvement of TAAR1 in the modulatory effect of TAs on D2 function. Notably, a mutual functional interaction between TAAR1 and D2 has been described,[26,33,34,52] supported by the evidence that TAAR1 and D2 can form constitutive functional heterodimers.[33] D2 activation appears to reduce TAAR1 signaling in heterologous expression systems,[26] while D2 antagonists increase TA-activated TAAR1 signaling and disrupts TAAR1–D2 heterodimer formation in vitro.[33]

Electrophysiological recordings of VTA DAergic neurons have demonstrated that TAAR1 signaling modulates D2 receptor function, tuning their desensitization rate and the potency of agonists.[52] Moreover the D2-mediated inhibition of DA release in the NAcc is potentiated by the stimulation of TAAR1.[34] The genetic deletion of TAAR1 by controlling D2 receptor expression and sensitivity in the striatum[27,51–53] produces a modulation of D2-dependent behaviors, such as haloperidol-induced catalepsy.[53]

Together, these evidences support a strong reciprocal inhibitory interaction between TAAR1 and D2 receptors. Overall, TAAR1 seems constitutively activated by ambient TAs to inhibit the DAergic mesencephalic system.

Remarkably, the TAAR1-dependent modulation of D2 function has been demonstrated only in VTA DAergic neurons but not in SNpc. Differently to what has been found in VTA, in nigral DAergic neurons we reported a TAAR1-independent inhibitory effect of TAs on D2 function.[37] Accordingly, a role of TAAR1 in regulating in vivo DA release has been reported in NAcc but not in the dorsal striatum,[34] demonstrating no clear TAAR1-dependent effects in SNpc DAergic neurons. Therefore, it might be possible that the TAAR1 contribution to TAs-induced effect could be distinct in different DAergic pathways. Perhaps, the TAAR1 effects are rather subtle in the SNpc, as they can only be seen under precise experimental conditions.

Moreover, since TAAR4 expression has been reported in SNpc DAergic neurons,[28] this receptor, rather than TAAR1, could play a role in DAergic neuron modulation and, consequently, regulate the nigrostriatal pathway. However, additional investigations are required to ascertain whether TAs differentially modulate VTA versus SNpc DAergic cells.

TAs-INDUCED MODULATION OF GLUTAMATERGIC AND GABAergic TRANSMISSION ON DAergic MESENCEPHALIC NEURONS

Recent evidence demonstrates that TAs modulate glutamatergic synaptic transmission on VTA DAergic cells, by means of TAAR1-dependent mechanisms. In particular, TAAR1 activation decreases the excitatory synaptic inputs on VTA DAergic cells.[39] This suggests the presence of TAAR1 on excitatory neurons impinging on DA cells, which regulates glutamate release.

Otherwise, in SNpc DAergic neurons, TAs modulate the GABAergic transmission mediated by the metabotropic $GABA_B$ receptors. Indeed, TAs reduce postsynaptic $GABA_B$-dependent responses through an interference with the G-protein-dependent coupling between $GABA_B$ and potassium channels.[35] A similar mechanism has also been suggested for reduction of the $GABA_B$-mediated presynaptic inhibition of the $GABA_A$ transmission.[36] Thus, under particular conditions, TAs might lead to a disinhibition of SNpc DAergic cells.

TAs INVOLVEMENT IN BRAIN DISORDERS ASSOCIATED WITH DAergic DYSFUNCTION

Abnormal levels of TAs have been implicated in several neurological and psychiatric disorders strongly associated with a dysfunction of DAergic transmission.

Regarding TAs dysregulation in schizophrenia, several reports documented altered levels of β-PEA, TYR, and TRYP or their metabolites in blood and/or urine in schizophrenic patients.[1,54,55] Notably, some TAAR genes localize to a region of human chromosome 6q23.1, which has been genetically associated with schizophrenia and bipolar disorder.[56] Indeed TAAR2 and TAAR4 are considered susceptibility genes for schizophrenia, since in schizophrenic patients single nucleotide polymorphisms in TAAR2 and TAAR4 genes have been identified.[17,29] Recently, in animal models of schizophrenia, a TAAR1 partial agonist demonstrated antipsychotic and proattentional effects.[57]

In relation to PD, it could be hypothesized that TAs and TAAR might regulate DA release and its effect on the residual DAergic neurons of PD patients.[58] Interestingly, the therapeutic use of the MAO_B inhibitors, selegiline and rasaligine, in PD by increasing TAs levels, should amplify their effects. Moreover, after a therapeutic loading with L-DOPA, the COMT-produced DA metabolite, 3-methoxytyramine, can affect movement by interacting with TAAR1.

Regarding ADHD, plasmatic and urinary levels of TAs and their metabolites are lower in patients with healthy children.[1,54,55] Moreover treatment with methylphenidate, representing the first choice drug for ADHD therapy, increases β-PEA urinary levels in patients.

The evidence that TAAR1 is the target of several drugs of abuse suggest its involvement in addiction. Moreover recent evidence supports the development of TAAR1 agonists as potential treatment for addiction since TAAR1 activation reduced various behavioral effects of cocaine and methamphetamine.[59]

CONCLUSION

In conclusion, growing evidence strongly supports a key role of TAs in the modulation of the DAergic mesencephalic system. The data present in the literature reveal complex interactions between TAs and the DAergic system, occurring at both presynaptic and postsynaptic loci (Fig. 6.1).

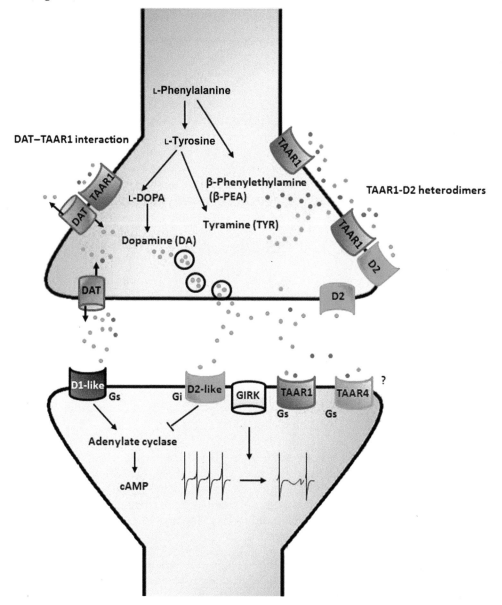

FIGURE 6.1   Modulatory effects of TAs on DAergic neurotransmission.

TAs are in the DA biosynthetic processes, thus their levels could directly determine DA amount in the presynaptic terminals, regulating DA-mediated neurotransmission and functions.

Moreover, TAs control DAT-mediated DA release by means of different mechanisms, both dependent or independent of the interplay with TAAR1. A modulatory role of TAs on D2 function, which controls firing activity and DA release, has also been described. This is also dependent on both TAAR1-mediated and -independent mechanisms. A specific role of TAAR1 in the regulation of GIRK channels, which affect spontaneous firing activity, has been reported in VTA DAergic neurons.[52]

In general, TAAR1-dependent effects seem to be predominant in VTA, whereas indirect TAAR1-independent actions appear to prevail in SNpc. Thus, a different behavior of TAs is probably occurring in the nigrostriatal versus the mesolimbic pathways. If the complex actions of TAs in the DAergic systems could be fully understood it might be possible to develop useful drugs to treat human diseases related to the dysfunction of the DAergic system.

References

1. Grandy DK. Trace amine-associated receptor 1-family archetype or iconoclast? *Pharmacol Ther* 2007;**116**:355–90.
2. Lindvall O, Björklund A. The organization of the ascending catecholamine neuron systems in the rat brain as revealed by the glyoxylic acid fluorescence method. *Acta Physiol Scand* 1974;**412**:1–48.
3. Haber SN. The primate basal ganglia: parallel and integrative networks. *J Chem Neuroanat* 2003;**26**:317–30.
4. Ikemoto S. Dopamine reward circuitry: two projection systems from the ventral midbrain to the nucleus accumbens-olfactory tubercle complex. *Brain Res Rev* 2007;**56**:27–78.
5. Ilango A, Kesner AJ, Keller KL, Stuber GD, Bonci A, Ikemoto S. Similar roles of substantia nigra and ventral tegmental dopamine neurons in reward and aversion. *J Neurosci* 2014;**34**:817–22.
6. Volkow ND, Wang GJ, Telang F, Fowler JS, Logan J, Childress AR, et al. Cocaine cues and dopamine in dorsal striatum: mechanism of craving in cocaine addiction. *J Neurosci* 2006;**26**:6583–8.
7. Perez-Costas E, Melendez-Ferro M, Roberts RC. Basal ganglia pathology in schizophrenia: dopamine connections and anomalies. *J Neurochem* 2010;**113**:287–302.
8. Winton-Brown TT, Fusar-Poli P, Ungless MA, Howes OD. Dopaminergic basis of salience dysregulation in psychosis. *Trends Neurosci* 2014;**37**:85–94.
9. Amara SG, Sonders MS, Zahniser NR, Povlock SL, Daniels GM. Molecular physiology and regulation of catecholamine transporters. *Adv Pharmacol* 1998;**42**:164–8.
10. Missale C, Nash SR, Robinson SW, Jaber M, Caron MG. Dopamine receptors: from structure to function. *Physiol Rev* 1998;**78**:189–225.
11. Lacey MG, Mercuri NB, North RA. Dopamine acts on D2 receptors to increase potassium conductance in neurones of rat substantia nigra zona compacta. *J Physiol* 1987;**392**:397–416.
12. Perreault ML, Hasbi A, O'Dowd BF, George SR. Heteromeric dopamine receptor signaling complexes:emerging neurobiology and disease relevance. *Neuropsychopharmacol Rev* 2014;**39**:156–68.
13. Beckstead MJ, Grandy DK, Wickman K, Williams JT. Vesicular dopamine release elicits an inhibitory postsynaptic current in midbrain dopamine neurons. *Neuron* 2004;**42**:939–46.
14. Grace AA, Bunney BS. The control of firing pattern in nigral dopamine neurons: single spike firing. *J Neurosci* 1984;**4**:2866–76.
15. Prisco S, Natoli S, Bernardi G, Mercuri NB. Group I metabotropic glutamate receptors activate burst firing in rat midbrain dopaminergic neurons. *Neuropharmacology* 2002;**42**:289–96.
16. Baldessarini RJ, Vogt M. Regional release of aromatic amines from tissues of the rat brain in vitro. *J Neurochem* 1972;**19**:755–61.
17. Boulton AA, Baker GB. The subcellular distribution of beta-phenylethylamine, p-tyramine and tryptamine in rat brain. *J Neurochem* 1975;**25**:477–81.

18. Dyck LE. Release of some endogenous trace amines from rat striatal slices in the presence and absence of a monoamine oxidase inhibitor. *Life Sci* 1989;**44**:1149–56.
19. Boulton AA, Juorio AV, Philips SR, Wu PH. The effects of resperpine and 6-hydroxydopamine on the concentrations of some arylalkylamines in rat brain. *Br J Pharmacol* 1977;**59**:209–14.
20. Berry MD. Mammalian central nervous system trace amines. Pharmacologic amphetamine, physiological neuromodulators. *J Neurochem* 2004;**90**:257–71.
21. Berry MD, Shitut MR, Almousa A, Alcorn J, Tomberli B. Membrane permeability of trace amines: evidence for a regulated, activity-dependent, nonexocytotic, synaptic release. *Synapse* 2013;**67**:656–67.
22. Bunzow JR, Sonders MS, Arttamangkul S, Harrison LM, Zhang G, Quigley DI, et al. Amphetamine, 3,4-methylenedioxymethamphetamine, lysergic acid diethylamide, and metabolites of the catecholamine neurotransmitters are agonists of a rat trace amine receptor. *Mol Pharmacol* 2001;**60**:1181–8.
23. Borowsky B, Adham N, Jones KA, Raddatz R, Artymyshyn R, Ogozalek KL, et al. Trace amines: identification of a family of mammalian G protein-coupled receptors. *Proc Natl Acad Sci USA* 2001;**98**:8966–71.
24. Maguire JJ, Parker WA, Foord SM, Bonner TA, Neubig RR, Davenport AP. International Union of Pharmacology. LXXII. Recommendations for trace amine receptor nomenclature. *Pharmacol Rev* 2009;**61**:1–8.
25. Miller GM, Verrico CD, Jassen A, Konar M, Yang H, Panas H. Primate trace amine receptor 1 modulation by the dopamine transporter. *J Pharmacol Exp Ther* 2005;**313**:983–94.
26. Xie Z, Westmoreland SV, Bahn ME, Chen G, Yang H, Vallender E, et al. Rhesus monkey trace amine-associated receptor 1 signaling: enhancement by monoamine transporters and attenuation by D2 autoreceptor in vitro. *J Pharmacol Exp Ther* 2007;**321**:116–27.
27. Lindemann L, Meyer CA, Jeanneau K, Bradaia A, Ozmen L, Bluethmann H, et al. Trace amine-associated receptor 1 modulates dopaminergic activity. *J Pharmacol Exp Ther* 2008;**324**:948–56.
28. Duan J, Martinez M, Sanders AR, Hou C, Saitou N, Kitano T, et al. Polymorphisms in the *Trace Amine Receptor 4 (TRAR4)* Gene on Chromosome 6q23.2 are associated with susceptibility to Schizophrenia. *Am J Hum Genet* 2004;**75**:624–38.
29. Bly M. Examination of the trace amine-associated receptor 2 (TAAR2). *Schizophr Res* 2005;**80**:367–8.
30. Greenshaw AJ, Juorio AV, Nguyen TV. Depletion of striatal beta-phenylethylamine following dopamine but not 5-HT denervation. *Brain Res Bull* 1986;**17**:477–84.
31. Vaccari A. High affinity binding of [3H]-tyramine in the central nervous system. *Br J Pharmacol* 1986;**89**:15–25.
32. Hauger RL, Skolnick P, Paul SM. Specific [3H]ISPhenylethylamine binding sites in rat brain. *Eur J Pharmacol* 1982;**83**:147–8.
33. Espinoza S, Salahpour A, Masri B, Sotnikova TD, Messa M, Barak LS, et al. Functional interaction between trace amine-associated receptor 1 and dopamine D2 receptor. *Mol Pharmacol* 2011;**80**:416–25.
34. Leo D, Mus L, Espinoza S, Hoener MC, Sotnikova TD, Gainetdinov RR. Taar1-mediated modulation of presynaptic dopaminergic neurotransmission: role of D2 dopamine autoreceptors. *Neuropharmacology* 2014;**81**:283–91.
35. Federici M, Geracitano R, Tozzi A, Longone P, Di Angelantonio S, Bengtson CP, et al. Trace amines depress GABA B response in dopaminergic neurons by inhibiting G-betagamma-gated inwardly rectifying potassium channels. *Mol Pharmacol* 2005;**67**:1283–90.
36. Berretta N, Giustizieri M, Bernardi G, Mercuri NB. Trace amines reduce GABA(B) receptor-mediated presynaptic inhibition at GABAergic synapses of the rat substantia nigra pars compacta. *Brain Res* 2005;**1062**:175–8.
37. Ledonne A, Federici M, Giustizieri M, Pessia M, Imbrici P, Millan MJ, et al. Trace amines depress D(2)-autoreceptor-mediated responses on midbrain dopaminergic cells. *Br J Pharmacol* 2010;**160**:1509–20.
38. Ledonne A, Berretta N, Davoli A, Rizzo GR, Bernardi G, Mercuri NB. Electrophysiological effects of trace amines on mesencephalic dopaminergic neurons. *Front Syst Neurosci* 2011;**5**:1–5.
39. Revel FG, Meyer CA, Bradaia A, Jeanneau K, Calcagno E, Andre CB. Brain-specific overexpression of trace amine-associated receptor 1 alters monoaminergic neurotransmission and decreases sensitivity to amphetamine. *Neuropsychopharmacology* 2012;**37**:2580–92.
40. Raiteri M, Del Carmine R, Bertollini A, Levi G. Effect of sympathomimetic amines on the synaptosomal transport of noradrenaline, dopamine, and 5-hydroxytryptamine. *Eur J Pharmacol* 1977;**41**:133–43.
41. Dyck LE. Release of monoamines from striatal slices by phenelzine and b-phenylethylamine. *Prog Neuropsychopharmacol Biol Psychiatry* 1983;**7**:797–800.
42. Philips SR, Robson AM. In vivo release of endogenous dopamine from rat caudate nucleus by phenylethylamine. *Neuropharmacology* 1983;**22**:1297–301.

43. Nakamura M, Ishii A, Nakahara D. Characterization of b-phenylethylamine induced monoamine release in rat nucleus accumbens: a microdioalysis study. *Eur J Pharmacol* 1998;**349**:163–9.

44. Barroso N, Rodriguez M. Action of beta-phenylethylamine and related amines on nigrostriatal dopamine neurotransmission. *Eur J Pharmacol* 1996;**29**:195–203.

45. Pinnock RD. Sensitivity of compacta neurones in the rat substantia nigra slice to dopamine agonists. *Eur J Pharmacol* 1983;**96**:269–76.

46. Rodriguez M, Barroso N. β-Phenylethylamine regulation of dopaminergic nigrostriatal cell activity. *Brain Res* 1995;**703**:201–4.

47. Geracitano R, Federici M, Prisco S, Bernardi G, Mercuri NB. Inhibitory effects of trace amines on rat midbrain dopaminergic neurons. *Neuropharmacology* 2004;**46**:807–14.

48. Sotnikova TD, Budygin EA, Jones SR, Dykstra LA, Caron MG, Gainetdinov RR. Dopamine transporter-dependent and -independent actions of trace amine beta-phenylethylamine. *J Neurochem* 2004;**9**:362–73.

49. Wolinsky TD, Swanson CJ, Smith KE, Zhong H, Borowsky B, Seeman P, et al. The trace amine 1 receptor knockout mouse: an animal model with relevance to schizophrenia. *Genes Brain Behav* 2007;**6**:628–39.

50. Xie Z, Westmoreland SV, Miller GM. Modulation of monoamine transporters by common biogenic amines via trace amine-associated receptor 1 and monoamine autoreceptors in human embryonic kidney 293 cells and brain synaptosomes. *J Pharmacol Exp Ther* 2008;**325**:629–40.

51. Revel FG, Moreau JL, Gainetdinov RR, Bradaia A, Sotnikova TD, Mory R, et al. TAAR1 activation modulates monoaminergic neurotransmission, preventing hyperdopaminergic and hypoglutamatergic activity. *Proc Natl Acad Sci USA* 2011;**108**:8485–90.

52. Bradaia A, Trube G, Stalder H, Norcross RD, Ozmen L, Wettstein JG, et al. The selective antagonist EPPTB reveals TAAR1-mediated regulatory mechanisms in dopaminergic neurons of the mesolimbic system. *Proc Natl Acad Sci USA* 2009;**106**:20081–6.

53. Espinoza S, Ghisi V, Emanuele M, Leo D, Sukhanov I, Sotnikova TD, et al. Postsynaptic D2 dopamine receptor supersensitivity in the striatum of mice lacking TAAR1. *Neuropharmacology* 2015;**93**:308–13.

54. Berry MD. The potential of trace amines and their receptors for treating neurological and psychiatric diseases. *Rev Recent Clin Trials* 2007;**2**:3–19.

55. Branchek TA, Blackburn TP. Trace amines receptors as target for novel therapeutics: legend, myth and fact. *Curr Opin Pharmacol* 2004;**3**:90–7.

56. Levinson DF, Holmans P, Straub RE, Owen MJ, Wildenauer DB, Gejman PV. Multicenter linkage study of schizophrenia candidate regions on chromosomes 5q, 6q, 10p, and 13q: schizophrenia linkage collaborative group III. *Am J Hum Genet* 2000;**67**:652–63.

57. Revel FG, Moreau JL, Gainetdinov RR, Ferragud A, Velázquez-Sánchez C, Sotnikova TD, et al. Trace amine-associated receptor 1 partial agonism reveals novel paradigm for neuropsychiatric therapeutics. *Biol Psychiatry* 2012;**72**:934–42.

58. Mercuri NB, Bernardi G. The 'magic' of L-dopa: why is it the gold standard Parkinson's disease therapy? *Trends Pharmacol Sci* 2005;**26**:341–4.

59. Jing L, Li JX. Trace amine-associated receptor1: a promising target for the treatment of psychostimulant addiction. *Eur J Pharmacol* 2015;**761**:345–52.

Trace Amine-Associated Receptors in the Cellular Immune System

G. Andersen and D. Krautwurst

German Research Center for Food Chemistry, Leibniz Institute,
Freising, Germany

O U T L I N E

INTRODUCTION

Biogenic amines are bioactive compounds. They are derived from enzymatic decarboxylation of, for example, certain amino acids,[1] and are metabolized by monoamine oxidases. Certain amines can be found in the plasma and in the central nervous system at low pico- to nanomolar concentrations,[2–7] and thus have been termed "trace amines."[8,9] Receptors, that specifically recognize trace amines (TAARs), were initially found in the brain.[8,10,11] TAAR are members of rhodopsin-like family A of G-protein-coupled receptors (GPCR). The impact of certain biogenic amines, such as β-phenylethylamine (2-PEA) and tyramine (TYR), on blood pressure, cardiac function, and brain monoaminergic systems has been demonstrated, mainly in studies with rodents.[3,11–14] Gene-targeted receptor knockout mice revealed Taar1 as a modulator of dopaminergic neurotransmission.[15,16] Moreover, mRNA for murine Taar1, Taar2, Taar6, and Taar9 was identified in gastrointestinal tissue.[17] TAAR have been identified in neurons of olfactory epithelia of mouse and zebrafish, where they may detect volatile amines such as odorants, pheromones, or kairomones, and mediate olfaction-guided behavior.[18–25] Gene-targeting supported the notion of TAARs as yet another family of olfactory receptors, at least in mice.[22,23] In line with this, gene expression for four out of the six intact human TAAR genes was reported in human nasal mucosa at trace levels.[26] However, TAAR gene expression has been reported for a variety of nonolfactory, peripheral tissues,[27] including human blood leukocytes.[4,28–31] Recently, our group has demonstrated a potent, amine-induced, and TAAR1/TAAR2-mediated activation of different, leukocyte-specific functions.[28]

Here we review the evidence for TAAR RNA and protein expression, signal transduction, and a functional role of TAAR in our cellular immune system.

RNA Expression of TAAR in Blood Leukocytes

For a variety of TAAR, gene-specific transcripts have been detected in blood leukocytes (Table 7.1). In murine leukocytes, RNA expression of Taar1, Taar2, Taar3, and Taar5 was shown for B cells as well as NK cells.[29] Human leukocytes have been shown to

TABLE 7.1   RNA expression of TAAR genes in leukocytes

RNA	Species	Blood Cell Type	Technique Used	References
Taar 1	Mouse	B cells, NK cells	RT-PCR	[29]
Taar 2	Mouse	B cells, NK cells	RT-PCR	[29]
Taar 3	Mouse	B cells, NK cells	RT-PCR	[29]
Taar 5	Mouse	B cells, NK cells	RT-PCR	[29]
TAAR1	Human	Leukocytes	RT-PCR	[4]
	Human	PBMC	RT-PCR	[29]
	Human	Monocytes, B cells, T cells, NK cells, neutrophils	RT-(q)PCR	[28]
TAAR2	Human	PBMC	RT-PCR	[29]
	Human	Monocytes, B cells, T cells, NK cells, neutrophils	RT-(q)PCR	[28]
TAAR5	Human	Monocytes, B cells, T cells, NK cells, neutrophils	RT-PCR	[28]
TAAR6	Human	Leukocytes	RT-PCR	[4]
	Human	Monocytes, B cells, T cells, NK cells, neutrophils	RT-PCR	[28]
TAAR8	Human	Leukocytes	RT-PCR	[4]
TAAR9	Human	Leukocytes	RT-PCR	[4]
	Human	Monocytes, B cells, T cells, NK cells, neutrophils	RT-PCR	[28]

Note: RT-(q)PCR, reverse transcriptase (quantitative) polymerase chain reaction; PBMC, peripheral mononuclear cells.

express gene-specific transcripts of TAAR1,[4,29] TAAR6, TAAR8, TAAR9,[4] and TAAR2.[29] Interestingly, stimulation of the leukocytes with phytohemagglutinin for 24h induced a notable increase in the TAAR1 and TAAR2 transcript levels.[29]

Recently, a comprehensive and more detailed study demonstrated that out of the six TAAR identified in humans to date, five TAAR are differentially expressed on the transcript level in neutrophils, monocytes, T, B, and NK cells.[28] In this study, the most abundant TAAR transcripts over all cell types investigated were TAAR1 and TAAR2, revealing the highest abundance in neutrophils, B, and T cells. Within these cell types TAAR2 expression always exceeded that of TAAR1. Although the abundance of TAAR5 and TAAR6 transcripts was generally lower than that of TAAR1 and TAAR2 transcripts, TAAR5 and TAAR6 were also most frequently expressed in B cells (TAAR5, TAAR6) and T cells (TAAR6). TAAR9 was equally expressed in all of the six cell types, whereas TAAR8-specific transcripts were generally not detectable.[28] These results clearly show that TAAR RNA expression is not restricted to one leukocyte cell type only, but rather seems to be differentially coordinated in immune cells, pointing to a specific role of these olfactory receptors in innate and adaptive immunity.

Protein Expression of TAAR in Blood Leukocytes

In line with the occurrence of a TAAR1-specific transcript in B cells, TAAR1 protein expression was demonstrated in immortalized rhesus monkey B-cell lines,[30] and both, in human nonmalignant B cells, and cells from different B-cell malignancies[31] (Table 7.2). Both groups reported on a regulation of TAAR1 protein in leukocytes: TAAR1 was upregulated following stimulation of isolated human B cells with phorbol myristate acetate/ionomycin for 4 days,[31] as well as after stimulation of rhesus monkey peripheral blood mononuclear cells with phytohemagglutinin for 48h,[30] although the latter cannot be attributed to B cells exclusively.

Very recently, TAAR1 expression has been shown in human T cells (Table 7.2), as well as in lymph nodes from HIV-1-infected patients.[32] Remarkably, treatment of resting T cells with methamphetamine, a ligand for TAAR1, significantly increased TAAR1 protein expression 24h poststimulation.

Babusyte et al.[28] demonstrated the expression of TAAR1 and TAAR2 in neutrophils (Table 7.2). Furthermore, the authors showed that both receptors are coexpressed in a

TABLE 7.2   TAAR protein expression in leukocytes

Protein	Species	Blood Cell Type	Technique Used	References
TAAR1	Rhesus monkey	B cells	IB	[30]
	Human	B cells	IB	[31]
	Human	T cells	IB	[32]
	Human	Neutrophils	IB, ICC	[28]
TAAR2	Human	Neutrophils	IB, ICC	[28]

Note: IB, immunoblot; ICC, immunocytochemistry.

subpopulation of neutrophils, whose distinctiveness has yet to be determined. Based on the finding that TAAR1 and TAAR2 are expressed in the same neutrophils, and both could be detected by coimmunoprecipitation and immunoblotting, it is likely that TAAR1 and TAAR2 form functional dimers. In fact, dimerization may be a general prerequisite for the functional expression of TAAR since, at least in thyroid epithelial cells, TAAR require heterodimerization in order to enter the secretory pathway and reach the cell membrane.[33] Heterodimerization of TAAR1 and the dopamine D2 receptor has been shown in HEK293 cells, and suggested for striatal dopaminergic neurons.[34]

Signal Transduction of TAAR in Blood Leukocytes

Initially, TAAR1 signaling via G protein G_s has been reported to result in an intracellular cAMP accumulation in *Xenopus* oocytes and COS-7 cells.[8] Likewise, several groups reported on gene expression of odorant receptor signal transduction components, such as $G_{\alpha olf}$, in olfactory sensory neurons,[21,23] suggesting that amine responses observed in TAAR-expressing olfactory sensory neurons are most likely mediated by the coupling of TAAR to the canonical odorant receptor cAMP signaling cascade.

Leukocytes express RNA for olfactory and taste signaling molecules, such as G proteins $G_{\alpha olf}$ and gustducin.[35,36] Methamphetamine increased cAMP levels via TAAR1 in human T cells.[32] In isolated human neutrophils, however, our group has recently shown that a 2-PEA-induced chemotaxis depended on signaling via pertussis toxin-sensitive G proteins of the G_i-type family, as did an odorant- and tastant-induced chemotaxis in these cells.[36,37] An amine-induced cAMP signaling was demonstrated for TAAR5 in *Xenopus* oocytes and HANA3A cells.[38] However, other signaling pathways have been suggested for TAAR. For instance, a basal $G_{q/11}$ (inositol-1,4,5-trisphosphate, IP3) and $G_{12/13}$ (mitogen-activated protein kinase) signaling activity has been reported for TAAR5 in HEK293 cells, recently,[39] and a basal G protein $G_{i/o}$ signaling activity was reported for TAAR8 in HEK293 cells.[40]

GPCR are pleiotropic with respect to their G protein coupling.[41] So-called "biased" ligands may produce cell-specific agonism by stabilizing unique GPCR conformations to selectively activate distinct signaling pathways.[42] It is therefore not surprising that TAAR, depending on the cellular context, may activate different G protein signaling pathways, which always has to be carefully controlled in the respective cell-based functional assay.

Biogenic Amine-Induced and TAAR-Mediated Cellular Functions in Isolated Blood Leukocytes

Biogenic amines have been shown to occur in various foods like fish, meat, cheese, vegetables, and wines.[43–47] TAAR agonists, such as 2-PEA, TYR, and 3-iodothyronamine (T1AM), have been identified in the plasma of healthy individuals at concentrations of 14–66 nM.[4–6] Biogenic amines may thus enter the bloodstream postprandially, and reach TAAR expressed on blood leukocytes. However, data on the impact of biogenic amines on leukocyte function are limited.

Babusyte et al.[28] demonstrated a role of TAAR during chemotaxis in neutrophils (Table 7.3). Neutrophils build a first line of defense in the innate immune system, based on their ability to detect chemotactic signals like *N*-formylmethionyl-leucyl-phenylalanine, and

TABLE 7.3 Demonstrated Functions of TAAR in Leukocytes

TAAR	Involved in	Ligand Used	Technique Used	References
TAAR1	Neutrophil migration	2-PEA	siRNA	[28]
		TYR		
		T1AM		
	IL-4 secretion of T cells	2-PEA, TYR, T1AM	siRNA	[28]
	IgE secretion of B cells	2-PEA, TYR, T1AM	siRNA	[28]
	RNA expression of T-cell markers (CCL5, SPP1)	2-PEA	siRNA	[28]
	IL-2 secretion of T cells	Methamphetamine	Gene silencing	[32]
TAAR2	Neutrophil migration	2-PEA, TYR, T1AM	siRNA	[28]
	IL-4 secretion of T cells	2-PEA, TYR, T1AM	siRNA	[28]
	IgE secretion of B cells	2-PEA, TYR, T1AM	siRNA	[28]
	RNA expression of T-cell markers (CCL5, SPP1)	2-PEA	siRNA	[28]

thereupon rapidly migrate to the sites of inflammation. In the study described, migration of human neutrophils expressing TAAR1- and TAAR2-specific transcripts as well as neutrophils that did not express TAAR1 and TAAR2 RNA was measured using the TAAR1 agonists 2-PEA, TYR, and T1AM in concentrations ranging from 0.01 to 10 nM. Only cells expressing TAAR1- and TAAR2-specific transcripts showed a dose-dependent migration towards the chemoattractants, with EC_{50} values of 0.066 ± 0.007 nM (2-PEA), 0.1 ± 0.009 nM (TYR), and 0.19 ± 0.08 nM (T1AM), whereas cells which were tested negative for the presence of TAAR1 and TAAR2 mRNA did not respond. This result was confirmed by using TAAR1- and TAAR2-specific siRNA, since chemotactic migration towards 2-PEA, TYR, and T1AM was strongly diminished in cells transfected with the specific siRNA compared to non-siRNA-transfected control cells.

By interacting with TAAR1 and TAAR2, biogenic amines also seem to affect the secretion of cytokines by human T cells (Table 7.3).[28] As a part of the adaptive immune system, T cells are activated by specific antigens presented on the surface of professional antigen-presenting cells. Once activated, T helper cells (CD3[+]CD4[+]) can differentiate mainly into two types of effector cell: Th1 cells and Th2 cells. Whilst differentiation into Th1 cells leads to the secretion of, for example, interleukin-2 (IL-2) and interferon γ, differentiation into Th2 cells results in the secretion of, for example, interleukin-4 (IL-4) and interleukin-5 (IL-5).[48] Stimulation of non-siRNA-transfected T cells with 2-PEA, TYR, and T1AM in a range from 0.01 to 1 nM concentration-dependently increased the secretion of IL-4 up to 300% of the control.[28] In this regard, 2-PEA revealed the highest efficacy of the three biogenic amines tested, whilst the respective EC_{50} values were comparable (2-PEA: 0.23 ± 0.008 nM, TYR: 0.24 ± 0.04 nM, T1AM: 0.1 ± 0.05 nM). In contrast, IL-4 secretion of T cells transfected with siRNA specific for TAAR1 and TAAR2 was barely affected by incubation of the cells with the biogenic amines.

A very recent study dealt with the impact of the TAAR1 agonist methamphetamine, a substance structurally related to some biogenic amines, on the cytokine secretion of human T cells[32] (Table 7.3). Here, after incubation with 100 μM methamphetamine and subsequent activation via CD3/CD28, the secretion of IL-2 was higher in cells transfected with a siRNA specific for TAAR1, compared to control cells transfected with scrambled siRNA. The authors also demonstrated that methamphetamine increased cAMP levels via TAAR1. Intracellular cAMP has been shown to reveal an inhibitory effect on IL-2 production and thus on T-cell activation.[49–55] In contrast, IL-4 production is not affected by cAMP.[56,57] Taken together, it is feasible that biogenic amines are able to modulate T-cell responses by activating TAAR1/TAAR2, leading to a decreased IL-2 and an increased IL-4 production, thereby participating in the induction of a shift towards Th2 responses. This is further supported by the fact that incubation of T cells with 1 nM 2-PEA for 24h led to a decreased RNA expression of the cytokine secreted phosphoprotein 1 (SPP1), which enhances Th1 cytokine production and inhibits Th2 cytokine production.[28]

Moreover, TAAR1- and TAAR2-mediated immunoglobulin E (IgE) secretion induced by biogenic amines in B cells (Table 7.3).[28] Like T cells, B cells are part of the adaptive immune system. B cells can be activated by T cells which will lead to the synthesis and secretion of antibodies. Babusyte et al.[28] showed that incubation of non-siRNA-transfected B cells with 2-PEA, TYR, and T1AM (1–30 nM for 12h) concentration-dependently induced the secretion of IgE. The respective EC_{50} values were 1.37 ± 0.08 nM (2-PEA), 2.58 ± 0.28 nM (TYR), and 2.22 ± 0.4 nM (T1AM). This amine-induced IgE secretion was strongly suppressed in B cells transfected with siRNA specific for TAAR1 and TAAR2. Since IgE induction requires IL-4 stimulation and T-cell–B-cell interaction,[58–60] the coincubation of purified T cells with B cells strongly increased the effect of 2-PEA to induce IgE synthesis by purified B cells, with 2-PEA and IL-4 being at least additive on IgE secretion.[28]

Altogether, since Th2 cytokines, such as IL-4, play a major role in allergic inflammation,[61,62] for example, food allergy responses are critically regulated by Th2 cytokines,[63] an amine-enhanced IL-4 signaling may be critical for individuals that are predisposed to allergic diseases by an altered signaling function of their IL-4 receptor.[64]

Interestingly, high concentrations (above 50 μM) of biogenic amines revealed a strong cytotoxic potential towards malignant and nonmalignant B cells.[31] Of the three biogenic amines tested in that study [o-phenyl-3-iodotyramine (o-PIT), T1AM, TYR], T1AM was the most potent, followed by o-PIT. TYR was identified as the least potent biogenic amine with regard to its cytotoxic potential. Notably, activation of tonsillar B cells seems to induce a resistance of the cells, since incubation of activated B cells strongly diminished the cytotoxicity of the biogenic amines, although the mechanism responsible for this finding still remains elusive.

CONCLUDING REMARKS

Whilst TAAR RNA expression has been demonstrated in several types of leukocytes, TAAR protein expression in these cells has not been as well studied, probably due to sparse information on TAAR-specific agonists, and the fact that TAAR-specific antibodies are difficult to obtain. However, this leaves basic questions unresolved to date. Knowledge on TAAR function in innate as well as adaptive immunity is just beginning to emerge, but already suggests TAAR as an important group of targets for receptor- and immune-cell-specific bioactives, such as foodborne amines, thus moving TAAR into the focus of amine intolerance, allergy, and food safety, and beyond, opening the perspective of a TAAR-specific pharmacological modulation of our cellular immune system. Research on TAAR in our cellular immune system is likely to be looming large in the very near future.

References

1. Zhang M. Aromatic L-amino acid decarboxylase cells in the spinal cord: a potential origin of monoamines. *Neural Regener Res* 2015;**10**(5):715–7.
2. Galli E, Marchini M, Saba A, et al. Detection of 3-iodothyronamine in human patients: a preliminary study. *J Clin Endocrinol Metab* 2012;**97**(1):E69–74.
3. Berry MD. Mammalian central nervous system trace amines. Pharmacologic amphetamines, physiologic neuromodulators. *J Neurochem* 2004;**90**(2):257–71.
4. D'Andrea G, Terrazzino S, Fortin D, Farruggio A, Rinaldi L, Leon A. HPLC electrochemical detection of trace amines in human plasma and platelets and expression of mRNA transcripts of trace amine receptors in circulating leukocytes. *Neurosci Lett* 2003;**346**(1–2):89–92.
5. Hoefig CS, Kohrle J, Brabant G, et al. Evidence for extrathyroidal formation of 3-iodothyronamine in humans as provided by a novel monoclonal antibody-based chemiluminescent serum immunoassay. *J Clin Endocrinol Metab* 2011;**96**(6):1864–72.
6. Miura Y. Plasma beta-phenylethylamine in Parkinson's disease. *Kurume Med J* 2000;**47**(4):267–72.
7. Scanlan TS, Suchland KL, Hart ME, et al. 3-Iodothyronamine is an endogenous and rapid-acting derivative of thyroid hormone. *Nat Med* 2004;**10**(6):638–42.
8. Borowsky B, Adham N, Jones KA, et al. Trace amines: identification of a family of mammalian G protein-coupled receptors. *Proc Natl Acad Sci USA* 2001;**98**(16):8966–71.
9. Boulton AA. Trace amines and the neurosciences. In: Boulton AA, Baker GB, Dewhurst WG, Sandler M, editors. *Neurobiology of the trace amines*. Clifton, NJ: Human Press; 1984. p. 13–24.
10. Lindemann L, Ebeling M, Kratochwil NA, Bunzow JR, Grandy DK, Hoener MC. Trace amine-associated receptors form structurally and functionally distinct subfamilies of novel G protein-coupled receptors. *Genomics* 2005;**85**(3):372–85.

11. Miller GM. The emerging role of trace amine-associated receptor 1 in the functional regulation of monoamine transporters and dopaminergic activity. *J Neurochem* 2011;**116**(2):164–76.

12. Chiellini G, Frascarelli S, Ghelardoni S, et al. Cardiac effects of 3-iodothyronamine: a new aminergic system modulating cardiac function. *FASEB J* 2007;**21**(7):1597–608.

13. Fehler M, Broadley KJ, Ford WR, Kidd EJ. Identification of trace-amine-associated receptors (TAAR) in the rat aorta and their role in vasoconstriction by beta-phenylethylamine. *Naunyn Schmiedebergs Arch Pharmacol* 2010;**382**(4):385–98.

14. Frascarelli S, Ghelardoni S, Chiellini G, et al. Cardioprotective effect of 3-iodothyronamine in perfused rat heart subjected to ischemia and reperfusion. *Cardiovasc Drugs Ther* 2011;**25**(4):307–13.

15. Lindemann L, Meyer CA, Jeanneau K, et al. Trace amine-associated receptor 1 modulates dopaminergic activity. *J Pharmacol Exp Ther* 2008;**324**(3):948–56. Epub 2007 Dec 2014.

16. Xie Z, Miller GM. Beta-phenylethylamine alters monoamine transporter function via trace amine-associated receptor 1: implication for modulatory roles of trace amines in brain. *J Pharmacol Exp Ther* 2008;**325**(2):617–28. Epub 2008 Jan 2008.

17. Ito J, Ito M, Nambu H, et al. Anatomical and histological profiling of orphan G-protein-coupled receptor expression in gastrointestinal tract of C57BL/6J mice. *Cell Tissue Res* 2009;**338**(2):257–69. Epub 2009 Sep 2010.

18. Fleischer J, Schwarzenbacher K, Breer H. Expression of trace amine-associated receptors in the Grueneberg ganglion. *Chem Senses* 2007;**32**(6):623–31.

19. Hussain A, Saraiva LR, Korsching SI. Positive Darwinian selection and the birth of an olfactory receptor clade in teleosts. *Proc Natl Acad Sci USA* 2009;**106**(11):4313–8.

20. Liberles SD. Trace amine-associated receptors are olfactory receptors in vertebrates. *Ann NY Acad Sci* 2009;**1170**:168–72.

21. Liberles SD, Buck LB. A second class of chemosensory receptors in the olfactory epithelium. *Nature* 2006;**442**(7103):645–50.

22. Dewan A, Pacifico R, Zhan R, Rinberg D, Bozza T. Non-redundant coding of aversive odours in the main olfactory pathway. *Nature* 2013;**497**(7450):486–9.

23. Zhang J, Pacifico R, Cawley D, Feinstein P, Bozza T. Ultrasensitive detection of amines by a trace amine-associated receptor. *J Neurosci* 2013;**33**(7):3228–39.

24. Ferrero DM, Lemon JK, Fluegge D, et al. Detection and avoidance of a carnivore odor by prey. *Proc Natl Acad Sci USA* 2011;**108**(27):11235–40.

25. Hussain A, Saraiva LR, Ferrero DM, et al. High-affinity olfactory receptor for the death-associated odor cadaverine. *Proc Natl Acad Sci USA* 2013;**110**(48):19579–84.

26. Carnicelli V, Santoro A, Sellari-Franceschini S, Berrettini S, Zucchi R. Expression of trace amine-associated receptors in human nasal mucosa. *Chem Percept* 2010;**3**(2):99–107.

27. Zucchi R, Chiellini G, Scanlan TS, Grandy DK. Trace amine-associated receptors and their ligands. *Br J Pharmacol* 2006;**149**(8):967–78.

28. Babusyte A, Kotthoff M, Fiedler J, Krautwurst D. Biogenic amines activate blood leukocytes via trace amine-associated receptors TAAR1 and TAAR2. *J Leukoc Biol* 2013;**93**:387–94.

29. Nelson DA, Tolbert MD, Singh SJ, Bost KL. Expression of neuronal trace amine-associated receptor (TAAR) mRNAs in leukocytes. *J Neuroimmunol* 2007;**192**(1–2):21–30.

30. Panas MW, Xie Z, Panas HN, Hoener MC, Vallender EJ, Miller GM. Trace amine associated receptor 1 signaling in activated lymphocytes. *J Neuroimmune Pharmacol* 2012;**7**(4):866–76.

31. Wasik AM, Millan MJ, Scanlan T, Barnes NM, Gordon J. Evidence for functional trace amine associated receptor-1 in normal and malignant B cells. *Leuk Res* 2011;**36**(2):245–9.

32. Sriram U, Cenna JM, Haldar B, et al. Methamphetamine induces trace amine-associated receptor 1 (TAAR1) expression in human T lymphocytes: role in immunomodulation. *J Leukoc Biol* 2016;**99**(1):213–23. Epub Aug 24, 2015.

33. Qatato M, Amoah A, Szumska J, et al. TAAR expression in thyroid epithelial cell lines as to establish an in vitro-model for signalling analysis. *Exp Clin Endocrinol Diabetes* 2014;**122**(3):P083.

34. Espinoza S, Salahpour A, Masri B, et al. Functional interaction between trace amine-associated receptor 1 and dopamine D2 receptor. *Mol Pharmacol* 2011;**80**(3):416–25.

35. Flegel C, Manteniotis S, Osthold S, Hatt H, Gisselmann G. Expression profile of ectopic olfactory receptors determined by deep sequencing. *PLoS One* 2013;**8**(2):e55368.

36. Malki A, Fiedler J, Fricke K, Ballweg I, Pfaffl MW, Krautwurst D. Class I odorant receptors, TAS1R and TAS2R taste receptors, are markers for subpopulations of circulating leukocytes. *J Leukoc Biol* 2015;**97**(3):533–45.

37. Geithe C, Andersen G, Malki A, Krautwurst D. A butter aroma recombinate activates human class-I odorant receptors. *J Agric Food Chem* 2015;**63**(43):9410–20.
38. Wallrabenstein I, Kuklan J, Weber L, et al. Human trace amine-associated receptor TAAR5 can be activated by trimethylamine. *PLoS One* 2013;**8**(2):e54950.
39. Dinter J, Muhlhaus J, Wienchol CL, et al. Inverse agonistic action of 3-iodothyronamine at the human trace amine-associated receptor 5. *PLoS One* 2015;**10**(2):e0117774.
40. Muhlhaus J, Dinter J, Nurnberg D, et al. Analysis of human TAAR8 and murine Taar8b mediated signaling pathways and expression profile. *Int J Mol Sci* 2014;**15**(11):20638–55.
41. Gudermann T, Kalkbrenner F, Schultz G. Diversity and selectivity of receptor-G protein interaction. *Annu Rev Pharmacol Toxicol* 1996;**36**:429–59.
42. Kenakin T. Functional selectivity and biased receptor signaling. *J Pharmacol Exp Ther* 2011;**336**(2):296–302.
43. Karovicova J, Kohajdova Z. Biogenic amines in food. *Chem Pap* 2005;**59**(1):70–9.
44. Lange J, Thomas K, Wittmann C. Comparison of a capillary electrophoresis method with high-performance liquid chromatography for the determination of biogenic amines in various food samples. *J Chromatogr B Analyt Technol Biomed Life Sci* 2002;**779**(2):229–39.
45. Mayr CM, Schieberle P. Development of stable isotope dilution assays for the simultaneous quantitation of biogenic amines and polyamines in foods by LC-MS/MS. *J Agric Food Chem* 2012;**60**(12):3026–32.
46. Oenal A. A review: current analytical methods for the determination of biogenic amines in foods. *Food Chem* 2007;**103**(4):1475–86.
47. Silla Santos MH. Biogenic amines: their importance in foods. *Int J Food Microbiol* 1996;**29**(2–3):213–31.
48. Smith-Garvin JE, Koretzky GA, Jordan MS. T cell activation. *Annu Rev Immunol* 2009;**27**:591–619.
49. Ruppelt A, Mosenden R, Gronholm M, et al. Inhibition of T cell activation by cyclic adenosine 5'-monophosphate requires lipid raft targeting of protein kinase A type I by the A-kinase anchoring protein ezrin. *J Immunol* 2007;**179**(8):5159–68.
50. Tsuruta L, Lee HJ, Masuda ES, et al. Cyclic AMP inhibits expression of the IL-2 gene through the nuclear factor of activated T cells (NF-AT) site, and transfection of NF-AT cDNAs abrogates the sensitivity of EL-4 cells to cyclic AMP. *J Immunol* 1995;**154**(10):5255–64.
51. Anastassiou ED, Paliogianni F, Balow JP, Yamada H, Boumpas DT. Prostaglandin E2 and other cyclic AMP-elevating agents modulate IL-2 and IL-2R alpha gene expression at multiple levels. *J Immunol* 1992;**148**(9):2845–52.
52. Wacholtz MC, Minakuchi R, Lipsky PE. Characterization of the 3',5'-cyclic adenosine monophosphate-mediated regulation of IL2 production by T cells and Jurkat cells. *Cell Immunol* 1991;**135**(2):285–98.
53. Minakuchi R, Wacholtz MC, Davis LS, Lipsky PE. Delineation of the mechanism of inhibition of human T cell activation by PGE2. *J Immunol* 1990;**145**(8):2616–25.
54. Averill LE, Stein RL, Kammer GM. Control of human T-lymphocyte interleukin-2 production by a cAMP-dependent pathway. *Cell Immunol* 1988;**115**(1):88–99.
55. Mary D, Aussel C, Ferrua B, Fehlmann M. Regulation of interleukin 2 synthesis by cAMP in human T cells. *J Immunol* 1987;**139**(4):1179–84.
56. Betz M, Fox BS. Prostaglandin E2 inhibits production of Th1 lymphokines but not of Th2 lymphokines. *J Immunol* 1991;**146**(1):108–13.
57. Novak TJ, Rothenberg EV. cAMP inhibits induction of interleukin 2 but not of interleukin 4 in T cells. *Proc Natl Acad Sci USA* 1990;**87**(23):9353–7.
58. Barnes PJ. Pathophysiology of allergic inflammation. *Immunol Rev* 2011;**242**(1):31–50.
59. Jin H, Malek TR. Redundant and unique regulation of activated mouse B lymphocytes by IL-4 and IL-21. *J Leukoc Biol* 2006;**80**(6):1416–23. Epub 2006 Aug 1430.
60. Mitchison NA. T-cell-B-cell cooperation. *Nat Rev Immunol* 2004;**4**(4):308–12.
61. Tan HP, Lebeck LK, Nehlsen-Cannarella SL. Regulatory role of cytokines in IgE-mediated allergy. *J Leukoc Biol* 1992;**52**(1):115–8.
62. Paul WE, Zhu J. How are T(H)2-type immune responses initiated and amplified? *Nat Rev Immunol* 2010;**10**(4):225–35.
63. Prescott VE, Forbes E, Foster PS, Matthaei K, Hogan SP. Mechanistic analysis of experimental food allergen-induced cutaneous reactions. *J Leukoc Biol* 2006;**80**(2):258–66.
64. Hershey GK, Friedrich MF, Esswein LA, Thomas ML, Chatila TA. The association of atopy with a gain-of-function mutation in the alpha subunit of the interleukin-4 receptor. *N Engl J Med* 1997;**337**(24):1720–5.

Trace Amines and Their Receptors in the Control of Cellular Homeostasis

M.D. Berry

Department of Biochemistry, Memorial University of Newfoundland,
St. John's, NL, Canada

INTRODUCTION

A role for trace amines in neurologic and psychiatric disorders, and by extension in maintaining neuronal homeostasis, has been hypothesized for close to 40 years.[1–3] Historically, this family of endogenous compounds has been thought to consist primarily of 2-phenylethylamine, *p*-tyramine, and tryptamine, and sometimes extended to also include *p*-octopamine and synephrine.[4,5] As described in detail elsewhere in this volume (see chapter: Synthesis and Neurochemistry of Trace Amines) in vertebrates, de novo *p*-tyramine, 2-phenylethylamine, and tryptamine are synthesized in neurons (and possibly other cell types) via aromatic L-amino acid decarboxylase (AADC; EC 4.1.1.28) from the precursor amino acids L-tyrosine, L-phenylalanine, and L-tryptophan, respectively.

Despite their long history, it was only with the 2001 discovery[6,7] of a family of G-protein-coupled receptors (GPCR), subsequently named trace amine-associated receptors (TAARs), at least some of which are selectively activated by trace amines, that a physiological and/or pathological relevance for trace amines in mammalian species began to be accepted by the mainstream neuroscience community. Even with the discovery of TAAR, the role of trace amine systems remains largely enigmatic, and the majority of studies have focused on a single family member, TAAR1, the only member for which endogenous trace amine ligands have been clearly established and recognized by IUPHAR.[8] This chapter will, therefore, largely focus on recent developments in the pharmacology and physiology of TAAR1, although other TAAR are briefly discussed with respect to their role in olfactory-mediated innate behaviors.

The TAAR family is currently thought to comprise 28 distinct subfamilies (TAAR1–28), with isoforms of many of these present.[9,10] TAAR10–28 have only been identified in teleosts,[10] often show a high degree of species-specificity,[11–13] may have lost the consensus amine binding domain,[10] and will not be discussed further here. Terrestrial vertebrates express members of the TAAR1–9 subfamilies, although again there is a high degree of variability between species.[9,14] In humans, a single member of each of these subfamilies has been identified, three of which (TAAR3, TAAR4, and TAAR7) are thought to be pseudogenes. TAAR2[15] and TAAR9[16] may also be pseudogenes in some individuals. In contrast, closely related primate species contain as few as three functional TAAR (TAAR1, TAAR5, and TAAR6) with species-specific pseudogenization having occurred in other TAAR isoforms.[17] In commonly used laboratory rodent species there is an expansion of the TAAR repertoire with 17 (rat) and 15 (mouse) functional receptors.[9,10,14] This expansion largely reflects the presence of multiple isoforms of TAAR7 and TAAR8. Strain differences in the functional TAAR complement may also be present in commercial stocks, with a recent report identifying a defunctionalizing mutation of TAAR1 in the DBA/2J mouse strain.[18] Interestingly TAAR1 also appears to have been pseudogenized in the entire *Canis* family.[17] Although loss of individual TAAR appears readily accommodated, and does not result in overt phenotypes in the absence of additional challenge,[19–21] TAAR appear to be under strong selection pressures,[10,17] suggesting they play a key role in species survival.

Neuronal synthesis and identified GPCR target(s) appears consistent with trace amines acting as neurotransmitters. Indeed, the so-called D-neurons of the central nervous system[22] would seem to be prime candidates for a trace aminergic neurotransmitter system, consisting of neurons that express AADC, but neither tyrosine hydroxylase, tryptophan hydroxylase, catecholamines, nor serotonin.[22] Available evidence, however, indicates that trace amines are not bona fide neurotransmitters; they do not appear to be stored in synaptic vesicles,[23,24] and their release from terminals does not occur in response to depolarization.[25–27] These latter observations strongly suggest a nonexocytotic release mechanism, and we,[27] and others,[28,29] have shown that trace amines readily diffuse across lipid bilayers. Consistent with this, as described in detail below, the primary role of TAAR1 appears to be in the regulation of basal aminergic tone, in particular dopaminergic tone, rather than mediating neurotransmission.

TRACE AMINE-ASSOCIATED RECEPTOR 1

Consistent with historical classification schemes of trace amine family members, p-tyramine and 2-phenylethylamine have been validated as endogenous full agonists at TAAR1.[8] Tryptamine also possesses agonistic activity[6,7] but has been much less studied in this regard. In addition to the classical trace amines, a variety of other endogenous compounds have been shown to be agonists at TAAR1, including 3-iodothyronamine, an endogenous metabolite of thyroid hormone,[30] the O-methyl metabolites of catecholamine neurotransmitters,[7] and the potent hallucinogen N,N-dimethyltryptamine,[7] and it may be useful to expand the trace amine family classification to include at least some of these. Although a number of highly selective synthetic agonists and partial agonists for TAAR1 have been described,[31–33] only a single antagonist compound (N-(3-ethoxy-phenyl)-4-pyrrolidin-1-yl-3-trifluoromethyl-benzamide, EPPTB) is known.[34] Although useful, the pharmacokinetic profile of EPPTB makes it difficult to use in vivo. The lack of a useable selective antagonist has severely hampered the elucidation of physiological roles for TAAR1, necessitating a reliance on gene-silencing techniques. TAAR1 has also been confirmed to be a receptor target of various psychotropic drugs of abuse including d-amphetamine, methamphetamine, 3,4-methylenedioxymethamphetamine (MDMA; "ecstasy"), and lysergic acid diethylamide.[7,35–37] In contrast cocaine does not appear to directly interact with TAAR1.[7]

Signal transduction at TAAR1 occurs through coupling to the G_s protein, with subsequent activation of adenylate cyclase, generation of cAMP, and activation of protein kinase A.[6,7] In addition, TAAR1 has been shown to tonically activate inwardly rectifying K^+-channels, which have the characteristics of the G protein $\beta\gamma$-subunit controlled GirK channels.[34,38] Despite this apparent classical coupling to G proteins, TAAR1 is thought to be poorly translocated to the plasma membrane,[7,39] possibly due to a lack of N-terminal glycosylation.[40] Rather TAAR1 appears to be primarily intracellular, possibly associated with as-yet unidentified intracellular membranes.[41] The lack of good-quality antibodies against TAAR1, along with inherent low expression levels, have prevented the direct determination of its synaptic location, but available evidence suggests that TAAR1 mediates both pre-[42] and postsynaptic[43] effects.

TAAR1 is distributed throughout the brain, primarily associated with areas containing monoaminergic cell bodies and projections, including the locus coeruleus, substantia nigra, raphe nuclei, limbic areas, prefrontal cortex, striatum, and ventral tegmental area (VTA).[6,20,44] In addition to neuronal localization, a single study has reported TAAR1 expression in astrocytes.[45] In addition TAAR1 has been identified in a number of peripheral tissues including adipocytes,[46] blood vessels,[47] gastrointestinal tract,[48–50] leukocytes,[51–55] pancreas,[50,56] stomach,[50,57,58] and possibly liver.[6]

TAAR1 Role in Dopaminergic Transmission

There is now good evidence that TAAR1 acts as an endogenous brake, preventing hyperactivity of monoaminergic systems. Below we provide a brief overview of recent advances in understanding the role of TAAR1 in controlling dopaminergic activity. The reader is also referred to chapters "Effects of Trace Amines on the Dopaminergic Mesencephalic System" and "Trace Amine-Associated Receptor 1 Modulation of Dopamine System" for further details.

Administration of the TAAR1 antagonist EPPTB causes an increase in the basal firing rate of dopaminergic neurons in the VTA,[34] while TAAR1 agonists decrease the firing rate.[20,34,38] A similar effect of TAAR1 agonists has also been reported on serotonergic neurons of the dorsal raphe nuclei.[38] The effect of EPPTB alone has been taken as indicative of either constitutive activity of TAAR1 or tonic activation of the receptor by endogenous ligands.[34] Alternatively EPPTB may also be an inverse agonist rather than an antagonist.[59] Tonic activation of TAAR1 is consistent with the ease with which the endogenous agonists *p*-tyramine and 2-phenylethylamine cross cell membranes,[27,29] and with release occurring in a nonexocytotic manner.[27] It is also consistent with earlier suggestions that trace amines function to maintain basal neuronal tone within defined physiological limits.[5] In further agreement with the constitutive activity/tonic activation of TAAR1, the selective partial agonist RO5203648 decreases the spontaneous firing rate of VTA dopaminergic and raphe serotonergic neurons,[60] and EPPTB alone decreases cAMP levels below baseline values.[34] These changes in dopaminergic neuronal activity in response to TAAR1 ligands are accompanied by the expected changes in extracellular dopamine levels.[42,60] A dampening effect of TAAR1 activation on dopaminergic activity is also seen at the whole animal level, with TAAR1 agonists preventing hyperactivity caused by cocaine-induced dopamine transporter (DAT; SLC6A3) inhibition, or following DAT knockout.[38,50]

In general, TAAR1 knockout animals show the same electrophysiological, neurochemical, and behavioral responses as those seen following administration of EPPTB. TAAR1 knockout results in an increased dopamine release in the nucleus accumbens, as well as increased levels of the extracellular dopamine metabolite homovanillic acid,[42] and show increased sensitivity to a wide variety of dopaminergic stimuli.[43] Somewhat paradoxically, the neuron-specific overexpression of TAAR1 also leads to an increase in the spontaneous firing rate of VTA, raphe, and locus coeruleus neurons, along with increased extracellular levels of dopamine and norepinephrine in the nucleus accumbens and 5-HT in the prefrontal cortex.[61] The reason for this discrepancy of TAAR1 overexpression and knockout causing apparently the same response is unknown, but may relate to compensatory mechanisms in response to long-term changes in gene expression, basal neuronal tone, or differences in response based on pre- and postsynaptic expression of TAAR1.

The molecular bases of the above effects have now begun to be elucidated. In particular the regulation of dopaminergic activity appears to be due to a TAAR1-mediated regulation of D2-like receptors. Knockout of TAAR1 causes increased striatal levels of D2-like receptors,[43] an increased proportion of D2-like receptors in the high-affinity state[19] and an enhanced potency of agonists at D2-like autoreceptors.[34] In contrast, Leo et al.[42] reported a loss of D2-like autoreceptor function in TAAR1 knockout animals, while in wildtype animals TAAR1 agonists increased D2-like receptor-mediated effects on dopamine release. Heterodimerization of TAAR1 and D2-like receptors has been demonstrated,[62,63] providing a physical process to explain the regulation. Recently, TAAR1-mediated effects on D2-like receptors were shown to also occur at the postsynaptic level,[43] with TAAR1 knockout augmenting postsynaptic D2-like receptor-mediated responses. Intriguingly this effect of TAAR1 appeared to be biased in nature, with the enhancement being due solely to an augmentation of the G-protein-independent β-arrestin 2 signaling pathway.[43,64] TAAR1 itself has also been shown to recruit β-arrestin 2[64] and these authors reported that heterodimerization with D2-like receptors enhanced TAAR1-mediated β-arrestin 2 signaling, while decreasing the recruitment of β-arrestin 2 to D2-like receptors.[64] Effects of TAAR1[43] or its ligands[65] on D1-like receptor activity do not appear to occur.

In addition to regulation of D2-like receptor function, TAAR1 activation has also been reported to regulate the activity of DAT. In a series of studies, Miller and colleagues[44,66,67] reported that activation of TAAR1 resulted in a decrease in dopamine reuptake by DAT, while also causing an increased outflow of dopamine,[35] which was suggested to occur via DAT following "inversion" of the transporter. Such effects could reasonably be expected to result in increased dopaminergic transmission, at least acutely, which is in marked contrast to the dampening effect of TAAR1 activation generally reported by others. In contrast, Leo et al.[42] reported no effect of TAAR1 activation on the extracellular clearance of dopamine, which was interpreted as a lack of effect of TAAR1 on DAT. The reason(s) for the differences in response is unknown. Different responses between heterologous expression systems and native preparations may certainly play a role. It should also be noted that it has recently become apparent that the polyspecific organic cation transporters (OCTs; SLC22 family) and plasma membrane monoamine transporter (PMAT; SLC29A4) play a much more active role in regulating neurotransmitter systems than was previously assumed.[68–72] In the past, these transporters were thought to be little more than a nonspecific overflow clearance mechanism.[73] None of the above studies examined for possible OCT/PMAT-mediated transport, and differential regulation of OCT- and/or PMAT-mediated transport could provide an explanation for some of the discrepancies between studies.

In summary, although there are conflicting reports as to the precise nature of the regulation, it is now clear that activation of TAAR1 regulates the responsivity of the dopaminergic neurotransmitter systems. The complexity of this regulation involving reciprocal interactions likely contributes to some of the conflicting details that have been reported. This crosstalk includes regulation of trace amine synthesis by dopamine receptors (described in detail in chapter: Synthesis and Neurochemistry of Trace Amines) coupled with tonic activation/constitutive activity of TAAR1; TAAR1 control of dopamine autoreceptor function (detailed above); D2-like receptor regulation of TAAR1 signaling[62,64]; TAAR1 regulation of DAT; and TAAR1 regulation of postsynaptic D2-like dopamine receptor function. In general, however, TAAR1 activation appears to act as a brake, preventing hyperdopaminergic states through regulation of D2-like receptors, such that decreased TAAR1 function results in an enhanced sensitivity to activation of dopaminergic systems. Further, this regulation of D2-like receptor activity may occur in a biased manner, with only the G-protein-independent (β-arrestin 2) signaling pathway being enhanced.

TAAR1 Regulation of Other Neurotransmitters

With respect to the regulation of serotonergic neuronal activity briefly described above, it would appear that similar molecular processes underlie the responses to those seen with the dopaminergic system; activation of TAAR1 results in increased agonist potency and increased desensitization at $5HT_{1a}$ autoreceptors.[38] This effect on agonist potency is opposite that seen with D2-like autoreceptors, where activation of TAAR1 generally decreases

D2-like receptor agonist potency. Again, evidence for constitutive activity/tonic activation of TAAR1, this time in dorsal raphe neurons, was present.[38] Interestingly tryptamine, a TAAR1 agonist, was previously reported to increase the proportion of $5HT_{2a}$ receptors in the high-affinity agonist state,[74] although a role for TAAR1 in regulating $5HT_{2a}$ receptors has not been directly demonstrated. In much the same way as the observed regulation of DAT, Miller and colleagues[66,67] reported a TAAR1-mediated regulation of the high-affinity serotonin transporter. Regulation of adrenoceptor function by TAAR ligands may also occur and is the subject of chapter "Differential Modulation of Adrenergic Receptor Signaling by Octopamine, Tyramine, Phenylethylamine, and 3-Iodothyronamine," and will not be discussed here.

Somewhat surprisingly TAAR1 has also been shown to regulate non-monoaminergic neurotransmission. TAAR1 knockout has been shown to result in an alteration of the subunit composition of prefrontal cortex N-methyl-D-aspartate (NMDA) receptors.[75] Both total and phosphorylated forms of the GluN1 subunit were decreased in knockout animals, while total GluN2B was also decreased. Although there was no change in basal neuronal activity, knockout animals showed neuronal membranes that were more depolarized as a result of an increased magnitude of excitatory postsynaptic potentials.[75] This again provides evidence of a postsynaptic site of action of TAAR1. In wildtype animals, as expected, TAAR1 agonists cause the opposite effects. As seen in the monoaminergic systems, regulation of glutamate transporters has also been suggested, with a TAAR1-mediated regulation of excitatory amino acid transporter (EAAT) 2 (SLC1A2).[45] Similarly methamphetamine and amphetamine, both potent TAAR1 agonists, have been reported to modulate glutamatergic transmission[76] and regulate EAAT3 (SLC1A1) transporter activity,[77] although the role of TAAR1 in these responses was not specifically addressed. Consistent with these apparent enhancements of glutamatergic transmission, selective TAAR1 activation prevents the hyperlocomotor responses and cognitive deficits induced by NMDA receptor blockade.[38,50] Enhanced AMPA receptor activity has also been reported following systemic 2-phenylethylamine administration.[78]

Functional Relevance

TAAR1 in Schizophrenia

As previously indicated, trace amines have long been implicated as playing an etiologic role in psychiatric illness in general,[3] and schizophrenia in particular, and this is reviewed in detail in chapter "β-Phenylethylamine-Class Trace Amines in Neuropsychiatric Disorders: A Brief Historical Perspective." Indeed, long before the discovery of TAAR, 2-phenylethylamine was proposed to be an endogenous amphetamine.[79,80] The discovery of TAAR further strengthened the putative relationship, with TAAR located to chromosome 6q23,[6,7] corresponding closely to the well-defined and oft-replicated SCZD5 schizophrenia susceptibility locus,[81–83] and a variety of psychotropic agents shown to be potent ligands at TAAR1.[7]

The spectrum of activity of TAAR1 described above has made it a particularly attractive target with respect to schizophrenia. Prevention of hyperdopaminergic states while also normalizing hypoglutamatergic states positions TAAR1 uniquely to address two of the major pathologies hypothesized to contribute to schizophrenia symptomatology, overactivity of midbrain dopamine systems and underactivity of frontal cortical glutamate neurons.[50] With respect to the dopaminergic regulation, the selective effect of TAAR1 on D2-like receptors is again attractive, with D2-like receptor blockade known to correlate with antipsychotic efficacy. Further, joint antagonism of D2-like and $5HT_{1a}$ receptors is a characteristic of the so-called atypical neuroleptics, and TAAR1 appears to interact with both of these receptor subtypes. Thus, TAAR1 agonists appear to be uniquely positioned for development as novel therapeutics to address some of the major underlying pathologies associated with schizophrenia, and the first such agents are now in clinical trials.[84]

Not only is the spectrum of TAAR1 activity promising for novel therapeutic development, it is hoped that compounds directed toward TAAR1 may have a lower propensity for adverse side effects. TAAR1 knockout animals are ostensibly normal, with only subtle behavioral changes present, and even then only following challenge.[19-21] This may relate to the very nature of the trace amine system, not being directly responsible for neurotransmission, but rather serving to maintain neuronal tone within defined limits. Further, the use of a TAAR1 partial agonist, such as RO5203648, may further decrease the likelihood of side effects. This compound has been shown to have a neuronal activation map very similar, although not identical, to the atypical neuroleptic olanzapine.[50] In contrast to olanzapine, RO5203648 does not induce weight gain in experimental animals, and in fact prevents olanzapine-induced weight gain.[50] Further, RO5203486 does not appear to share the propensity of a number of other clinically used antipsychotics for inducing narcolepsy.[50] The putative role of dedicated trace aminergic neurons in schizophrenia is the subject of chapter "Involvement of So-Called D-Neuron (Trace Amine Neuron) in the Pathogenesis of Schizophrenia: D-Cell Hypothesis," and readers are referred there for further discussion.

TAAR1 in Drug Abuse

Central monoamine systems in general, and the dopamine system in particular, are implicated in reward systems, drug-seeking behavior, and abuse.[85,86] Even prior to TAAR identification, trace amines had been proposed to play a role in reward and addiction behaviors.[3,87] The identification of a number of psychotropic drugs of abuse as potent ligands at TAAR1 immediately raised interest in this receptor as a novel target for the pharmacotherapy of drug dependence.[88] This interest has been further strengthened as the interplay between TAAR1 and dopaminergic activity has become apparent.

TAAR1 knockout animals show greater locomotor activation in response to either amphetamine or methamphetamine,[89] consistent with TAAR1 knockout animals being more susceptible to dopamine hyperactivity. On the surface this may seem a little surprising, given that amphetamines are potent agonists at TAAR1. Amphetamines are promiscuous ligands, however, interacting with a number of other sites that promote dopamine hyperactivity, and they can cause both aversive and rewarding responses. In mice selectively bred for methamphetamine consumption, a defunctionalizing mutation in TAAR1 was identified in the DBA/2J background strain that selected for high methamphetamine consumption.[18] Further study showed that high-consuming animals showed lower taste aversion and reduced hypothermic responses to methamphetamine, which the authors interpreted as TAAR1 playing a major role in the aversive properties of amphetamines, while the pleasurable, rewarding aspects, were associated with interaction at dopaminergic targets.[18] A similar loss of MDMA-induced hypothermic responses has also been seen in TAAR1 knockout mice.[36] Whether this indicates a role for TAAR1 in thermoregulation is an open question. It does, however, suggest that activation of TAAR1 may be effective in preventing the self-administration of amphetamine-like drugs. Consistent with this, TAAR1 knockout animals acquire methamphetamine-conditioned place preference (CPP) quicker, and retain it for longer, than do their wildtype littermates,[89] further suggesting beneficial effects of TAAR1 agonists on both drug-seeking behavior and relapse.

Consistent with the above, selective TAAR1 agonists have been shown to prevent methamphetamine-induced hyperactivity,[90] behavioral sensitization,[90,91] self-administration,[90,91] and reinstatement,[91] although methamphetamine-induced dopamine release and inhibition of dopamine uptake in vitro were not affected.[90] Similarly, brain-selective overexpression of TAAR1 decreases amphetamine-induced responses.[61] In contrast, an amphetamine-induced, TAAR1-mediated, regulation of dopamine uptake and release through DAT has been reported,[35] while MDMA-induced dopamine release was enhanced following TAAR1 knockout.[36] The discrepancy between studies on the effects of TAAR1 on amphetamine-induced dopamine release likely reflects differences in response between in vitro and in vivo systems, particularly with respect to the promiscuous nature of amphetamine pharmacology. Notwithstanding these differences, the combined evidence is convincing that selective TAAR1 activation is beneficial in preventing both amphetamine self-administration and relapse behaviors in animal models.

Although cocaine is not a ligand for TAAR1, similar beneficial effects of TAAR1 activation have also been observed in various cocaine self-administration protocols. TAAR1 agonists decrease cocaine-induced sensitization,[92,93] cocaine-induced hyperactivity,[38,50] cocaine-seeking behavior following withdrawal,[94] prevent cue-,[92] and cocaine-primed[92,94] reinstatement, and cocaine-induced CPP.[92] Further, agonists prevent the cocaine-induced elevations in extracellular dopamine, but do not alter dopamine half-life,[94] suggesting that TAAR1-mediated effects do not involve inhibition of dopamine reuptake processes. Although a lack of effect of TAAR1 agonists on cocaine-induced hyperlocomotion[93] and cocaine-induced CPP[60] has been reported, the results generally indicate a potential utility of TAAR1 agonists in the treatment of cocaine, as well as amphetamine-related drug abuse. Given that the pharmacology of both of these drugs of abuse is primarily dopaminergic in nature, it would seem likely that the previously described role of TAAR1 in preventing dopaminergic hyperactivity underlies these effects. In this regard it is interesting to note that TAAR1 knockout does not affect morphine-induced CPP,[89] and neither TAAR1 agonists[60,90,91,94] nor TAAR1 knockout[95] affect sucrose self-administration, behaviors that are not primarily dopaminergic in origin. TAAR1 knockout animals do, however, show increased ethanol consumption and greater effects of acute alcohol exposure[95] and dopaminergic systems are implicated in alcohol-related behaviors.[96] Interestingly, the DBA/2J mouse strain reported to contain a defunctionalizing mutation in TAAR1[18] is also reported to show enhanced reward responses to both ethanol and cocaine.[97]

In summary, TAAR1 activity appears to be important in the suppression of abuse-related behaviors for drugs that act primarily through activation of dopaminergic systems. Importantly from a clinical perspective, the TAAR1 partial agonist RO5203648 alone does not support self-administration,[90] suggesting that it does not have abuse potential in its own right.

Peripheral TAAR1

The role of TAAR1 in the periphery has received very little attention. A role in the control of energy metabolism appears likely, although whether this is mediated through peripheral effects, central effects, or a combination of both requires systematic study. TAAR1 has been identified in adipocytes,[46] insulin-secreting pancreatic β cells,[50,56] although not the glucagon secreting α-cells,[50] the stomach,[50,57,58] and possibly hepatocytes,[7,50] all of which have well-established roles in the control of nutrient storage and use. In the stomach TAAR1 is preferentially localized to the somatostatin-producing D-cells that play a role in the control of nutrient absorption and postfeeding secretion of peptide hormones.[58] Somatostatin release from these cells is increased following selective activation of TAAR1, and it was suggested that trace amines produced by decarboxylation of precursor amino acids in the gastrointestinal tract may act as a signal for nutrient presence.[58] Daily administration of RO5263397 for 2 weeks prevents olanzapine-induced increases in food intake, weight gain, and fat accumulation without affecting weight gain when administered alone,[50] unless higher doses were administered which were associated with decreased food intake and decreased weight gain in comparison to controls.[50] Such effects suggest that the trace amine system plays a role in the control of nutrient intake and energy metabolism, although again whether such effects are mediated centrally, peripherally, or through a combination of the two remains an open question.

These studies with selective TAAR1 agonists build on previous observations of the effects of the endogenous thyroid hormone metabolite 3-iodothyronamine. Systemic administration of 3-iodothyronamine has been reported to increase food intake,[98] cause ketonuria,[99] loss of body weight due to increased fat utilization,[99,100] hyperglycemia,[101] and changes in both insulin and glucagon secretion.[56] Although promiscuous,[102] 3-iodothyronamine is a potent agonist at TAAR1,[30] and some of these reported effects bear a striking resemblance to the effects of RO5263397. Indeed, a TAAR1-mediated increase in insulin secretion from cultured β-cells has been reported in response to 3-iodothyronamine, while 3-iodothyronamine acting via α_{2a}-adrenoceptors caused a decrease in insulin secretion.[56] Interestingly, methamphetamine was previously suggested to increase insulin secretion through interaction with an unknown target in the pancreas.[103] Centrally mediated effects of 3-iodothyronamine on plasma glucose, insulin, and glucagon levels have also been reported.[101,104] There is, therefore, growing evidence to support a role for TAAR1 in the regulation of glucose and lipid homeostasis, although the exact nature of this role requires further investigation.

It has also become increasingly apparent in recent years that TAAR1 also plays a role in the immune system, with multiple groups now reporting that TAAR1 is involved in leukocyte activation. TAAR1 expression has been identified in various leukocytes[51,105] including granulocytes,[54] B cells,[51–54] T cells,[54,55] and NK cells,[51] and is upregulated following leukocyte activation.[51–53] In granulocytes TAAR1 mediates their chemotactic response toward the classic trace amines 2-phenylethylamine and p-tyramine.[54] In this respect it is interesting to note that trace amines have been suggested to be released from platelets following their activation.[105] In T cells and B cells TAAR1 was shown to be required for trace amine-induced secretion of cytokines and immunoglobulins.[54] In contrast, Wasik et al.[53] observed agonist-induced apoptosis in B cells expressing TAAR1. Further, it was recently reported that methamphetamine increases TAAR1 expression in T cells, and causes a TAAR1-dependent decrease in their interleukin-2 levels.[55] Immune dysfunction resulting in an increased susceptibility to infection is associated with a number of drugs of abuse, including amphetamines.[106,107] Interestingly, both metabolic disorder and immune dysfunction also occur with increased prevalence in schizophrenia,[108] although the basis of this is unknown. The combination of the above studies raises the possibility that TAAR1 may provide a single molecular target for the coexistence of such apparently disparate disorders.

TRACE AMINES AND OLFACTION

In 2006 TAAR were identified as a new family of odorant receptor, distinct from the olfactory receptor and vomeronasal receptor families.[109] With the exception of TAAR1, all functional TAAR isoforms are expressed in the olfactory epithelium of all vertebrate species examined, including humans.[9,110–114] Within the olfactory epithelium, each TAAR isoform is only expressed in a distinct, limited, subset of olfactory sensory neurons, which project to distinct glomeruli in the olfactory bulb.[21,114–116] Of note, olfactory TAAR appear to exhibit exquisite sensitivity to their ligands (where known), with limits of detection in the picomolar range, several orders of magnitude lower in the native systems than in commonly used expression systems.[117]

Selective activation of individual TAAR by various volatile, endogenous, amines has been reported and is summarized in Table 8.1. Many of these ligands also induce innate behavioral responses, at least some of which appear to be mediated by TAAR. Pan-TAAR knockout causes a loss of innate avoidance responses to various volatile amines including 2-phenylethylamine, isoamylamine, cadaverine, trimethylamine, and N-methylpiperidine.[21] TAAR4 has

TABLE 8.1   Known Endogenously Produced Ligands of TAAR

	Selective Ligands	Amino Acid Precursor	Notes
TAAR1	2-Phenylethylamine	L-Phenylalanine	Stress increases urinary PE levels
	p-Tyramine	L-Tyrosine	
TAAR3	Isoamylamine	L-Leucine	Putative mouse pheromone. Enriched in male urine and induces female puberty
TAAR4	2-Phenylethylamine	L-Phenylalanine	Stress increases urinary PE levels
	p-Tyramine	L-Tyrosine	
TAAR5	Trimethylamine	Produced by constitutive microbes from amino acid precursors	Enriched in male mouse urine
			Found in spoiled fish
			Metabolized by flavin monooxygenase 3 in most species
TAAR7e	5-Methoxy-N,N-dimethyltryptamine	L-Tryptophan	Identified in urine
TAAR13c	Cadaverine[a]	L-Lysine	Component of decaying fish

[a]Receptor activated in mouse unknown.

been reported to be required for the detection and subsequent innate avoidance of both predator urine and 2-phenylethylamine,[21,117] with 2-phenylethylamine shown to be enriched in carnivore urine and responsible for the innate avoidance shown by prey species.[118] Interestingly in this respect, urinary 2-phenylethylamine levels are also increased by stress,[119] and 2-phenylethylamine has been proposed to be an endogenous anxiogen.[120] As such, 2-phenylethylamine may serve as both a conspecific and heterospecific urinary cue. In addition to TAAR4, mouse TAAR5 and TAAR7f, and rat TAAR8c and TAAR9 are activated by urine,[121] although the component(s) responsible for the activation have not yet been identified.

There is some evidence that species-specificity in TAAR-mediated olfactory responses may also occur, with one report that ligands for a given TAAR in one species do not necessarily activate the ortholog in other species.[122] Further, at least some TAAR ligands mediate species-specific responses; trimethylamine, which selectively activates TAAR5,[112,117,123,124] is an attractant in mice,[123] but causes avoidance behavior in other species including rats[112,123]; cadaverine, which activates TAAR13c in teleosts, is an attractant to some teleost species, but mediates avoidance in others[125] and in mice.[21] Differential expression of olfactory TAAR isoforms during development has also been observed in migratory eels,[126] suggesting that they play a role in one or more aspects of migration. TAAR have also been proposed to play a role in the migratory behavior of Atlantic salmon.[12] The brain area(s) subsequently activated following stimulation of olfactory TAAR, and hence how these various innate behaviors are initiated at the cellular level, is still unknown. For a more thorough discussion of TAAR-mediated olfactory behaviors the reader is referred to the recent review by Liberles[121] and references therein.

Taken together the above findings support TAAR playing a role in species-specific responses to chemical cues related to navigation, conspecific signaling, and/or predator avoidance. It is also worth noting that while the ligands in Table 8.1 were identified on the basis of their activation of olfactory TAAR, many of these are produced endogenously and could play a role elsewhere in the body. In particular many of these ligands are produced by either the decarboxylation of common amino acids by endogenous decarboxylase enzymes such as ornithine decarboxylase[125] and AADC (see chapter: Synthesis and Neurochemistry of Trace Amines), or due to metabolism of other dietary constituents by gut microbes.[127] Such a production of TAAR ligands from dietary sources may be an important aspect of the growing recognition of the role played by the gastrointestinal tract and its constituent microbial flora in health and disease.

CONCLUDING STATEMENTS

Over the past decade there has been a gradual accumulation of studies which provide convincing evidence for trace amines and their receptors playing an important role in cellular homeostasis. In the central nervous system there is especially strong evidence for TAAR1 activation playing a role in the control of neuronal tone. In particular TAAR1 acts to prevent hyperdopaminergic states through heterodimerization with D2-like receptors. This interaction initiates a biased signaling cascade, with the selective recruitment of the G-protein-independent β-arrestin 2 pathway. This control of hyperdopaminergic states makes TAAR1 an attractive candidate for the development of new therapeutics for schizophrenia and drug abuse/addiction and such agents have recently entered clinical trials. From this perspective the selective recruitment of β-arrestin 2 is of interest as this pathway has been implicated in both the pathology of schizophrenia[128,129] and in drug abuse behaviors.[130] There is also now very little doubt that TAAR play a role in olfaction-mediated behaviors. In particular TAAR activation induces innate behaviors, although the neuronal connectivity and molecular events that underlie such behavioral changes remains to be elucidated.

Outside of the CNS, the role and function of TAAR has been much less studied. There is sufficient circumstantial evidence of TAAR1 playing a role in energy metabolism that a systematic study of TAAR1 effects in the insulin-secreting β-cells of the pancreas, adipocytes, and hepatocytes is warranted. The recent development of selective TAAR1 agonists will considerably aid in this regard. There is also now very good evidence that TAAR1 is involved in the process of leukocyte activation and the relevance of this to pathological conditions should be an area of increased study in future years.

References

1. Sandler M, Reynolds GP. Does phenylethylamine cause schizophrenia? *Lancet* 1976;**307**(7950):70–1.
2. Reynolds GP. Phenylethylamine—a role in mental illness? *Trends Neurosci* 1979;**2**:265–8.
3. Berry MD. The potential of trace amines and their receptors for treating neurological and psychiatric diseases. *Rev Recent Clin Trials* 2007;**2**(1):3–19.
4. Philips SR, Rozdilsky B, Boulton AA. Evidence for the presence of *m*-tyramine, *p*-tyramine, tryptamine, and phenylethylamine in the rat brain and several areas of the human brain. *Biol Psychiatry* 1978;**13**(1):51–7.
5. Berry MD. Mammalian central nervous system trace amines. Pharmacologic amphetamines, physiologic neuromodulators. *J Neurochem* 2004;**90**(2):257–71.
6. Borowsky B, Adham N, Jones KA, et al. Trace amines: identification of a family of mammalian G protein-coupled receptors. *Proc Natl Acad Sci USA* 2001;**98**(16):8966–71.
7. Bunzow JR, Sonders MS, Arttamangkul S, et al. Amphetamine, 3,4-methylenedioxymethamphetamine, lysergic acid diethylamide, and metabolites of the catecholamine neurotransmitters are agonists of a rat trace amine receptor. *Mol Pharmacol* 2001;**60**(6):1181–8.
8. Alexander SP, Benson HE, Faccenda E, et al. The concise guide to pharmacology 2013/14: G protein-coupled receptors. *Br J Pharmacol* 2013;**170**(8):1459–581.
9. Hashiguchi Y, Nishida M. Evolution of trace amine associated receptor (TAAR) gene family in vertebrates: lineage-specific expansions and degradations of a second class of vertebrate chemosensory receptors expressed in the olfactory epithelium. *Mol Biol Evol* 2007;**24**(9):2099–107.
10. Hussain A, Saraiva LR, Korsching SI. Positive Darwinian selection and the birth of an olfactory receptor clade in teleosts. *Proc Natl Acad Sci USA* 2009;**106**(11):4313–8.
11. Hashiguchi Y, Furuta Y, Nishida M. Evolutionary patterns and selective pressures of odorant/pheromone receptor gene families in teleost fishes. *PLoS One* 2008;**3**(12):e4083.
12. Tessarolo JA, Tabesh MJ, Nesbitt M, Davidson WS. Genomic organization and evolution of the trace amine-associated receptor (TAAR) repertoire in Atlantic Salmon (*Salmo salar*). *G3 (Bethesda)* 2014;**4**(6):1135–41.
13. Azzouzi N, Barloy-Hubler F, Galibert F. Identification and characterization of cichlid TAAR genes and comparison with other teleost TAAR repertoires. *BMC Genom* 2015;**16**(1):335.
14. Gloriam DE, Bjarnadottir TK, Schioth HB, Fredriksson R. High species variation within the repertoire of trace amine receptors. *Ann NY Acad Sci* 2005;**1040**:323–7.
15. Bly M. Examination of the trace amine-associated receptor 2 (TAAR2). *Schizophr Res* 2005;**80**(2–3):367–8.
16. Vanti WB, Muglia P, Nguyen T, et al. Discovery of a null mutation in a human trace amine receptor gene. *Genomics* 2003;**82**(5):531–6.
17. Vallender EJ, Xie Z, Westmoreland SV, Miller GM. Functional evolution of the trace amine associated receptors in mammals and the loss of TAAR1 in dogs. *BMC Evol Biol* 2010;**10**:51.
18. Harkness JH, Shi X, Janowsky A, Phillips TJ. Trace amine-associated receptor 1 regulation of methamphetamine intake and related traits. *Neuropsychopharmacology* 2015;**40**(9):2175–84.

19. Wolinsky TD, Swanson CJ, Smith KE, et al. The trace amine 1 receptor knockout mouse: an animal model with relevance to schizophrenia. *Genes Brain Behav* 2007;**6**(7):628–39.

20. Lindemann L, Meyer CA, Jeanneau K, et al. Trace amine-associated receptor 1 modulates dopaminergic activity. *J Pharmacol Exp Ther* 2008;**324**(3):948–56.

21. Dewan A, Pacifico R, Zhan R, Rinberg D, Bozza T. Non-redundant coding of aversive odours in the main olfactory pathway. *Nature* 2013;**497**(7450):486–9.

22. Jaeger CB, Ruggiero DA, Albert VR, Park DH, Joh TH, Reis DJ. Aromatic L-amino acid decarboxylase in the rat brain: immunohistochemical localization in neurons of the brain stem. *Neuroscience* 1984;**11**(3):691–713.

23. Boulton AA, Juorio AV, Philips SR, Wu PH. The effects of reserpine and 6-hydroxydopamine on the concentrations of some arylalkylamines in rat brain. *Br J Pharmacol* 1977;**59**:209–14.

24. Juorio AV, Greenshaw AJ, Wishart TB. Reciprocal changes in striatal dopamine and b-phenylethylamine induced by reserpine in the presence of monoamine oxidase inhibitors. *Naunyn Schmiedebergs Arch Pharmacol* 1988;**338**:644–8.

25. Henry DP, Russell WL, Clemens JA, Plebus LA. Phenylethylamine and *p*-tyramine in the extracellular space of the rat brain: quantification using a new radioenzymatic assay and in situ microdialysis. In: Boulton AA, Juorio AV, Downer RGH, editors. *Trace amines; comparative and clinical neurobiology*. Clifton, NJ: Humana Press; 1988. p. 239–50.

26. Dyck LE. Release of some endogenous trace amines from rat striatal slices in the presence and absence of a monoamine-oxidase inhibitor. *Life Sci* 1989;**44**(17):1149–56.

27. Berry MD, Shitut MR, Almousa A, Alcorn J, Tomberli B. Membrane permeability of trace amines: evidence for a regulated, activity-dependent, nonexocytotic, synaptic release. *Synapse* 2013;**67**(10):656–67.

28. Boulton AA, Juorio AV, Paterson IA. Phenylethylamine in the CNS: effects of monoamine oxidase inhibiting drugs, deuterium substitution and lesions and its role in the neuromodulation of catecholaminergic neurotransmission. *J Neural Transm Suppl* 1990;**29**:119–29.

29. Mosnaim AD, Callaghan OH, Hudzik T, Wolf ME. Rat brain-uptake index for phenylethylamine and various monomethylated derivatives. *Neurochem Res* 2013;**38**(4):842–6.

30. Scanlan TS, Suchland KL, Hart ME, et al. 3-Iodothyronamine is an endogenous and rapid-acting derivative of thyroid hormone. *Nat Med* 2004;**10**(6):638–42.

31. Galley G, Stalder H, Goergler A, Hoener MC, Norcross RD. Optimisation of imidazole compounds as selective TAAR1 agonists: discovery of RO5073012. *Bioorg Med Chem Lett* 2012;**22**(16):5244–8.

32. Cichero E, Espinoza S, Franchini S, et al. Further insights into the pharmacology of the human trace amine-associated receptors: discovery of novel ligands for TAAR1 by a virtual screening approach. *Chem Biol Drug Des* 2014;**84**:712–20.

33. Chiellini G, Nesi G, Digiacomo M, et al. Design, synthesis, and evaluation of thyronamine analogues as novel potent mouse trace amine associated receptor 1 (mTAAR1) agonists. *J Med Chem* 2015;**58**(12):5096–107.

34. Bradaia A, Trube G, Stalder H, et al. The selective antagonist EPPTB reveals TAAR1-mediated regulatory mechanisms in dopaminergic neurons of the mesolimbic system. *Proc Natl Acad Sci USA* 2009;**106**(47):20081–6.

35. Xie Z, Miller GM. A receptor mechanism for methamphetamine action in dopamine transporter regulation in brain. *J Pharmacol Exp Ther* 2009;**330**(1):316–25.

36. Di Cara B, Maggio R, Aloisi G, et al. Genetic deletion of trace amine 1 receptors reveals their role in auto-inhibiting the actions of ecstasy (MDMA). *J Neurosci* 2011;**31**(47):16928–40.

37. Reese EA, Norimatsu Y, Grandy MS, Suchland KL, Bunzow JR, Grandy DK. Exploring the determinants of trace amine-associated receptor 1's functional selectivity for the stereoisomers of amphetamine and methamphetamine. *J Med Chem* 2014;**57**(2):378–90.

38. Revel FG, Moreau JL, Gainetdinov RR, et al. TAAR1 activation modulates monoaminergic neurotransmission, preventing hyperdopaminergic and hypoglutamatergic activity. *Proc Natl Acad Sci USA* 2011;**108**(20):8485–90.

39. Miller GM, Verrico CD, Jassen A, et al. Primate trace amine receptor 1 modulation by the dopamine transporter. *J Pharmacol Exp Ther* 2005;**313**(3):983–94.

40. Barak LS, Salahpour A, Zhang X, et al. Pharmacological characterization of membrane-expressed human trace amine-associated receptor 1 (TAAR1) by a bioluminescence resonance energy transfer cAMP biosensor. *Mol Pharmacol* 2008;**74**(3):585–94.

41. Xie Z, Vallender EJ, Yu N, et al. Cloning, expression, and functional analysis of rhesus monkey trace amine-associated receptor 6: evidence for lack of monoaminergic association. *J Neurosci Res* 2008;**86**(15):3435–46.

42. Leo D, Mus L, Espinoza S, Hoener MC, Sotnikova TD, Gainetdinov RR. TAAR1-mediated modulation of presynaptic dopaminergic neurotransmission: role of D2 dopamine autoreceptors. *Neuropharmacology* 2014;**81**:283–91.

43. Espinoza S, Ghisi V, Emanuele M, et al. Postsynaptic D2 dopamine receptor supersensitivity in the striatum of mice lacking TAAR1. *Neuropharmacology* 2015;**93**:308–13.

44. Xie Z, Westmoreland SV, Bahn ME, et al. Rhesus monkey trace amine-associated receptor 1 signaling: enhancement by monoamine transporters and attenuation by the D2 autoreceptor in vitro. *J Pharmacol Exp Ther* 2007;**321**(1):116–27.

45. Cisneros IE, Ghorpade A. Methamphetamine and HIV-1-induced neurotoxicity: role of trace amine associated receptor 1 cAMP signaling in astrocytes. *Neuropharmacology* 2014;**85C**:499–507.

46. Regard JB, Sato IT, Coughlin SR. Anatomical profiling of G protein-coupled receptor expression. *Cell* 2008;**135**(3):561–71.

47. Broadley KJ, Fehler M, Ford WR, Kidd EJ. Functional evaluation of the receptors mediating vasoconstriction of rat aorta by trace amines and amphetamines. *Eur J Pharmacol* 2013;**715**(1–3):370–80.

48. Kidd M, Modlin IM, Gustafsson BI, Drozdov I, Hauso O, Pfragner R. Luminal regulation of normal and neoplastic human EC cell serotonin release is mediated by bile salts, amines, tastants, and olfactants. *Am J Physiol Gastrointest Liver Physiol* 2008;**295**(2):G260–72.

49. Ito J, Ito M, Nambu H, et al. Anatomical and histological profiling of orphan G-protein-coupled receptor expression in gastrointestinal tract of C57BL/6J mice. *Cell Tissue Res* 2009;**338**(2):257–69.

50. Revel FG, Moreau JL, Pouzet B, et al. A new perspective for schizophrenia: TAAR1 agonists reveal antipsychotic- and antidepressant-like activity, improve cognition and control body weight. *Mol Psychiatry* 2013;**18**(5):543–56.

51. Nelson DA, Tolbert MD, Singh SJ, Bost KL. Expression of neuronal trace amine-associated receptor (TAAR) mRNAs in leukocytes. *J Neuroimmunol* 2007;**192**(1–2):21–30.

52. Panas MW, Xie Z, Panas HN, Hoener MC, Vallender EJ, Miller GM. Trace amine associated receptor 1 signaling in activated lymphocytes. *J Neuroimmune Pharmacol* 2012;**7**(4):866–76.

53. Wasik AM, Millan MJ, Scanlan T, Barnes NM, Gordon J. Evidence for functional trace amine associated receptor-1 in normal and malignant B cells. *Leuk Res* 2012;**36**(2):245–9.

54. Babusyte A, Kotthoff M, Fiedler J, Krautwurst D. Biogenic amines activate blood leukocytes via trace amine-associated receptors TAAR1 and TAAR2. *J Leukoc Biol* 2013;**93**(3):387–94.

55. Sriram U, Cenna JM, Haldar B, et al. Methamphetamine induces trace amine-associated receptor 1 (TAAR1) expression in human T lymphocytes: role in immunomodulation. *J Leukoc Biol* 2015;**99**(1):213–23.

56. Regard JB, Kataoka H, Cano DA, et al. Probing cell type-specific functions of Gi in vivo identifies GPCR regulators of insulin secretion. *J Clin Invest* 2007;**117**(12):4034–43.

57. Chiellini G, Erba P, Carnicelli V, et al. Distribution of exogenous [125I]-3-iodothyronamine in mouse in vivo: relationship with trace amine-associated receptors. *J Endocrinol* 2012;**213**(3):223–30.

58. Adriaenssens A, Yee Hong Lam B, Billing L, et al. A transcriptome-led exploration of molecular mechanisms regulating somatostatin-producing D-cells in the gastric epithelium. *Endocrinology* 2015;**156**(11):3924–36.

59. Stalder H, Hoener MC, Norcross RD. Selective antagonists of mouse trace amine-associated receptor 1 (mTAAR1): discovery of EPPTB (RO5212773). *Bioorg Med Chem Lett* 2011;**21**(4):1227–31.

60. Revel FG, Moreau JL, Gainetdinov RR, et al. Trace amine-associated receptor 1 partial agonism reveals novel paradigm for neuropsychiatric therapeutics. *Biol Psychiatry* 2012;**72**(11):934–42.

61. Revel FG, Meyer CA, Bradaia A, et al. Brain-specific overexpression of trace amine-associated receptor 1 alters monoaminergic neurotransmission and decreases sensitivity to amphetamine. *Neuropsychopharmacology* 2012;**37**(12):2580–92.

62. Espinoza S, Salahpour A, Masri B, et al. Functional interaction between trace amine-associated receptor 1 and dopamine D2 receptor. *Mol Pharmacol* 2011;**80**(3):416–25.

63. Salahpour A, Espinoza S, Masri B, Lam V, Barak LS, Gainetdinov RR. BRET biosensors to study GPCR biology, pharmacology, and signal transduction. *Front Endocrinol* 2012;**3**:105.

64. Harmeier A, Obermueller S, Meyer CA, et al. Trace amine-associated receptor 1 activation silences GSK3β signaling of TAAR1 and D2R heteromers. *Eur Neuropsychopharmacol* 2015;**25**(11):2049–61.

65. Berry MD. The effects of pargyline and 2-phenylethylamine on D1-like dopamine receptor binding. *J Neural Transm* 2011;**118**(7):1115–8.

66. Xie Z, Miller GM. Beta-phenylethylamine alters monoamine transporter function via trace amine-associated receptor 1: implication for modulatory roles of trace amines in brain. *J Pharmacol Exp Ther* 2008;**325**(2):617–28.

67. Xie Z, Westmoreland SV, Miller GM. Modulation of monoamine transporters by common biogenic amines via trace amine-associated receptor 1 and monoamine autoreceptors in human embryonic kidney 293 cells and brain synaptosomes. *J Pharmacol Exp Ther* 2008;**325**(2):629–40.

68. Vialou V, Balasse L, Callebert J, Launay JM, Giros B, Gautron S. Altered aminergic neurotransmission in the brain of organic cation transporter 3-deficient mice. *J Neurochem* 2008;**106**(3):1471–82.

69. Baganz N, Horton R, Martin K, Holmes A, Daws LC. Repeated swim impairs serotonin clearance via a corticosterone-sensitive mechanism: organic cation transporter 3, the smoking gun. *J Neurosci* 2010;**30**(45):15185–95.

70. Bacq A, Balasse L, Biala G, et al. Organic cation transporter 2 controls brain norepinephrine and serotonin clearance and antidepressant response. *Mol Psychiatry* 2012;**17**(9):926–39.

71. Graf EN, Wheeler RA, Baker DA, et al. Corticosterone acts in the nucleus accumbens to enhance dopamine signaling and potentiate reinstatement of cocaine seeking. *J Neurosci* 2013;**33**(29):11800–10.

72. Marcinkiewcz CA, Devine DP. Modulation of OCT3 expression by stress, and antidepressant-like activity of decynium-22 in an animal model of depression. *Pharmacol Biochem Behav* 2015;**131**:33–41.

73. Courousse T, Gautron S. Role of organic cation transporters (OCTs) in the brain. *Pharmacol Ther* 2015;**146**: 94–103.

74. Frenken M, Kaumann AJ. Effects of tryptamine mediated through 2 states of the 5-HT2 receptor in calf coronary artery. *Naunyn Schmiedebergs Arch Pharmacol* 1988;**337**(5):484–92.

75. Espinoza S, Lignani G, Caffino L, et al. TAAR1 modulates cortical glutamate NMDA receptor function. *Neuropsychopharmacology* 2015;**40**(9):2217–27.

76. Zhang S, Jin Y, Liu X, et al. Methamphetamine modulates glutamatergic synaptic transmission in rat primary cultured hippocampal neurons. *Brain Res* 2014;**1582**:1–11.

77. Underhill SM, Wheeler DS, Li M, Watts SD, Ingram SL, Amara SG. Amphetamine modulates excitatory neurotransmission through endocytosis of the glutamate transporter EAAT3 in dopamine neurons. *Neuron* 2014;**83**(2):404–16.

78. Ishida K, Murata M, Kato M, Utsunomiya I, Hoshi K, Taguchi K. Phenylethylamine stimulates striatal acetylcholine release through activation of the AMPA glutamatergic pathway. *Biol Pharm Bull* 2005;**28**(9):1626–9.

79. Borison RL, Mosnaim AD, Sabelli HC. Brain 2-phenylethylamine as a major mediator for the central actions of amphetamine and methylphenidate. *Life Sci* 1975;**17**:1331–44.

80. Janssen PAJ, Leysen JE, Megens AAHP, Awouters FHL. Does phenylethylamine act as an endogenous amphetamine in some patients? *Int J Neuropsychopharmacol* 1999;**2**:229–40.

81. Cao Q, Martinez M, Zhang J, et al. Suggestive evidence for a schizophrenia susceptibility locus on chromosome 6q and a confirmation in an independent series of pedigrees. *Genomics* 1997;**43**(1):1–8.

82. Kaufmann CA, Suarez B, Malaspina D, et al. NIMH genetics initiative millenium schizophrenia consortium: linkage analysis of African-American pedigrees. *Am J Med Genet* 1998;**81**(4):282–9.

83. Levinson DF, Holmans P, Straub RE, et al. Multicenter linkage study of schizophrenia candidate regions on chromosomes 5q, 6q, 10p, and 13q: schizophrenia linkage collaborative group III. *Am J Hum Genet* 2000;**67**(3):652–63.

84. Jing L, Li JX. Trace amine-associated receptor 1: a promising target for the treatment of psychostimulant addiction. *Eur J Pharmacol* 2015;**761**:345–52.

85. Sulzer D. How addictive drugs disrupt presynaptic dopamine neurotransmission. *Neuron* 2011;**69**(4):628–49.

86. Volkow ND, Morales M. The brain on drugs: from reward to addiction. *Cell* 2015;**162**(4):712–25.

87. Shannon HE, Thompson WA. Behavior maintained under fixed-interval and second-order schedules by intravenous injections of endogenous noncatecholic phenylethylamines in dogs. *J Pharmacol Exp Ther* 1984;**228**(3):691–5.

88. Revel FG, Hoener MC, Renau-Piqueras J, Canales JJ. Targeting trace-amine associated receptors in the treatment of drug addiction. In: Canales JJ, editor. *Emerging targets for drug addiction treatment*. New York: Nova Science Publishers; 2012.

89. Achat-Mendes C, Lynch LJ, Sullivan KA, Vallender EJ, Miller GM. Augmentation of methamphetamine-induced behaviors in transgenic mice lacking the trace amine-associated receptor 1. *Pharmacol Biochem Behav* 2012;**101**(2):201–7.

90. Cotter R, Pei Y, Mus L, et al. The trace amine-associated receptor 1 modulates methamphetamine's neurochemical and behavioral effects. *Front Neurosci* 2015;**9**:39.

91. Jing L, Zhang Y, Li JX. Effects of the trace amine associated receptor 1 agonist RO5263397 on abuse-related behavioral indices of methamphetamine in rats. *Int J Neuropsychopharmacol* 2014;**18**:1–7.
92. Thorn DA, Jing L, Qiu Y, et al. Effects of the trace amine-associated receptor 1 agonist RO5263397 on abuse-related effects of cocaine in rats. *Neuropsychopharmacology* 2014;**39**(10):2309–16.
93. Thorn DA, Zhang C, Zhang Y, Li JX. The trace amine associated receptor 1 agonist RO5263397 attenuates the induction of cocaine behavioral sensitization in rats. *Neurosci Lett* 2014;**566**:67–71.
94. Pei Y, Lee J, Leo D, Gainetdinov RR, Hoener MC, Canales JJ. Activation of the trace amine-associated receptor 1 prevents relapse to cocaine seeking. *Neuropsychopharmacology* 2014;**39**(10):2299–308.
95. Lynch LJ, Sullivan KA, Vallender EJ, Rowlett JK, Platt DM, Miller GM. Trace amine associated receptor 1 modulates behavioral effects of ethanol. *Subst Abuse Res Treat* 2013;**7**:117–26.
96. Nutt DJ, Lingford-Hughes A, Erritzoe D, Stokes PR. The dopamine theory of addiction: 40 years of highs and lows. *Nat Rev Neurosci* 2015;**16**(5):305–12.
97. Fish EW, Riday TT, McGuigan MM, Faccidomo S, Hodge CW, Malanga CJ. Alcohol, cocaine, and brain stimulation-reward in C57Bl6/J and DBA2/J mice. *Alcohol Clin Exp Res* 2010;**34**(1):81–9.
98. Dhillo WS, Bewick GA, White NE, et al. The thyroid hormone derivative 3-iodothyronamine increases food intake in rodents. *Diabetes Obes Metab* 2009;**11**(3):251–60.
99. Haviland JA, Reiland H, Butz DE, et al. NMR-based metabolomics and breath studies show lipid and protein catabolism during low dose chronic T(1)AM treatment. *Obesity (Silver Spring)* 2013;**21**(12):2538–44.
100. Braulke LJ, Klingenspor M, DeBarber A, et al. 3-Iodothyronamine: a novel hormone controlling the balance between glucose and lipid utilisation. *J Comp Physiol [B]* 2008;**178**(2):167–77.
101. Manni ME, De Siena G, Saba A, et al. 3-Iodothyronamine: a modulator of the hypothalamus-pancreas-thyroid axes in mice. *Br J Pharmacol* 2012;**166**(2):650–8.
102. Zucchi R, Accorroni A, Chiellini G. Update on 3-iodothyronamine and its neurological and metabolic actions. *Front Physiol* 2014;**5**:402.
103. McMahon EM, Andersen DK, Feldman JM, Schanberg SM. Methamphetamine-induced insulin release. *Science* 1971;**174**(4004):66–8.
104. Klieverik LP, Foppen E, Ackermans MT, et al. Central effects of thyronamines on glucose metabolism in rats. *J Endocrinol* 2009;**201**(3):377–86.
105. D'Andrea G, Terrazzino S, Fortin D, Farruggio A, Rinaldi L, Leon A. HPLC electrochemical detection of trace amines in human plasma and platelets and expression of mRNA transcripts of trace amine receptors in circulating leukocytes. *Neurosci Lett* 2003;**346**(1–2):89–92.
106. Boyle NT, Connor TJ. Methylenedioxymethamphetamine ('Ecstasy')-induced immunosuppression: a cause for concern? *Br J Pharmacol* 2010;**161**(1):17–32.
107. Sriram U, Haldar B, Cenna JM, Gofman L, Potula R. Methamphetamine mediates immune dysregulation in a murine model of chronic viral infection. *Front Microbiol* 2015;**6**:793.
108. Steiner J, Bernstein HG, Schiltz K, et al. Immune system and glucose metabolism interaction in schizophrenia: a chicken-egg dilemma. *Prog Neuropsychopharmacol Biol Psychiatry* 2014;**48**:287–94.
109. Liberles SD, Buck LB. A second class of chemosensory receptors in the olfactory epithelium. *Nature* 2006;**442**(7103):645–50.
110. Gliem S, Schild D, Manzini I. Highly specific responses to amine odorants of individual olfactory receptor neurons in situ. *Eur J Neurosci* 2009;**29**(12):2315–26.
111. Carnicelli V, Santoro A, Sellari-Franceschini S, Berrettini S, Zucchi R. Expression of trace amine-associated receptors in human nasal mucosa. *Chemosens Percept* 2010;**3**(2):99–107.
112. Horowitz LF, Saraiva LR, Kuang D, Yoon KH, Buck LB. Olfactory receptor patterning in a higher primate. *J Neurosci* 2014;**34**(37):12241–52.
113. Kanageswaran N, Demond M, Nagel M, et al. Deep sequencing of the murine olfactory receptor neuron transcriptome. *PLoS One* 2015;**10**(1):e0113170.
114. Syed AS, Sansone A, Roner S, Bozorg Nia S, Manzini I, Korsching SI. Different expression domains for two closely related amphibian TAARs generate a bimodal distribution similar to neuronal responses to amine odors. *Sci Rep* 2015;**5**:13935.
115. Johnson MA, Tsai L, Roy DS, et al. Neurons expressing trace amine-associated receptors project to discrete glomeruli and constitute an olfactory subsystem. *Proc Natl Acad Sci USA* 2012;**109**(33):13410–5.

116. Pacifico R, Dewan A, Cawley D, Guo C, Bozza T. An olfactory subsystem that mediates high-sensitivity detection of volatile amines. *Cell Rep* 2012;**2**(1):76–88.

117. Zhang J, Pacifico R, Cawley D, Feinstein P, Bozza T. Ultrasensitive detection of amines by a trace amine-associated receptor. *J Neurosci* 2013;**33**(7):3228–39.

118. Ferrero DM, Lemon JK, Fluegge D, et al. Detection and avoidance of a carnivore odor by prey. *Proc Natl Acad Sci USA* 2011;**108**(27):11235–40.

119. Paulos MA, Tessel RA. Excretion of beta-phenylethylamine is elevated in humans after profound stress. *Science* 1982;**215**:1127–9.

120. Lapin IP. Beta-phenylethylamine (PEA): an endogenous anxiogen? Three series of experimental data. *Biol Psychiatry* 1990;**28**(11):997–1003.

121. Liberles SD. Trace amine-associated receptors: ligands, neural circuits, and behaviors. *Curr Opin Neurobiol* 2015;**34C**:1–7.

122. Staubert C, Boselt I, Bohnekamp J, Rompler H, Enard W, Schoneberg T. Structural and functional evolution of the trace amine-associated receptors TAAR3, TAAR4 and TAAR5 in primates. *PLoS One* 2010;**5**(6):e11133.

123. Li Q, Korzan WJ, Ferrero DM, et al. Synchronous evolution of an odor biosynthesis pathway and behavioral response. *Curr Biol* 2013;**23**(1):11–20.

124. Wallrabenstein I, Kuklan J, Weber L, et al. Human trace amine-associated receptor TAAR5 can be activated by trimethylamine. *PLoS One* 2013;**8**(2):e54950.

125. Hussain A, Saraiva LR, Ferrero DM, et al. High-affinity olfactory receptor for the death-associated odor cadaverine. *Proc Natl Acad Sci USA* 2013;**110**(48):19579–84.

126. Churcher AM, Hubbard PC, Marques JP, Canario AV, Huertas M. Deep sequencing of the olfactory epithelium reveals specific chemosensory receptors are expressed at sexual maturity in the European eel *Anguilla anguilla*. *Mol Ecol* 2015;**24**(4):822–34.

127. Williams BB, Van Benschoten AH, Cimermancic P, et al. Discovery and characterization of gut microbiota decarboxylases that can produce the neurotransmitter tryptamine. *Cell Host Microbe* 2014;**16**(4):495–503.

128. Oda Y, Kanahara N, Kimura H, Watanabe H, Hashimoto K, Iyo M. Genetic association between G protein-coupled receptor kinase 6/beta-arrestin 2 and dopamine supersensitivity psychosis in schizophrenia. *Neuropsychiatr Dis Treat* 2015;**11**:1845–51.

129. Park SM, Chen M, Schmerberg CM, et al. Effects of beta-arrestin-biased dopamine D2 receptor ligands on schizophrenia-like behavior in hypoglutamatergic mice. *Neuropsychopharmacology* 2016;**41**(3):704–15.

130. Beaulieu JM, Gainetdinov RR, Caron MG. Akt/GSK3 signaling in the action of psychotropic drugs. *Annu Rev Pharmacol Toxicol* 2009;**49**:327–47.

Trace Amine-Associated Receptor 1 Modulation of Dopamine System

D. Leo and S. Espinoza

Department of Neuroscience and Brain Technologies, Italian Institute of
Technology, Genova, Italy

INTRODUCTION

From their first description in 2001,[1,2] trace amine-associated receptors (TAARs) attracted an enduring interest among physiologists and pharmacologists, since they have been shown to be involved in many different physiological processes ranging from regulation of brain functions to olfaction and, more recently, to the immune system. The TAARs family consists of nine genes in human (including three pseudogenes) and in chimpanzee genomes (including six pseudogenes), while 19 and 16 genes (including two and one pseudogenes) are present in the rat and mouse genomes, respectively.[3,4] Among them, only two TAAR receptors

(TAAR1 and TAAR4) have shown activation by trace amines (TAs).[1,2] At the moment, TAAR1 is the best-studied member of the TAARs family. All these gene clusters are in a narrow region of approximately 100–200 kilobases on the same chromosome, reminiscent of similar chromosomal organization of some members of the olfactory receptors. The gene for TAAR1 is located on chromosome 6q23.2 in humans and encodes for a 332-amino-acid protein generated from a single exon.[5]

Expression

TAARs expression has been debated between different groups, likely due to the different techniques used to detect TAARs; all the TAARs (except TAAR1) have been shown to be expressed in the olfactory epithelium in mice,[6–8] as putative olfactory receptors. The first two groups that cloned the TAARs family found that TAAR1 was expressed in many brain regions and in peripheral organs such as the liver, kidney, spleen, pancreas, and heart.[1,2,9] There are also indications that other TAAR members are expressed in the brain (ie, TAAR5, TAAR6, TAAR9).[10]

TAAR1 and TAAR2 are also expressed in human blood leukocytes, particularly in polymorphonuclear T and B cells. Conversely, lower expression was found for TAAR5, TAAR6, and TAAR9.[11–13]

TAAR1 expression and localization have been extensively studied in mouse,[1,3,14] rat,[2] and monkey.[15–17] Although sometimes conflicting results were reported, most likely due to the different techniques used to study TAAR1 expression, there is general agreement about TAAR1 expression in the brain, especially in monoaminergic nuclei and their projecting areas. By using RT-PCR and in situ hybridization, Borowsky and colleagues identified TAAR1 expression in many mouse brain areas (including the amygdala, cerebellum, hypothalamus, dorsal root ganglia, and hippocampus) and, most interestingly, in key regions of the monoamine system [including substantia nigra (SN), ventral tegmental area (VTA), locus coeruleus (LC), and dorsal raphe (DR)]. TAAR1 expression was also confirmed by Lindermann and colleagues by using a transgenic mouse model where LacZ gene was inserted into the TAAR1 gene to have a specific expression of the β-galactosidase driven by the TAAR1 promoter.[18] In rhesus monkey, TAAR1 mRNA is expressed in monoaminergic regions including the SN/VTA, LC, DR, amygdala, caudate nucleus, putamen, and nucleus accumbens (NAc),[16] whereas in human fewer studies were performed, but high expression of TAAR1 was found at least in the amygdala.[1] Taken together, it is convincing that TAAR1 is widely expressed in the primary monoaminergic areas of the brain and may therefore be implicated in the modulation of dopamine, serotonin, and potentially norepinephrine neurotransmission modulating locomotor, emotional, and motivated behaviors traditionally associated with monoaminergic activity.

Cellular Signaling

TAAR1 is a Gαs-coupling receptor and its stimulation increases cAMP levels.[1,2,19] Moreover, TAAR1 is able to increase extracellular signal-regulated kinase (ERK) and cAMP response element-binding (CREB) phosphorylation in vitro and in vivo in the striatum,[20] and protein kinase A and protein kinase C phosphorylation in HEK-293 cells.[12] Recently, evidence on TAAR1 involvement with the β-arrestin2/AKT/GSK3 signaling cascade was reported.[21,22] Many groups in the last few years have been struggling with TAAR1 in vitro pharmacology because of its difficulty to be expressed at the plasma membrane, at least in heterologous cell systems. Still is not clear whether TAAR1 in its natural environment, such as neurons, is expressed at the membrane or in the intracellular compartments. Technical limitations, such as the lack of specific antibodies, are the main reasons for the low reliable evaluation of cellular distribution of TAAR1.[23] Many modifications of the receptor structure have been made to improve the TAAR1 surface expression in order to study its cellular signaling. Bunzow et al.[2] built a human/rat chimera, while Liberles and Buck[7] added a peptide from bovine rhodopsin at the N-terminus of the receptor. In our lab, we fused the first nine amino acids of the β2-adrenergic receptor to the N-terminus, leading to a significant level of membrane expression sufficient to reliably study TAAR1 pharmacology.[19,24]

TAAR1 Regulation of Dopamine System

Because TAAR1 is expressed in areas such as VTA and DR, many studies focus their attention on TAAR1 modulation of monoaminergic systems, particularly the dopamine system. In VTA dopaminergic neurons, TAAR1 is able to regulate the firing rate of these neurons, with TAAR1 knockout (TAAR1-KO) mice showing an increase in firing frequency.[25] The application of the selective antagonist N-(3-ethoxy-phenyl)-4-pyrrolidin-1-yl-3-trifluoromethyl-benzamide (EPPTB) produces the same result while the full agonist decreases the firing frequency of these neurons.[26] Interestingly, the partial agonist behaves as an antagonist in this experimental paradigm, suggesting that there is a tonic activation of TAAR1.[27] It has been proved, both in vivo and in vitro, that TAAR1 interacts with D2 dopamine receptor. Bradaia et al.[25] demonstrated that acute application of the selective antagonist EPPTB increased the potency of DA at D2 dopamine receptors in DA neurons. In our laboratory, we used fast-scan cyclic voltammetry (FSCV) to evaluate the inhibitory effect of the D2 dopamine autoreceptors using the D2 agonist quinpirole and the TAAR1 agonist RO5166017 (a generous gift from F. Hoffmann, La Roche Ltd) in the NAc.[28] A significant quinpirole-induced reduction in DA release was observed in both TAAR1-KO and wildtype (WT) mice, indicating a preserved D2 dopamine autoreceptor function in the absence of TAAR1. However, the D2 dopamine receptor agonist showed an additive effect when combined with the TAAR1 agonist RO5166017 in WT mice but not in TAAR1-deficient animals. Presumably, there is an increase in the D2 dopamine receptor-mediated autoinhibition of DA neurons under tonic activation of TAAR1 that supports the role of TAAR1 as a homeostatic regulatory mechanism preventing the excess activity of DA neurons. Finally, to further investigate the relationship between the level of autoreceptor stimulation and the resulting inhibition of DA release, we used paired stimuli to evoke DA release in TAAR1-KO and control NAc slices. In this paradigm, DA released by the first pulse activates D2 dopamine autoreceptors and thus results in less amount of DA release evoked by the following pulse. In this test, TAAR1-KO animals had higher amplitude of DA release following the second pulse of stimulation, thus directly indicating that D2 dopamine receptor-mediated autoinhibition was less active in the absence of TAAR1.

On the other hand, there is evidence indicating a TAAR1 modulation on D2 dopamine postsynaptic receptors. Our in vitro work using a BRET-based assay on HEK-293 cells showed the formation of heterodimers between the long isoform of the D2 dopamine receptor (mainly expressed at postsynaptic sites) and TAAR1.[29] This complex was sensitive to D2 dopamine receptor conformation, since the treatment with haloperidol (a D2 dopamine receptor antagonist) was able to decrease the complex formation. Similarly, haloperidol treatment enhanced TAAR1 signaling in these cells.[24] Moreover, this interaction between D2 dopamine receptors (D2R) and TAAR1 was evident in vivo since in TAAR1-KO mice haloperidol was less effective in producing cataleptic behavior and to induce c-fos expression in the dorsal striatum. Our in vivo studies also showed that in the striatum of TAAR1-KO mice the D2 dopamine receptors were overexpressed and supersensitive, since the D2-dependent/β-arrestin2-dependent AKT/GSK3 pathway was activated. Moreover, quinpirole-induced locomotor activation was also more pronounced in TAAR1-KO mice.[22]

More recently, using a TAAR1-KO- and TAAR1-overexpressing rat model, the functional link between TAAR1 and D2R has been analyzed.[21] Coimmunoprecipitation experiments supported the idea of a functional interaction between the two receptors in both the heterologous cell system and in brain tissue. The authors also demonstrated, for the first time, that TAAR1 signals via β-arrestin2 recruitment shifting TAAR1 signaling from cAMP accumulation towards β-arrestin2 signaling.

Using rhesus monkey TAAR1 sequence, it has also been shown that TAAR1 seems to be able to bind many monoamine transporters, that is, serotonin transporter, dopamine transporter (DAT), and norepinephrine transporter.[15,17,30] Coexpression studies showed the presence of TAAR1 and DAT in a subset of SN neurons, while synaptosomal preparation revealed that TAAR1 can modulate DAT functions. It has been hypothesized that TAAR1 interaction with these transporters might provide a mechanism by which TAAR1 ligands can enter the cytoplasm and bind TAAR1 in intracellular compartments. However, TAAR1 agonist was not able to influence DAT functions in FSCV slices experiments.[28]

TAAR1 seems to regulate important functions also outside the central nervous system. In a very recent article, Raab and colleagues showed TAAR1 expression in human pancreatic islets, duodenum and jejunum, and pylorus of the stomach where it acts on gastrointestinal and pancreatic islet hormone secretion. Thus TAAR1 qualifies also as a novel and promising target for the treatment of type 2 diabetes and obesity.[9]

TAAR1 and its Ligands

TAs are endogenous compounds found at low levels in the brain and include molecules such as β-phenylethylamine (PEA), p-tyramine, octopamine, synephrine, and tryptamine. Structurally related to classic monoamines, for many years they have been considered as side products with little physiological significance. However, with the cloning of TAARs, their role has been reconsidered. TAs are not selective for TAAR1 and interact with other important biological targets such as monoamine transporters, even if at high, nonphysiological concentrations. As demonstrated by many laboratories under different experimental conditions, TAAR1 can be activated by PEA and p-tyramine, with PEA being more potent against mouse and human TAAR1 and p-tyramine being more potent against rat

TAAR1.[1,2,13,15,18,19,31–33] A relatively weaker activity of octopamine and tryptamine at the rat and human TAAR1 has been described.

TAAR1 has low affinity for classic monoamine neurotransmitters such as dopamine, serotonin, and norepinephrine but high affinity for adrenergic drugs, apomorphine, the dopamine metabolite 3-methoxytyramine (3-MT), ractopamine (RAC), and thyroid derivatives such as thyronamine.[2,20,34–36] One interesting class of compounds that activates TAAR1 is amphetamines.[2] This evidence suggested that part of the physiological actions of these compounds may be mediated by TAAR1 and many studies focused their attention on the evaluation of TAAR1 as a novel target for drug abuse disorders (see section below).

Other interesting TAAR1 ligands are thyronamines, compounds structurally related to thyroid hormones.[35,37,38] 3-Iodothyronine (T1AM) and its deiodinated relative thyronamine (T0AM) are potent full agonists of human, rat, and mouse TAAR1, and when administered to rats, they induce profound physiological effects such as hypothermia, alteration of metabolism, cardiac effects, and behavioral suppression. However, these compounds are not selective and can interact, for example, also with vesicular monoamine transporter.

3-MT, 4-MT, normetanephrine, and metanephrine can exert potent TAAR1 agonistic activity. Sotnikova et al.[20] demonstrated that 3-MT can induce behavioral effects in a dopamine-independent manner and these effects are partially mediated by TAAR1, suggesting a possible role for TAAR1 in Parkinson's disease (PD) (see section below).

A recent intriguing study also suggested that the food additive RAC used to feed livestock in the United States is a full agonist of TAAR1. Using human cystic fibrosis transmembrane conductance regulator (hCFTR) chloride channels as a sensor for intracellular cAMP, they found that RAC and p-tyramine produced concentration-dependent increases in chloride conductance in oocytes coexpressing hCFTR and mouse TAAR1, which was completely reversed by the TAAR1 selective antagonist EPPTB.[39]

An important step in understanding TAAR1 physiology came from the discovery of the first selective TAAR1 ligands.[25,26] Several studies have been performed with these compounds in various preclinical animal models and they all supported the idea that TAAR1 may represent a novel target for treating psychiatric disorders such as schizophrenia, bipolar disorder, and addiction. Since only one antagonist was described, and with poor solubility and poor brain–blood barrier penetration,[25] different groups worked to understand the molecular determinants responsible for ligand–receptor interaction in order to advance the pharmacological innovation.[40–43] These studies could help in the drug discovery process to find new TAAR1 ligands, such as a systemically available antagonist that would improve the comprehension of TAAR1 physiology.

TAAR1 and Psychiatric Disorders

The fact that TAs, and in particular PEA, can activate TAAR1 raised immediately the idea that TAAR1 could be involved in the pathophysiology of psychiatric disorders. Many studies in the last 50 years linked the dysregulation of TA levels to several human disorders, such as depression, schizophrenia, PD, migraine, and ADHD.[44–47] For example, increased PEA plasma levels were found in schizophrenic patients and increased PEA urinary excretion in paranoid schizophrenics.[48,49] Moreover, altered PEA levels have been linked to the pathogenesis of depression with the so-called "PEA hypothesis," suggesting that a decreased level of endogenous PEA may be responsible for the etiology of depression, while an excess of PEA could cause manic episodes.[50,51] Interestingly, monoamine oxidase (MAO) inhibitors have been largely used as antidepressant drugs and it is believed that part of this action may be due to an increase in TA levels, since MAO enzymes are responsible for TA degradation in the brain. In fact, in MAO B-KO mice the levels of PEA are largely increased compared to WT animals and these mice display a phenotype resembling normal animals treated with antidepressant drugs. Intriguingly, dopamine levels in these mice are unaffected.[52]

While interesting connections exist between TAAR1 and psychiatric diseases, up to now a direct causal linkage is missing, in part due to poor genetic studies and in part to the complexity and poor knowledge of the etiology of these disorders. The only genetic association that has been found refers to another TAAR, TAAR6, and mutations of this receptor have been studied in response to antidepressant treatment and suicidal behavior.[53] Moreover, a single nucleotide polymorphism in the TAAR6 gene has been associated with schizophrenia.[54] However, more studies are needed, especially in the view of recent interesting evidence on the involvement of TAAR1 in the pathophysiology of addiction.[55]

Until the recent development of TAAR1 selective ligands, the first experimental evidence of an involvement of TAAR1 in brain physiology came from the study of the mouse line lacking this receptor. Three independent groups generated TAAR1-KO mice and all of them reported a substantially similar phenotype as regards to a supersensitive dopaminergic system and to monoamine-related dysregulation.[14,18,33] In fact, TAAR1-KO animals do not display an overt phenotype when tested for the most common behavioral and neurological assays, but they start displaying a difference when challenged with dopaminergic drugs. Amphetamine, when injected acutely i.p., is able to produce an enhanced response in TAAR1-KO mice compared to control animals, increasing both locomotor activity and dopamine release in the striatum.[18,33] 3,4-Methylenedioxymethamphetamine (MDMA), another amphetamine-related compound, produces a similar effect on locomotion and dopamine release. In addition, MDMA can also increase serotonin release in both dorsal and ventral striatum.[14] Since the ability of many amphetamines to activate TAAR1 and the property of TAAR1 to modulate the dopaminergic system, TAAR1 has also been studied as a new target for addiction (see the section below). The first study on TAAR1-KO mice described this novel mouse line as a model resembling certain aspects of schizophrenia.[33] The enhanced sensitivity to amphetamine is connected to the positive symptomatology and also mimics the altered response to this drug in patients. Another feature that links TAAR1 to schizophrenia is the D2 dopamine receptor supersensitivity.[56] In TAAR1-KO mice, a higher proportion of D2 dopamine receptor in the high-affinity states has been found.[33] A recent study also show that the D2-related signaling, in particular the G-protein-independent/β-arrestin2-dependent pathway is dysregulated in TAAR1-KO mice, indicating that D2 dopamine receptors in the striatum of TAAR1-KO mice are more sensitive to both endogenous dopamine and to the stimulation with a D2 dopamine receptor agonist.[22] Moreover, TAAR1-KO mice showed a deficit in prepulse inhibition test, indicating an impairment in sensorimotor gating that is known to be deficient in schizophrenic patients.[33]

Since all the TAs that can bind TAAR1 are not selective but can bind other receptors, the recent development of the first selective TAAR1 ligands has been of extreme importance to better understanding TAAR1 physiology. The availability of both full and partial TAAR1 agonists provided the first opportunity to study the consequence of TAAR1 activation in experimental animal models. As the absence of TAAR1 led to a supersensitive dopaminergic system, the activation of TAAR1 is able to reduce an excess of dopaminergic activation, obtained in both pharmacological or genetic animal models.[57] The development of both full and partial agonists and their use in several experimental paradigms, from in vitro assays to animal model of diseased states, also gave important information on the difference between these compounds. As it happened for other neurotransmitters, the TAAR1 partial agonist, in some conditions, behaves as an antagonist. For example, it increases the firing frequency in VTA dopaminergic neurons, while the full agonist decreases it.[26,27] This evidence suggests that there exists a tonic TAAR1 activation at resting conditions, which may be due to the basal endogenous TAs levels. In behavioral experiments, both full and partial agonists share similar properties, even if some differences exist. It will be interesting to study their detailed in vivo pharmacology in order to better understand TAAR1 physiology and in particular TAAR1 modulation of monoamine systems.

As described before, TAAR1 can also influence the serotonergic system, at least by modulation the firing frequency and the 5-HT1A receptor activity in the DR.[26] Serotonin, and in particular 5-HT1A autoreceptors, is important in the regulation of mood, cognition, and for its antidepressant response.[58–60] Thus, since their ability to modulate both dopaminergic and serotonergic system, TAAR1 agonists have been proposed as possible drugs for the treatment of different disorders, including schizophrenia, bipolar disorders, depression, and drug abuse.

In preclinical animal models, TAAR1 activation can reduce the hyperlocomotion induced by the DAT inhibitor cocaine or by the N-methyl-D-aspartate (NMDA) receptor antagonists L-687,414 and phencyclidine (PCP).[26,27] In the same way, TAAR1 agonists reduce the hyperlocomotion naturally present in DAT-KO mice and in NR1-knockdown mice, two models of hyperdopaminergia and hypoglutamatergia, that recapitulate some schizophrenia endophenotypes (in this case the positive symptoms) and represent the two main hypothesis of the etiology of this disease. Interestingly, both partial and full agonists showed an addictive effect in these behavioral paradigms if used in combination with two atypical antipsychotics, olanzapine and risperidone.[61] This suggests that TAAR1 agonists could be used also as add-on treatment to current antipsychotics.

Regarding side effects, TAAR1 agonists do not show extrapyramidal side effects and the partial agonist can reduce the haloperidol-induced catalepsy,[61] as seen in TAAR1-KO animals.[24] This effect also demonstrates that the partial TAAR1 activation can led to a dual effect. When the dopamine system is overactivated, such as in the case of cocaine treatment, it may act as an agonist to reduce the excess of dopaminergic activation. When there is a deficiency in the dopamine transmission, such as in the case of D2 receptor blockade with haloperidol treatment, it may act as an antagonist to increase the dopaminergic tone.

Regarding the TAAR1 influence on serotonergic system, the potential use of the full and partial agonists as antidepressants and anxiolytics has been tested. In rats, only the partial agonist was effective in the forced swim test, while in monkeys both the partial and full agonists showed antidepressant activity in the differential reinforcement of low-rate behavior paradigm.[27,61] Moreover, TAAR1 activation was shown to induce anxiolytic-like behaviors in the stress-induced hyperthermia assay, further suggesting a potential role of TAAR1 ligands for mood disorders.[27]

A study using phMRI revealed that both full and partial agonists activate several brain regions with a pattern similar to olanzapine, although with some differences, including the prefrontal area, suggesting a possible role of TAAR1 in cognition.[61] The first report about this aspect showed that TAAR1 agonists could improve cognitive performances in the object retrieval paradigm in monkeys, increasing the percentage of correct responses. Moreover, in rats, TAAR1 agonists could revert the deficit induced by PCP administration in the attentional set-shifting test.[61] In our lab, we demonstrated that TAAR1 is expressed in pyramidal neurons of layer V of prefrontal cortex (PFC).[62] In TAAR1-KO mice, these glutamatergic neurons displayed a deficient functionality of NMDA-mediated current and an altered subunit composition of NMDA receptors. Behaviorally, TAAR1-KO mice presented aberrant cognitive behaviors, including an impulsive and perseverative phenotype.[62] These data indicate that TAAR1 plays an important role in the modulation of NMDA receptor-mediated glutamate transmission in the PFC and in certain aspects of cognition. Interestingly, impulsivity is a key feature of the addiction process, and TAAR1 agonists have been found to be effective in different models of addiction (see section below). It should be noted that more studies are necessary to understand the precise domain of cognition that TAAR1 can influence, in order to focus TAAR1 agonists in animal models of disorders with a cognitive symptomatology.

TAAR1 and Addiction

When the TAAR family was cloned in 2001, among the agonists found to activate TAAR1, one interesting class of compounds was amphetamines.[2] This evidence immediately led to the idea that some physiological actions of these addictive compounds could be mediated by TAAR1. As described above, TAAR1-KO mice have a supersensitive dopaminergic system, while TAAR1 activation can reduce an excess of dopamine transmission. Altogether, it is now accepted that TAAR1 can represent a "brake" in amphetamine actions, like an endogenous negative feedback, since amphetamines, by activating TAAR1, reduce their own activity on the dopamine system. In fact, it has been demonstrated with amphetamine, MDMA and methamphetamine that all these compounds were more active in TAAR-KO animals compared to WT animals.[14,18,33] Achat-Mendes et al.[63] demonstrated that the conditioned place preference (CPP) induced by methamphetamine was acquired earlier in TAAR1-KO mice and retained longer, as evaluated by extinction training. An interesting study showed that a mouse line obtained by selective breeding that displayed a high voluntary consumption of methamphetamine carried a nonfunctional allele for TAAR1.[55] These animals behave like TAAR1-KO mice as regards to their increase sensitivity to the rewarding effect of methamphetamine. Moreover, TAAR1-KO mice were also shown to be more sensitive to the rewarding effect of ethanol and to its sedative effects.[64]

This evidence suggests a potential role of TAAR1 ligands as a possible treatment for addiction for different drugs of abuse. Addictive drugs can act in different ways, but all of them seem to share the ability to enhance the dopaminergic mesolimbic neurotransmission.[65] There are several ways to modulate dopamine transmission, either influencing the neuronal firing of dopaminergic neurons, interfering with dopamine reuptake through DAT, or altering the presynaptic regulation at the level of terminals.[65] As described earlier, TAAR1 seems to affect all these processes, thus representing an intriguing novel target for addiction.

Some interesting studies explored the potential use of TAAR1 selective ligands for the treatment of cocaine addiction. The first evidence came from Revel et al.[27] that demonstrated that the administration of a TAAR1 partial agonist reduced the cocaine intake in rats with a history of cocaine self-administration. Two other groups, independently, studied TAAR1 agonists in cocaine-abuse-related effect in rats. Both full and partial agonists were able to reduce the context-induced renewal of drug seeking, with no alteration in the lever-pressing task maintained by food.[66] Similarly, the partial agonist was able to completely block the cocaine-primed reinstatement of cocaine seeking. Both these two behavioral assays are well validated models of drug relapse, an important and not well managed aspect of addiction. The same group confirmed the ability of TAAR1 partial and full agonists to reduce cocaine self-administration and expanded this evidence to intracranial self-stimulation.[67] Regarding the mechanism, it has been shown that TAAR1 activation can reduce the dopamine release induced by cocaine in the NAc without altering DAT function but involving other mechanisms, such as D2 presynaptic receptors.[66] In another study, a TAAR1 partial agonist was shown to be effective in reducing cocaine sensitization and the expression of CPP.[68,69] However, it was not able to modify the development of the CPP. Altogether, these data demonstrated that TAAR1 activation reduces the sensitizing, reinforcing and rewarding effect of cocaine in experimental animal models, hoping that TAAR1 selective ligands could be used for future clinical trials to have a proof of concept in human drug addicts.

TAAR1 and Parkinson's Disease

One recent study tested the hypothesis that TAAR1 ligands could be beneficial in Parkinson's disease.[70] PD is a neurodegenerative disorder characterized by the loss of the dopaminergic neurons of the SN. All the approved treatments affect the dopamine system in different ways, with the dopamine precursor L-DOPA being the most used. For this reason, TAAR1 has been thought to be a possible target also for this disorder.[71,72] Alvarsson and colleagues showed that TAAR1-KO mice with a unilateral lesion with 6-OH-DA displayed an enhanced rotational behavior as well as L-DOPA-induced dyskinesia (LID). Conversely, the subchronic treatment with a full TAAR1 agonist reduced both the L-DOPA-induced rotational behavior and LID. This behavioral evidence was accompanied by an opposite effect on the phosphorylation of the subunit GluA1 of the AMPA receptor, known to be a biochemical indicator of LID. Interestingly, TAAR1 was able to reduce the evoked glutamate release in striatum (that was enhanced in two models of PD) with a mechanism that may be involving presynaptic D2 dopamine receptors.[70]

Another evidence for the possible involvement of TAAR1 in pathophysiology of PD came from the first report on TAAR1 where among the ligands found to activate TAAR1 was 3-MT.[2,19,73] 3-MT is a dopamine metabolite produced by the activity of the catechol-o-methyltransferase (COMT) and traditionally considered an inactive metabolite with no biological activity and which was used as a reflection of dopamine extracellular levels.[74] However, there were already studies that showed an increase in 3-MT levels in PD patients developing dyskinesia after chronic L-DOPA treatment, both in the brain and in urine.[75–77] In our lab, we demonstrated that 3-MT can induce behavioral effects in a dopamine-independent manner and these effects were partially mediated by TAAR1.[20] In normal mice, the central administration of 3-MT caused a temporary mild hyperactivity with a concomitant set of abnormal movements. Furthermore, 3-MT induced significant ERK and CREB phosphorylation in the mouse striatum, in a TAAR1-dependent manner.[20] These data suggest a possible role for 3-MT in L-DOPA-induced side effects and indicate COMT as the rate-limiting step for the production of an active chemical active on TAAR1.

CONCLUSION

The discovery of TAAR1 receptor attracted the attention of the research community and many studies have been performed to comprehend its physiology. The recent development of selective TAAR1 ligands and knockout mouse line were essential tools to understand the ability of TAAR1 in regulating the monoaminergic signaling. The TAAR1 ability to act as a "regulator" of dopamine tone opens new prospects in targeting psychiatric disorders via TAAR1 modulation but a proof of concept in human physiology is still needed. The emerging role of TAAR1 in cognition-related functions could extend the interest of this receptor as a new target for several human pathologies.

References

1. Borowsky B, Adham N, Jones KA, et al. Trace amines: identification of a family of mammalian G protein-coupled receptors. *Proc Natl Acad Sci USA* 2001;**98**(16):8966–71.
2. Bunzow JR, Sonders MS, Arttamangkul S, et al. Amphetamine, 3,4-methylenedioxymethamphetamine, lysergic acid diethylamide, and metabolites of the catecholamine neurotransmitters are agonists of a rat trace amine receptor. *Mol Pharmacol* 2001;**60**(6):1181–8.
3. Lindemann L, Ebeling M, Kratochwil NA, Bunzow JR, Grandy DK, Hoener MC. Trace amine-associated receptors form structurally and functionally distinct subfamilies of novel G protein-coupled receptors. *Genomics* 2005;**85**(3):372–85.
4. Lindemann L, Hoener MC. A renaissance in trace amines inspired by a novel GPCR family. *Trends Pharmacol Sci* 2005;**26**(5):274–81.
5. Zucchi R, Chiellini G, Scanlan TS, Grandy DK. Trace amine-associated receptors and their ligands. *Br J Pharmacol* 2006;**149**(8):967–78.
6. Dewan A, Pacifico R, Zhan R, Rinberg D, Bozza T. Non-redundant coding of aversive odours in the main olfactory pathway. *Nature* 2013;**497**(7450):486–9.
7. Liberles SD, Buck LB. A second class of chemosensory receptors in the olfactory epithelium. *Nature* 2006;**442**(7103):645–50.
8. Zhang J, Pacifico R, Cawley D, Feinstein P, Bozza T. Ultrasensitive detection of amines by a trace amine-associated receptor. *J Neurosci* 2013;**33**(7):3228–39.
9. Raab S, Wang H, Uhles S, et al. Incretin-like effects of small molecule trace amine-associated receptor 1 agonists. *Mol Metab* 2016;**5**(1):47–56.
10. Grandy DK. Trace amine-associated receptor 1-family archetype or iconoclast? *Pharmacol Ther* 2007;**116**(3):355–90.
11. Babusyte A, Kotthoff M, Fiedler J, Krautwurst D. Biogenic amines activate blood leukocytes via trace amine-associated receptors TAAR1 and TAAR2. *J Leukoc Biol* 2013;**93**(3):387–94.
12. Panas MW, Xie Z, Panas HN, Hoener MC, Vallender EJ, Miller GM. Trace amine associated receptor 1 signaling in activated lymphocytes. *J Neuroimmune Pharmacol* 2012;**7**(4):866–76.
13. Wasik AM, Millan MJ, Scanlan T, Barnes NM, Gordon J. Evidence for functional trace amine associated receptor-1 in normal and malignant B cells. *Leuk Res* 2012;**36**(2):245–9.
14. Di Cara B, Maggio R, Aloisi G, et al. Genetic deletion of trace amine 1 receptors reveals their role in auto-inhibiting the actions of ecstasy (MDMA). *J Neurosci* 2011;**31**(47):16928–40.

15. Miller GM, Verrico CD, Jassen A, et al. Primate trace amine receptor 1 modulation by the dopamine transporter. *J Pharmacol Exp Ther* 2005;**313**(3):983–94.

16. Xie Z, Westmoreland SV, Bahn ME, et al. Rhesus monkey trace amine-associated receptor 1 signaling: enhancement by monoamine transporters and attenuation by the D2 autoreceptor in vitro. *J Pharmacol Exp Ther* 2007;**321**(1):116–27.

17. Xie Z, Miller GM. Trace amine-associated receptor 1 is a modulator of the dopamine transporter. *J Pharmacol Exp Ther* 2007;**321**(1):128–36.

18. Lindemann L, Meyer CA, Jeanneau K, et al. Trace amine-associated receptor 1 modulates dopaminergic activity. *J Pharmacol Exp Ther* 2008;**324**(3):948–56.

19. Barak LS, Salahpour A, Zhang X, et al. Pharmacological characterization of membrane-expressed human trace amine-associated receptor 1 (TAAR1) by a bioluminescence resonance energy transfer cAMP biosensor. *Mol Pharmacol* 2008;**74**(3):585–94.

20. Sotnikova TD, Beaulieu J-M, Espinoza S, et al. The dopamine metabolite 3-methoxytyramine is a neuromodulator. *PLoS One* 2010;**5**(10):e13452.

21. Harmeier A, Obermueller S, Meyer CA, et al. Trace amine-associated receptor 1 activation silences GSK3β signaling of TAAR1 and D2R heteromers. *Eur Neuropsychopharmacol* 2015;**25**(11):2049–61.

22. Espinoza S, Ghisi V, Emanuele M, et al. Postsynaptic D2 dopamine receptor supersensitivity in the striatum of mice lacking TAAR1. *Neuropharmacology* 2015;**93**:308–13.

23. Miller GM. The emerging role of trace amine-associated receptor 1 in the functional regulation of monoamine transporters and dopaminergic activity. *J Neurochem* 2011;**116**(2):164–76.

24. Espinoza S, Salahpour A, Masri B, et al. Functional interaction between trace amine-associated receptor 1 and dopamine D2 receptor. *Mol Pharmacol* 2011;**80**(3):416–25.

25. Bradaia A, Trube G, Stalder H, et al. The selective antagonist EPPTB reveals TAAR1-mediated regulatory mechanisms in dopaminergic neurons of the mesolimbic system. *Proc Natl Acad Sci USA* 2009;**106**(47): 20081–6.

26. Revel FG, Moreau J-L, Gainetdinov RR, et al. TAAR1 activation modulates monoaminergic neurotransmission, preventing hyperdopaminergic and hypoglutamatergic activity. *Proc Natl Acad Sci USA* 2011;**108**(20):8485–90.

27. Revel FG, Moreau J-L, Gainetdinov RR, et al. Trace amine-associated receptor 1 partial agonism reveals novel paradigm for neuropsychiatric therapeutics. *Biol Psychiatry* 2012;**72**(11):934–42.

28. Leo D, Mus L, Espinoza S, Hoener MC, Sotnikova TD, Gainetdinov RR. Taar1-mediated modulation of presynaptic dopaminergic neurotransmission: role of D2 dopamine autoreceptors. *Neuropharmacology* 2014;**81**:283–91.

29. Espinoza S, Masri B, Salahpour A, Gainetdinov RR. BRET approaches to characterize dopamine and TAAR1 receptor pharmacology and signaling. *Methods Mol Biol Clifton NJ* 2013;**964**:107–22.

30. Xie Z, Miller GM. A receptor mechanism for methamphetamine action in dopamine transporter regulation in brain. *J Pharmacol Exp Ther* 2009;**330**(1):316–25.

31. Navarro HA, Gilmour BP, Lewin AH. A rapid functional assay for the human trace amine-associated receptor 1 based on the mobilization of internal calcium. *J Biomol Screen* 2006;**11**(6):688–93.

32. Reese EA, Bunzow JR, Arttamangkul S, Sonders MS, Grandy DK. Trace amine-associated receptor 1 displays species-dependent stereoselectivity for isomers of methamphetamine, amphetamine, and para-hydroxyamphetamine. *J Pharmacol Exp Ther* 2007;**321**(1):178–86.

33. Wolinsky TD, Swanson CJ, Smith KE, et al. The trace amine 1 receptor knockout mouse: an animal model with relevance to schizophrenia. *Genes Brain Behav* 2007;**6**(7):628–39.

34. Liu IS, Kusumi I, Ulpian C, Tallerico T, Seeman P. A serotonin-4 receptor-like pseudogene in humans. *Brain Res Mol Brain Res* 1998;**53**(1–2):98–103.

35. Scanlan TS, Suchland KL, Hart ME, et al. 3-Iodothyronamine is an endogenous and rapid-acting derivative of thyroid hormone. *Nat Med* 2004;**10**(6):638–42.

36. Sukhanov I, Espinoza S, Yakovlev DS, Hoener MC, Sotnikova TD, Gainetdinov RR. TAAR1-dependent effects of apomorphine in mice. *Int J Neuropsychopharmacol* 2014;**17**(10):1683–93.

37. Chiellini G, Nesi G, Digiacomo M, et al. Design, synthesis, and evaluation of thyronamine analogues as novel potent mouse trace amine associated receptor 1 (mTAAR1) agonists. *J Med Chem* 2015;**58**(12):5096–107.

38. Hart ME, Suchland KL, Miyakawa M, Bunzow JR, Grandy DK, Scanlan TS. Trace amine-associated receptor agonists: synthesis and evaluation of thyronamines and related analogues. *J Med Chem* 2006;**49**(3):1101–12.

39. Liu X, Grandy DK, Janowsky A. Ractopamine, a livestock feed additive, is a full agonist at trace amine-associated receptor 1. *J Pharmacol Exp Ther* 2014;**350**(1):124–9.

40. Cichero E, Espinoza S, Franchini S, et al. Further insights into the pharmacology of the human trace amine-associated receptors: discovery of novel ligands for TAAR1 by a virtual screening approach. *Chem Biol Drug Des* 2014;**84**(6):712–20.

41. Cichero E, Espinoza S, Gainetdinov RR, Brasili L, Fossa P. Insights into the structure and pharmacology of the human trace amine-associated receptor 1 (hTAAR1): homology modelling and docking studies. *Chem Biol Drug Des* 2013;**81**(4):509–16.

42. Reese EA, Norimatsu Y, Grandy MS, Suchland KL, Bunzow JR, Grandy DK. Exploring the determinants of trace amine-associated receptor 1's functional selectivity for the stereoisomers of amphetamine and methamphetamine. *J Med Chem* 2014;**57**(2):378–90.

43. Tallman KR, Grandy DK. A decade of pharma discovery delivers new tools targeting trace amine-associated receptor 1. *Neuropsychopharmacology* 2012;**37**(12):2553–4.

44. Berry MD. Mammalian central nervous system trace amines. Pharmacologic amphetamines, physiologic neuromodulators. *J Neurochem* 2004;**90**(2):257–71.

45. Branchek TA, Blackburn TP. Trace amine receptors as targets for novel therapeutics: legend, myth and fact. *Curr Opin Pharmacol* 2003;**3**(1):90–7.

46. D'Andrea G, Terrazzino S, Fortin D, Cocco P, Balbi T, Leon A. Elusive amines and primary headaches: historical background and prospectives. *Neurol Sci* 2003;**24**(Suppl. 2):S65–7.

47. Sandler M, Reynolds GP. Does phenylethylamine cause schizophrenia? *Lancet Lond Engl* 1976;**1**(7950):70–1.

48. Potkin SG, Karoum F, Chuang LW, Cannon-Spoor HE, Phillips I, Wyatt RJ. Phenylethylamine in paranoid chronic schizophrenia. *Science* 1979;**206**(4417):470–1.

49. Shirkande S, O'Reilly R, Davis B, Durden D, Malcom D. Plasma phenylethylamine levels of schizophrenic patients. *Can J Psychiatry* 1995;**40**(4):221.

50. Davis BA, Boulton AA. The trace amines and their acidic metabolites in depression—an overview. *Prog Neuropsychopharmacol Biol Psychiatry* 1994;**18**(1):17–45.

51. Sabelli HC, Mosnaim AD. Phenylethylamine hypothesis of affective behavior. *Am J Psychiatry* 1974;**131**(6):695–9.

52. Grimsby J, Toth M, Chen K, et al. Increased stress response and beta-phenylethylamine in MAOB-deficient mice. *Nat Genet* 1997;**17**(2):206–10.

53. Pae C-U, Drago A, Kim J-J, et al. TAAR6 variations possibly associated with antidepressant response and suicidal behavior. *Psychiatry Res* 2010;**180**(1):20–4.

54. Duan J, Martinez M, Sanders AR, et al. Polymorphisms in the trace amine receptor 4 (TRAR4) gene on chromosome 6q23.2 are associated with susceptibility to schizophrenia. *Am J Hum Genet* 2004;**75**(4):624–38.

55. Harkness JH, Shi X, Janowsky A, Phillips TJ. Trace amine-associated receptor 1 regulation of methamphetamine intake and related traits. *Neuropsychopharmacology* 2015;**40**(9):2175–84.

56. Seeman P, Weinshenker D, Quirion R, et al. Dopamine supersensitivity correlates with D2High states, implying many paths to psychosis. *Proc Natl Acad Sci USA* 2005;**102**(9):3513–8.

57. Espinoza S, Gainetdinov RR. Neuronal functions and emerging pharmacology of TAAR1 *Topics in medicinal chemistry*. Berlin Heidelberg: Springer; 2014;1–20 <http://link.springer.com/chapter/10.1007/7355_2014_78> [accessed 17.11.15].

58. Barnes NM, Sharp T. A review of central 5-HT receptors and their function. *Neuropharmacology* 1999;**38**(8):1083–152.

59. Gardier AM, Malagié I, Trillat AC, Jacquot C, Artigas F. Role of 5-HT1A autoreceptors in the mechanism of action of serotoninergic antidepressant drugs: recent findings from in vivo microdialysis studies. *Fundam Clin Pharmacol* 1996;**10**(1):16–27.

60. Millan MJ. Improving the treatment of schizophrenia: focus on serotonin (5-HT)(1A) receptors. *J Pharmacol Exp Ther* 2000;**295**(3):853–61.

61. Revel FG, Moreau J-L, Pouzet B, et al. A new perspective for schizophrenia: TAAR1 agonists reveal antipsychotic- and antidepressant-like activity, improve cognition and control body weight. *Mol Psychiatry* 2013;**18**(5):543–56.

62. Espinoza S, Lignani G, Caffino L, et al. TAAR1 modulates cortical glutamate NMDA receptor function. *Neuropsychopharmacology* 2015;**40**(9):2217–27.

63. Achat-Mendes C, Lynch LJ, Sullivan KA, Vallender EJ, Miller GM. Augmentation of methamphetamine-induced behaviors in transgenic mice lacking the trace amine-associated receptor 1. *Pharmacol Biochem Behav* 2012;**101**(2):201–7.

64. Lynch LJ, Sullivan KA, Vallender EJ, Rowlett JK, Platt DM, Miller GM. Trace amine associated receptor 1 modulates behavioral effects of ethanol. *Subst Abuse Res Treat* 2013;**7**:117–26.

65. Sulzer D. How addictive drugs disrupt presynaptic dopamine neurotransmission. *Neuron* 2011;**69**(4):628–49.

66. Pei Y, Lee J, Leo D, Gainetdinov RR, Hoener MC, Canales JJ. Activation of the trace amine-associated receptor 1 prevents relapse to cocaine seeking. *Neuropsychopharmacology* 2014;**39**(10):2299–308.

67. Pei Y, Mortas P, Hoener MC, Canales JJ. Selective activation of the trace amine-associated receptor 1 decreases cocaine's reinforcing efficacy and prevents cocaine-induced changes in brain reward thresholds. *Prog Neuropsychopharmacol Biol Psychiatry* 2015;**63**:70–5.

68. Thorn DA, Jing L, Qiu Y, et al. Effects of the trace amine-associated receptor 1 agonist RO5263397 on abuse-related effects of cocaine in rats. *Neuropsychopharmacology* 2014;**39**(10):2309–16.

69. Thorn DA, Zhang C, Zhang Y, Li J-X. The trace amine associated receptor 1 agonist RO5263397 attenuates the induction of cocaine behavioral sensitization in rats. *Neurosci Lett* 2014;**566**:67–71.

70. Alvarsson A, Zhang X, Stan TL, et al. Modulation by trace amine-associated receptor 1 of experimental parkinsonism, L-DOPA responsivity, and glutamatergic neurotransmission. *J Neurosci* 2015;**35**(41):14057–69.

71. Sotnikova TD, Caron MG, Gainetdinov RR. Trace amine-associated receptors as emerging therapeutic targets. *Mol Pharmacol* 2009;**76**(2):229–35.

72. Sotnikova TD, Zorina OI, Ghisi V, Caron MG, Gainetdinov RR. Trace amine associated receptor 1 and movement control. *Parkinsonism Relat Disord* 2008;**14**(Suppl. 2):S99–S102.

73. Wainscott DB, Little SP, Yin T, et al. Pharmacologic characterization of the cloned human trace amine-associated receptor1 (TAAR1) and evidence for species differences with the rat TAAR1. *J Pharmacol Exp Ther* 2007;**320**(1):475–85.

74. Kehr W. 3-Methoxytyramine as an indicator of impulse-induced dopamine release in rat brain in vivo. *Naunyn Schmiedebergs Arch Pharmacol* 1976;**293**(3):209–15.

75. Muskiet FA, Thomasson CG, Gerding AM, Fremouw-Ottevangers DC, Nagel GT, Wolthers BG. Determination of catecholamines and their 3-O-methylated metabolites in urine by mass fragmentography with use of deuterated internal standards. *Clin Chem* 1979;**25**(3):453–60.

76. Rajput AH, Fenton ME, Di Paolo T, Sitte H, Pifl C, Hornykiewicz O. Human brain dopamine metabolism in levodopa-induced dyskinesia and wearing-off. *Parkinsonism Relat Disord* 2004;**10**(4):221–6.

77. Siirtola T, Sonninen V, Rinne UK. Urinary excretion of monoamines and their metabolites in patients with Parkinson's disease. Response to long-term treatment with levodopa alone or in combination with a dopa decarboxylase inhibitor and clinical correlations. *Clin Neurol Neurosurg* 1975;**78**(2):77–88.

10

Trace Amines as Intrinsic Monoaminergic Modulators of Spinal Cord Functional Systems

S. Hochman and E.A. Gozal

Department of Physiology, Emory University School of Medicine,
Atlanta, GA, United States

O U T L I N E

INTRODUCTION

The trace amines (TAs) comprise a class of neuroactive monoamines that are synthesized from the same precursor amino acids and essential synthesis enzyme as the classical monoamine modulatory transmitters. This chapter reappraises their role in relation to recent findings on their unique motor facilitatory actions,[1] and is divided into four sections. (1) We first provide an overview of the TAs. Their detectability in trace amounts due to lack of storage led not only to their name, but also to an expression lability that made it difficult to ascribe a role in CNS function but easy to dismiss as metabolic byproducts. The 2001 discovery of G-protein-coupled trace amine-associated receptors (TAARs) preferentially activated by TAs established mechanisms by which TAs can produce effects of their own,[2,3] and inspired a new wave of investigation that placed the TAAR1 receptor as a component of brain monoaminergic signaling. Still, without identification of discrete CNS trace aminergic neuronal circuits, a role for the TAs in CNS modulation remained uncertain. (2) A population of neurons intrinsic to spinal cord expresses the TA synthesis enzyme aromatic-L-amino-acid decarboxylase (AADC) but not monoamines, and may represent a specific TA-ergic neuronal system.[4] This section describes our recent results in the isolated neonatal rat spinal cord showing that the TAs can facilitate expression of spinal pattern generating circuits independent of, but with comparable ability to, the major descending monoaminergic pathways.[1] Experimental results support a role for the TAs as a distinct class of intrinsic spinal monoaminergic neuromodulators acting intracellularly, putatively on TAARs. (3) We then explore the hypothetical functional relevance of the TAs both as a sympathomimetic intracellular metabolic effector, and in tonically setting spinal circuit excitability. (4) Finally, emphasis on the need for additional experiments including in the adult segues into possible neurotherapeutic approaches that could preferentially act on the TA-ergic system to improve motor performance via facilitating expression of locomotor circuits.

The Trace Amines and Their Receptors

A brief description of the classical monoamines and conventional conception of neuromodulation is first provided to compare and contrast to the TAs. The monoamines serotonin (5-HT), norepinephrine (noradrenaline), epinephrine (adrenaline), and dopamine are important CNS neuromodulatory transmitters made from precursor aromatic amino acids in neurons containing their synthesis enzymes. Like all classical transmitters, these monoamines are stored in vesicles and released in response to depolarization. They bind to G-protein-coupled receptors (except 5-HT$_3$) and modulate cellular responsiveness via alterations in signal transduction pathways. After release, they are rapidly sequestered into the presynaptic neurons through specific high-affinity transporters and are degraded by the monoamine oxidases. In spinal cord, as in brain, these neuromodulatory transmitters reconfigure neural networks to express various behavioral "states,"[5,6] one being facilitated expression of the spinal circuitry generating locomotion.

The trace amines (TAs) are structurally, metabolically, physiologically, and pharmacologically similar to classical monoamine neurotransmitters (Fig. 10.1).[7] Like the classical monoamines, the TAs are synthesized via enzymatic decarboxylation of their precursor aromatic amino acids with aromatic-L-amino acid decarboxylase (AADC; also called DOPA decarboxylase) and are metabolized primarily via monoamine oxidases (MAOs).[8] Tyramine, β-phenylethylamine (PEA), and tryptamine are synthesized from tyrosine, phenylalanine, and tryptophan, respectively. Endogenous TA synthesis has been well documented in all vertebrate and invertebrate species studied[9,10] including in the mammalian CNS.[7,11] While a large earlier literature asserted their role as endogenous neuromodulators of

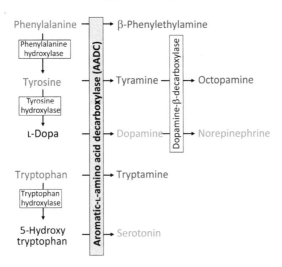

FIGURE 10.1   Comparison of monoamine synthesis pathways. The trace amines (TAs) are a group of endogenous monoamines that include tryptamine, tyramine, octopamine, and β-phenylethylamine (PEA; blue). The TAs have structural, metabolic, physiologic, and pharmacologic similarities to the classical monoamine transmitters (green) and are synthesized from the same precursor aromatic amino acids (red). Unlike the classical monoamines, aromatic-L-amino acid decarboxylase (AADC; also called dopa decarboxylase) is the only enzyme required to produce the TAs tryptamine, tyramine, and PEA. Conversion from the TAs to the monoamines does not appear to occur.

monoaminergic excitability and neurotransmission, the more conventional view of the TAs has been as metabolic byproducts.[10,12]

The TAs have synthesis rates comparable to those of dopamine and norepinephrine.[13,14] As MAO inhibitors lead to their rapid buildup, TAs levels are clearly metabolically regulated.[10] TAs in the CNS are generally not thought to be stored as a reserve pool in vesicles like the classical monoamines. An important consequence is that the TAs have an especially short half-life of approximately 30 s.[14] Consequently, measured TA levels are far below those of the sequestered classical monoamines and thus relatively "trace" in quantity. Nevertheless, the TAs circulate in cerebrospinal fluid at levels comparable to the classical monoamines. They also have a heterogeneous CNS distribution.[13–15] For example, tryptamine concentrations in spinal cord are much higher than in brain.[16–18]

In 2001, G-protein-coupled TAARs were discovered and included receptors preferentially activated by TAs.[2,3] Tyramine and PEA were shown to activate TAAR1, and PEA and tryptamine activated TAAR4, founding a mechanism by which TAs can produce effects of their own.[2] More recent observations using selective TAAR1 pharmacology and TAAR1 knockout mice identified a role for CNS TAAR1 receptors. In brief, TAAR1 activity was shown to depress monoamine transport and spiking in dopaminergic and serotonergic neurons via interactions with their presynaptic D_2 and 5-HT_{1A} autoreceptors, respectively.[19–25] While likely,[22] a demonstrably clear role for TAs in TAAR-mediated actions in intact circuits remains elusive.

Relating a Putative Intrinsic Spinal Cord TA-ergic System to Modulation of Intrinsic Spinal Locomotor Activity

In the 1980s, 16 anatomically distinct clusters of CNS neuronal populations were identified in rat that contain the essential TA synthesis enzyme AADC but no other monoamine synthesis enzymes and were called D cells.[4,26,27] One population, D1, is intrinsic to spinal cord. D1 cells associate with the spinal cord central canal, and ultrastructural identification of synapses and secretory vesicles confirmed D1 cells as neurons.[4] At least one of their processes projects into the lumen of the central canal, which makes them part of a group of cerebrospinal fluid (CSF)-contacting neurons.[27,28] Spinal CSF-contacting nerve cells situated around the central canal are found from cyclostomes to mammals including monkeys.[28,29] That CNS TAAR1 expression was also reported in spinal cord[2] introduced the possibility that D1 cells are an endogenous spinal cord trace aminergic transmitter system.

We first confirmed in neonatal rat that the spinal cord contains the substrates for TA biosynthesis (AADC) and for receptor-mediated actions via TAAR1 and TAAR4, including in cells associated with the central canal.[1] Fig. 10.2 provides examples of the distribution observed for AADC immunolabeling in the neonatal spinal cord. AADC labeling is clearly preferentially in the motor circuitry-containing ventral half of the spinal cord with particularly strong labeling associated with a ventral neuronal stream emanating from the central canal and terminating in a midline white matter tract in the ventral funiculus. Sporadic labeling of other spinal interneurons and weaker labeling of motor neurons were also seen.

After spinal cord injury in the adult rat, AADC-expressing D cells facilitate spinal motor excitability by increasing their expression of monoamines.[30] As a morphologically similar population of neurons activates motor circuits in larval zebrafish,[31] D1 cells may function to monitor CSF-related events and relay the information into a modulatory recruitment of motor circuits. We therefore explored a role for the TAs in the neuromodulation of rat spinal cord motor circuits using the in vitro isolated neonatal rat spinal cord.[1] We examined the actions of the TAs tyramine, tryptamine, and PEA on motor activity and found that tyramine and tryptamine consistently increased motor activity and both had prominent direct actions on motoneurons[1] (also see Ref. 32).

We then examined whether the TAs could play a role in recruitment of neural circuits comprising the central pattern generator (CPG) for locomotion. Locomotor CPGs are intrinsic to spinal cord and present at birth in the rat and mouse.[33] Interneuronal populations responsible for generating coordinated locomotion are preferentially located in the ventromedial gray matter of the caudal thoracic and rostral lumbar spinal cord (Fig. 10.2B).[34–38] When the isolated spinal cord is maintained in vitro and activated by bath-applied neurochemicals, a motor pattern consistent with locomotion emerges and can be measured using suction electrodes on lumbar ventral motor nerve roots that correspond to dominating activity in flexors and extensors (Fig. 10.3).[39–41] When the hindlimbs are left intact, motor coordination is remarkably comparable to that seen in the adult,[39,42–47] albeit at considerably slower frequency.[48] While numerous bath-applied neuroactive substances can generate a patterned motor output consistent with locomotion, a combination of serotonin (5-HT) and N-methyl-D-aspartate (NMDA) is the most widely used "cocktail" to generate a stable and long-lasting locomotor rhythm.[39–41]

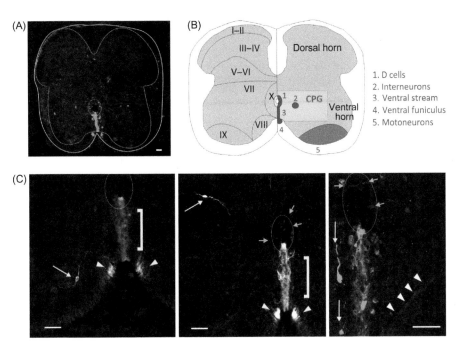

FIGURE 10.2 Distribution of AADC in the neonatal spinal cord. (A) Low power of AADC immunolabeling in spinal cord. Superimposed on this section is an outline of the spinal cord with interior white lines approximately demarcating gray matter. Epithelial cell layer surrounding central canal is outlined. (B) Summary of anatomical organization of lumbar spinal cord in cross-section divided into symmetrical halves. The left half divides the spinal cord into Rexed's laminae. Laminae I–VI comprise anatomically divisible regions of the predominantly sensory-processing dorsal horn while laminae VII–X comprise anatomically divisible regions of the predominantly motor-processing ventral horn. The right half identifies the boxed region where the most prominent locomotor central pattern-generating (CPG) circuits are located. The right half also identifies the primary locations of AADC labeled neural elements. Numbers correspond to elements defined at right. (C) Three panels focusing on AADC labeling of D cells intermingled with epithelial cells surrounding the central canal (blue arrows), ventral stream of AADC labeling projecting ventrally (yellow brackets), and white matter labeling in the ventral funiculus (yellow arrowheads). Some interneurons are also labeled (yellow arrows) and there is also weak AADC immunolabeling in putative motoneurons encircled and in a blood vessel in far right panel (white arrowheads). Scale bars are 50 μm in panels A and C.

In the presence of sublocomotor doses of bath-applied NMDA, we observed that the TAs could lead to the emergence of a locomotor pattern indistinguishable from that ordinarily observed with 5-HT (Fig. 10.4A).[1] These actions could be maintained following block of all Na^+-dependent monoamine transporters or the vesicular monoamine transporter, demonstrating that TA actions can occur independent of interactions with descending monoaminergic projections. Interestingly, we observed that bath-applied TAs always took much longer to recruit locomotor circuits than 5-HT, and this corresponded with the observation that tryptamine and tyramine-evoked locomotor actions depended on intracellular uptake via pentamidine-sensitive Na^+-independent membrane transporters. A need for intracellular

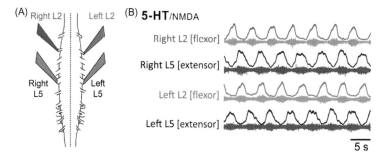

FIGURE 10.3 Experimental setup (A) For in vitro studies on locomotor-like motor rhythms. Suction electrodes are placed on lumbar L2 and L5 ventral roots bilaterally to monitor population motoneuron flexor and extensor activity. Drugs are applied to the bath. (B) Locomotor-like activity (LLA) is recruited with the addition of various neurochemicals with the combination of NMDA and 5-HT being the most common and illustrated here. Activity recorded from L2 ventral roots typically report flexor motoneuron activity while L5 ventral roots predominantly reflect extensor motoneuron activity. Rectified and low-pass filtered (top traces) and raw records (lower traces) are displayed for each recorded root.

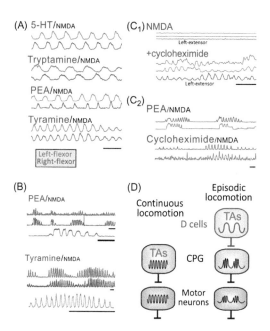

FIGURE 10.4 Example activity patterns and hypothetical network for trace amine-mediated actions. (A) Locomotor rhythms generated in the presence of 5-HT and the TAs. Activity patterns for left (red) and right (blue) flexors. (B) Episodic bouts of locomotor-like rhythms are shown for PEA and tyramine at slow (top two traces) and at an expanded timescale (bottom). (C) Cycloheximide-induced accumulation of intracellular amino acids can lead to the expression of LLA. (C_1) Cycloheximide subsequent to applied NMDA elicits coordinated locomotor-like activity. An additional root reporting extensor activity is shown to verify locomotor circuit coordination. (C_2) Cycloheximide-induced episodic bursting patterns are similar to those observed with applied TAs. (D) Proposed circuitry responsible for TA-induced emergence of continuous and episodic locomotor rhythms. (Left) Direct actions on CPG neurons comparable to those seen with the descending neuromodulatory transmitter serotonin. (Right) Distinct actions on neurons that drive the CPG could lead to the episodic waxing and waning of CPG output to motor neurons. Candidate neurons are the lamina X AADC$^+$ D cells. Scale bars are 10 s.

transport was consistent with the anatomical observation that TAAR1 and TAAR4 were intracellularly localized[1] and earlier observations of a prominent cytoplasmic location for TAAR1.[3,49]

The TAs also generated distinctive higher-order motor rhythms characterized by variable periods of quiescence interspersed with episodic bouts of locomotor-like activity, suggesting that the TAs also have discrete actions elsewhere (Fig. 10.4B). To test whether endogenous TA biosynthesis has the capacity to recruit locomotor circuits, we increased intracellular amino acid levels with cycloheximide.[50] Locomotor-like activity emerged. That emergent rhythms had episodic bouts that were uniquely TA-like, suggested recruitment following endogenous TA synthesis (Fig. 10.4C).

Overall, both our anatomical and functional evidence support the TAs as bona fide endogenous monoamine neuromodulators with their own unique neuromodulatory status. Observations suggest that the TAs excite motoneurons, locomotor CPG neurons, and also have unique actions on circuits that episodically modulate the locomotor CPG. Presumably these actions are at D1 cells (Fig. 10.4D). TA-evoked actions may occur via endogenous synthesis but cytoplasmic amino acid substrate availability may be limited during this anabolic period of development in the neonate.

Possible Physiological Relevance of the TAs

TAs as Effectors of Dietary and Metabolic Status

If intracellular localization of TAARs is a prominent feature of TA-mediated actions, effector coupling may be strongly linked to TA synthesis. Consequently, substrate precursor aromatic amino acid availability may be a decisive event for TA-based TAAR activation in AADC-expressing cells (Fig. 10.5). As diurnal rhythms in plasma aromatic amino acid levels relate to patterns of food consumption,[51] TA/TAAR events may link directly to dynamics in diet. For example, while even large doses of administered tyrosine do not transiently alter brain norepinephrine or dopamine levels,[52,53] rapid transient increases in brain tyrosine are seen.[54] A direct link between substrate precursor levels and CNS function would represent an uncommon form of neuromodulation,[55] yet CSF-contacting D1 cells with capacity for TA synthesis via AADC would be perfectly placed to serve such a role.

The TA/TAAR system may also serve to link to activation of amino acid mobilizing catabolic pathways to enhanced motor output as an integral part of survival responses—during

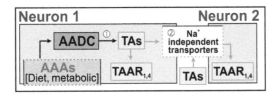

FIGURE 10.5 Putative transport and intracellular signaling mechanisms for observed TA actions. TAAR1 or TAAR4-containing neurons may be activated by the TAs following endogenous TA synthesis from their precursor aromatic amino acids (AAAs) in cells containing the essential synthesis enzyme AADC ①, or following exogenous application or transcellular transport via widely expressed Na$^+$-independent membrane transporters ②.

starvation to promote food-seeking, or in systemic autonomic sympathetic stress responses (eg, fight or flight). A role in amplification of motor behaviors associated with urgency is consistent with TAARs being important, as olfactory odorant receptors in innate survival responses,[56] in sympathomimetic cardiovascular actions,[57,58] and in activation of blood leukocytes.[59] It is also consistent with observed analgesic actions of PEA.[60,61] Last, TA transport via organic cation transporters is consistent with these transporters being associated with stress-related brain circuits.[62]

The Role of an Exquisitely Regulated Trace Aminergic System in Metamodulation

The previously described high TA synthesis rates may be an important clue to their function,[7] as it supports rapidly adjustable shifts in TA levels that are exquisitely attuned to moment-to-moment fluctuations in substrate. If the primary role for the TAs was TAAR-mediated modulation of classical monoaminergic circuit function,[63] with TAAR activity levels setting endogenous modulatory tone,[64] metabolic alterations in AADC activity or amino acid substrate availability could rapidly adjust endogenous TA levels and actions on TAARs and thus function as a metamodulator via changes in biochemical integration.[65,66]

While this would apply anywhere in AADC-expressing monoaminergic-modulated circuits with TAAR1 and/or TAAR4 expression, it may also apply to TAARs in AADC⁻ neurons. If the TAs are bidirectionally transported across membranes by pentamidine-sensitive organic cation transporters,[67] or the plasma membrane monoamine transporter,[68] which are widely expressed in spinal cord,[69] neurons with TAAR1 or TAAR4 could be modulated by nearby TA-synthesizing AADC⁺ cells via volume transmission (Fig. 10.5). Importantly, as AADC is expressed in spinal cord vasculature,[1,70,71] and AADC expression in microvasculature appears to be strongly modifiable (eg, after spinal cord injury),[71] the circulatory system can be a viable diffuse delivery system for broad TA-based metamodulation of neural circuit function.

Future Considerations

To strengthen our initial observations, it will be essential to mechanistically link changes in substrate availability to endogenous TA synthesis, TA synthesis to activation of TAARs, and activation of TAARs to facilitation of locomotor activity. Combined use of transgenic approaches that selectively target and control activity of AADC-expressing spinal neurons (eg, D1 cells) and newer-generation TAAR ligands would greatly aid such efforts.[19,25,72] Additional studies are also necessary to extend findings in the neonate to motor behaviors in the adult. It appears that the TAs can integrate and interact with conventional monoamines in spinal cord, as we have observed that the TAs can broadly facilitate ongoing 5-HT-induced locomotion (Gozal and Hochman, unpublished observations). A more detailed understanding of these interactions is clearly required, including cellular locus and monoamine receptors involved.

As the TA/TAAR modulatory system may also help set the baseline excitability of spinal motor systems, therapeutic elevations in TAAR signaling may help elevate hypoactive motor circuits in various disease states. Unfortunately, TA synthesis and degradation enzymes are common to the classical monoamines. For example, elevating TA expression levels via block of degradation (eg, MAO inhibitors) would also alter classical monoamine content and actions. However, assuming TA synthesis is rate-limiting for the TA actions but not for the classical monoamines (with reserve pools in synaptic vesicles), signal transduction-mediated alterations in AADC activity could have preferential short-term influence on the TAs over classical monoamine concentrations.[10,73] As experimental studies in animals show that diets rich in TAs (abundant in chocolate, aged cheeses, and wines) can lead to excitability changes in locomotor circuits,[74,75] dietary supplementation of aromatic amino acid precursors could also have short-term functional consequences on motor circuits.[76]

The identification of the TAs as an intrinsic spinal cord system that can modulate motor systems[1] may be particularly well-suited for the management of spinal cord injuries with compromised descending monoaminergic systems. Monoamine receptor agonists improve functional outcome after spinal cord injury in animal models.[77–79] Perhaps strategically timed delivery of aromatic amino acid precursors and/or TA dietary supplements could complement and further improve motor performance during rehabilitative training, including via improved motor endurance by their cardiovascular sympathomimetic actions.[58] Moreover, as the TAs have been shown to depress reflexes,[60,61,80] they may concomitantly be used to reduce hyperreflexia and/or pain.

CONCLUSION

We conclude that the mammalian spinal cord contains an anatomical substrate for the TAs to modulate spinal locomotor function independent of known classical monoamine-containing circuits. D1 cells represent one likely CNS TA-ergic neuronal system that expresses intracellular-located TAARs to modulate motor function. As CSF-contacting neurons, D1 cells may serve to translate CSF levels of aromatic amino acid levels to alter motor function. Further studies on their actions and potential neurotherapeutic relevance are clearly needed.

Acknowledgments

Support: NSF IOS-0745164 (SH); Paralyzed Veterans of America (SH); Craig H Neilsen Foundation (SH); NINDS F31 NS057911-02 (EAG).

References

1. Gozal EA, O'Neill BE, Sawchuk MA, Zhu H, Halder M, et al. Anatomical and functional evidence for trace amines as unique modulators of locomotor function in the mammalian spinal cord. *Front Neural Circuits* 2014;**8**:134.
2. Borowsky B, Adham N, Jones KA, Raddatz R, Artymyshyn R, et al. Trace amines: identification of a family of mammalian G protein-coupled receptors. *Proc Natl Acad Sci USA* 2001;**98**(16):8966–71. Epub 2001 Jul 8917.
3. Bunzow JR, Sonders MS, Arttamangkul S, Harrison LM, Zhang G, et al. Amphetamine, 3,4-methylene-dioxymethamphetamine, lysergic acid diethylamide, and metabolites of the catecholamine neurotransmitters are agonists of a rat trace amine receptor. *Mol Pharmacol* 2001;**60**(6):1181–8.
4. Jaeger CB, Teitelman G, Joh TH, Albert VR, Park DH, et al. Some neurons of the rat central nervous system contain aromatic-L-amino-acid decarboxylase but not monoamines. *Science* 1983;**219**(4589):1233–5.
5. Hochman S, Garraway SM, Machacek DW, Shay BL. 5-HT receptors and the neuromodulatory control of spinal cord function. *Motor Neurobiol Spinal Cord* 2001:47–87.
6. Jacobs BL, Fornal CA. 5-HT and motor control: a hypothesis. *Trends Neurosci* 1993;**16**(9):346–52.

7. Saavedra JM. β-phenylethylamine, phenylethanoamine, tyramine and octopamine. In: Trendelenburg U, Weiner N, editors. *Catecholamines II*. Berlin: Springer-Verlag; 1989. p. 181–210.

8. Shimazu S, Miklya I. Pharmacological studies with endogenous enhancer substances: beta phenylethylamine, tryptamine, and their synthetic derivatives. *Prog Neuropsychopharmacol Biol Psychiatry* 2004;**28**(3):421–7.

9. Axelrod J, SJM, Usdin E. Trace amines in the brain. In: UeaS M, editor. *Trace amines and the brain*. New York: Marcel Dekker Inc.; 1976. p. 1–20.

10. Berry MD. Mammalian central nervous system trace amines. Pharmacologic amphetamines, physiologic neuromodulators. *J Neurochem* 2004;**90**(2):257–71.

11. Boulton AA, Juorio AV, Downer RGH. Trace amines *Comparative and clinical neurobiology*. Clifton, NJ: Humana Press; 1988.

12. Boulton AA. Phenylethylaminergic modulation of catecholaminergic neurotransmission. *Prog Neuropsychopharmacol Biol Psychiatry* 1991;**15**(2):139–56.

13. Durden DA, Philips SR. Kinetic measurements of the turnover rates of phenylethylamine and tryptamine in vivo in the rat brain. *J Neurochem* 1980;**34**(6):1725–32.

14. Paterson IA, Juorio AV, Boulton AA. 2-Phenylethylamine: a modulator of catecholamine transmission in the mammalian central nervous system? *J Neurochem* 1990;**55**(6):1827–37.

15. Henry DP, Russell WL, Clemens JA, Plebus LA. Phenylethylamine and p-tyramine in the extracellular space of rat brain: quantification using a new radioenzymatic assay and in situ microdialysisBoulton A.A.Juorio AV, Downer RGH, editors. *Trace amines: comparative and clinical neurobiology*, **1988**. New Jersey: Humana Press; 1988. p. 239–50.

16. Martin WR, Sloan JW, Buchwald WF, Clements TH. Neurochemical evidence for tryptaminergic ascending and descending pathways in the spinal cord of the dog. *Psychopharmacologia* 1975;**43**(2):131–4.

17. Sloan JW, Martin WR, Clements TH, Buchwald WF, Bridges SR. Factors influencing brain and tissue levels of tryptamine: species, drugs and lesions. *J Neurochem* 1975;**24**(3):523–32.

18. Snodgrass SR, Horn AS. An assay procedure for tryptamine in brain and spinal cord using its (3H)dansyl derivative. *J Neurochem* 1973;**21**(3):687–96.

19. Wolinsky TD, Swanson CJ, Smith KE, Zhong H, Borowsky B, et al. The trace amine 1 receptor knockout mouse: an animal model with relevance to schizophrenia. *Genes Brain Behav* 2007;**6**(7):628–39.

20. Lindemann L, Meyer CA, Jeanneau K, Bradaia A, Ozmen L, et al. Trace amine-associated receptor 1 modulates dopaminergic activity. *J Pharmacol Exp Ther* 2008;**324**(3):948–56.

21. Xie Z, Vallender EJ, Yu N, Kirstein SL, Yang H, et al. Cloning, expression, and functional analysis of rhesus monkey trace amine-associated receptor 6: evidence for lack of monoaminergic association. *J Neurosci Res* 2008;**86**(15):3435–46.

22. Xie Z, Miller GM. Beta-phenylethylamine alters monoamine transporter function via trace amine-associated receptor 1: implication for modulatory roles of trace amines in brain. *J Pharmacol Exp Ther* 2008;**325**(2):617–28.

23. Bradaia A, Trube G, Stalder H, Norcross RD, Ozmen L, et al. The selective antagonist EPPTB reveals TAAR1-mediated regulatory mechanisms in dopaminergic neurons of the mesolimbic system. *Proc Natl Acad Sci USA* 2009;**106**(47):20081–6.

24. Revel FG, Moreau JL, Gainetdinov RR, Bradaia A, Sotnikova TD, et al. TAAR1 activation modulates monoaminergic neurotransmission, preventing hyperdopaminergic and hypoglutamatergic activity. *Proc Natl Acad Sci USA* 2011;**108**(20):8485–90.

25. Leo D, Mus L, Espinoza S, Hoener MC, Sotnikova TD, et al. TAAR1-mediated modulation of presynaptic dopaminergic neurotransmission: role of D2 dopamine autoreceptors. *Neuropharmacology* 2014;**81**:283–91.

26. Jaeger CB, Ruggiero DA, Albert VR, Joh TH, Reis DJ. Immunocytochemical localization of aromatic-L-amino acid decarboxylase. In: Bjorkland A, Hokfelt T, editors. *Handbook of chemical neuroanatomy, Vol. 2, classical transmitters in the CNS, Part 1*. Amsterdam: Elsevier; 1984. p. 387–408.

27. Nagatsu I, Sakai M, Yoshida M, Nagatsu T. Aromatic L-amino acid decarboxylase-immunoreactive neurons in and around the cerebrospinal fluid-contacting neurons of the central canal do not contain dopamine or serotonin in the mouse and rat spinal cord. *Brain Res* 1988;**475**(1):91–102.

28. Vigh B, Manzano e Silva MJ, Frank CL, Vincze C, Czirok SJ, et al. The system of cerebrospinal fluid-contacting neurons. Its supposed role in the nonsynaptic signal transmission of the brain. *Histol Histopathol* 2004;**19**(2):607–28.

29. LaMotte CC. Vasoactive intestinal polypeptide cerebrospinal fluid-contacting neurons of the monkey and cat spinal central canal. *J Comp Neurol* 1987;**258**(4):527–41.

30. Wienecke J, Ren LQ, Hultborn H, Chen M, Moller M, et al. Spinal cord injury enables aromatic L-amino acid decarboxylase cells to synthesize monoamines. *J Neurosci.* 2014;**34**(36):11984–2000.

31. Wyart C, Del Bene F, Warp E, Scott EK, Trauner D, et al. Optogenetic dissection of a behavioural module in the vertebrate spinal cord. *Nature* 2009;**461**(7262):407–10.

32. Kitazawa T, Saito K, Ohga A. Effects of catecholamines on spinal motoneurones and spinal reflex discharges in the isolated spinal cord of the newborn rat. *Brain Res* 1985;**351**(1):31–6.

33. Nishimaru H, Kudo N. Formation of the central pattern generator for locomotion in the rat and mouse. *Brain Res Bull* 2000;**53**(5):661–9.

34. Kiehn O. Development and functional organization of spinal locomotor circuits. *Curr Opin Neurobiol* 2011;**21**(1):100–9.

35. Kjaerulff O, Kiehn O. Distribution of networks generating and coordinating locomotor activity in the neonatal rat spinal cord in vitro: a lesion study. *J Neurosci* 1996;**16**(18):5777–94.

36. Cazalets JR, Borde M, Clarac F. Localization and organization of the central pattern generator for hindlimb locomotion in newborn rat. *J Neurosci* 1995;**15**(7 Pt 1):4943–51.

37. Cowley KC, Schmidt BJ. Regional distribution of the locomotor pattern-generating network in the neonatal rat spinal cord. *J Neurophysiol* 1997;**77**(1):247–59.

38. Kremer E, Lev-Tov A. Localization of the spinal network associated with generation of hindlimb locomotion in the neonatal rat and organization of its transverse coupling system. *J Neurophysiol* 1997;**77**(3):1155–70.

39. Hochman S, Gozal EA, Hayes HB, Anderson JT, DeWeerth SP, et al. Enabling techniques for in vitro studies on mammalian spinal locomotor mechanisms. *Front Biosci (Landmark Ed)* 2012;**17**:2158–80.

40. Kiehn O. *Neuronal mechanisms for generating locomotor activity.* New York, NY: New York Academy of Sciences; 1998.

41. Whelan P, Bonnot A, O'Donovan MJ. Properties of rhythmic activity generated by the isolated spinal cord of the neonatal mouse. *J Neurophysiol* 2000;**84**(6):2821–33.

42. Smith JC, Feldman JL, Schmidt BJ. Neural mechanisms generating locomotion studied in mammalian brain stem-spinal cord in vitro. *FASEB J* 1988;**2**(7):2283–8.

43. Hayes HB, Chang YH, Hochman S. An in vitro spinal cord-hindlimb preparation for studying behaviorally relevant rat locomotor function. *J Neurophysiol* 2009;**101**(2):1114–22.

44. Juvin L, Simmers J, Morin D. Propriospinal circuitry underlying interlimb coordination in mammalian quadrupedal locomotion. *J Neurosci* 2005;**25**(25):6025–35.

45. Kiehn O, Kjaerulff O. Spatiotemporal characteristics of 5-HT and dopamine-induced rhythmic hindlimb activity in the in vitro neonatal rat. *J Neurophysiol* 1996;**75**(4):1472–82.

46. Klein DA, Patino A, Tresch MC. Flexibility of motor pattern generation across stimulation conditions by the neonatal rat spinal cord. *J Neurophysiol* 2010;**103**(3):1580–90.

47. Klein DA, Tresch MC. Specificity of intramuscular activation during rhythms produced by spinal patterning systems in the in vitro neonatal rat with hindlimb attached preparation. *J Neurophysiol* 2010;**104**(4):2158–68.

48. Akay T, Acharya HJ, Fouad K, Pearson KG. Behavioral and electromyographic characterization of mice lacking EphA4 receptors. *J Neurophysiol* 2006;**96**(2):642–51.

49. Miller GM. The emerging role of trace amine-associated receptor 1 in the functional regulation of monoamine transporters and dopaminergic activity. *J Neurochem* 2011;**116**(2):164–76.

50. Beugnet A, Tee AR, Taylor PM, Proud CG. Regulation of targets of mTOR (mammalian target of rapamycin) signalling by intracellular amino acid availability. *Biochem J* 2003;**372**(Pt 2):555–66.

51. Fernstrom JD, Wurtman RJ, Hammarstrom-Wiklund B, Rand WM, Munro HN, et al. Diurnal variations in plasma concentrations of tryptophan, tryosine, and other neutral amino acids: effect of dietary protein intake. *Am J Clin Nutr* 1979;**32**(9):1912–22.

52. Alonso R, Agharanya JC, Wurtman RJ. Tyrosine loading enhances catecholamine excretion by rats. *J Neural Transm* 1980;**49**(1–2):31–43.

53. Agharanya JC, Wurtman RJ. Effect of acute administration of large neutral and other amino-acids on urinary-excretion of catecholamines. *Life Sci* 1982;**30**(9):739–46.

54. Morre MC, Hefti F, Wurtman RJ. Regional tyrosine levels in rat brain after tyrosine administration. *J Neural Transm* 1980;**49**(1–2):45–50.

55. Cansev M, Wurtman RJ. Aromatic amino acids in the brain 3rd ed.Lajtha A, editor. *Handbook of neurochemistry and molecular neurobiology*, **vol. 6**. Berlin, Heidelberg: Springer-Verlag; 2007. p. 60–97.

56. Liberles SD. Trace amine-associated receptors: ligands, neural circuits, and behaviors. *Curr Opin Neurobiol* 2015;**34C**:1–7.

57. Broadley KJ, Fehler M, Ford WR, Kidd EJ. Functional evaluation of the receptors mediating vasoconstriction of rat aorta by trace amines and amphetamines. *Eur J Pharmacol* 2013;**715**(1–3):370–80.
58. Broadley KJ. The vascular effects of trace amines and amphetamines. *Pharmacol Ther* 2010;**125**(3):363–75.
59. Babusyte A, Kotthoff M, Fiedler J, Krautwurst D. Biogenic amines activate blood leukocytes via trace amine-associated receptors TAAR1 and TAAR2. *J Leukoc Biol* 2013;**93**(3):387–94.
60. Mosnaim AD, Hudzik T, Wolf ME. Analgesic effects of beta-phenylethylamine and various methylated derivatives in mice. *Neurochem Res* 2014;**39**(9):1675–80.
61. Reddy SV, Maderdrut JL, Yaksh TL. Spinal cord pharmacology of adrenergic agonist-mediated antinociception. *J Pharmacol Exp Ther* 1980;**213**(3):525–33.
62. Courousse T, Bacq A, Belzung C, Guiard B, Balasse L, et al. Brain organic cation transporter 2 controls response and vulnerability to stress and GSK3 beta signaling. *Mol Psychiatry* 2015;**20**(7):889–900.
63. Xie Z, Miller GM. Trace amine-associated receptor 1 as a monoaminergic modulator in brain. *Biochem Pharmacol* 2009;**78**(9):1095–104.
64. Revel FG, Meyer CA, Bradaia A, Jeanneau K, Calcagno E, et al. Brain-specific overexpression of trace amine-associated receptor 1 alters monoaminergic neurotransmission and decreases sensitivity to amphetamine. *Neuropsychopharmacology* 2012;**37**(12):2580–92.
65. Katz P, Edwards D. Metamodulation: the control and modulation of neuromodulation *Beyond neurotransmission: neuromodulation and its importance for information processing*. New York: Oxford University Press; 1999.349.81
66. Katz PS, Clemens S. Biochemical networks in nervous systems: expanding neuronal information capacity beyond voltage signals. *Trends Neurosci* 2001;**24**(1):18–25.
67. Schomig E, Lazar A, Grundemann D. Extraneuronal monoamine transporter and organic cation transporters 1 and 2: a review of transport efficiency. *Handb Exp Pharmacol* 2006;**175**:151–80.
68. Engel K, Wang J. Expression and immunolocalization of the plasma membrane monoamine transporter in the brain. *Mol Pharmacol* 2005;**68**(5):1397–407.
69. *Allen_Spinal_Cord_Atlas. Allen Institute for Brain Science [Online]*. Seattle, <http://mousespinal.brain-map.org>; 2009, WAoa.
70. Hardebo JE, Falck B, Owman C, Rosengren E. Studies on the enzymatic blood-brain barrier: quantitative measurements of DOPA decarboxylase in the wall of microvessels as related to the parenchyma in various CNS regions. *Acta Physiol Scand* 1979;**105**(4):453–60.
71. Li Y, Li L, Stephens MJ, Zenner D, Murray KC, et al. Synthesis, transport, and metabolism of serotonin formed from exogenously applied 5-HTP after spinal cord injury in rats. *J Neurophysiol* 2014;**111**(1):145–63.
72. Ferrero DM, Wacker D, Roque MA, Baldwin MW, Stevens RC, et al. Agonists for 13 trace amine-associated receptors provide insight into the molecular basis of odor selectivity. *ACS Chem Biol* 2012;**7**(7):1184–9.
73. Hadjiconstantinou M, Neff NH. Enhancing aromatic L-amino acid decarboxylase activity: implications for L-DOPA treatment in Parkinson's disease. *CNS Neurosci Ther* 2008;**14**(4):340–51.
74. Jackson DM. Beta-phenylethylamine and locomotor activity in mice. Interaction with catecholaminergic neurones and receptors. *Arzneimittelforschung* 1975;**25**(4):622–6.
75. Marsden CA, Curzon G. The role of tryptamine in the behavioural effects of tranylcypromine+ L-tryptophan. *Neuropharmacology* 1979;**18**(2):159–64.
76. Thurmond JB, Lasley SM, Kramarcy NR, Brown JW. Differential tolerance to dietary amino acid-induced changes in aggressive behavior and locomotor activity in mice. *Psychopharmacology (Berl)* 1979;**66**(3):301–8.
77. Musienko P, van den Brand R, Marzendorfer O, Roy RR, Gerasimenko Y, et al. Controlling specific locomotor behaviors through multidimensional monoaminergic modulation of spinal circuitries. *J Neurosci* 2011;**31**(25):9264–78.
78. Courtine G, Gerasimenko Y, van den Brand R, Yew A, Musienko P, et al. Transformation of nonfunctional spinal circuits into functional states after the loss of brain input. *Nat Neurosci* 2009;**12**(10):1333–42.
79. Guertin PA. Recovery of locomotor function with combinatory drug treatments designed to synergistically activate specific neuronal networks. *Curr Med Chem* 2009;**16**(11):1366–71.
80. Bowman WC, Callingham BA, Osuide G. Effects of tyramine on a spinal reflex in the anaesthetized chick. *J Pharm Pharmacol* 1964;**16**:505–15.

Trace Amine-Associated Receptors: Ligands and Putative Role in the Central Nervous System

A. Accorroni[1] and R. Zucchi[2]

[1]Scuola Superiore St. Anna, Pisa, Italy [2]Department of Pathology,
University of Pisa, Pisa, Italy

INTRODUCTION

It has long been known that trace amines produce significant neurological effects and may be regarded as neuromodulators or even as proper neurotransmitters. A breakthrough in this field has been the discovery of a family of G-protein-coupled receptors that interact specifically with trace amines, now known as trace amine-associated receptors (TAARs).

It was later ascertained that some TAARs interact also with different endogenous substances (eg, thyronamines) and pharmacological agents (eg, psychotropic drugs). In this chapter we are going to discuss the tissue distribution of TAARs, with special regard to the central nervous system, and we will review the different classes of TAAR ligands presently known. Particular emphasis is going to be placed on the putative roles of TAARs in the physiology of the central nervous system and in neurological diseases. Furthermore, the implications derived from available transgenic models of TAAR knockout or overexpression will be summarized.

TRACE AMINE-ASSOCIATED RECEPTORS AND THEIR DISTRIBUTION

Trace amine-associated receptors (TAARs) are a large family of G-protein-coupled receptors. The first member of this family, now known as TAAR1, was identified in 2001 by two different groups of investigators,[1,2] through a degenerate PCR approach, that is, using primers based on the known sequences of serotonin or dopamine receptors on rat cDNA and genomic DNA. When expressed in heterologous cells, the novel receptor induced cAMP production upon exposure to p-tyramine or β-phenylethylamine, while classical biogenic amines (namely dopamine, norepinephrine, epinephrine, serotonin, and histamine) were either ineffective or much less potent. The name "trace amine receptor" was therefore proposed.

The next step was an extensive effort to analyze the genomic sequences of several vertebrate and invertebrate species, looking for homologies to the putative hallmarks of the new receptor.[3–5] A large number of genes have been reported, with nine of them in humans, nine in chimpanzee, 19 in rat, 16 in mouse, 112 in zebrafish, and several dozens in other fish species.[6] Some sequences apparently corresponded to pseudogenes, for example, three in humans, six in chimpanzee, and two in rat. It turned out that only a few members of this family show a clear response to trace amines, so it was suggested to modify their denomination to "trace amine-*associated* receptors." A nomenclature for the whole gene family has been developed, based on the physical location on the chromosome relative to TAAR1 (in mammals all TAARs are located on the same chromosome), and on the putative origin of the different genes in evolution (see chapter: Trace Amines and Their Receptors in the Control of Cellular Homeostasis). Orthologous genes are characterized by different numbers, while paralogue genes, that is, genes which were probably generated by duplication events within the lineage of each species, are identified by a letter suffix added after the number.

TAAR1 has a broad tissue distribution. Through in situ hybridization histochemistry[1], TAAR1 mRNA was detected in many regions of the mouse brain and the use of a *LacZ* reporter[7] suggested a more circumscribed distribution of TAAR1 in dopaminergic and serotoninergic brain areas.

Quantitative RT-PCR showed that TAAR1 is expressed at moderate levels in the stomach and at low levels in the kidney, lung, and small intestine.[1] By a similar technique TAAR1 cDNA was detected in rat heart and in mouse stomach, intestine, testis, and thyroid,[8,9] although expression levels were generally low. More recently, significant TAAR1 expression has been reported in neonatal rat spinal cord[10] and in rhesus monkey and human leukocytes.[11–14] Interestingly, evidence of intracellular expression was reported both in brain and in spinal cord.[1,10]

TAARs different from TAAR1 are expressed at high levels in mouse olfactory epithelium.[15] Double labeling experiments showed that different TAAR subtypes are expressed in different cells, and that TAARs are not coexpressed with odorant receptors. It was therefore suggested that olfactory TAARs represent a specific class of odorant receptors, probably involved in behavioral responses.[6]

In spite of their preferential localization in the olfactory epithelium, extraolfactory expression of TAAR subtypes different from TAAR1 has been reported by several investigators. With regard to the central nervous system, TAAR subtypes 2–9 are expressed in neonatal rat spinal cord,[10] where TAAR4 shows an intracellular location.[10] TAAR6 expression was detected in several human brain regions, namely basal ganglia, frontal cortex, substantia nigra, amygdala, and hippocampus,[16] although this pattern was not reproduced in rhesus monkey.[17] Notably, in basal ganglia, frontal cortex, and substantia nigra, TAAR6 expression was over 10-fold higher than TAAR1 expression. TAAR6 and TAAR8 expression were observed in mouse amygdala,[1] while very low levels of TAAR8a expression were reported[18] in several locations of the mouse brain, namely cerebellum, cortex, hippocampus, striatum, and olfactory bulb.

Apart from the central nervous system, there is consistent evidence of the presence of TAARs in white blood cells and phagocytes: mouse B cells, macrophages, and dendritic cells have been reported to express all TAAR subtypes,[14] while human leukocytes apparently express TAAR2, TAAR6, TAAR8, and TAAR9.[13,14,19]

TAARs different from TAAR1 have been occasionally detected in several mouse tissues, although these reports need confirmation by focused investigations, since expression levels were usually very low, and in most cases proper positive and negative controls have not been provided. This applies to TAAR2 expression in the gastrointestinal tract;[20] TAAR6 expression in kidney;[1] TAAR8 expression in heart, liver, kidney, intestine, spleen, testis, and thyroid;[1,8,21] and TAAR9 expression in kidney, pituitary, and skeletal muscle.[22] Rat hearts contain transcripts for TAAR8a, and, at a lower level, TAAR1, TAAR2, TAAR3, and TAAR4.[9]

On the whole, while most TAARs appear to have a primary role as olfactory receptors, it is possible that during vertebrate evolution they may have assumed additional function, particularly in the central nervous system.

TAAR LIGANDS: TRACE AMINES

Trace amines, namely *para*-tyramine, *meta*-tyramine, tryptamine, β-phenylethylamine, *para*-octopamine, and *meta*-octopamine[23] were the first compounds to be characterized as ligands for TAAR1. In COS-7 cells and in HEK293 cells expressing recombinant human TAAR1, trace amines were much more powerful than dopamine, histamine, and serotonin in inducing cAMP production. In particular, p-tyramine and β-phenylethylamine showed the lowest EC_{50}, ranging, respectively, between 69–214 nM and 240–324 nM.[1,2] In general, TAAR1 orthologues obtained from different species had EC_{50} values for p-tyramine and β-phenylethylamine in the range of 0.1–1.4 μM,[15,24] while EC_{50} values for octopamine and tryptamine averaged 2–10 μM and 1.5–45 μM, respectively.[24] Dopamine and serotonin were 5- to 25-fold less potent than *p*-tyramine and β-phenylethylamine, and the maximum response was about half, suggesting a partial agonist action.[24]

Investigations performed in heterologous cells transfected with other TAAR subtypes failed to show any response to trace amines, with the exception of COS-7 cells expressing the rat TAAR4, where cAMP production was induced by β-phenylethylamine and *p*-tyramine with EC_{50} in the micromolar range.[1] Notably, TAAR4 is a pseudogene in human.

Administration of exogenous trace amines or increased levels of endogenous trace amines have long been known to produce "sympathomimetic" effects, such as tachycardia, nausea, sweating, hyperthermia, and headache. These effects have been attributed to competition with catecholamine- or serotonin-binding sites, leading to increased catecholamine or serotonin availability. However, interaction with these molecular targets occurs at micromolar concentrations, while trace amines are present in tissues at much lower concentrations, on the order of 0.1 nM to 10 nM, representing less than 1% of total biogenic amines.[5,23] Therefore, it seems more likely that the effects of endogenous trace amines, if any, are mediated by higher-affinity receptors, potentially including TAAR1.

In the central nervous system, trace amines have been initially thought to act as neuromodulators. However, more recent reports suggest that they may also produce direct effects on neuronal excitability and physiological responses.

A neuromodulator can be defined as a chemical that is released from a neuron and is able to potentiate or inhibit the action of a coexisting neurotransmitter. Some trace amines have been shown to modulate neuronal responses to dopamine, in areas involved in brain reward; indeed, Paterson et al.[25] and Berry et al.[23] showed that physiological concentrations of β-phenylethylamine increase neuronal sensitivity to either dopamine or synthetic dopamine agonists. In addition, β-PEA inhibits the electrically evoked release of acetylcholine in rat striatum through a mechanism that requires the integrity of the striatonigral dopaminergic system.[26] β-Phenylethylamine also increases neuronal responses to norepinephrine[27–29] and potentiates cortical neuron responses to electrical stimulation of the locus coeruleus,[28] while it seems to produce no effect on neuronal responses to serotonin.[29] Moreover in rat midbrain slices, the exogenous application of β-phenylethylamine suppresses the inward rectifying currents operated by the $GABA_B$ receptor.[30]

Consistent results have been reported for tyramines, which potentiate both dopamine[31,32] and norepinephrine-mediated effects on neuronal physiology,[31] but seems to induce no significant effects on serotonin and glutamate transmission.[32] Octopamines on the other hand potentiate responses to norepinephrine but not to dopamine and serotonin.[33]

The effects that trace amines produce per se mimic those of amphetamine, namely, they induce increased alertness, irritability, euphoria, and insomnia. However, it has to be underlined that the amphetamine-like effects described above were only observed at concentrations that largely exceed endogenous trace amine levels.

Also, trace amines have recently been implicated in the regulation of locomotor activity. In particularly, the in vitro work by Gozal et al.[10] suggested that tryptamine and tyramine may represent a system that interacts directly with the central pattern-generating neurons and is capable of recruiting specific locomotor circuits.

Given the putative function of trace amines as neuromodulators and neurotransmitters, it is not surprising that a pathophysiological role for trace amines in the genesis of several neurological and psychiatric disorders has been advocated. This issue is discussed elsewhere (see Section III. Trace Amines and Neurological Disorders), but it is interesting to point out that several genome-wide association studies demonstrated an association between TAAR genes located on the region 6q23.2 and genes known to be involved in schizophrenia.[34–37] This latter result is supported by reports suggesting that multiple single nucleotide polymorphism identified in TAAR2[38] and TAAR6 (previously known as trace amine receptor 4)[16] genes have been associated with an increased susceptibility to schizophrenia. However, the relationship between TAAR genes and schizophrenia is not completely straightforward. Indeed, a linkage between TAAR polymorphisms and this psychiatric disorder has not been confirmed in all ethnicities.[16,39–41]

Collectively, the data discussed above, even if still controversial, uncover potential neuromodulatory effects of trace amines in schizophrenia, and suggest that targeting trace amine activity might represent a novel therapeutic approach to integrate with already-validated protocols.

TAAR LIGANDS: THYRONAMINES

Thyronamines are chemically related to thyronines,[42,43] and to thyroid hormone, indeed, thyronamines represent iodinated thyronine derivatives. If compared to thyroid hormone, all thyronamines lack the carboxyl group, while they may retain one or more iodine atoms on the aromatic rings. Many different thyronamines were synthetized in Scanlan's laboratory, and tested in HEK293 expressing either the rat or mouse TAAR1.[44,45] Several compounds turned out to produce a concentration-dependent increase in cAMP concentration, and the most effective were the following: 3-iodothyronamine (T_1AM) >3,5-diiodothyronamine >3,5,3'-triiodothyronamine >thyronamine. In particular, the EC_{50} for T_1AM was in the range 10–100 nM for the rat and mouse receptors, respectively.

T_1AM is an endogenous compound, since it has been detected in virtually every tissue in rodents, as well as in human blood (reviewed in Ref. 46). Its assay is technically difficult, and some discrepancy exists between published results. In particular, serum concentrations determined by chemiluminescence immunoassay or by HPLC coupled to mass spectrometry differ by nearly two orders of magnitude.[47,48] These problems may be related to the extensive preanalytical processing needed for mass spectrometry-based techniques, and/or to the high-affinity binding of T_1AM to plasma proteins like apo-B100,[49] which complicates the calibration of competition-based immunoassays. In any case, the bulk of evidence suggests that tissue and serum T_1AM levels are in the nanomolar range, and therefore they are not far from the EC_{50} for TAAR1, as estimated in vitro.

Thyronamine metabolism has not been definitely unravelled, but liver T_1AM levels were reduced in mice subjected to experimental hypothyroidism,[50] suggesting that it is either produced in the thyroid or by peripheral thyroid hormone metabolism. Recently, in an isolated gut preparation, thyroxine conversion into T_1AM has been shown by successive deiodination and decarboxylation steps.[51] Aromatic amino acid decarboxylase, which is involved in the synthesis of trace amines, does not appear to be able to act on thyronines,[52] whose decarboxylation is thought to be catalyzed by ornithine decarboxylase.

Administration of exogenous T_1AM in vivo showed that its metabolism may include different pathways: deiodination to thyronine, oxidative deamination to 3-iodothyroacetic acid, acetylation to N-acetyl-T_1AM, or esterification with glucoronate or sulfate.[53–55] Thyronamine and 3-iodothyroacetic acid have been detected in rat liver and brain, respectively,[56,57] most likely as a result of endogenous T_1AM metabolism.

Treatment with exogenous T_1AM produced several functional effects in vitro or in vivo, namely: reduced body temperature,[44] negative cardiac inotropic and chronotropic effects,[58] reduced body weight,[59,60] reduction of insulin and stimulation of glucagon secretion,[61,62] prolearning and antiamnestic action,[63] reduced pain threshold,[63] and reduction of non-REM sleep.[64] Many effects were observed using very high T_1AM concentrations, in the micromolar range, but some metabolic and neurologic actions occurred with dosages increasing endogenous T_1AM levels by about an order of magnitude, and might therefore have physiological or pathophysiological relevance.[46] However, it should not be assumed that the effects of T_1AM are necessarily mediated by TAAR1, because recent evidence suggest that TAAR1 is not the only target of T_1AM. T_1AM has been reported to behave as an inverse agonist on TAAR5,[65] and it is a biased agonist of α_{2A} adrenergic receptor, since it specifically activates the MAP kinase pathway.[66] In addition, some functional responses may be mediated by T1AM catabolites, particularly by 3-iodothyroacetic acid[57,67] (see chapter: β-Phenylethylamine-Class Trace Amines in Neuropsychiatric Disorders: A Brief Historical Perspective).

A major goal of ongoing research is determining which receptor is responsible for the specific effects of thyronamines. TAAR1 does not appear to be involved in the hypothermic effect, which was reproduced in TAAR1 knockout mice.[68] Cardiac effects were shared by several trace amines in the isolated rat heart, but the order of potency did not match the potency observed on rat TAAR1, suggesting that a different receptor was involved.[69] Modulation of glucose and insulin secretion, as well as the consequent hyperglycemia, have been attributed to interaction with pancreatic α_{2A} receptors, while stimulation of pancreatic TAAR1 apparently produced an opposite effect.[70] Preliminary evidence that some electrophysiological effects of T_1AM in the entorhinal cortex are mediated by TAAR1 was recently reported, through the use of specific TAAR1 antagonists.[71]

PHARMACOLOGICAL AGENTS AS TAAR1 LIGANDS

Several psychoactive agents have been shown to interact with TAARs, particularly with TAAR1. In addition, different synthetic substances have been devised as TAAR agonists or antagonists.

Amphetamine and its derivatives stimulated cAMP production in heterologous cells expressing rat TAAR1.[2] The most powerful agonist was 4-hydroxyamphetamine, whose EC_{50} at rat TAAR1 averaged 51 nM, followed by amphetamine, methamphetamine, 3,4-methylenedioxymetamphetamine (MDMA, known as "ecstasy"), and the hallucinogenic amphetamine 2-amino,1-[2,5-dimethoxy-4-iodophenyl]-propane. Similar results were obtained with mouse and primate TAAR1, although species differences in the stereoselectivity for isomers were reported.[72,73]

Another group of TAAR1 agonists is represented by ergot alkaloids and ergoline derivatives, namely ergometrine, dihydroergotamine, d-lysergic acid diethylamide, and the antiparkinsonian agents bromocriptine and lisuride. At micromolar concentrations,

TAAR1-mediated cAMP production was stimulated also by inhibitors of dopamine transporter, such as nomifensine and 1-methyl-4-phenyl-1,2,3,6-tetrahydropyridine (MTPT).[2]

It should be stressed that the drugs mentioned above have other molecular targets too, so the contribution of TAAR1-mediated responses to their effects remains uncertain. However, the chemical structure of these molecules and of endogenous TAAR1 agonists has guided the efforts to develop novel synthetic TAAR ligands.

Several phenethylamine derivatives have been tested on human TAAR1.[74,75] The initial investigations showed that derivatives with small substituents at ring position 2 were as effective or even slightly more effective than the parent compound, while analogues with multiple substituents and analogues in which the primary amino group was converted to a secondary or a tertiary amino group were partial agonists. Hart et al.[45] adopted the strategy of modifying the T_1AM molecule. A large series of derivatives was obtained by chemical interventions aimed at replacing the 3-iodine with an alkyl group, removing the phenol hydroxyl, modifying or substituting the amine, increasing the distance between the two aryl rings. The most potent derivatives on rat and mouse TAAR1 were 3-methyl-thyronamine, N-methyl-O-(p-trifluoromethyl)benzyl-tyramine, O-phenyl-3-iodotyramine, and O-(p-fluoro)phenyl-3-iodotyramine. In vivo, these substances proved to be able to induce hypothermia in mice, with an EC_{50} equal to or even lower than T1AM.

Tan et al.[76,77] introduced further developments. On the basis of the consideration that TAAR1 is targeted by phenethylamine and amphetamines, they tested different modifications in the ethylamine portion of the phenoxyphenylethylamine scaffold. In addition, they observed that transforming the inner ring of the phenoxyphenethylamine scaffold into a naphthyl group yielded high affinity for both rat and mouse TAAR1.

Chiellini et al.[78] applied to the T1AM molecule a few chemical interventions which had proved to be effective in the synthesis of thyroid hormone analogues: linking the two aromatic rings by a methylene group, replacing the hydroxyl group with an amino group (NH2), and introducing an oxy-ethylamine side chain in position 1. The most effective compound (4-(4-(2-aminoethoxy)-2-methylbelnzyl)aniline) had a potency on mouse TAAR1 similar to T1AM, as evaluated on the basis of cAMP production in heterologous expression models. In functional experiments, it increased plasma glycemia and reduced cardiac output. Since these effects are unlikely to be mediated by TAAR1, it seems that these derivatives share the multitarget properties of T1AM.

A different strategy was followed by Revel et al.[79] They produced an iterative series of structural modifications on an amino-oxazoline α_{2A}-adrenergic receptor agonist, and discovered a novel compound, RO5166017 [(S)-4-[(ethyl-phenyl-amino)-methyl]-4,5-dihydro-oxazol-2-ylamine], which showed high potency and efficacy at mouse, rat, monkey, and human TAAR1. The maximal effects were in a range similar to that achieved by β-phenylethylamine stimulation, and the estimated EC_{50} was 3, 14, and 55 nM with mouse, rat, and human TAAR1, respectively. Further developments yielded additional derivatives,[79] among which two compounds, known as RO5256390 and RO5263397, showed a very high affinity for TAAR1, as well as highly selectivity versus other aminergic receptors. The former appears to be a full agonist and the latter a partial agonist.

These derivatives, and other compounds of this class, have been used in several functional studies to try to unravel the physiological role of TAAR1 in the modulation of the monoaminergic system and its possible applications as a drug discovery target for cocaine addiction and schizophrenia. In particular, the use of RO5166017 in rodents demonstrated that TAAR1, through a putative interaction with D2 receptors, decreases electrically evoked dopamine release at the level of the nucleus accumbens and dorsal striatum,[80] inhibits the firing frequency of dopamine neurons in the ventral tegmental area, and blocks dopamine-dependent hyperlocomotion in cocaine-treated and in dopamine transporter KO mice.[79] In addition, TAAR1 also appeared to be involved in the inhibition of firing of serotoninergic neurons in the dorsal raphe nucleus. With regard to the possible role of TAAR1 in the treatment of cocaine addiction, RO5256390-mediated TAAR1 activation[81] suppressed the cocaine-seeking behavior after a 2-week period of withdrawal from cocaine self-administration. The same compound was demonstrated to decrease cocaine reinforcing efficacy.[82] Moreover, using RO5263397 it has been found that TAAR1 activation promotes vigilance in rodents and produces procognitive and antidepressant effects in both rodents and primate models.[83] The results employing the TAAR1 full agonist RO5256390 in rodents also supported a possible role of this receptor in the treatment of schizophrenia. Indeed, TAAR1 activation blocks the hyperactivity induced by psychostimulant compounds and produces a brain activation pattern that closely resembles that of the antipsychotic drug olanzapine. RO5256390 significantly reduced haloperidol-induced catalepsy and was also able to prevent the increase in body weight and fat accumulation typically associated with olanzapine treatment.[83]

Extensive screening of the Roche compound library on a chimeric human/rat TAAR1 also allowed the identification of several benzanilides, which behaved as TAAR1 antagonists. Further work aimed at optimizing this effect yielded a selective TAAR1 antagonist, N-(3-ethoxy-phenyl)-4-pyrrolidin-1-yl-3-trifluoromethyl-benzamide (EPPTB). EPPTB has an IC$_{50}$ of 28 nM at mouse TAAR1, and it has been used in several functional investigations that shed light into the close interaction existing between TAAR1 and the dopaminergic system. In particular, using EPPTB, Bradaia et al.[84] demonstrated that TAAR1 may reduce the firing frequency of the dopaminergic neurons in the mesolimbic system and may also activate an inwardly rectifying K$^+$ current in the same system. On the whole, the research on synthetic TAAR ligands has provided valuable tools for experimental investigations on the functional role of TAAR1, and has opened new perspectives in the study of psychotropic agents.

ODOROUS AMINES AS TAAR LIGANDS

In 2006, Liberles and Buck[15] proposed that all TAARs other than TAAR1 should be regarded as olfactory receptors. This hypothesis was based on the observation that most TAAR subtypes were expressed in the mouse olfactory epithelium, and that exposure to a large number (over 300) of odorous compounds induced cAMP production in HEK cells transfected with vectors encoding for individual mouse or human TAARs. These observations have opened the way to further investigations on the olfactory role of TAARs, which appear to have major importance in rodents, but might be relevant also in human. This issue is discussed in detail in Section II. Trace Amines and Olfaction.

LESSONS FROM TRANSGENIC MOUSE MODELS

Transgenic mice lacking TAAR1 have been generated using standard molecular biology techniques by Wolinsky et al.[85] Homozygous TAAR1 KO mice appeared normal and showed no difference in body weight, body size, and basal body temperature from wildtype littermates. A normal response was observed in two standard behavioral tests, namely the elevated plus maze, that is used as a test of anxiety, and the Y-maze, that is used as a test for working memory. A significant difference was however observed in prepulse inhibition of the acoustic startle response. Prepulse inhibition is the phenomenon in which a weak stimulus reduces the response to a subsequent strong "startling" stimulus, and the extent of this phenomenon was reduced in TAAR1 KO mice.

In pharmacological studies, KO mice showed an increased sensitivity to amphetamine, measured on the basis of enhanced locomotor activity.[7,85] This behavioral effect was associated with substantial neurochemical changes, since in dorsal striatum the amphetamine-induced increase of extracellular dopamine, norepinephrine, and serotonin levels was significantly enhanced. Electrophysiological recordings from dopaminergic neurons in the ventral tegmental area revealed higher firing rate, and, contrary to wildtype littermates, KO mice showed no inhibition of dopaminergic neurons by tyramine. The latter finding suggests that under normal conditions TAAR1 is either constitutively active or tonically activated by an endogenous ligand. It was also observed that the proportion of high-affinity D2 dopamine receptors was increased in TAAR KO mice versus their littermates. Recently, overexpression of D2, but not of D1 dopamine receptors has been confirmed in the striatum of TAAR1 KO mice. G-protein-independent activation of the AKT/GSK3 signaling pathway has also been reported.[86] The mechanisms of TAAR-dependent modulation of dopaminergic neurotransmission are further analyzed in Chapter "Trace Amine-Associated Receptor 1 Modulation of Dopamine System".

These findings are consistent with the hypothesis that TAAR1 is involved in neuromodulation. They also raise pathophysiological interest, since TAAR1 KO mice show features of dopamine supersensitivity, namely defect in sensorimotor gating and enhanced response to amphetamine, which have been consistently observed in schizophrenia.

Experiments performed with 3,4-methylenedioxyamphetamine (NMDA, or "ecstasy") also revealed a specific role for TAAR1.[87] In wildtype animals, NMDA produced time- and dose-dependent hypothermia and hyperthermia, while in TAAR1 KO mice only hyperthermia was produced. NMDA increased dorsal striatum dialysate levels of dopamine and serotonin, which were amplified in TAAR1 KO mice. On the other hand, the hypothermia induced in vivo by high T1AM dosages was independent of TAAR1 expression.[68]

The locomotor response to morphine, as evaluated by conditioned place preference studies, was not affected in TAAR1 KO mice,[88] and the locomotor responses induced by low doses of apomorphine were also unaffected.[89] However, TAAR1 KO showed a significant reduction in some behavioral responses induced by high doses of apomorphine, such as climbing behavior and certain types of stereotypes.[89]

The effects of several antipsychotic agents, particularly haloperidol and clozapine, were also altered in TAAR1 KO mice,[90,91] and differences in ethanol-related behavior have also been reported.[92] TAAR1 KO mice showed a significantly greater preference for ethanol in a two-bottle choice test, higher sedative-like effects of acute ethanol, and a lower locomotor activity in response to an acute ethanol challenge.

A mouse line showing generalized brain TAAR1 overexpression has been produced.[93] Transgenic mice did not show overt differences in general motor function and behavior. In females only, a small but significant increase in body weight and a small but significant reduction in body temperature were detected. As expected, reduced locomotor responses were observed after amphetamine administration, and extracellular dopamine and norepinephrine concentration (measured in accumbens nucleus) were not affected. However, electrophysiological recordings provided unexpected findings. The spontaneous firing rates of dopaminergic neurons in the ventral tegmental area were increased, as observed in TAAR1 KO mice. This was attributed to decreased activity of a population of GABA neurons, which physiologically inhibit dopaminergic neurons. Increased firing of serotoninergic neurons in the dorsal raphe nucleus and of adrenergic neurons in nucleus coeruleus were also reported. In addition, in vivo microdialysis showed significantly higher baseline levels of norepinephrine and dopamine in accumbens nucleus, and elevated serotonin in medial prefrontal cortex.

These results underscore the complexity of TAAR1 interaction with different neurotransmitters. Apparently, opposite modulatory effects are possible, and the conclusions derived from KO animals may be an oversimplification. In general, the hypothesis that TAAR1 plays a neuromodulatory role is supported by transgenic models, but much additional experimental work is necessary before any physiological and pathophysiological implications are drawn.

While most research on transgenic models has been focused on TAAR1, a mouse strain in which all the so-called olfactory TAAR genes (ie, TAAR2–TAAR9) were deleted has been produced.[94] This model was used to show the behavioral relevance of TAAR-mediated odor perception. Aside from olfactory-related changes, homozygous mice had a normal weight and did not show any significant difference in locomotor activity versus wildtype littermates. However, more detailed behavioral, electrophysiological or neurochemistry observations have not been reported so far. Therefore, no additional information on the potential nonolfactory functions of these TAAR subtypes has been obtained.

CONCLUSION

TAARs represent a large class of receptors recently discovered that may play important biological functions, since they have been preserved in vertebrate evolution. The widespread expression of TAAR1 in the central nervous system, its activation by psychoactive drugs, its interaction with the dopaminergic and serotoninergic systems, and the observation that TAAR1 knockout mice show a neurological phenotype, make it a candidate for the neurological effects of trace amines and thyronamines. In spite of these findings, it should be acknowledged that there is still very little evidence of a causal relationship between TAAR1 activation by endogenous compounds and specific physiological responses. Even less is known for the extraolfactory role of the other TAAR subtypes, which are not activated in vitro by trace amines, and whose brain expression is less extensive. However, a significant association of specific TAAR2 and/or TAAR6 polymorphisms and psychiatric diseases has been observed at least in some pedigrees. On the whole, further investigations are highly needed, and they might provide intriguing implications for central nervous system physiology and pathophysiology.

References

1. Borowsky B, Adham N, Jones KA, et al. Trace amines: identification of a family of mammalian G protein-coupled receptors. *Proc Natl Acad Sci USA* 2001;**98**(16):8966–71.
2. Bunzow JR, Sonders MS, Arttamangkul S, et al. Amphetamine, 3,4-methylenedioxymethamphetamine, lysergic acid diethylamide, and metabolites of the catecholamine neurotransmitters are agonists of a rat trace amine receptor. *Mol Pharmacol* 2001;**60**(6):1181–8.
3. Gloriam DEI, Bjarnadottir TK, Schioet HB, Fredriksson R. High species variation within the repertoire of trace amine receptors. *Ann NY Acad Sci* 2005;**1040**(1):323–7.
4. Lindemann L, Hoener MC. A renaissance in trace amines inspired by a novel GPCR family. *Trends Pharmacol Sci* 2005;**26**(5):274–81.
5. Zucchi R, Chiellini G, Scanlan TS, Grandy DK. Trace amine-associated receptors and their ligands. *Br J Pharmacol* 2006;**149**(8):967–78.
6. Liberles SD. Trace amine-associated receptors: ligands, neural circuits, and behaviors. *Curr Opin Neurobiol* 2015;**34**:1–7.
7. Lindemann L, Meyer CA, Jeanneau K, et al. Trace amine-associated receptor 1 modulates dopaminergic activity. *J Pharmacol Exp Ther* 2008;**324**(3):948–56.
8. Chiellini G, Erba P, Carnicelli V, et al. Distribution of exogenous [125I]-3-iodothyronamine in mouse in vivo: relationship with trace amine-associated receptors. *J Endocrinol* 2012;**213**(3):223–30.
9. Chiellini G, Grzywacz P, Plum LA, Barycki R, Clagett-Dame M, DeLuca HF. Synthesis and biological properties of 2-methylene-19-nor-25-dehydro-1alpha-hydroxyvitamin D(3)-26,23-lactones – weak agonists. *Bioorg Med Chem* 2008;**16**(18):8563–73.
10. Gozal EA, O'Neill BE, Sawchuk MA, et al. Anatomical and functional evidence for trace amines as unique modulators of locomotor function in the mammalian spinal cord. *Front Neural Circuits* 2014;**8**:134.
11. Panas MW, Xie Z, Panas HN, Hoener MC, Vallender EJ, Miller GM. Trace amine associated receptor 1 signaling in activated lymphocytes. *J Neuroimmune Pharmacol* 2012;**7**(4):866–76.
12. Wasik AM, Millan MJ, Scanlan T, Barnes NM, Gordon J. Evidence for functional trace amine associated receptor-1 in normal and malignant B cells. *Leuk Res* 2012;**36**(2):245–9.
13. Babusyte A, Kotthoff M, Fiedler J, Krautwurst D. Biogenic amines activate blood leukocytes via trace amine-associated receptors TAAR1 and TAAR2. *J Leukoc Biol* 2013;**93**(3):387–94.
14. Nelson DA, Tolbert MD, Singh SJ, Bost KL. Expression of neuronal trace amine-associated receptor (TAAR) mRNAs in leukocytes. *J Neuroimmunol* 2007;**192**(1–2):21–30.
15. Liberles SD, Buck LB. A second class of chemosensory receptors in the olfactory epithelium. *Nature* 2006;**442**(7103):645–50.
16. Duan J, Martinez M, Sanders AR, et al. Polymorphisms in the trace amine receptor 4 (TRAR4) gene on chromosome 6q23.2 are associated with susceptibility to schizophrenia. *Am J Hum Genet* 2004;**75**(4):624–38.
17. Xie Z, Vallender EJ, Yu N, et al. Cloning, expression, and functional analysis of rhesus monkey trace amine-associated receptor 6: evidence for lack of monoaminergic association. *J Neurosci Res* 2008;**86**(15):3435–46.
18. Dinter J, Gourdain P, Lai NY, et al. Different antigen-processing activities in dendritic cells, macrophages, and monocytes lead to uneven production of HIV epitopes and affect CTL recognition. *J Immunol* 2014;**193**(9):4322–34.
19. D'Andrea G, Terrazzino S, Fortin D, Farruggio A, Rinaldi L, Leon A. HPLC electrochemical detection of trace amines in human plasma and platelets and expression of mRNA transcripts of trace amine receptors in circulating leukocytes. *Neurosci Lett* 2003;**346**(1–2):89–92.
20. Ito J, Ito M, Nambu H, et al. Anatomical and histological profiling of orphan G-protein-coupled receptor expression in gastrointestinal tract of C57BL/6J mice. *Cell Tissue Res* 2009;**338**(2):257–69.
21. Mühlhaus J, Dinter J, Nürnberg D, et al. Analysis of human TAAR8 and murine TAAR8B mediated signaling pathways and expression profile. *Int J Mol Sci* 2014;**15**(11):20638–55.

22. Vanti WB, Muglia P, Nguyen T, et al. Discovery of a null mutation in a human trace amine receptor gene. *Genomics* 2003;**82**(5):531–6.

23. Berry MD. Mammalian central nervous system trace amines. Pharmacologic amphetamines, physiologic neuromodulators. *J Neurochem* 2004;**90**(2):257–71.

24. Lindemann L, Ebeling M, Kratochwil NA, Bunzow JR, Grandy DK, Hoener MC. Trace amine-associated receptors form structurally and functionally distinct subfamilies of novel G protein-coupled receptors. *Genomics* 2005;**85**(3):372–85.

25. Paterson IA, Juorio AV, Boulton AA. 2-Phenylethylamine: a modulator of catecholamine transmission in the mammalian central nervous system? *J Neurochem* 1990;**55**(6):1827–37.

26. Baud P, Arbilla S, Cantrill RC, Scatton B, Langer SZ. Trace amines inhibit the electrically evoked release of [3H] acetylcholine from slices of rat striatum in the presence of pargyline: similarities between beta-phenylethylamine and amphetamine. *J Pharmacol Exp Ther* 1985;**235**(1):220–9.

27. Paterson IA. The potentiation of cortical neurone responses to noradrenaline by beta-phenylethylamine: effects of lesions of the locus coeruleus. *Neurosci Lett* 1988;**87**(1–2):139–44.

28. Paterson IA. The potentiation of cortical neuron responses to noradrenaline by 2-phenylethylamine is independent of endogenous noradrenaline. *Neurochem Res* 1993;**18**(12):1329–36.

29. Paterson IA, Boulton AA. β-Phenylethylamine enhances single cortical neurone responses to noradrenaline in the rat. *Brain Res Bull* 1988;**20**(2):173–7.

30. Federici M, Geracitano R, Tozzi A, et al. Trace amines depress GABA B response in dopaminergic neurons by inhibiting G-βγ-gated inwardly rectifying potassium channels. *Mol Pharmacol* 2005;**67**(4):1283–90.

31. Jones RS, Boulton AA. Interactions between p-tyramine, m-tyramine, or beta-phenylethylamine and dopamine on single neurones in the cortex and caudate nucleus of the rat. *Can J Physiol Pharmacol* 1980;**58**(2):222–7.

32. Jones RS, Juorio AV, Boulton AA. Changes in levels of dopamine and tyramine in the rat caudate nucleus following alterations in impulse flow in the nigrostriatal pathway. *J Neurochem* 1983;**40**(2):396–401.

33. Jones RS. Noradrenaline-octopamine interactions on cortical neurones in the rat. *Eur J Pharmacol* 1982;**77**(2–3):159–62.

34. Cao Q, Martinez M, Zhang J, et al. Suggestive evidence for a schizophrenia susceptibility locus on chromosome 6q and a confirmation in an independent series of pedigrees. *Genomics* 1997;**43**(1):1–8.

35. Levinson DF, Holmans P, Straub RE, et al. Multicenter linkage study of schizophrenia candidate regions on chromosomes 5q, 6q, 10p, and 13q: schizophrenia linkage collaborative group III. *Am J Hum Genet* 2000;**67**(3):652–63.

36. Schwab SG, Hallmayer J, Albus M, et al. A genome-wide autosomal screen for schizophrenia susceptibility loci in 71 families with affected siblings: support for loci on chromosome 10p and 6. *Mol Psychiatry* 2000;**5**(6):638–49.

37. Mowry BJ, Nancarrow DJ. Molecular genetics of schizophrenia. *Clin Exp Pharmacol Physiol* 2001;**28**(1–2):66–9.

38. Bly M. Examination of the trace amine-associated receptor 2 (TAAR2). *Schizophr Res* 2005;**80**(2–3):367–8.

39. Amann D, Avidan N, Kanyas K, et al. The trace amine receptor 4 gene is not associated with schizophrenia in a sample linked to chromosome 6q23. *Mol Psychiatry* 2005;**11**(2):119–21.

40. Ikeda M, Iwata N, Suzuki T, et al. No association of haplotype-tagging SNPs in TRAR4 with schizophrenia in Japanese patients. *Schizophr Res* 2005;**78**(2–3):127–30.

41. Venken T, Alaerts M, Adolfsson R, Broeckhoven CV, Del-Favero J. No association of the trace amine-associated receptor 6 with bipolar disorder in a northern Swedish population. *Psychiatr Genet* 2006;**16**(1):1–2.

42. Piehl S, Hoefig CS, Scanlan TS, Köhrle J. Thyronamines—past, present, and future. *Endocr Rev* 2011;**32**(1):64–80.

43. Zucchi R, Ghelardoni S, Chiellini G. Cardiac effects of thyronamines. *Heart Fail Rev* 2010;**15**(2):171–6.

44. Scanlan TS, Suchland KL, Hart ME, et al. 3-Iodothyronamine is an endogenous and rapid-acting derivative of thyroid hormone. *Nat Med* 2004;**10**(6):638–42.

45. Hart ME, Suchland KL, Miyakawa M, Bunzow JR, Grandy DK, Scanlan TS. Trace amine-associated receptor agonists: synthesis and evaluation of thyronamines and related analogues. *J Med Chem* 2006;**49**(3):1101–12.

46. Zucchi R, Accorroni A, Chiellini G. Update on 3-iodothyronamine and its neurological and metabolic actions. *Front Physiol* 2014;**5**:402.

47. Hoefig CS, Kohrle J, Brabant G, et al. Evidence for extrathyroidal formation of 3-iodothyronamine in humans as provided by a novel monoclonal antibody-based chemiluminescent serum immunoassay. *J Clin Endocrinol Metab* 2011;**96**(6):1864–72.

48. Galli E, Marchini M, Saba A, et al. Detection of 3-iodothyronamine in human patients: a preliminary study. *J Clin Endocrinol Metab* 2012;**97**(1):E69–74.

49. Roy G, Placzek E, Scanlan TS. ApoB-100-containing lipoproteins are major carriers of 3-iodothyronamine in circulation. *J Biol Chem* 2012;**287**(3):1790–800.

50. Hackenmueller SA, Marchini M, Saba A, Zucchi R, Scanlan TS. Biosynthesis of 3-iodothyronamine (T1AM) is dependent on the sodium-iodide symporter and thyroperoxidase but does not involve extrathyroidal metabolism of T4. *Endocrinology* 2012;**153**(11):5659–67.

51. Hoefig CS, Wuensch T, Rijntjes E, et al. Biosynthesis of 3-iodothyronamine from L-thyroxine in murine intestinal tissue. *Endocrinology* 2015;**156**(11):4356–64.

52. Hoefig CS, Renko K, Piehl S, et al. Does the aromatic L-amino acid decarboxylase contribute to thyronamine biosynthesis? *Mol Cell Endocrinol* 2012;**349**(2):195–201.

53. Wood WJL, Geraci T, Nilsen A, DeBarber AE, Scanlan TS. Iodothyronamines are oxidatively deaminated to iodothyroacetic acids in vivo. *Chembiochem* 2009;**10**(2):361–5.

54. Pietsch CA, Scanlan TS, Anderson RJ. Thyronamines are substrates for human liver sulfotransferases. *Endocrinology* 2007;**148**(4):1921–7.

55. Hackenmueller SA, Scanlan TS. Identification and quantification of 3-iodothyronamine metabolites in mouse serum using liquid chromatography–tandem mass spectrometry. *J Chromatogr A* 2012;**1256**:89–97.

56. Saba A, Chiellini G, Frascarelli S, et al. Tissue distribution and cardiac metabolism of 3-iodothyronamine. *Endocrinology* 2010;**151**(10):5063–73.

57. Musilli C, De Siena G, Manni ME, et al. Histamine mediates behavioural and metabolic effects of 3-iodothyroacetic acid, an endogenous end product of thyroid hormone metabolism. *Br J Pharmacol* 2014;**171**(14):3476–84.

58. Chiellini G, Frascarelli S, Ghelardoni S, et al. Cardiac effects of 3-iodothyronamine: a new aminergic system modulating cardiac function. *FASEB J* 2007;**21**(7):1597–608.

59. Braulke LJ, Klingenspor M, DeBarber A, et al. 3-Iodothyronamine: a novel hormone controlling the balance between glucose and lipid utilisation. *J Comp Physiol B* 2008;**178**(2):167–77.

60. Haviland JA, Reiland H, Butz DE, et al. NMR-based metabolomics and breath studies show lipid and protein catabolism during low dose chronic T(1)AM treatment. *Obesity (Silver Spring)* 2013;**21**(12):2538–44.

61. Manni ME, De Siena G, Saba A, et al. 3-Iodothyronamine: a modulator of the hypothalamus-pancreas-thyroid axes in mice. *Br J Pharmacol* 2012;**166**(2):650–8.

62. Klieverik LP, Foppen E, Ackermans MT, et al. Central effects of thyronamines on glucose metabolism in rats. *J Endocrinol* 2009;**201**(3):377–86.

63. Manni ME, De Siena G, Saba A, et al. Pharmacological effects of 3-iodothyronamine (T1AM) in mice include facilitation of memory acquisition and retention and reduction of pain threshold. *Br J Pharmacol* 2013;**168**(2):354–62.

64. James TD, Moffett SX, Scanlan TS, Martin JV. Effects of acute microinjections of the thyroid hormone derivative 3-iodothyronamine to the preoptic region of adult male rats on sleep, thermoregulation and motor activity. *Horm Behav* 2013;**64**(1):81–8.

65. Dinter J, Mühlhaus J, Wienchol CL, et al. Inverse agonist action of 3-iodothyronamine at the human trace amine-associated receptor 5. *PLoS One* 2015;**10**(2):e0117774.

66. Dinter J, Mühlhaus J, Jacobi SF, et al. 3-Iodothyronamine differentially modulates α-2A-adrenergic receptor-mediated signaling. *J Mol Endocrinol* 2015;**54**(3):205–16.

67. Laurino A, De Siena G, Resta F, et al. 3-Iodothyroacetic acid, a metabolite of thyroid hormone, induces itch and reduces threshold to noxious and to painful heat stimuli in mice. *Br J Pharmacol* 2015;**172**(7):1859–68.

68. Panas HN, Lynch LJ, Vallender EJ, et al. Normal thermoregulatory responses to 3-iodothyronamine, trace amines and amphetamine-like psychostimulants in trace amine associated receptor 1 knockout mice. *J Neurosci Res* 2010;**88**(9):1962–9.

69. Frascarelli S, Ghelardoni S, Chiellini G, et al. Cardiac effects of trace amines: Pharmacological characterization of trace amine-associated receptors. *Eur J Pharmacol* 2008;**587**(1–3):231–6.

70. Regard JB, Kataoka H, Cano DA, et al. Probing cell type-specific functions of Gi in vivo identifies GPCR regulators of insulin secretion. *J Clin Invest* 2007;**117**(12):4034–43.

71. Accorroni A, Criscuolo C, Sabatini M, Donzelli R, Saba A, Zucchi R. 3-Iodothyronamine effect on long-term potentiation: rescuing β-amyloid-induced neuronal dysfunction. *Thyroid* 2015;**25**(Suppl. 1):P159 [abstract].

72. Reese EA, Bunzow JR, Arttamangkul S, Sonders MS, Grandy DK. Trace amine-associated receptor 1 displays species-dependent stereoselectivity for isomers of methamphetamine, amphetamine, and para-hydroxyamphetamine. *J Pharmacol Exp Ther* 2007;**321**(1):178–86.

73. Lewin AH, Miller GM, Gilmour B. Trace amine-associated receptor 1 is a stereoselective binding site for compounds in the amphetamine class. *Bioorg Med Chem* 2011;**19**(23):7044–8.

74. Wainscott DB, Little SP, Yin T, et al. Pharmacologic characterization of the cloned human trace amine-associated receptor 1 (TAAR1) and evidence for species differences with the rat TAAR1. *J Pharmacol Exp Ther* 2006;**320**(1):475–85.

75. Lewin AH, Navarro HA, Mascarella SW. Structure-activity correlations for beta-phenethylamines at human trace amine receptor 1. *Bioorg Med Chem* 2008;**16**(15):7415–23.

76. Tan ES, Groban ES, Jacobson MP, Scanlan TS. Toward deciphering the code to aminergic G protein-coupled receptor drug design. *Chem Biol* 2008;**15**(4):343–53.

77. Tan ES, Miyakawa M, Bunzow JR, Grandy DK, Scanlan TS. Exploring the structure–activity relationship of the ethylamine portion of 3-iodothyronamine for rat and mouse trace amine-associated receptor 1. *J Med Chem* 2007;**50**(12):2787–98.

78. Chiellini G, Nesi G, Digiacomo M, et al. Design, synthesis, and evaluation of thyronamine analogues as novel potent mouse trace amine associated receptor 1 (mTAAR1) agonists. *J Med Chem* 2015;**58**(12):5096–107.

79. Revel FG, Moreau J-L, Gainetdinov RR, et al. TAAR1 activation modulates monoaminergic neurotransmission, preventing hyperdopaminergic and hypoglutamatergic activity. *Proc Natl Acad Sci USA* 2011;**108**(20):8485–90.

80. Leo D, Mus L, Espinoza S, Hoener MC, Sotnikova TD, Gainetdinov RR. TAAR1-mediated modulation of presynaptic dopaminergic neurotransmission: role of D2 dopamine autoreceptors. *Neuropharmacology* 2014;**81**:283–91.

81. Pei Y, Lee J, Leo D, Gainetdinov RR, Hoener MC, Canales JJ. Activation of the trace amine-associated receptor 1 prevents relapse to cocaine seeking. *Neuropsychopharmacology* 2014;**39**(10):2299–308.

82. Pei Y, Mortas P, Hoener MC, Canales JJ. Selective activation of the trace amine-associated receptor 1 decreases cocaine's reinforcing efficacy and prevents cocaine-induced changes in brain reward thresholds. *Prog Neuropsychopharmacol Biol Psychiatry* 2015;**63**:70–5.

83. Revel FG, Moreau J-L, Pouzet B, et al. A new perspective for schizophrenia: TAAR1 agonists reveal antipsychotic- and antidepressant-like activity, improve cognition and control body weight. *Mol Psychiatry* 2013;**18**(5):543–56.

84. Bradaia A, Trube G, Stalder H, et al. The selective antagonist EPPTB reveals TAAR1-mediated regulatory mechanisms in dopaminergic neurons of the mesolimbic system. *Proc Natl Acad Sci USA* 2009;**106**(47):20081–6.

85. Wolinsky TD, Swanson CJ, Smith KE, et al. The trace amine 1 receptor knockout mouse: an animal model with relevance to schizophrenia. *Genes Brain Behav* 2007;**6**(7):628–39.

86. Espinoza S, Lignani G, Caffino L, et al. TAAR1 modulates cortical glutamate NMDA receptor function. *Neuropsychopharmacology* 2015;**40**(9):2217–27.

87. Di Cara B, Maggio R, Aloisi G, et al. Genetic deletion of trace amine 1 receptors reveals their role in auto-inhibiting the actions of ecstasy (MDMA). *J Neurosci* 2011;**31**(47):16928–40.

88. Achat-Mendes C, Lynch LJ, Sullivan KA, Vallender EJ, Miller GM. Augmentation of methamphetamine-induced behaviors in transgenic mice lacking the trace amine-associated receptor 1. *Pharmacol Biochem Behav* 2012;**101**(2):201–7.

89. Sukhanov I, Espinoza S, Yakovlev DS, Hoener MC, Sotnikova TD, Gainetdinov RR. TAAR1-dependent effects of apomorphine in mice. *Int J Neuropsychopharmacol* 2014;**17**(10):1683–93.

90. Karmacharya R, Lynn SK, Demarco S, et al. Behavioral effects of clozapine: Involvement of trace amine pathways in C. elegans and M. musculus. *Brain Res* 2011;**1393**:91–9.

91. Espinoza S, Salahpour A, Masri B, et al. Functional interaction between trace amine-associated receptor 1 and dopamine D2 receptor. *Mol Pharmacol* 2011;**80**(3):416–25.

92. Lynch LJ, Sullivan KA, Vallender EJ, Rowlett JK, Platt DM, Miller GM. Trace amine associated receptor 1 modulates behavioral effects of ethanol. *Subst Abuse* 2013;**7**:117–26.

93. Revel FG, Meyer CA, Bradaia A, et al. Brain-specific overexpression of trace amine-associated receptor 1 alters monoaminergic neurotransmission and decreases sensitivity to amphetamine. *Neuropsychopharmacology* 2012;**37**(12):2580–92.

94. Dewan A, Pacifico R, Zhan R, Rinberg D, Bozza T. Non-redundant coding of aversive odours in the main olfactory pathway. *Nature* 2013;**497**(7450):486–9.

β-Phenylethylamine Requires the Dopamine Transporter to Release Dopamine in *Caenorhabditis elegans*

L. Carvelli

Department of Biomedical Sciences School of Medicine & Health Sciences,
University of North Dakota, Grand Forks, ND, United States

OUTLINE

INTRODUCTION

Although a number of studies have been performed to elucidate the role of β-phenylethylamine (βPEA) in the nervous system, we still do not know how and where this endogenous amine exactly affects neuronal transmission. It is well established that βPEA is synthetized and broadly expressed in the mammalian brain, but the highest levels are found in those cerebral areas containing dopaminergic neurons, that is, the nucleus accumbens, caudate putamen, and olfactory tubercles.[1] In fact, in-depth studies demonstrated that βPEA coexists with dopamine in neurons containing tyrosine hydroxylase.[2] This suggests that βPEA might affect and/or play a role in the dopaminergic system and the reward pathway in particular. The involvement of βPEA with the dopaminergic system was somehow predictable since βPEA shares chemical and structural similarities with dopamine and drugs of abuse, such as amphetamines, which directly interacts with key elements of the dopaminergic system.

The concept that βPEA is involved in the reward pathway and acts as an endogenous psychostimulant, although supported by recent data, is not new but goes back to early studies performed in 1941 in which Schulte et al. demonstrated that βPEA produced stimulation of locomotor activity in mice.[3] These results were supported by a number of studies showing that βPEA, and in general all phenylethylamine compounds including amphetamine, induces characteristic stereotyped behaviors and increases locomotor activity.[4,5] For example, Jackson[6] and Dourish[7] performed a detailed analysis of the behaviors induced by βPEA in mice and found that βPEA exerts a biphasic stimulant effect on locomotor activity. At low drug concentrations (50 mg/kg), βPEA produced a short-lived locomotor stimulation which lasts 15 min, whereas at higher doses (75–150 mg/kg) the drug produced an earlier phase of stereotyped behaviors followed by a late phase, 20–60 min postinjection, in which increased locomotor activity appeared. Also in this late phase, the increase in motor activity lasted only a few minutes. Further analysis suggested that the biphasic effect of βPEA on motor activity may be closely related to the emergence of various stereotyped behaviors: compulsive grooming, forepaw padding, sniffing, and head weaving in the earlier phase and compulsive rearing, licking, and hyperactivity in the late phase. Interestingly, these data highlighted a major difference between βPEA and amphetamine. Although these two compounds are structurally almost identical, βPEA causes behavioral effects that extinguish within few minutes, whereas amphetamine generates behaviors that last longer. It was suggested that degradation of βPEA by the monoamine oxidase-type B (MAO-B) underlies the short-lasting behavioral effects caused by βPEA with respect to amphetamine as the inclusion of the MAO-B inhibitor deprenyl enhanced the discriminative and reinforcing stimulus effects caused by βPEA in monkeys.[8]

As mentioned above, the assumption that βPEA acts mainly in the dopaminergic system originated from behavioral experiments showing that βPEA enhances locomotion and induces stereotyped behaviors similarly to amphetamine. However, this hypothesis has been proved to be true in several in vitro experiments using a variety of ex vivo preparations.[9–11] A pioneer study in this field was performed by Raiteri et al.[9] In this study, 1 μM βPEA caused dopamine release in synaptosomes prepared from rat striatal and preloaded with [^{3}H]-dopamine.[10] Interestingly, the authors also found that the βPEA-induced dopamine release was blocked by nomifensine, which is a specific and well-characterized dopamine and norepinephrine transporter inhibitor. The inhibition by nomifensine of the release of dopamine indicated that the dopamine transporter is involved in this type of dopamine release, and it provided the first direct evidence that βPEA promotes dopamine release via the dopamine transporter. This type of drug-induced neurotransmitter release via the transporter was called efflux. Since βPEA is structurally similar to dopamine, it could be argued that nomifensine blocks the βPEA-induced dopamine efflux by preventing βPEA from entering into the cell through the transporter. However, this explanation seems unlikely since βPEA may enter into the cell more easily by diffusion than through the carrier. In fact, βPEA is much more lipophilic than dopamine, with a log P value of 1.39 and 0.03, respectively. Thus the inhibition of βPEA-induced dopamine release by nomifensine can be interpreted as being largely due to inhibition of the transport-mediated efflux of cytoplasmic dopamine. The authors also suggested that this process probably involves a step in which βPEA displaced dopamine from its storage vesicles. In this case, the exit of the dopamine displaced from the storage vesicles is accelerated through the exchange at the plasma membrane with βPEA present extracellularly. This is a reasonable explanation since βPEA competes with dopamine for reuptake via the dopamine transporter.[12]

Besides promoting dopamine efflux through the dopamine transporter, βPEA also affects dopamine release through vesicle fusion. Yamada et al.[13] demonstrated that while low concentrations (300 nM) of βPEA have minimal effects on spontaneous or evoked vesicle-mediated dopamine release, it had a significant inhibitory effect on the quinpirole-induced reduction and sulpiride-induced enhancement of evoked and vesicle-mediated dopamine release in rat striatal slides. Since quinpirole is an agonist of the dopamine D2 receptors and sulpiride is an antagonist of the same receptors, it is tempting to assume that βPEA displaces quinpirole and sulpiride from their binding to the dopamine D2 receptors to prevent their effects. However no study, as of today, has shown that βPEA binds to the D2 receptors. On the other hand, several studies suggested the effects of βPEA on the D2 receptors are mediated by dopamine: βPEA induces dopamine efflux and the increased extracellular levels of dopamine act on the D2 receptors.[14–16] In this case, the higher levels of extracellular dopamine generated by βPEA would prevent sulpiride from binding to the D2 receptors and thus inhibit the sulpiride-induced enhancement of vesicle-mediated dopamine release. However, this model does not explain why βPEA prevents the quinpirole-induced reduction of vesicle-mediated dopamine release. It would be expected that the higher extracellular levels of dopamine reached following βPEA treatment would potentiate the effects of quinpirole rather than inhibit them. Therefore, it is possible that βPEA involves other targets. For instance, two independent groups demonstrated that βPEA binds and activates a specific class of trace amine receptors, the TAAR1.[17,18] These receptors are located intracellularly and have been shown to modulate the psychostimulant effects generated by phenylethylamine compounds.

Taken together, these data prove that βPEA acts at multiple targets within the dopaminergic system, making it particularly difficult to discriminate and pinpoint the role of βPEA in an already complex system such as the central nervous system.

The ability of βPEA to increase extracellular dopamine has also been demonstrated in in vivo studies.[19] These studies showed that systemic or local administration of βPEA causes a remarkable increase in dopamine concentration of dialysate samples collected from the striatum and nucleus accumbens of freely moving or anesthetized rats, while it caused only a small increase in the dialysate levels of serotonin.[20–23] The results of these studies were highly appreciated in the field because they proved that the capability of phenylethylamine compounds to induce dopamine release is not an artifact generated by synaptosomes or other in vitro preparations but is a reliable process that takes place in intact cells and living animals. Moreover, these studies also provided a direct correlation between the increased levels of extracellular dopamine caused by βPEA and the βPEA-induced stereo-typed behaviors. For example, Kuroki et al.[20] demonstrated that rats receiving daily administration of βPEA for 14 or 28 days exhibited behavioral sensitization such that the intensity of βPEA-induced stereotyped behaviors increased during the treatment. Interestingly, the authors also reported that a challenge dose after a drug-free period of 7 days reinstated the enhanced stereotyped behaviors and significantly increased the extracellular dopamine concentration in striatal perfusates collected from βPEA-treated rats with respect to saline-treated animals.

In summary, a number of studies have recognized that βPEA affects behaviors in animals, but the molecular mechanism underlying these effects has not yet been identified. Most of the studies performed to elucidate the mechanism of action of βPEA lead to two major findings: βPEA blocks the reuptake of dopamine by binding to the dopamine transporter,[24,25] and it causes dopamine efflux by inducing reverse transport of dopamine through the transporter.[10,12,26–28] The βPEA-induced dopamine efflux is carrier-mediated and Ca^{2+}- and vesicle-independent.[29]

βPEA CAUSES BEHAVIORAL CHANGES IN CAENORHABDITIS ELEGANS

βPEA generates behavioral effects in different model systems including the simple nematode Caenorhabditis elegans (C. elegans). With only 959 somatic cells, 300 of which are neurons, C. elegans possesses a well-defined nervous system capable of executing various behaviors and rudimentary learning. Although it is a primitive organism, C. elegans shares many of the essential physiological characteristics of human biology. Thus, it provides the ideal compromise between complexity and tractability. When treated with βPEA, C. elegans exhibits a unique behavior named swimming-induced paralysis (SWIP). Normally, nematodes placed in water swim vigorously for several hours. However, if βPEA is included in the solution C. elegans sinks to the bottom of the well and stops moving within a few seconds (Fig. 12.1A).[30] In other words, C. elegans exhibits SWIP. The βPEA-induced paralysis is not a permanent paralysis. As soon as βPEA is washed out or simply diluted, the animals start to swim again as in control solution. Therefore, contrary to what is seen in mammals, in C. elegans βPEA has an inhibitory rather than stimulant effect on locomotion. This is not surprising since in C. elegans the dopaminergic system evolved to slow down the animals after they encounter food.[31] The behavioral effects caused by βPEA occur within a few seconds after the drug administration, are dose-dependent and, as previously shown in mammals, are short-lasting (Fig. 12.1A). In fact, most of the animals treated with 0.3 and 0.5 mM βPEA recovered from SWIP within 6 min, 75% and 71%, respectively (Fig. 12.1A). However, at 1 mM βPEA only $36 \pm 4\%$ of animals recovered from SWIP. These results have one direct implication: the SWIP recovery is specifically linked to βPEA treatment. Specifically, it is inversely proportional to the concentration of βPEA used. Therefore, while the βPEA-induced maximal values of SWIP increases as we increase the concentration of βPEA applied, the number of animals recovering from SWIP decreases under these same conditions. Another important implication of these results is that the recover/transient effects generated by βPEA in C. elegans are most likely caused by factors other than βPEA degradation. In fact, although MAO homologs

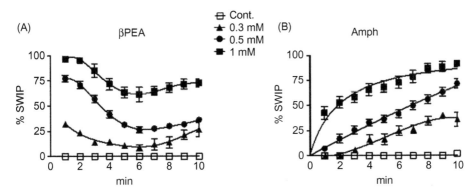

FIGURE 12.1   βPEA- and amphetamine-induced behaviors in *C. elegans*. (A) Animals placed in a solution containing 0, 0.3, 0.5, and 1mM βPEA exhibited 0%, 32%, 77%, and 96% SWIP after 1min treatment and 0%, 26%, 36%, and 73% SWIP after 10min. (B) 0, 0.3, 0.5, and 1mM amphetamine caused 0%, 36%, 65%, and 92% SWIP after 10-min treatment. *Source: These results are reproduced with permission from Safratowich BD, Hossain M, Bianchi L, Carvelli L. Amphetamine potentiates the effects of β-phenylethylamine through activation of an amine-gated chloride channel. J Neurosci 2014;***34***:4686–91.*

have been identified in *C. elegans*,[32] it is unlikely that the transient SWIP is caused by faster βPEA degradation as animals are continually immersed in a solution containing βPEA.

Previously, Carvelli et al.[33] demonstrated that SWIP can also be induced by treating *C. elegans* with another phenylethylaminic compound and well-known dopamine releaser, amphetamine (Fig. 12.1B). By combining behavioral and in vitro studies in wildtype and mutant animals, the authors suggested that amphetamine-induced SWIP is caused by an increase of extracellular concentrations of dopamine mediated by the dopamine transporter, which ultimately overstimulate the *C. elegans* D2-like dopaminergic receptors DOP3. In fact, behavioral data showed that mutants that do not produce dopamine, that is, lacking the tyrosine hydroxylase which is the key enzyme in the synthesis of dopamine, the dopamine transporter (*dat-1*) or the dop-3 receptors exhibited highly reduced SWIP with respect to wildtype animals (Fig. 12.2C). Because of the similar chemical-structural properties between amphetamine and βPEA, it would be expected that βPEA utilizes the same pathway used by amphetamine to generate SWIP. However, the comparison between the βPEA- and amphetamine-induced SWIP results highlighted a remarkable difference between these two compounds. Similarly to βPEA, amphetamine generated SWIP in a dose-dependent manner but maximal SWIP values were obtained only after prolonged treatments of about 10min (Fig. 12.1B). These results suggested that βPEA utilizes a different pathway and/or involves different proteins to generate SWIP.

The SWIP behavioral experiments performed using amphetamine clearly indicate that dopamine, which is released in the synaptic cleft by amphetamine, is the main neurotransmitter required to generate SWIP, most likely by acting directly at the DOP3 receptors. As the βPEA-induced SWIP happens within a few seconds following drug application, it is reasonable to imagine that βPEA acts as a neurotransmitter and thus binds to the DOP3 receptors causing rapid SWIP, whereas amphetamine requires a longer period because there needs to be enough released dopamine to trigger the postsynaptic DOP3 receptors.

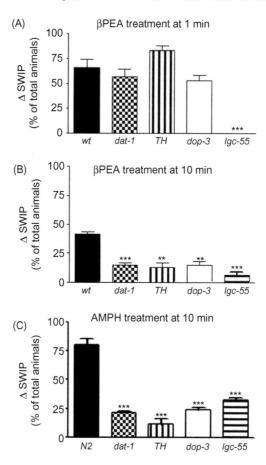

FIGURE 12.2   Proteins required for βPEA- and amphetamine-induced SWIP at different time points. (A) The dopamine transporter (*dat-1*), tyrosine hydroxylase (*TH*), or the dopamine receptor (*dop-3*) knockout animals did not exhibit statistical differences in SWIP after 1 min of 0.5 mM βPEA treatment, whereas the LGC-55 knockouts (*lgc-55*) showed 100% reduction in SWIP. (B) After 10 min, *dat-1*, *TH*, *dop-3*, and *lgc-55* knockouts exhibited 66%, 68%, 66%, and 85% SWIP reduction with respect to wt animals. (C) Ten minutes of amphetamine treatment caused the same reduction in SWIP measured after 10 min of βPEA treatment in the *dat-1*, *TH*, and *dop-3* knockout animals (***$p = 0.0001$; **$p = 0.001$, two-way ANOVA). *Source: These results are reproduced with permission from Safratowich BD, Hossain M, Bianchi L, Carvelli L. Amphetamine potentiates the effects of β-phenylethylamine through activation of an amine-gated chloride channel. J Neurosci 2014;34:4686–91.*

However, this assumption was not supported by experimental data. In fact, animals lacking expression of the DOP3 receptors still exhibited βPEA-induced SWIP after 1 min of βPEA treatment (Fig. 12.2A).[30] Moreover, mutant animals lacking the tyrosine hydroxylase or the dopamine transporter, which are both needed to generate SWIP after 10-min treatment with amphetamine (Fig. 12.2C), exhibited the same levels of βPEA-induced SWIP as the wildtype animals after 1-min treatment (Fig. 12.2A). Therefore, these results suggest that within 1 min of treatment βPEA does not induce dopamine efflux and does not require the DOP3 receptors to generate SWIP.

βPEA ACTIVATES THE LIGAND-GATED CHLORIDE CHANNEL LGC-55 TO GENERATE BEHAVIORAL EFFECTS IN C. ELEGANS

The fact that βPEA does not need dopamine, the dopamine transporter, and the DOP3 receptors to generate SWIP after 1-min treatment, suggests that it might act through one or some of the trace amine receptors identified both in C. elegans and mammals. However, animals knocked out for the trace amine receptors, such as the three isoforms of the tyramine receptor ser-2, the octopamine receptor ser-3, and the octopamine or tyramine receptor tyra-3 exhibited SWIP at the same rate as the wildtype animals.[30]

In 2009, two independent groups identified a new class of ionotropic receptors that are activated by biogenic amines.[34,35] Among these receptors, the LGC-55 has higher affinity for trace amines, such as tyramine, than for dopamine. Thus, the LGC-55 receptors could be activated by βPEA. And indeed, animals knocked out for the LGC-55 receptors completely lost the ability to display SWIP following 1-min βPEA treatment (Fig. 12.2A). These results demonstrate that the LGC-55 is the only receptor required by βPEA to generate SWIP within 1 min. After 10 min, the percentage of lgc-55 knockout animals exhibiting SWIP was still very low with respect to wildtype animals (Fig. 12.2B). However, contrary to what was seen after 1 min of βPEA treatment, after 10 min of treatment animals knocked out for tyrosine hydroxylase, the dopamine transporter, or the dop3 receptors also exhibited highly reduced SWIP (Fig. 12.2B). These results demonstrate that βPEA promotes two distinct mechanisms of action to generate SWIP: (1) within a few seconds, it rapidly activates the LGC-55 channels and (2) it indirectly acts on the DOP3 receptors by promoting dopamine release through the dopamine transporter. The latter is measurable only after prolonged treatments (5–10 min) because it requires time to build up enough extracellular dopamine levels to overstimulate the DOP3 receptors, just like amphetamine does (compare Fig. 12.2B and C). Moreover, these results demonstrate for the first time that βPEA activates a channel to alter locomotor behaviors.

The LGC-55 channels are members of the cys-loop ligand-gated channels superfamily which includes $GABA_A$, $GABA_{A-\rho}$, glycine, nicotinic acetylcholine, AMPA, NMDA, and $5-HT_3$ receptors. When activated, these receptors allow large movement of ions which can hyper- or depolarize the cell membrane. Specifically, $GABA_A$, $GABA_{A-\rho}$, and glycine receptors are Cl^--permeable, thus they hyperpolarize the cell membrane; whereas the nicotinic acetylcholine, AMPA, NMDA, and $5-HT_3$ receptors in most of the cases depolarize the cell membrane as they are permeable to cations (Na^+, K^+, Ca^{2+}). When expressed in Xenopus oocytes and activated by tyramine, the LGC-55 channels generate Cl^- currents,[34,35] suggesting therefore that the LGC-55 may hyperpolarize the cell membrane potential.

The behavioral experiments discussed above demonstrated that βPEA requires the LGC-55 to generate fast SWIP in C. elegans within few seconds. However, these experiments did not prove that βPEA is an actual substrate of these channels. For instance, βPEA could induce release of another neurotransmitter which in turn activates the LGC-55. The capability of βPEA to directly activate the channel was demonstrated in vitro by performing two-electrode voltage clamp experiments in LGC-55 expressing Xenopus oocytes. When the oocyte's membrane potential was clamped at –60 mV, βPEA generated inward Cl^- currents that were proportional to the concentration of βPEA applied.[30] These results demonstrate that βPEA directly activates the LGC-55 channels, which was not a surprise since other endogenous amines, for example, tyramine,[34,35] and less efficiently dopamine[35] and amphetamine,[36] were previously shown to directly activate the LGC-55. However, it was interesting to discover that the endogenous amines tyramine and βPEA generated larger currents and activated the LGC-55 channels more efficiently than amphetamine (K_m = 4, 9, and 152 μM, respectively). This might suggest that the LGC-55 evolved to be specifically activated by endogenous trace amines like βPEA, rather than dopamine (EC_{50} = 159 μM)[35] or amphetamine (K_m = 152 μM).[36]

Taken together, these results suggest that βPEA generates Cl^--mediated currents through the activation of the LGC-55 channels and likely hyperpolarize the cell membrane which ultimately causes SWIP. This would explain the rapid effect of βPEA in generating SWIP with respect to amphetamine.

Since βPEA and amphetamine both activate the LGC-55 channels but with different efficacy it would be interesting to test whether amphetamine prevents the activation of the LGC-55 by the endogenous βPEA or whether the two compounds act synergistically on the LGC-55 channels. To investigate this hypothesis Safratowich et al.[30] performed in vitro experiments in which Xenopus oocytes expressing the LGC-55 channels were perfused either with low concentrations of amphetamine (1–10 μM) alone or together with 1 μM βPEA, or with 1 μM βPEA alone (Fig. 12.3A–C). Interestingly, they found that the currents generated by 1 μM βPEA were increased by three times if 1 μM amphetamine was coperfused along with 1 μM βPEA (Fig. 12.3A–C). Similarly, behavioral experiments demonstrated that amphetamine enhanced the SWIP response caused by βPEA within 1 min of treatment (Fig. 12.3D). However, the potentiation of the LGC-55 channels induced by amphetamine decreased as higher concentrations of amphetamine were applied together with βPEA (Fig. 12.3B,C). These results suggest that amphetamine binds to an allosteric binding site within the LGC-55 channel. If this is the case, we could image a scenario in which, when amphetamine binds to the allosteric site and βPEA occupies the binding site, the channel adopts a more stable open conformation than if only βPEA is bound to the channel. This would explain the larger currents measured when both amphetamine and βPEA are applied (Fig. 12.3C, gray bars). It is not clear however why at higher concentrations amphetamine loses its ability to potentiate the βPEA-induced currents. One possible explanation might be found in the fact that amphetamine itself, at high concentrations, activates the LGC-55 channels although less efficiently than βPEA, K_m = 152 and 9 μM, respectively. For example, if we assume that amphetamine causes conformational changes of the channel which are distinct from those induced by βPEA, then the amphetamine-induced conformational changes allow the channel to operate at low efficacy as only small amount of currents are measured under these conditions (Fig. 12.3C, black bars). On the other hand, the βPEA-induced conformational changes would switch the channels in a state where larger amounts of current are tolerated (Fig. 12.3C, white bar), and while amphetamine stabilizes this larger-current supporting state by binding to the allosteric site, higher concentrations of amphetamine compete with βPEA for the same substrate-binding site and therefore reduce the amount of current generated. Undoubtedly, these ideas are mainly hypothetical considerations and a number of experiments need to be done to support these for-now pure speculations. But they might be true based on the fact that multiple binding sites exist at the channel. In fact, the LGC-55 channels are comprised of five subunits and each subunit contains a substrate-binding site. Therefore, in theory, five substrates could bind and activate each channel.

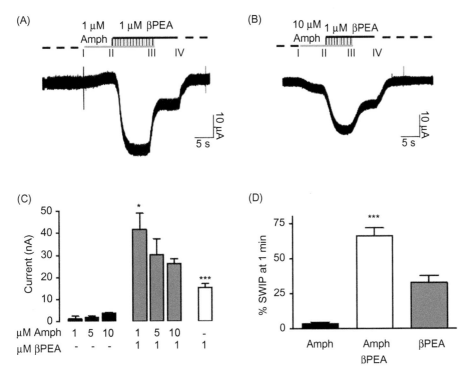

FIGURE 12.3   Amphetamine potentiates the activation of the LGC-55 channels by βPEA. (A,B) Representative recordings of LGC-55 expressing oocytes perfused with 1 μM amphetamine (Amph) alone (I–II). Subsequently, 1 μM βPEA was added together with 1 or 10 μM Amph (II–III). Amph was removed and βPEA was perfused alone (III–IV). (C) Average of currents recorded as in A and B. (D) After 1 min, cotreatment of 0.5 mM Amph and 0.3 mM βPEA induced higher SWIP with respect to Amph or βPEA treatments alone (*p=0.001, two-way ANOVA). *Source: These results are reproduced with permission from Safratowich BD, Hossain M, Bianchi L, Carvelli L. Amphetamine potentiates the effects of β-phenylethylamine through activation of an amine-gated chloride channel. J Neurosci 2014;34:4686–91.*

However, experimental data demonstrated that the probability to switch in the open state is higher if only two or three binding sites of the channel are occupied.[37] We could imagine a situation in which one or two binding sites are occupied by βPEA and only one occupied by amphetamine. This will create a more stable conformation of the channel with respect to that produced if all binding sites are occupied by βPEA which allows the channel to stay open longer and therefore to generate larger currents (Fig. 12.3C). On the other hand, if the majority of the binding sites are occupied by amphetamine, the channel adopts a conformation that permits less current to go through (Fig. 12.3C), likely because the channel is very unstable in this open state.

Although mammalian homologs of the LGC-55 channels have not yet been identified, the existence of amine-gated channels in mammals has long been suggested.[38] For instance, Hatton and Yang[39] demonstrated that in the brain, histamine generates fast inhibitory postsynaptic potentials through the activation of not-yet-identified Cl⁻ channels which are different from the well-known ionotropic GABA and glycine receptors as they are insensitive to bicuculline or strychnine. These data suggest that LGC-55 channels might indeed exist in mammals.

βPEA REQUIRES THE DOPAMINE TRANSPORTER TO INCREASE THE EXTRACELLULAR LEVELS OF DOPAMINE IN C. *ELEGANS*

The capability of βPEA to cause dopamine efflux has been documented in mammalian model systems both in vivo, by performing microdialysis, and in in vitro experiments.[14,23,40] Moreover, behavioral studies suggest that also in C. *elegans* βPEA generate dopamine efflux.[30] In fact, similarly to amphetamine, βPEA induces SWIP behaviors after 10 min of treatment only if animals express (1) a functional tyrosine hydroxylase, (2) an active dopamine transporter, and (3) DOP3 receptors (Fig. 12.2B); suggesting therefore that by inducing dopamine efflux, βPEA increases the extracellular levels of dopamine which ultimately activate the DOP3 receptors and cause SWIP. However, these data did not provide direct evidence that the C. *elegans* dopamine transporter releases dopamine in the synaptic cleft following βPEA treatment. On the other hand, Hossain et al.[41] demonstrated that using both mammalian cells expressing the C. *elegans* dopamine transporter and cultured neurons isolated from C. *elegans* embryos, βPEA increases the extracellular concentration of dopamine via the dopamine transporter. Both neurons and transfected cells were preloaded with radiolabeled dopamine and then treated with βPEA to induce efflux. The amount of radioactivity counted in the extracellular media was therefore representative of the dopamine released. In transfected cells, the amount of extracellular dopamine measured after 1 min incubation with βPEA was 120% higher than control-treated samples. This increase was even higher in neuronal cultures (160%). To prove the increase in extracellular dopamine was mediated by the dopamine transporter, the authors performed parallel experiments in which βPEA was co-applied with the dopamine transporter blocker RTI-55.[41] As expected, RTI-55 prevented the βPEA-induced dopamine efflux (Fig. 12.4). Interestingly though, while RTI-55 completely blocked the βPEA-induced dopamine efflux in the transfected cells (Fig. 12.4A), the same concentration of RTI-55 diminished the βPEA-induced dopamine release only in part in cultured neurons (Fig. 12.4B). It is possible therefore that in native systems the mechanism activated by βPEA to induce dopamine release includes other targets than just the dopamine transporter.

The βPEA-induced dopamine efflux was also detected by performing single-cell amperometry recordings.[41] In this technique, a carbon-fiber electrode is held at 700 mV, that is the potential at which dopamine is oxidized to dopamine *o*-quinone. The two electrodes released during this reaction can be detected by the same carbon fiber as an electric signal. These electric traces or oxidative currents recorded are therefore representative of the amount of dopamine in the solution. When C. *elegans* dopaminergic neurons engineered to express cytosolic GFP were perfused with 50 μM amphetamine, a transient increase in oxidative currents was measured in 44% of neurons tested.[41] As previous data showed that the dopamine transporter is expressed only in a subpopulation of dopaminergic neurons,[42] this result suggested that βPEA induces dopamine efflux only in those dopaminergic neurons expressing the dopamine transporter. This hypothesis was supported by studies showing that βPEA did not increase the basal oxidative currents in neurons isolated from animals lacking expression of the dopamine transporter.[41]

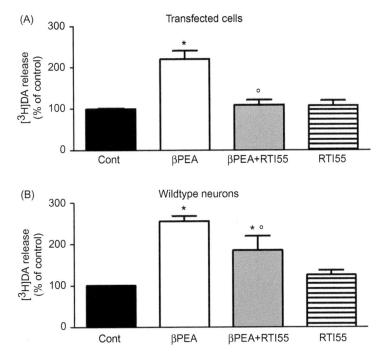

FIGURE 12.4   βPEA increases the amount of extracellular dopamine in LLC-pk1 cells expressing the C. *elegans* dopamine transporter and in cultures of C. *elegans* dopaminergic neurons. In both cell preparations, 100 µM βPEA applied for 1 min caused a significant increase in extracellular [³H]-dopamine with respect to control-treated cells. This effect was completely blocked in transfected cells by RTI-55, whereas in neuronal cultures RTI-55 only in part prevented the βPEA-induced dopamine release. In both cell preparations, RTI-55 alone did not alter the extracellular concentration of [³H]-dopamine. (*$p < 0.001$, °$p < 0.01$; two-way ANOVA). *Source: These results are reproduced with permission from Hossain M, Wickramasekara RN, Carvelli L. β-Phenylethylamine requires the dopamine transporter to increase extracellular dopamine in Caenorhabditis elegans dopaminergic neurons. Neurochem Int 2014;73:27–31.*

IN C. *ELEGANS* βPEA DOES NOT REQUIRE THE VESICULAR MONOAMINE TRANSPORTER TO GENERATE DOPAMINE EFFLUX

Although βPEA has been the subject of intensive investigation for many years, the intraneuronal pools of dopamine that contribute to βPEA-induced dopamine efflux have not been elucidated. Previous studies suggested that βPEA as well as amphetamine, release dopamine from a single cytoplasmic pool that is maintained by continuous dopamine synthesis and by spontaneous and drug-induced release of dopamine from the vesicular store.[14,28] This result was supported by the fact that βPEA-induced release of dopamine is Ca²⁺-independent, carrier-mediated, and is not affected by pretreatment with the vesicular monoamine transporter (VMAT) blocker reserpine. In line with these data, reserpine did not affect the stereotyped behaviors induced by βPEA in rats as well.[43] The observation that the increase of extracellular dopamine by βPEA does not require Ca²⁺ but is carrier-dependent excludes the possibility that dopamine is released in the synaptic cleft through vesicle fusion but rather suggest that dopamine is released through the transporter. This conclusion was further supported by data obtained in *C. elegans* mutants. As this model system offers the advantage to measure the activity of the dopamine transporter in cultured dopaminergic neurons, Hossain et al.[41] investigated whether genetic ablation of the VMAT impacts the ability of βPEA to induce dopamine efflux. They found that βPEA increases the extracellular levels of dopamine in neurons lacking expression of VMAT as much as it does in wildtype neurons, proving therefore that vesicular release of dopamine is not involved in the mechanism of action of βPEA.

CONCLUSION

In summary, the use of *C. elegans* as a model organism has generated data showing that βPEA acts primarily at the dopaminergic system to generate behavioral effects.[30] Moreover, thanks to this model, it was possible to identify a new target, the ligand-gated ion channel LGC-55, required by βPEA to induce fast and transient behaviors within 1 min of treatment. If LGC-55 homologs are proven to exist in mammals, these data might explain the biphasic behavioral effects observed in rodents when treated with βPEA.[6]

References

1. Paterson IA, Juorio AV, Boulton AA. 2-Phenylethylamine: a modulator of catecholamine transmission in the mammalian central nervous system? *J Neurochem* 1990;**55**:1827–37.
2. Juorio AV, Paterson IA, Zhu MY, Matte G. Electrical stimulation of the substantia nigra and changes of 2-phenylethylamine synthesis in the rat striatum. *J Neurochem* 1991;**6**:213–20.
3. Schulte J, Reif EC, Bader JA, Lawrence WS, Tainter ML. Further study of central stimulation from sympathomimetic amines. *J Pharmacol Exp Ther* 1941;**71**:62–72.
4. Dourish C. Behavioural effects of acute and chronic beta-phenylethylamine administration in the rat: evidence for the involvement of 5-hydroxytryptamine. *Neuropharmacology* 1981;**20**:1067–72.
5. Lapin IP. Antagonism by CPP (+/−)-3-(2-carboxypiperazin-4-yl)-propyl-1-phosphonic acid, of beta-phenylethylamine (PEA)-induced hypermotility in mice of different strains. *Pharmacol Biochem Behav* 1996;**55**:175–8.
6. Jackson D. The effect of β-phenylethylamine upon spontaneous motor activity in mice: a dual effect o locomotor activity. *J Pharm Pharmacol* 1972;**24**:383–9.
7. Dourish CT. A pharmacological analysis of the hyperactivity syndrome induced by beta-phenylethylamine in the mouse. *Br J Pharmacol* 1982;**77**:129–39.
8. Bergman J, Yasar S, Winger G. Psychomotor stimulant effects of b-phenylethylamine in monkeys treated with MAO-B inhibitors. *Psychopharmacology (Berl)* 2001;**159**:21–30.
9. Raiteri M, Bertollini A, del Carmine R, Levi G. Effects of phenethylamine derivatives on the release of biogenic amines from synaptosomes. *Biochem Soc Trans* 1976;**4**:121–4.
10. Raiteri M, Cerrito F, Cervoni AM, Levi G. Dopamine can be released by two mechanisms differentially affected by the dopamine transport inhibitor nomifensine. *J Pharmacol Exp Ther* 1979;**208**:195–202.
11. Dyck L. Release of monoamines from striatal slices by phenelzine and beta-phenylethylamine. *Prog Neuropsychopharmacol Biol Psychiatry* 1983;**7**:797–800.
12. Raiteri M, Del Carmine R, Bertollini A, Levi G. Effect of sympathomimetic amines on the synaptosomal transport of noradrenaline, dopamine and 5-hydroxytryptamine. *Eur J Pharmacol* 1977;**41**:133–43.
13. Yamada S, Harano M, Tanaka M. Antagonistic effects of beta-phenylethylamine on quinpirole- and (−)-sulpiride-induced changes in evoked dopamine release from rat striatal slices. *Eur J Pharmacol* 1998;**343**:145–50.
14. Ishida K, Murata M, Katagiri N, Ishikawa M, Abe K, Kato M, et al. Effects of beta-phenylethylamine on dopaminergic neurons of the ventral tegmental area in the rat: a combined electrophysiological and microdialysis study. *J Pharmacol Exp Ther* 2005;**314**:916–22.

15. Geracitano R, Federici M, Prisco S, Bernardi G, Mercuri NB. Inhibitory effects of trace amines on rat midbrain dopaminergic neurons. *Neuropharmacology* 2004;**46**:807–14.
16. Paetsch P, Greenshaw AJ. Down-regulation of beta-adrenergic and dopaminergic receptors induced by 2-phenylethylamine. *Cell Mol Neurobiol* 1993;**13**:203–15.
17. Borowsky B, Adham N, Jones KA, Raddatz R, Artymyshyn R, Ogozalek KL, et al. Trace amines: identification of a family of mammalian G protein-coupled receptors. *Proc Natl Acad Sci USA* 2001;**98**:8966–71.
18. Bunzow J, Sonders MS, Arttamangkul S, Harrison LM, Zhang G, Quigley DI, et al. Amphetamine, 3,4-methylenedioxymethamphetamine, lysergic acid diethylamide, and metabolites of the catecholamine neurotransmitters are agonists of a rat trace amine receptor. *Mol Pharmacol* 2001;**60**:1181–8.
19. Philips SR, Robson AM. In vivo release of endogenous dopamine from rat caudate nucleus by phenylethylamine. *Neuropharmacology* 1983;**22**:1297–301.
20. Kuroki T, Tsutsumi T, Hirano M, Matsumoto T, Tatebayashi Y, Nishiyama K, et al. Behavioral sensitization to beta-phenylethylamine (PEA): enduring modifications of specific dopaminergic neuron systems in the rat. *Psychopharmacology (Berl)* 1990;**102**:10.
21. Nakamura M, Ishii A, Nakahara D. Characterization of beta-phenylethylamine-induced monoamine release in rat nucleus accumbens: a microdialysis study. *Eur J Pharmacol* 1998;**349**:163–9.
22. Ishida K, Murata M, Katagiri N, Ishikawa M, Abe K, Kato M, et al. Effects of beta-phenylethylamine on dopaminergic neurons of the ventral tegmental area in the rat: a combined electrophysiological and microdialysis study. *J Pharmacol Exp Ther* 2005;**314**:916–22.
23. Murata M, Katagiri N, Ishida K, Abe K, Ishikawa M, Utsunomiya I, et al. Effect of beta-phenylethylamine on extracellular concentrations of dopamine in the nucleus accumbens and prefrontal cortex. *Brain Res* 2009;**1269**:40–6.
24. Horn AS, Snyder SH. Steric requirements for catecholamine uptake by rat brain synaptosomes: studies with rigid analogs of amphetamine. *J Pharmacol Exp Ther* 1972;**180**:523–30.
25. Sulzer D, Sonders MS, Poulsen NW, Galli A. Mechanisms of neurotransmitter release by amphetamines: a review. *Prog Neurobiol* 2005;**75**:406–33.
26. Dyck L. Release of monoamines from striatal slices by phenelzine and beta-phenylethylamine. *Prog Neuropsychopharmacol Biol Psychiatry* 1983;**7**:797–800.
27. Bailey BA, Philips SR, Boulton AA. In vivo release of endogenous dopamine, 5-hydroxytryptamine and some of their metabolites from rat caudate nucleus by phenylethylamine. *Neurochem Res* 1987;**12**:173–8.
28. Parker E, Cubeddu LX. Comparative effects of amphetamine, phenylethylamine and related drugs on dopamine efflux, dopamine uptake and mazindol binding. *J Pharmacol Exp Ther* 1988;**245**:199–210.
29. Liang NY, Rutledge CO. Evidence for carrier-mediated efflux of dopamine from corpus striatum. *Biochem Pharmacol* 1982;**31**:2479–84.
30. Safratowich BD, Hossain M, Bianchi L, Carvelli L. Amphetamine potentiates the effects of β-phenylethylamine through activation of an amine-gated chloride channel. *J Neurosci* 2014;**34**:4686–91.
31. Sawin E, Ranganathan R, Horvitz HR. C. elegans locomotory rate is modulated by the environment through a dopaminergic pathway and by experience through a serotonergic pathway. *Neuron* 2000;**26**:619–31.
32. Weyler W. Evidence for monoamine oxidase in *Caenorhabditis elegans*. *Worm Breeder's Gazette* 1992;**12**:52.
33. Carvelli L, Matthies DS, Galli A. Molecular mechanisms of amphetamine actions in *Caenorhabditis elegans*. *Mol Pharmacol* 2010;**78**:151–6.
34. Ringstad N, Abe N, Horvitz HR. Ligand-gated chloride channels are receptors for biogenic amines in *C. elegans*. *Science* 2009;**325**:96–100.
35. Pirri JK, McPherson AD, Donnelly JL, Francis MM, Alkema MJ. A tyramine-gated chloride channel coordinates distinct motor programs of a *Caenorhabditis elegans* escape response. *Neuron* 2009;**62**:526–38.
36. Safratowich BD, Lor C, Bianchi L, Carvelli L. Amphetamine activates an amine-gated chloride channel to generate behavioral effects in *Caenorhabditis elegans*. *J Biol Chem* 2013;**288**:21630–7.
37. Sine SM, Engel AG. Recent advances in Cys-loop receptor structure and function. *Nature* 2006;**440**:448–55.
38. Yang Q, Hatton GI. Histamine mediates fast synaptic inhibition of rat supraoptic oxytocin neurons via chloride conductance activation. *Neuroscience* 1994;**61**:955–64.
39. Hatton GI, Yang QZ. Ionotropic histamine receptors and H_2 receptors modulate supraoptic oxytocin neuronal excitability and dye coupling. *J Neurosci* 2001;**21**:2974–82.
40. Sotnikova TD, Budygin EA, Jones SR, Dykstra LA, Caron MG, Gainetdinov RR. Dopamine transporter-dependent and -independent actions of trace amine beta-phenylethylamine. *J Neurochem* 2004;91362–73.

41. Hossain M, Wickramasekara RN, Carvelli L. β-Phenylethylamine requires the dopamine transporter to increase extracellular dopamine in *Caenorhabditis elegans* dopaminergic neurons. *Neurochem Int* 2014;**73**:27–31.

42. Li X, Qi J, Yamaguchi T, Wang HL, Morales M. Heterogeneous composition of dopamine neurons of the rat A10 region: molecular evidence for diverse signaling properties. *Brain Struct Funct* 2012;**218**:1159–76.

43. Barroso N, Rodriguez M. Action of beta-phenylethylamine and related amines on nigrostriatal dopamine neurotransmission. *Eur J Pharmacol* 1996;**297**:195–203.

SECTION II

TRACE AMINES AND OLFACTION

13

Trace Amine-Mediated Olfactory Learning and Memory in Mammals and Insects: A Brief Comparative Review

T. Farooqui

Department of Entomology, The Ohio State University, Columbus, OH, United States

Trace Amines and Neurological Disorders
DOI: http://dx.doi.org/10.1016/B978-0-12-803603-7.00013-6

INTRODUCTION

Olfaction is a sense, in which external chemical information is carried and transformed into the central nervous system in patterns of brain activity, which is involved in mediating odor perception. This ancient sensory modality plays a crucial role in most animals by providing an ability to detect and discriminate among a wide variety of odors in the external environment. Thus most animals can approach or avoid various odor sources by judging their quality in a variety of behavioral contexts involving their survival and reproduction. Insects rely heavily on olfaction to guide various important activities, including foraging, mating, as well as allowing social interactions, oviposition, and predator avoidance. Such a heavy reliance on olfactory information has made insects attractive model systems for dissecting olfactory memory formation.[1] Moreover, the olfactory nervous systems of insects are remarkably homologous in structure and function to those of vertebrates, suggesting that the mechanisms for olfactory learning may be shared.[2] For example, after an animal smells an odor, olfactory sensory neurons generate an activity pattern across olfactory glomeruli of the first sensory neuropil, the vertebrate olfactory bulb (OB), or insect antennal lobe (AL), where they make synapses with secondary neurons, local neurons, and output neurons; and information is then relayed to higher-order brain centers by mitral/tufted cells in vertebrates,[3] or by projection neurons in insects.[4] Insects offer unique advantages to the olfaction field, starting from the olfactory conditioning in the honeybee *Apis mellifera*,[5–8] molecular genetics tractability in fruit fly *Drosophila melanogaster*,[9–12] fast acquisition, long retention, and easy rewriting of memory in cricket *Gryllus bimaculatus*,[13,14] to broad expression of the olfactory system and phenotypic transformation affecting associative learning in desert locust *Schistocerca gregaria*.[15,16] Among insects, the honeybee *A. mellifera* worker has emerged to be a robust, influential, and central insect model to study olfactory learning and memory due to its sophisticated cognitive abilities in both foraging and task performance within the colony,[8,17–19] and ease of detecting the behavioral,[20–22] neuroanatomical and neurophysiological,[23,24] and pharmacological and reverse genetics approaches,[25,26] as well as the ability to create iron-mediated oxidative stress in the honeybee brain.[27] These findings collectively support that this social insect may serve as a potential valid model for olfactory dysfunction in humans with aging and neurological disorders.

Trace amines (tyramine, β-phenylethylamine, tryptamine, and octopamine) are biogenic amines, which are found in trace levels (0.1–10 nm), representing <1% of total biogenic amines in the mammalian central nervous system.[28] However, the prominent biogenic amines found in insects are octopamine and its biochemical precursor tyramine, which are thought to be the chief insect neurotransmitters,[29–32] because they fulfill similar functions in invertebrates as epinephrine and norepinephrine do in vertebrates. Tyramine and octopamine also play important neuromodulatory roles in invertebrates, particularly in insects.[33–36] Octopamine shows arousing effect and leads to higher sensitivity for sensory inputs, induced learning performance and increased foraging behavior, whereas tyramine acts antagonistically to octopamine.[33,34]

Age-related olfactory dysfunction is very common in humans.[37] Olfactory and gustatory disorders increase with advancing age.[38] Olfactory disorders are common characteristics in several neurological disorders.[39–53] Extensive investigations have confirmed the association

TABLE 13.1   Olfactory Dysfunction is a Common Symptom of Aging and in Patients with Neurodegenerative and Neuropsychiatric Disorders

Neurological Disorders	Olfactory Dysfunction	References
Aging	√	37–39
Parkinson disease	√	40–42
Alzheimer disease	√	40,41,43,44
Huntington disease	√	45,46
Multiple sclerosis	√	47,48
Amyotrophic lateral	√	49,50
Schizophrenia	√	51–53

between several neurological disorders (particularly neurodegenerative diseases, such as Parkinson disease (PD) and other synucleinopathies, Alzheimer disease (AD), and mild cognitive impairment progressing to dementia), and in neuropsychiatric diseases (such as schizophrenia) and olfactory dysfunction (Table 13.1), implicating its high diagnostic significance in humans. Moreover, dramatically altered trace amine levels have been reported in several neurological disorders (as described in chapter: Trace Amines in Neuropsychiatric Disorders), suggesting the involvement of trace amines in the pathophysiology of these diseases.[28,54,55]

This chapter will be presented in five main sections:

1. In the first section, I discuss trace amines as neuromodulators that play crucial roles for animal behavior.
2. In the second section, I compare the types of chemosensory receptors in mammals and insects olfactory systems.
3. In the third section, I describe the trace amine-associated receptors (TAARs) in vertebrates.
4. In the fourth section, I describe octopamine and tyramine receptors in invertebrates.
5. The fifth section concludes whether insects have something profound to offer if used as a potential model to understand the molecular mechanism underlying olfactory dysfunction during aging and neurological disorders.

TRACE AMINES: AS NEUROTRANSMITTER, NEUROMODULATOR, AND NEUROHORMONE

Trace amines belong to a group of endogenous amines, which are structurally, metabolically, physiologically, and pharmacologically similar to the monoamine neurotransmitters, such as dopamine, norepinephrine, and serotonin (5-hydroxytryptamine, 5-HT) in the mammalian brain.[28,54] Unlike the classical monoamines, aromatic-L-amino acid decarboxylase (AADC) is the sole enzyme required for the synthesis of trace amines.[28] Moreover, trace amines cannot be converted into monoamines.[55] Their uniqueness is mainly based on their trace levels (several hundred fold lower than classical neurotransmitters), high rates of metabolism (rapid turnover rate), and heterogeneous distribution in the mammalian brain and peripheral nervous tissues.[55,56] In contrast to vertebrates, trace amines (such as octopamine and tyramine) are present in high concentration and serve as well-established neurotransmitters, neuromodulators, and neurohormones in many invertebrate species, including insects.[29–36]

Both neurotransmitters and neuromodulators play important roles in animal cognition and behavior. Similarly, *neurohormones* are involved in regulating the expression of behavior. Neurotransmitter-mediated synaptic transmission (classic transmission) and neuromodulator-mediated synaptic transmission differ from each other in both mechanism and function:

- A neurotransmitter is a messenger released from a neuron at an anatomically specialized junction, which diffuses across a narrow cleft to affect one or sometimes two postsynaptic neurons, a muscle cell, or another effector cell in an excitatory or inhibitory manner, depending on the neurotransmitter and on the postsynaptic ionotropic receptor. A neurotransmitter's postsynaptic response is relatively fast and short-lasting (milliseconds).
- A neuromodulator is released from a neuron in the central nervous system or peripheral nervous system, affecting other neurons or effector cells by binding to appropriate metabotropic postsynaptic receptors. A neuromodulator modulates (enhance or suppress) the effects of neurotransmitters at the synapse, producing a slow but lasting response (seconds to hours).
- A neurohormone is released into the circulatory system (hemolymph, blood) by specialized cells (neurosecretory cells). By binding to the metabotropic receptors or cytoplasmic receptors, it leaves a long-lasting response (months to years to life-long). A neurohormone may exert its effects on distant peripheral targets and may differ only in degree from a neuromodulator in the extent of its action.

Trace Amine-Mediated Neuromodulatory Actions in Vertebrates

The central and peripheral systems of vertebrates traditionally include β-phenyl ethylamine, *para* tyramine, octopamine, and tryptamine.[57–60] Trace amines exert their effect by activating a class of G-protein-coupled receptors (GPCRs), termed trace amine-associated receptors (TAARs). TAAR1 is the most widely studied of these receptors, and is the only receptor that is expressed and widely distributed in human brain monoaminergic systems. TAAR1 binds to endogenous trace amines, a wide spectrum of biogenic amines, and psychoactive compounds, and is known to modulate monoaminergic activity. Under normal conditions, trace amines are present in the mammalian brain and peripheral nervous tissues at very low (<10 nM) concentrations that are several hundred-fold below those for the classical monoaminergic neurotransmitters (dopamine, norepinephrine, epinephrine, and serotonin). However, these levels are dramatically altered (unusually high or low) with an imbalance in their functions in several neurological disorders.[28,61–64] A substantial body of evidence is available to suggest that trace amines may play significant roles in the coordination of biogenic amine-based synaptic physiology.[28,54] Trace amines act as endogenous

TABLE 13.2   Trace Amines and Their Modulatory Roles in Mammals

Trace Amines	TAARS	Neuromodulatory Role
Tyramine and tryptamine	TAAR1, TAAR4	Modulates locomotor function in the mammalian spinal cord[68]
β-PE	TAAR1	TAAR1 activity depresses monoamine transport and limits dopaminergic and serotonergic neuronal firing rates via interactions with presynaptic D2 and 5-HT1A autoreceptors, respectively[65,66]
β-PE	TAAR1	Modulates catecholaminergic neurotransmission[65,66]
β-PE	TAAR1	Modulatory effects on monoamine transporters by interacting with TAAR1[69]
RO5166017	TAAR1	Modulates not only DA but also 5-HT neurotransmission[70]
Tyramine, β-PE, and amphetamine	TAAR1	↑ Locomotion in rodents and primates[71–74]
Amphetamine	TAAR1	*Taar1*$^{-/-}$ mice are hypersensitive to the psychostimulant amphetamine[69,75]
RO5073012	TAAR1	*Taar1* Tg mice are hyposensitive to d-amphetamine, but attenuating TAAR1 activity with RO5073012 restored the stimulating effects of amphetamine on locomotion[76]

β-PE, β-phenylethylamine; *DA*, dopamine; *5-HT*, serotonin; *RO5073012*, a partial TAAR1 agonist; *RO5166017*, a selective TAAR1 agonist; *Taar1−/−* mice, mutant mice lacking the *Taar1* gene; *Taar1* Tg mice, overexpression of *Taar1* gene.

neuromodulators of monoaminergic excitability and neurotransmission.[28,65] Trace amines have been suggested to contribute to several physiological processes, including modulation of aminergic neurotransmission (Table 13.2).[65–76] It was earlier proposed that TAAR1 inhibits locomotor activity via a down-modulating dopamine neurotransmission,[66] which is now supported by another study demonstrating TAAR1-mediated activation of K$^+$ channels in dopamine neurons.[67] TAAR1 activates K$^+$ channels in dopamine tonically, the overruling effect of blocking TAAR1 is a net increase in the firing rate of dopamine neurons. Antagonizing TAAR1 with its selective antagonist N-(3-ethoxyphenyl)-4-(pyrrolidin-1-yl)-3-(trifluoromethyl)benzamide (EPPTB) in this study reveals TAAR1-mediated regulatory mechanisms in dopaminergic neurons of the mesolimbic system. Thus EPPTB-induced increase in the potency of dopamine at D2 receptors may be a part of a homeostatic feedback mechanism compensating for the lack of inhibitory TAAR1 tone, which could provide therapeutic potential for treating PD.[67] In contrast, TAAR1 has been demonstrated to negatively modulate the dopaminergic system, suggesting that it can also serve as a novel target for treating neuropsychiatric disorders, such as schizophrenia.[72] Studies in mutant mice lacking the *Taar1* gene have reported that subjects were hypersensitive to amphetamine (a psychostimulant), as evidenced by increased locomotor activity response and striatal release of monoamines following an acute challenge.[66,75] Revel et al.[76] generated a line of transgenic mice that overexpresses *Taar1* gene in the brain. In contrast to *Taar1*$^{-/-}$ mice, they reported that *Taar1* Tg mice are hyposensitive to d-amphetamine, suggesting that TAAR1 overexpression alters monoaminergic neurotransmission. Collectively, these studies suggest that modulation of monoaminergic neurotransmission by TAAR1 may impact on a broad variety of neurophysiological functions in vertebrates.

Trace Amine-Mediated Neuromodulatory Actions in Invertebrates

In invertebrates, biogenic amine neurotransmitters (norepinephrine and epinephrine) are essentially lacking. However, they are replaced by related biogenic amines (*p*-tyramine and *p*-octopamine) that act as bonafide neurotransmitters, neuromodulators, and neurohormones, mediating diverse complex behaviors and many crucial physiological processes.[29–36,77–86] Both octopamine and tyramine are considered as trace amines in vertebrates, but they are present in high concentration in the peripheral nervous system, central nervous system, and various other tissues in invertebrates, including insects. The neuromodulatory role of octopamine in regulating diverse insect behaviors has been known for some time.[35,77,78] Octopamine is known to modulate the activity of flight muscles, peripheral organs (such as fat body, oviduct, and hemocytes), and almost all sense organs in the peripheral nervous system; whereas it is essential for the regulation of a plethora of behaviors, including motivation, desensitization of sensory inputs, initiation, and maintenance of various rhythmic behaviors, locomotion, and learning and memory in the central nervous system (Table 13.3). However, the neuromodulatory role of tyramine is relatively new, because until recently tyramine was mainly considered to be a precursor of octopamine.[79]

Octopamine and tyramine exert their actions by binding to specific octopamine and tyramine receptors, which belong to the GPCR family, allowing insects to respond to external stimuli with a fine-tuned adequate response. Both octopamine and tyramine are synthesized from tyrosine, found ubiquitously in insects, and play independent but opposite roles in a wide spectrum of behaviors ranging from locomotion, aggression, mating, and egg-laying, to learning and memory.[34,36,80–86]

Collective evidence strongly suggests that octopamine and tyramine are involved in regulating different physiological functions, including modulation of sensory perception. The next section reviews recent progress on chemosensory receptors involved in the mammalian and insect olfactory systems.

CHEMOSENSORY RECEPTORS IN MAMMAL AND INSECT OLFACTORY SYSTEMS

Many animal species depend on chemosensory systems to investigate their environments for food, mates, danger, predators, and pathogens, which is essential for their survival. Such chemosensory systems have also been evolved in mammals and insects to detect and discriminate among odors. In the mammalian nose, a variety of odor receptor families are expressed in olfactory receptor neurons (ORNs) (Table 13.4). These chemosensory GPCRs receptors belong to five classes of mammalian receptors: (1) odorant receptors (ORs);[87–89] (2) trace amine-associated receptors (TAARs);[86,88–91] (3) two distinct vermonasal

TABLE 13.3 Neuromodulatory Roles of Octopamine in Insects

Trace Amine	Neuromodulatory Roles	Insect Species
Octopamine	Olfactory learning and memory	*A. mellifera, D. melanogaster*, and *G. bimaculatus*
Octopamine	Vision	Locusts and *A. mellifera*
Octopamine	Motivation	*Locusta migratoria*
Octopamine	Feeding behavior	*A. mellifera* and *Phormia regina*
Octopamine	Discrimination of nestmates from non-nestmates	*A. mellifera* and *Solenopsis invicta*
Octopamine	Dance behavior	*A. mellifera*
Octopamine	Sting response	*A. mellifera*
Octopamine	Motor control	*L. migratoria*
Octopamine	Rhythmic behavior	*S. gregaria* and *Manduca sexta*
Octopamine	Division of labor	*A. mellifera*
Octopamine	Activity and energy metabolism of flight muscles, visceral muscle, peripheral organs, and sense organs	*L. migratoria* and *Acheta domesticus*
Octopamine	Conditional courtship	*D. melanogaster*
Octopamine	Sensitization and dishabituation	*L. migratoria*
Octopamine	Locomotion and grooming	*D. melanogaster*
Octopamine	Lipid and carbohydrate metabolism	*Acheta domesticus* and *L. migratoria*
Octopamine	Ovulation	*D. melanogaster*

*Information has been adapted from Roeder T. Octopamine in invertebrates. Prog Neurobiol 1999;**59**:533–54, Farooqui T. Octopamine-mediated neuromodulation of insect senses. Neurochem Res 2007; **32**(9):1511–29, and Farooqui T. Review of octopamine in insect nervous systems. Open Access Insect Physiol 2012;**4**:1–17.*

receptors (V1Rs and V2Rs);[86,92,93] (4) formyl peptide receptors (FPRs);[86,94,95] and (5) two distinct guanylyl cyclase receptors (GC-D and GC-G).[88,96–99] A minority of ORNs express TAARs, some of which respond to volatile amines found in urine, which likely act in the detection of social cues. All five classes of receptors are predicted to contain 7-transmembrane (7-TM) domains and have either been shown to signal via guanine nucleotide-binding proteins (so-called G proteins) or are likely to do so based on their sequence similarity to known GPCRs. In addition to these receptors, odorant-binding proteins (OBPs) have been observed in a variety of vertebrate species. These soluble proteins (150–160 amino acids long) of vertebrates are present in the perireceptor area, belong to the superfamily of lipocalins, and reversibly bind odorant molecules with μM affinities. OBPs are carrier proteins, folded in typical β-barrel shape, with eight β-sheets and one short segment of α-helix close to the C-terminus (Table 13.4).[100,101]

TABLE 13.4   Types of Chemosensory Receptors in Mammals and Insects

I. Mammalian Olfactory System	Ligands	Receptor Expression and Functional Properties
1. ORs[87–89]	Odorants	• On the cilia of OSNs in the MOE • Largest gene family in mammals • GPCRs (7-α-helical TMD), monomer • OR interaction with odorants is followed by coupling to a G-protein, activation of AC, and opening of cyclic nucleotide-gated channels • One OR recognizes multiple odorants
2. TAARs[88,90–93]	Volatile and highly aversive amines	• On the cilia of OSNs in the MOE and GG • A smaller gene family found in diverse vertebrates • GPCRs (7 α-helical TMD), monomer • Not related to ORs but distantly related to BARs • Mediate aversion or attraction towards these amines
3. V1Rs[88,94,102]	Small, volatile molecules and sulfated steroids	• VSNs in VNO in nasal septum express VRs • V1Rs expressed at the apical layer of VNO and MOE • V1Rs are GPCRs, monomer • Distantly related to the receptors of MOE • Coexpressed with $G_{\alpha i2}$ in neurons in an apical zone • Mainly associated with pheromones recognition
V2Rs[88,94,102]	Peptides, MUPs and sulfated steroids	• V2Rs expressed in the basal layer of VNO and GG • Coexpressed with $G_{\alpha o}$ in neurons in a basal zone • V2Rs are GPCRs, monomer and heteromer • Mainly associated with pheromone recognition
4. FPRs[88,94,95]	Pathogen- and inflammation-related compounds	• FPRs are expressed at the apical layer of VNO • GPCR, oligomeric state unknown • Consistently express $G_{\alpha i2}$ or $G_{\alpha o}$ • VNO FPRs possible role: detection of conspecifics or other species based on variations in normal bacterial flora or mitochondrial proteins • Other FPRs function in the immune system
5. GC-D[88,96,97]	Extracellular: uroguanylin and guanylin Intracellular: bicarbonate, Ca^{2+}, and neurocalcin-δ	• A subset of MOE neurons express GC-D receptor and the cyclic nucleotide-gated channel subunit (CNGA3) • Mediate excitatory cGMP-dependent transduction mechanism for odor recognition • Monotopic receptors (RTK type), dimer • OSNs use carbonic anhydrase to catalyze the conversion of CO_2 to bicarbonate and the opening of cGMP-sensitive ion channels • Bicarbonate activates cGMP-producing ability GC-D • The molecular mechanism of GC-D activation is distinct from other membrane GCs • T1Rs recognize sweet and umami stimuli • T2Rs detect bitter taste

(Continued)

TABLE 13.4   (Continued)

I. Mammalian Olfactory System	Ligands	Receptor Expression and Functional Properties
GC-G[88,98,99]	2,3-DMP and TMT may be a potential ligand	• Expressed in GG neurons • Olfactory subsystem coexpresses cGMP and other signaling components (cGMP-regulated PDE2A and cGMP-gated CNGA3) • GC-G is activated by CO_2/bicarbonate • TMT may elicit innate defensive response • 2,3-DMP-sensitive GG neurons express signaling proteins associated with the second messenger cGMP (CNGA3 and GC-G), suggesting that cGMP may play a crucial role for odorant-induced electrical responses
6. OBPs[100,101]	2-Isobutyl-3-methoxypyrazine and odorants	• Secreted in nasal olfactory mucosa • Length: (~150–160 residues long), with β-sheets and one short segment of α-helix close to C-terminus • Carrier proteins belong to lipocalins family • Low detection threshold

II. Insect Olfactory System	Ligands	Receptor Expression and Functional Properties
1. ORs[88,103–106]	Food odors, general odors, pheromones	• Each OSN expresses between one and three ligand-binding members of the OR gene family • ORs are expressed in the olfactory organs (antenna and maxillary palp) and act as ligand-gated ion channels • A distinct 7-TMD topology with intracellular amino-termini compared with mammalian ORs • Heteromeric complexes of subunits (odor-specific OrX protein and a coreceptor Orco) • Lack homology to GPCRs or vertebrate ORs • OR/Orco complex forms odorant-activated ionotropic cation channels • Activated by volatile chemical cues • Can be sensitized by repeated subthreshold odor stimulation, involving metabotropic signaling • Primary transduction mechanism ionotropic
2. GRs[88,97,107]	CO_2	• Expressed on antenna (basiconic sensilla), proboscis, legs, wings, and genitalia • Hetrodimer (Gr21a and Gr63a) mediate CO_2 response • 7-TMDs, but adopt distinct membrane topology with intracellular amino-termini • $G_{\alpha q}$ signaling pathway to detect CO_2 • Distantly related to OR genes (Or83b) • Detect sweet and bitter tastants • G proteins ($G_{\alpha q}$ and $G_{\gamma 30A}$) play role in the response of CO_2-sensing neurons, but are not required for odor-mediated signaling • GRs play a role in accepting and rejecting different foods, and in the process of selecting an oviposition site

(Continued)

TABLE 13.4   (Continued)

II. Insect Olfactory System	Ligands	Receptor Expression and Functional Properties
3. IRs[88,108,109]	Ammonia, water, amines, vapor, and alcohols	• Expressed on antenna (coeloconic sensilla) • 7-TMDs, form heterocomplex assemblies • IRs are not related to insect ORs, but rather have evolved from iGluRs • IRs lack glutamate-interacting residues • Specifically respond to amines or acid-based odorants, which are ignored by ORs
4. OBPs[110,111]	Pheromone delivery	• Secreted in the lymph of chemosensilla • Length: (~130–140 residues long), made of six α-helical domains • Characterized by a pattern of six conserved cysteines paired into three interlocked disulfide bridges, assembled in strongly constrained folding • OBP genes # in different insect species: highly variable • Reversibly bind odorants and pheromones • Dual role: release of odorants and activation of ORs • OBPs bind semiochemicals, and activate specific chemoreceptors • Role of OBPs in chemodetection
5. CSPs[111–113]	Pheromone delivery	• Found in olfactory and gustatory organs of insects • Length: (~100–120 residues long), rich in α-helical domains • Possesses a conserved pattern of four cysteines forming two independent loops, assembled in a more flexible folding • # CSP genes: highly variable in different insect species • CSPs bind semiochemicals and activate specific chemoreceptors • Role of CSPs in chemodetection

GPCRs, G-protein-coupled receptors; ORs, olfactory receptors; BARs, biogenic amine receptors; TAARs, trace amine-associated receptors; VNO, vomeronasal organ; V1R, V2R, vomeronasal receptors; VSNs, vomeronasal sensory neurons; FPRs, formyl peptide receptors; GC-D and GC-G, guanylyl cyclase types D and G; phosphodiesterase 2A, cGMP-regulated PDE2A; cyclic nucleotide-gated cation channel α-3, cGMP-gated ion channel CNGA3; RTK, receptor tyrosine kinases; GRs, gustatory receptors; IRs, ionotropic receptors; iGluRs, ionotropic glutamate receptors; LIGC, ligand-gated ion channels; Orc, OR coreceptor; tuning Ors, other OR subunit(s) are highly divergent; OBPs, odorant-binding proteins; CSPs, chemosensory proteins; TMD, transmembrane spanning domain; MOE, main olfactory epithelium; GG, Grüneberg ganglion; 2,3-DMP, 2,3-dimethylpyrazine; TMT, 2,4,5-trimethylthiozoline; OSNs, olfactory sensory neurons; cGMP, cyclic guanosine monophosphate; MUP, major urinary protein.

In insects, chemosensory receptors include: (1) olfactory receptors (ORs),[88,103–106] (2) gustatory receptors (GRs),[88,97,107] and ionotropic receptors (IRs),[88,108,109] which play a central role in sensing chemical signals (Table 13.4). Odorant-binding proteins (OBPs) and chemosensory proteins (CSPs) also play a key role in insect chemical communication. Insect OBPs are soluble binding proteins, small (130–140 amino acid residues long), and are made of six α-helical domains assembled in a compact and stable structure. They are expressed in the lymph of chemosensilla on the antennae as well as in other chemosensory organs, such as the mouthparts and wings, and are considered to be crucial for insect-specific and -sensitive olfaction.[110,111] Like OBPs, the CSPs are another class of soluble binding proteins, enriched in the sensillum lymph, and found in other olfactory and gustatory organs of insects.[111] CSPs are smaller (100–120 amino acids) than OBPs, and present a conserved pattern of four cysteines forming two independent loops made of six α-helical domains, assembled in a compact and stable structure, and show no sequence similarity with OBPs, but exert similar properties in binding and transporting pheromones and other ligands.[112,113]

Both vertebrates and insects use their chemosensory receptors for detecting and discriminating among odors. Vertebrate chemosensory receptors belong to the family of GPCRs that initiate a cascade of cellular signaling events and thereby electrically excite the neuron; whereas insect ORs are odorant-gated ion channels that lack homology to GPCRs in vertebrates and possess a distinct 7-TM topology with the amino terminus located intracellularly. In vertebrates, binding of odorant molecules to an odorant receptor (OR) activates an olfactory-specific excitatory G-protein, $G_{\alpha olf}$, whose structure is similar to that of other types of G-proteins (consisting of three subunits, α, β, and γ, at the cytoplasmic surface of the ciliary membrane). It is likely that after the odorant binds to a receptor, $G_{\alpha olf}$ exchanges guanosine 5'-diphosphate (GDP) for guanosine 5'-triphosphate (GTP), and the GTP-bound $G_{\alpha olf}$ subunit dissociates from the β and γ subunits and activates AC, resulting in increased intraciliary concentration of cyclic adenosine monophosphate (cAMP). Cyclic nucleotide-gated channels located in the ciliary membrane are directly activated by cytoplasmic cAMP, and cause depolarizing influx of Na^+ and Ca^+ ions.[102] In insects, two models have been suggested for the molecular mechanisms responsible for insect OR-based signal transduction. According to first model, OR/Orco complexes form odorant activated ionotropic cation channels,[103] whereas in the second model, Orco itself is an ionotropic cation channel that is activated by fast, odor-dependent pathways as well as being affected by a slow, metabotropic mechanism involving G-proteins and AC.[104]

In *Drosophila*, IRs are expressed in coeloconic sensilla where OR/Orco complexes are usually not present, suggesting another model that IRs specifically respond to amines or acid-based odorants that are largely ignored by ORs.[106]

Collectively, the insect olfactory system exhibits remarkable organizational similarities with mammals, however insect ORs are structurally and functionally unrelated to mammalian ORs.

TRACE AMINE-ASSOCIATED RECEPTORS AND OLFACTION IN VERTEBRATES

In mammals, olfaction is mediated by two distinct organs that are located in the nasal cavity: (1) the main olfactory epithelium (MOE) binds volatile odorants, which is responsible for the conscious perception of odors, and (2) the vomeronasal organ (VNO) binds pheromones, which is responsible for various behavioral and neuroendocrine responses between individuals of a same species. Odorants and pheromones bind to 7-TM domains GPCRs that permit signal transduction. These receptors are encoded by large multigene families that evolved in mammal species in function of specific olfactory needs.[114]

TAARs are found in all vertebrates, including fish, amphibians, rodents, and humans, and form structurally and functionally distinct subfamilies of novel GPCRs. The mammalian TAARs family includes nine different subtypes. Most of these TARR subtypes retain amine recognition motifs, conserved in biogenic amine receptors, including an aspartic acid in TM helix III that forms a salt bridge with the ligand amino group. To date the best-characterized subtype is TAAR1, which activates the Gs protein/adenylyl cyclase pathway upon stimulation by trace amines and psychoactive substances like 3,4-methylenedioxy-methamphetamine (MDMA), popularly known as ecstasy, or lysergic acid diethylamide (LSD). This subtype is expressed in the brain (amygdala) and several peripheral organs, as well as in circulating lymphocytes, but not expressed in the human or mouse OE.[115–120] Eight out of the nine subtypes (TAAR2–TAAR9) were primarily found in the olfactory epithelium (OE) (Fig. 13.1), in similar patterns as ORs.[90,91] TAAR1 receptor gene is widely distributed throughout the human brain, shows low levels of expression in the amygdala, and also has trace levels in the cerebellum, hippocampus, and hypothalamus, as well as in some peripheral tissues.[71,72] TARR1 is potently activated by endogenous trace amines (such as tyramine and β-phenylethylamine), displays low affinity for tryptamine, and octopamine, activated by other amine compounds and psychostimulant drugs (such as amphetamine, 1-methyl-2-phenethylamine), and interacts with thyronamines (metabolites of thyroid hormones),[115–120] implying that TAAR1 may play a broader role in the brain. However, TAAR2–TAAR9 are considered as orphan receptors because they do not interact with trace amines. Not all TAARs show high affinity for trace amines. Based on their expression in

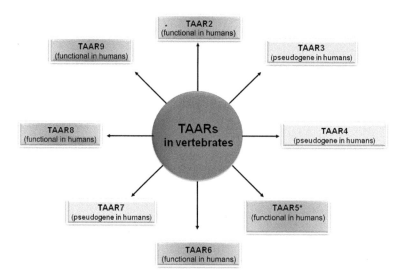

FIGURE 13.1   Mammalian trace amine-associated receptors (TAARs, 2–9 subtypes) expressed in the olfactory sensory neurons in the olfactory epithelium are considered as vertebrate olfactory receptors. Functional TAARs recognize ligands, and convert chemical signals from the environment into electrical signals, which are then transmitted to the brain. Most TAARs retain amine recognition motifs conserved in biogenic amine receptors. TARRs, existing as pseudogenes, have sequences similar to normal genes but due to gene mutation this pseudogene is not functional, and usually not transcribed.

the olfactory epithelium it is suggested that only some of these TAARs are olfactory receptors.[90] Both TAAR1 and TAAR2 are involved in mediating amine-induced blood leukocyte functions.[118] TAAR3 receptor in humans is encoded by the *TAAR3* gene, which is a pseudogene.[73,90,121] TAAR4 is extraordinarily sensitive and exhibits robust responses to subpicomolar odorant concentrations that are comparable to mammalian pheromone receptors.[90] TAAR4 serves as high-affinity amine detectors in mammalian olfactory systems. Its unprecedented sensitivity is mediated through receptor coupling to the canonical odorant transduction cascade.[90,119,121,122] Humans can smell volatile amines despite carrying open reading frame (ORF) disruptions in TAAR3 and TAAR4.[123] TAAR5 is of substantial interest because its gene is highly expressed in the human OE, and it is specifically activated by trimethylamine (TMA).[90,119,121–124] Agonist profiles of TAAR3, TAAR4, and TAAR5 vary significantly among mammals.[123] Human TAAR5 and TAAR8 have been demonstrated to be specifically expressed within the olfactory area of the human nasal mucosa. Several studies have reported association of the trace amine-associated receptor 6 gene (*TAAR6*) with susceptibility to schizophrenia and bipolar disorder, but results have not been consistent. However, the findings of Pae et al.[125] strongly support such an association in Korean patients. Moreover, this group has also reported the impact of a set of variations located in the heat shock protein 70 (HSP-70) and *TAAR6* in a sample of bipolar patients, implicating that HSP-70 may play a role in the disrupted mechanisms leading to bipolar disorder.[126] TAAR7 is not expressed in humans, but the TARR7 family of receptors is found in rodents. During evolution by gene duplication and subsequent mutation, the TAAR7 family in rodents has been expanded, resulting in a rapid and functional expansion of the olfactory receptor repertoire.[127] However, TAAR8 is detected in trace levels in human OE. Recently, Mühlhaus et al.[128] inevstigated the expression patterns of human TAAR8 and its murine orthologue *Taar8b* using TAAR1 agonists (PEA and 3-T_1AM), and both receptors were characterized by a basal $G_{i/o}$ signaling activity. However, the expression of *Taar8b* in young male and female C57BL6/J mice was reported at most marginal, close to the detection limit in the analyzed tissues.[128] TAAR9 is a functional receptor in most of the human population; however, it shows a polymorphism with a premature stop codon in 10–30% based on the population subgroup. The predominant TAAR-mediated signaling by receptors for which ligands are available utilizes a heterotrimeric G protein subunit ($G_{\alpha s}$), which activates adenylyl cyclase (AC) and converts ATP into cyclic AMP (cAMP), at least in rodent and recombinant systems.[73,90,91,121] Current understanding of TAAR-mediated signaling and functions is incomplete. Most of the information is available on TAAR1. For further extrapolation to understand trace amine systems, we need to develop selective agonists and antagonists for other functional TAAR subtpes.

TRACE AMINE-ASSOCIATED RECEPTORS AND OLFACTION IN INVERTEBRATES

Contrary to the situation in vertebrates, some trace amines (octopamine and tyramine) are the chief amines found in many invertebrate species. Invertebrate genomes lack sequences belonging to the TAAR family of GPCRs; however, both *p*-octopamine and *p*-tyramine receptors occur in insects (such as honeybee *A. mellifera* and fruit fly *D. melanogaster*) as well as in nematodes *Caenorhabditis elegans*. Invertebrate octopamine receptors (OAR) and tyramine receptors (TYR) are activated by octopamine and tramine, respectively. These receptors show close homology to vertebrate adrenergic receptors and dopamine receptors, but show no phylogenetic relation with vertebrate TAARs.[91,129] This suggests that octopamine and tyramine receptors in invertebrates have evolved independently.

Evans and Maqueira[130] proposed a receptor classification scheme, which was on the structural and signaling similarities between cloned *D. melanogaster* octopaminergic receptors and vertebrate adrenergic receptors. According to this scheme, octopaminergic receptors were grouped into three classes: (1) α-adrenergic-like (OCTα-R); (2) β-adrenergic-like (OCTβ-R); and (3) octopaminergic/tyraminergic (OCT/TYR-R) or tyraminergic (TYR-R) (Fig. 13.2).[130] The OCTα-R class has sequence homology with vertebrate α1-adrenergic receptors. Receptors of this class show a higher affinity for octopamine than tyramine, and are coupled with an increase in intracellular Ca^{2+} concentration as well as a small increase in intracellular cAMP levels.[131–135] The OCTβ-R class has sequence similarities with vertebrate

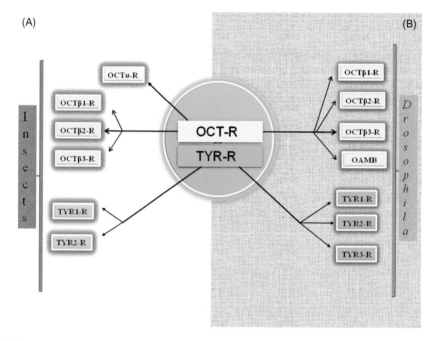

FIGURE 13.2 Octopamine and tyramine receptor classification: (A) in insects there are four different kinds of octopamine receptors (Octα-R, OCTβ1-R, OCTβ2-R, and OCTβ3) and two different kinds of tyramine receptors (TYR1-R and TYR2-R), whereas (B) in fruit fly *D. melanogaster*, there are four different octopamine receptors (Oamb, Octß1R, Octß2R, and Octß3R) and three different tyramine receptors (TYR1-R, TYR2-R, and TYR3-R). *Source: Information is adapted from references Farooqui T. Review of octopamine in insect nervous systems. Open Access Insect Physiol 2012;4:1–17, Evans PD, Maqueira B. Insect octopamine receptors: a new classification scheme based on studies of cloned Drosophila G-protein coupled receptors. Invert Neurosci 2005;5:111–8, Bayliss A, Roselli G, Eans PD. A comparison of the signalling properties of two tyramine receptors from Drosophila. J Neurochem 2013;12(1):37–48, and El-Kholy S, Stephano F, Li Y, Bhandari A, Flink C, Roeder T. Expression analysis of octopamine and tyramine receptors in Drosophila. Cell Tissue Res 2015;361(3):669–84.*

β-adrenergic receptors, and activation in this class of receptors after octopamine binding results in increased intracellular cAMP levels.[129,130] Based on pharmacological differences, the OCβ-R class is subdivided into three subclasses (OCTβ1-R, OCTβ2-R, and OCTβ3-R).[129] The OCT/TYR-R or TYR-R class of receptors has structural and pharmacological similarities with vertebrate α2-adrenergic receptors, and receptor activation can occur by binding to either tyramine or octopamine.[130] Tyramine-mediated activation of OCT/TYR-R class of receptors is coupled with inhibition of adenylyl cyclase via G_i protein, resulting in reduction of intracellular cAMP levels.[136–140] However, receptor activation in response to octopamine is coupled with an increase in intracellular Ca^{2+} release.[141,142] Moreover, a phylogenetic tree based on the comparison of 25 complete nucleotide sequences homology of insect octopaminergic and tyraminergic receptor genes supported applicability of the Evans and Maqueira classification scheme, except that there were two subclasses in the tyraminergic receptor class.[78] The TYR2-R class is specifically activated by tyramine, but not by other biogenic amines, and is selectively coupled to intracellular Ca^{2+} mobilization.[143,144] Bayliss et al.[145] proposed a new class of tyramine receptors (TYR3-R), which is structurally related to CG7431. The orthologues of CG7431 can be identified in the genomes of a number of insect species, whereas species orthologues of CG16766 are only found in *Drosophila* species.[145] Recently, four different octopamine receptors (Oamb, Octß1R, Octß2R, and Octß3R) and three different tyramine receptors (TyrR, TyrRII, and TyrRIII) are present in the fruit fly *D. melanogaster*.[146] Taken together, an updated insect classification of octopamine and tyramine receptors is shown in Fig. 13.2.

Moreover, octopamine and tyramine signaling is mediated through binding to distinct OAR and TYR receptor subtypes that belong to a family of seven transmembrane domain (7-TM) receptors. After ligand binding, these receptors may interact with either of the three types of guanine nucleotide-binding proteins: (1) a heterotrimeric G protein subunit (G_s) that activates AC, resulting in the stimulation of cAMP-dependent pathway; (2) a heterotrimeric G protein subunit ($G_{\alpha i}$) that inhibits the production of cAMP from ATP by inhibiting AC; and (3) another heterotrimeric G protein subunit ($G_{\alpha q}$) that activates phospholipase C (PLC), which hydrolyzes phosphatidylinositol 4,5-bisphosphate (PIP_2) to diaclglycerol (DAG) and inositol triphosphate (IP_3). DAG activates protein kinase C (PKC). Both PKC and IP_3 help in phosphorylation of some signaling proteins, ion channels, and transcription factors, which are responsible for the cellular response.[77,78] Based on species, tissue source, receptor type, and cell line used for the expression of cloned receptors, the second messengers may include Ca^{2+}, cAMP, inositol-1,4,5-trisphosphate (IP_3), and diacylglycerol (DAG).

Moreover, tryptamine (a plant-produced amine), and tyramine (an endogenous insect amine) have been recently reported to inhibit the odorant activation function of insect ORs via antagonizing the coreceptor subunit, thereby modulating insect olfaction.[147] Both trace amines blocked Orco activation by serving as Orco antagonists, suggesting that trace amines modulate insect olfaction.

CONCLUSION

Olfactory dysfunction (in particular, alteration in the ability to identify and discriminate the odors, as well as the odor threshold) is a common and early feature of many neurological disorders. These changes often occur as early manifestations of the pathology but are not always diagnosed on time. There are fundamental differences in transduction machineries between insects and vertebrates. However, the anatomical/morphological and functional features of insect odor-coding strategy are similar to those in vertebrates. Therefore, this review compares the trace amine-mediated olfactory systems of mammals and insects. Trace amines play a major role in insect physiology by interacting with specific plasma membrane GPCRs (OAR, TYR). In mammals, trace amines play a neuromodulatory role by interacting with a novel class of plasma membrane GPCRs (TAARS). Several TAARs have been shown to mediate aversion or attraction towards volatile amines. However, mammalian TAARs do not show a close phylogenetic relationship to their invertebrate counterparts (OAR, TYR), suggesting that TAARs evolved independently of invertebrate receptors, but they have the ability to interact with different amines, volatile amines, and possibly other unidentified endogenous compounds.

The olfactory system is important for human life quality and dramatic changes in trace amine levels and olfactory deficits have been observed in neurological disorders. The insect olfactory sensory system plays a predominant role in insect behavior, such as nest mate recognition, selection of oviposition sites, utilization in orientation, food foraging, mating, and escaping from predators and toxic compounds. The study of olfaction in insects is easily amenable to the laboratory, since dedicated protocols have been developed in which insects show rapid and robust odor learning abilities. The insect brain is easily accessible to neurophysiological (electrophysiological or optical imaging recordings), pharmacological and molecular genetics, and social manipulations. Moreover, insects are inexpensive, easy to work with, have short lifespans, and have less ethical or methodological limitations than studying in rodents or mammals. While the olfactory system functions may be quite different from those in humans, there are common principles in the two phyla. Many signaling pathways may be conserved and display similar activities. Despite having the above advantages of invertebrate models, transgenic insect model systems may suffer from several unphysiological features, including high protein levels due to integration of multiple transgene copies into the genomes, alterations in brain area specificity, and subcellular expression pattern of the transgene compared to endogenous genes. Thus genetic models fail to phenocopy the human diseases in that they generally lack a behavioral phenotype and/or the characteristic pathological feature of human disease. Secondly, compared to insects, research in rodents will also not provide an ideal model system for neurological disorders because of low order of cognition in mice, suggesting that models can never capture the full complexity of the human condition.

Among insects, in particular, the honeybee *A. mellifera* worker has been accepted as a great model in cognitive neuroscience because of its sophisticated cognitive abilities. Moreover other characteristics, such as its brain containing 1 million neurons (five orders of magnitude less than the human brain but four times greater than *Drosophila*), and ability to successfully induce oxidative stress in honeybee worker brain in the laboratory, favor *A. mellifera* to serve as an excellent tool for studying trace amine-mediated molecular mechanisms underlying olfactory dysfunction, which is among the first signs in several neurological disorders and aging.

References

1. Liu X, Davis RL. Insect olfactory memory in time and space. *Curr Opin Neurobiol* 2006;**16**(6):679–85.
2. Davis RL. Olfactory learning. *Neuron* 2004;**44**:31–8.
3. Wilson RI, Mainen ZF. Early events in olfactory processing. *Annu Rev Neurosci* 2006;**29**:163–201.
4. Stocker RF. The organization of the chemosensory system in *Drosophila melanogaster*: a review. *Cell Tissue Res* 1994;**275**:3–26.
5. Takeda K. Classical conditioned response in the honey bee. *J Insect Physiol* 1961;**6**:168–79.
6. Bitterman ME, Menzel R, Fietz A, Schäfer S. Classical conditioning of proboscis extension in honeybees (*Apis mellifera*). *J Comp Psychol* 1983;**1983**(97):107–19.
7. Menzel R, Müller U. Learning and memory in honeybees: from behavior to neural substrates. *Annu Rev Neurosci* 1996;**19**:379–404.
8. Menzel R. Memory dynamics in the honeybee. *J Comp Physiol A* 1999;**185**:323–40.
9. Dudai Y. Properties of learning and memory in *Drosophila melanogaster*. *J Comp Physiol A* 1977;**114**:69–89.
10. Tully T, Quinn WG. Classical conditioning and retention in normal and mutant *Drosophila melanogaster*. *J Comp Physiol A* 1985;**157**:263–77.
11. Davis RL. Olfactory memory formation in Drosophila: from molecular to systems neuroscience. *Annu Rev Neurosci* 2005;**28**:275–302.
12. McGuire SE, Deshazer M, Davis RL. Thirty years of olfactory learning and memory research in *Drosophila melanogaster*. *Prog Neurobiol* 2005;**76**:328–47.
13. Matsumoto Y, Mizunami M. Olfactory learning in the cricket *Gryllus bimaculatus*. *J Exp Biol* 2000;**203**:2581–8.
14. Matsumoto Y, Mizunami M. Olfactory memory capacity of the cricket *Gryllus bimaculatus*. *Biol Lett* 2006;**2**(4):608–10.
15. SimÕes P, Niven JE, Ott SR. Phenotypic transformation affects associative learning in the desert locust. *Curr Biol* 2013;**23**(23):2407–12.
16. SimÕes P, Ott SR, Niven JE. Associative olfactory learning in the desert locust, *Schistocerca gregaria*. *J Exp Biol* 2011;**214**:2495–503.
17. Menzel R, Giurfa M. Cognitive architecture of amini-brain: the honeybee. *Trends Cogn Sci* 2001;**5**:62–71.
18. Frasnelli E, Haase A, Rigosi E, Anfora G, Rogers LJ. The bee as a model to investigate brain and behavioural asymmetries. *Insects* 2014;**5**:120–38.
19. Smith BH, Cobey S. The olfactory memory of the honeybee *Apis mellifera*. II. Blocking between odorants in binary mixtures. *J Exp Biol* 1994;**195**:91–108.
20. Giurfa M, Sandoz JC. Invertebrate learning and memory: fifty years of olfactory conditioning of the proboscis extension response in honeybees. *Learn Mem* 2012;**19**(2):54–66.
21. Wright GA, Carlton M, Smith BH. A honeybee's ability to learn, recognize, and discriminate odors depends upon odor sampling time and concentration. *Behav Neurosci* 2009;**123**:36–43.
22. Smith BH. The olfactory memory of the honeybee *Apis mellifera* I. Odorant modulation of short- and intermediate- term memory after single-trial conditioning. *J Exp Biol* 1991;**161**:367–82.
23. Sandoz JC. Behavioral and neurophysiological study of olfactory perception and learning in honeybees. *Front Syst Neurosci* 2011;**5**:98.
24. Galizia CG, Kimmerle B. Physiological and morphological characterization of honeybee olfactory neurons combining electrophysiology, calcium imaging and confocal microscopy. *J Comp Physiol A* 2004;**190**:21–38.
25. Farooqui T, Robinson K, Vaessin H, Smith BH. Modulation of early olfactory processing by an octopaminergic reinforcement pathway in the honeybee. *J Neurosci* 2003;**23**:5370–80.
26. Farooqui T. Gene silencing by double-stranded RNA in *Apis mellifera*: a useful reverse genetics approach for studying complex biological phenomena. In: Lyland RT, Browning IB, editors. *RNA interference and research progress*. : Nova Science Publishers, Inc.; 2008. p. 19–210.
27. Farooqui T. Iron-induced oxidative stress modulates olfactory learning and memory in honeybees. *Behav Neurosci* 2008;**122**(2):433–47.
28. Berry MD. Mammalian central nervous system trace amines. Pharmacologic amphetamines, physiologic neuromodulators. *J Neurochem* 2004;**90**:257–71.
29. Evans PD, Robb S. Octopamine receptor subtypes and their modes of action. *Neurochem Res* 1993;**18**:869–74.
30. Axelrod J, Saavedra JM. Octopamine. *Nature* 1977;**265**:501–4.
31. David JC, Coulon JF. Octopamine in invertebrates and vertebrates. A review. *Prog Neurobiol* 1985;**24**:141–85.

32. Bleanau W, Baumann A. Aminergic signal transduction in invertebrates: focus on tyramine and octopamine receptors. *Recent Res Dev Neurochem* 2003;**6**(2003):225–40.

33. Scheiner R, Baumann A, Blenau W. Aminergic control and modulation of honeybee behaviour. *Curr Neuropharmacol* 2006;**4**(4):259–76.

34. Roeder T, Seifert M, Kähler C, Gewecke M. Tyramine and octopamine: antagonistic modulators of behavior and metabolism. *Arch Insect Biochem Physiol* 2003;**4**(1):1–13.

35. Roeder T. Octopamine in invertebrates. *Prog Neurobiol* 1999;**59**:533–41.

36. Saraswati S, Fox LE, Soll DR, Wu CF. Tyramine and octopamine have opposite effects on the locomotion of *Drosophila* larvae. *J Neurobiol* 2004;**58**:425–41.

37. Doty RL, Kamath V. The influences of age on olfaction: a review. *Front Psychol* 2014;**5**:20.

38. Welge-Lüssen A. Ageing, neurodegeneration, and olfactory and gustatory loss. *B-ENT* 2009;**5**(Suppl. 13):129–32.

39. Hüttenbrink K-B, Hummel T, Berg D, Gasser T, Hähner A. Olfactory dysfunction: common in later life and early warning of neurodegenerative disease. *Dtsch Arztebl Int* 2013;**110**(1–2):1–7.

40. Hawkes C. Olfaction in neurodegenerative disorder. *Mov Disord* 2003;**18**:364–72.

41. Hawkes C. Olfaction in neurodegenerative disorder. *Adv Otorhinolaryngol* 2006;**63**:133–51.

42. Haehner A, Hummel T, Hummel C, Sommer U, Junghanns S, Reichmann H. Olfactory loss may be a first sign of idiopathic Parkinson's disease. *Mov Disord* 2007;**22**(6):839–42.

43. Stamps JJ, Bartoshuk LM, Heilman KM. A brief olfactory test for Alzheimer's disease. *J Neurol Sci* 2013;**333**(1–2):19–24.

44. Graves AB, Bowen JD, Rajaram L, McCormick WC, McCurry SM, et al. Impaired olfaction as a marker for cognitive decline: interaction with apolipoprotein E epsilon4 status. *Neurology* 1999;**53**(7):1480–7.

45. Moberg PJ, Pearlson GD, Speedie LJ, Lipsey JR, Strauss ME, et al. Olfactory recognition: differential impairments in early and late Huntington's and Alzheimer's diseases. *J Clin Exp Neuropsychol* 1987;**9**:650–64.

46. Barrios FA, Gonzalez L, Favila R, Alonso ME, Salgado PM, et al. Olfaction and neurodegeneration in HD. *Neuroreport* 2007;**18**:73–6.

47. Rolet A, Magnin E, Millot JL, Berger E, Vidal C, et al. Olfactory dysfunction in multiple sclerosis: evidence of a decrease in different aspects of olfactory function. *Eur Neurol* 2013;**69**(3):166–70.

48. Fleiner F, Dahlslett SB, Schmidt F, Harms L, Goektas O. Olfactory and gustatory function in patients with multiple sclerosis. *Am J Rhinol Allergy* 2010;**24**(5):e93–7.

49. Ahlskog JE, Waring SC, Petersen RC, Esteban-Santillan C, Craig UK, O'Brien PC, et al. Olfactory dysfunction in Guamanian ALS, Parkinsonism, and dementia. *Neurology* 1998;**51**(6):1672–7.

50. Takeda T, Uchihara T, Kawamura S, Ohashi T. Olfactory dysfunction related to TDP-43 pathology in amyotrophic lateral sclerosis. *Clin Neuropathol* 2014;**33**(1):65–7.

51. Moberg PJ, Agrin R, Gur RE, Gur RC, Turetsky BI, et al. Olfactory dysfunction in schizophrenia: a qualitative and quantitative review. *Neuropsychopharmacology* 1999;**21**:325–40.

52. Moberg PJ, Kamath V, Marchetto DM, Calkins ME, Doty RL, et al. Meta-analysis of olfactory function in schizophrenia, first-degree family members, and youths at-risk for psychosis. *Schizophr Bull* 2014;**40**(1):50–9.

53. Moberg PJ, Arnold SE, Doty RL, Gur RE, Balderston CC, et al. Olfactory functioning in schizophrenia: relationship to clinical, neuropsychological, and volumetric MRI measures. *J Clin Exp Neuropsychol* 2006;**28**:1444–61.

54. Burchett SA, Hicks TP. The mysterious trace amines: protean neuromodulators of synaptic transmission in mammalian brain. *Prog Neurobiol* 2006;**79**(5–6):223–46.

55. Berry MD. The potential of trace amines and their receptors for treating neurological and psychiatric diseases. *Rev Recent Clin Trials* 2007;**2**(1):3–19.

56. Philips SR, Rozdilsky B, Boulton AA. Evidence for the presence of m-tyramine, p-tyramine, tryptamine, and phenylethylamine in the rat brain and several areas of the human brain. *Biol Psychiatry* 1978;**13**(1):51–7.

57. Schmitt R, Nasse O. Beitrag zur Keuntnifs des tyrosins. *Liebigs Ann Chem* 1865;**133**:211–6.

58. Boulton AA. Identification, distribution, metabolism, and function of meta and para tyramine, phenylethylamine and tryptamine in brain. *Adv Biochem Psychopharmacol* 1976;**15**:57–67.

59. Boulton AA, Dyck LE. Biosynthesis and excretion of *meta* and *para* tyramine in the rat. *Life Sci* 1974;**14**:2497–506.

60. Tallman JF, Saavedra JM, Axelrod J. Biosynthesis and metabolism of endogenous tyramine and its normal presence in sympathetic nerves. *J Pharmacol Exp Ther* 1976;**199**:216–21.

61. Paterson IA, Juorio AV, Boulton AA. 2-Phenylethylamine: a modulator of catecholamine transmission in the mammalian central nervous system? *J Neurochem* 1990;**1990**(55):1827–37.

62. Boulton AA. Trace amines and mental disorders. *Can J Neurol Sci* 1980;**7**:261–3.

63. Branchek TA, Blackburn TP. Trace amine receptors as targets for novel therapeutics: legend, myth and fact. *Curr Opin Pharmacol* 2003;**3**:90–7.

64. O'Reilly RL, Davis BA. Phenylethylamine and schizophrenia. *Prog Neuropsychopharmacol Biol Psychiatry* 1994;**18**:63–75.

65. Boulton AA. Phenylethylaminergic modulation of catecholaminergic neurotransmission. *Prog Neuropsychopharmacol Biol Psychiatry* 1991;**15**:139–56.

66. Lindemann L, Meyer CA, Jeanneau K, Bradaia A, Ozmen L, et al. Trace amine-associated receptor 1 modulates dopaminergic activity. *J Pharmacol Exp Ther* 2008;**324**:948–56.

67. Bradaia A, Trube G, Stalder H, Norcross RD, Ozmen L, et al. The selective antagonist EPPTB reveals TAAR1-mediated regulatory mechanisms in dopaminergic neurons of the mesolimbic system. *Proc Natl Acad Sci USA* 2009;**106**(47):20081–6.

68. Gozal A, O'Neill BE, Sawchuk MA, Zhu H, Halder M, et al. Anatomical and functional evidence for trace amines as unique modulators of locomotor function in the mammalian spinal cord. *Front Neural Circuits* 2014;**8**:134.

69. Xie Z, Miller GM. Beta-phenylethylamine alters monoamine transporter function via trace amine-associated receptor 1: implication for modulatory roles of trace amines in brain. *J Pharmacol Exp Ther* 2008;**325**(2):617–28.

70. Revel FG, Moreau JL, Gainetdinov RR, Bradaia A, Sotnikova TD, et al. TAAR1 activation modulates monoaminergic neurotransmission, preventing hyperdopaminergic and hypoglutamatergic activity. *Proc Natl Acad Sci USA* 2011;**108**:8485–90.

71. Borowsky B, Adham N, Jones KA, Raddatz R, Artymyshyn R, et al. Trace amines: identification of a family of mammalian G protein coupled receptors. *Proc Natl Acad Sci USA* 2001;**98**(16):8966–71.

72. Bunzow JR, Sonders MS, Arttamangkul S, Harrison LM, Zhang G, et al. Amphetamine, 3,4-methylenedioxymethamphetamine, lysergic acid diethylamide, and metabolites of the catecholamine neurotransmitters are agonists of a rat trace amine receptor. *Mol Pharmacol* 2001;**60**:1181–8.

73. Lindemann L, Ebeling M, Kratochwil NA, Bunzow JR, Grandy DK, et al. Trace amine-associated receptors form structurally and functionally distinct subfamilies of novel G protein-coupled receptors. *Genomics* 2005;**85**:372–85.

74. Miller GM, Verrico CD, Jassen AK, Konar M, Yang H, et al. Primate trace amine receptor 1 modulation by the dopamine transporter. *J Pharmacol Exp Ther* 2005;**313**:983–94.

75. Wolinsky TD, Swanson CJ, Smith KE, Zhong H, Borowsky B, et al. The trace amine 1 receptor knockout mouse: an animal model with relevance to schizophrenia. *Genes Brain Behav* 2007;**6**:628–39.

76. Revel FG, Meyer CA, Bradaia A, Jeanneau K, Calcagno E, et al. Brain-specific overexpression of trace amine-associated receptor 1 alters monoaminergic neurotransmission and decreases sensitivity to amphetamine. *Neuropsychopharmacology* 2012;**37**(12):2580–92.

77. Farooqui T. Octopamine-mediated neuromodulation of insect senses. *Neurochem Res* 2007;**32**(9):1511–29.

78. Farooqui T. Review of octopamine in insect nervous systems. *Open Access Insect Physiol* 2012;**4**:1–17.

79. Lange AB. Tyramine: from octopamine precursor to neuroactive chemical in insects. *Gen Comp Endocrinol* 2009;**162**:18–26.

80. Roeder T. Tyramine and octopamine: ruling behavior and metabolism. *Annu Rev Entomol* 2005;**50**:447–77.

81. Ohta H, Ozoe T. Molecular signaling, pharmacology, and physiology of octopamine and tyramine receptors as potential insect pest control agents. In: Cohen E, editor. *Advances in insect physiology – target receptors in the control of insect pests: part II.* Oxford (UK): Elsevier Ltd.; 2014. p. 73–166.

82. Vierk R, Pflueger HJ, Duch C. Differential effects of octopamine and tyramine on the central pattern generator for *Manduca* flight. *J Comp Physiol A Neuroethol Sens Neural Behav Physiol* 2009;**195**:265–77.

83. Hardie SL, Zhang JX, Hirsh J. Trace amines differentially regulate adult locomotor activity, cocaine sensitivity, and female fertility in *Drosophila melanogaster*. *Dev Neurobiol* 2007;**67**:1396–405.

84. Ma Z, Guo X, Lei H, Li T, Hao S, Kang L. Octopamine and tyramine respectively regulate attractive and repulsive behavior in locust phase changes. *Sci Rep* 2015;**5**:8036.

85. Fussnecker BL, Smith BH, Mustard JA. Octopamine and tyramine influence the behavioral profile of locomotor activity in the honey bee (*Apis mellifera*). *J Insect Physiol* 2006;**52**(10):1083–92.

86. Salomon M, Malka O, Meer RK, Heftz A. The role of tyramine and octopamine in the regulation of reproduction in queenless worker honeybees. *Naturwissenschaften* 2012;**99**(2):123–31.

87. Buck L, Axel R. A novel multigene family may encode odorant receptors: a molecular basis for odor recognition. *Cell* 1991;**65**:175–87.

88. Kaupp UB. Olfactory signaling in vertebrates and insects: differences and commonalities. *Nat Rev Neurosci* 2010;**11**(3):188–200.

89. Yoshihito Niimura Y. Olfactory receptor multigene family in vertebrates: from the viewpoint of evolutionary genomics. *Curr Genomics* 2012;**13**(2):103–14.

90. Liberles SD, Buck LB. A second class of chemosensory receptors in the olfactory epithelium. *Nature* 2006;**442**(7103):645–50.

91. Liberles SD. Trace amine-associated receptors are olfactory receptors in vertebrates. *Ann NY Acad Sci* 2009;**1170**:168–72.

92. Liberles SD. Trace amine-associated receptors: ligands, neural circuits, and behaviors. *Curr Opin Neurobiol* 2015;**34**:1–7.

93. Johnson MA, Tsai L, Roy DS, Valenzuela DH, Mosley C, et al. Neurons expressing trace amine-associated receptors project to discrete glomeruli and constitute an olfactory subsystem. *Proc Natl Acad Sci USA* 2012;**109**(33):13410–5.

94. Liberles SD, Horowitza LF, Kuanga D, Contosa JM, Wilsona KL, et al. Formyl peptide receptors are candidate chemosensory receptors in the vomeronasal organ. *Proc Natl Acad Sci USA* 2009;**1066**(24):9842–7.

95. Bufe B, Schumann T, Zufall F. Formyl peptide receptors from immune and vomeronasal system exhibit distinct agonist properties. *J Biol Chem* 2012;**287**(40):33644–55.

96. Leinders-Zufall T, Cockerham RE, Michalakis S, Biel M, Garbers DL, et al. Contribution of the receptor guanylyl cyclase GC-D to chemosensory function in the olfactory epithelium. *Proc Natl Acad Sci USA* 2007;**104**(36):14507–12.

97. Yao CA, Carlson JR. Role of G-proteins in odor-sensing and CO_2-sensing neurons in Drosophila. *Neurosci* 2010;**30**(13):4562–72.

98. Sun L, Wang H, Hu J, Han J, Matsunami H, et al. Guanylyl cyclase-D in the olfactory CO_2 neurons is activated by bicarbonate. *Proc Natl Acad Sci USA* 2009;**106**(6):2041–6.

99. Chao YC, Chang CJ, Hsieh HT, Lin CC, Yang RB. Guanylate cyclase-G, expressed in the *Grueneberg ganglion* olfactory subsystem, is activated by bicarbonate. *Biochem J* 2010;**432**(2):267–73.

100. Pelosi P. The role of perireceptor events in vertebrate olfaction. *Cell Mol Life Sci* 2001;**58**:503–9.

101. Tegoni M, Pelosi P, Vincent F, Spinelli S, Campanacci V, et al. Mammalian odorant binding proteins. *Biochim Biophys Acta* 2000;**1482**:229–40.

102. (a)Francia S, Pifferi S, Menini A, Tirindelli R. Chapter 10, vomeronasal receptors and signal transduction in the vomeronasal organ of mammals. In: Mucignat-Caretta C, editor. *Neurobiology of chemical communication*. Boca Raton (FL): CRC Press, Taylor & Francis; 2014.(b)Pifferi S, Menini A, Kurahashi T. Chapter 8, signal transduction in vertebrate olfactory cilia. In: Menini A, editor. *The neurobiology of olfaction*. Boca Raton (FL): CRC press/Taylor & Francis; 2010.

103. Sato K, Pellegrino M, Nakagawa T, Nakagawa T, Vosshall LB, et al. Insect olfactory receptors are heteromeric ligand-gated ion channels. *Nature* 2008;**452**:1002–6.

104. Wicher D, Schafer R, Bauernfeind R, Stensmyr MC, Heller R, et al. Drosophila odorant receptors are both ligand-gated and cyclic-nucleotide-activated cation channels. *Nature* 2008;**452**:1007–11.

105. Silbering AF, Benton R. Ionotropic and metabotropic mechanisms in chemoreception: 'chance or design'? *EMBO Rep* 2010;**11**(3):173–9.

106. Suh E, Bohbot J, Zwiebel LJ. Peripheral olfactory signaling in insects. *Curr Opin Insect Sci* 2014;**6**:86–92.

107. Proudel S, Kim Y, Kim YT, Lee Y. Gustatory receptors required for sensing umbelliferone in *Drosophila melanogaster*. *Insect Biochem Mol Biol* 2015;**23**(66):110–8.

108. Benton R, Vannice KS, Gomez-Diaz C, Vosshall LB. Variant ionotropic glutamate receptors as chemosensory receptors in Drosophila. *Cell* 2009;**136**:149–62.

109. Rytz R, Croset V, Benton R. Ionotropic receptors (IRs): chemosensory ionotropic glutamate receptors in Drosophila and beyond. *Insect Biochem Mol Biol* 2013;**43**(9):8888–97.

110. Fan J, Francis F, Liu Y, Chen JL, Cheng DF. An overview of odorant-binding protein functions in insect peripheral olfactory reception. *Genet Mol Res* 2011;**10**(4):3056–69.

111. Pelosi P, Iovvinella I, Felicioli A, Dani FR. Soluble proteins of chemical communication: an overview across arthropods. *Front Physiol* 2014;**5**:320.

112. Angeli S, Ceron F, Scaloni A, Monti M, Monteforti G, et al. Purification, structural characterization, cloning and immunocytochemical localization of chemoreception proteins from *Schistocerca gregaria*. *Eur J Biochem* 1999;**262**:745–54.

113. Tomaselli S, Crescenzi O, Sanfelice D, Ab E, Wechselberger R, et al. Solution structure of a chemosensory protein from the desert locust *Schistocerca gregaria*. *Biochemistry* 2006;**45**:10606–13.

114. Rouquier S, Giorgi D. Olfactory receptor gene repertoires in mammals. *Mutat Res* 2007;**616**(1–2):95–102.

115. Xie Z, Miller GM. Trace amine-associated receptor 1 as a monoaminergic modulator in brain. *Biochem Pharmacol* 2009;**78**(9):1095–104.

116. Miller GM. The emerging role of trace amine-associated receptor 1 in the functional regulation of monoamine transporters and dopaminergic activity. *J Neurochem* 2011;**116**(2):164–76.

117. Scanlan TS, Suchland KL, Hart ME, Chiellini G, Huang Y, et al. 3-Iodothyronamine is an endogenous and rapid-acting derivative of thyroid hormone. *Nat Med* 2004;**10**(6):638–42.

118. Babusyte A, Kotthoff M, Fiedler J, Krautwurst D. Biogenic amines activate blood leukocytes via trace amine-associated receptors TAAR1 and TAAR2. *J Leukoc Biol* 2013;**93**(3):387–94.

119. Carnicelli V, Santoro A, Sellari-Franceschini S, Berrettini S, Zucchi R. Expression of trace amine-associated receptors in human nasal mucosa. *Chemosensory Percept* 2010;**3**:99–107.

120. Zucchi R, Chiellini G, Scanlan TS, Grandy DK. Trace amine-associated receptors and their ligands. *Br J Pharmacol* 2006;**149**:967–78.

121. Maguire JJ, Parker WA, Foord SM, Bonner TI, Neubig RR, et al. International union of pharmacology. LXXII. Recommendations for trace amine receptor nomenclature. *Pharmacol Rev* 2009;**61**(1):1–8.

122. Wallrabenstein I, Kuklan J, Weber L, Zborala S, Werner M, et al. Human trace amine-associated receptor TAAR5 can be activated by trimethylamine. *PLoS One* 2013;**8**(2):e54950.

123. Staubert C, Boselt I, Bohnekamp J, Rompler H, Enard W, et al. Structural and functional evolution of the trace amine-associated receptors TAAR3, TAAR4 and TAAR5 in primates. *PLoS One* 2010;**5**:e11133.

124. Dinter J, Mühlhaus J, Wienchol CL, Yi CX, Nürnberg D, et al. Inverse agonistic action of 3-iodothyronamine at the human trace amine-associated receptor 5. *PLoS One* 2015;**10**(2):e0117774.

125. Pae CU, Yu HS, Amann D, Kim JJ, Lee CU, et al. Association of the trace amine associated receptor 6 (TAAR6) gene with schizophrenia and bipolar disorder in a Korean case control sample. *J Psychiatr Res* 2008;**42**(1):35–40.

126. Pae CU, Drago A, Mandelli L, De Ronchi D, Serretti A. TAAR 6 and HSP-70 variations associated with bipolar disorder. *Neurosci Lett* 2009;**465**(3):257–61.

127. Ferrero DM, Wacker D, Roque MA, Baldwin MW, Stevens RC, et al. Agonists for 13 trace amine-associated receptors provide insight into the molecular basis of odor selectivity. *ACS Chem Biol* 2012;**7**(7):1184–9.

128. Mühlhaus J, Dinter J, Nürnberg D, Rehders M, Depke M, et al. Analysis of human TAAR8 and murine TAAR8B mediated signaling pathways and expression profile. *Int J Mol Sci* 2014;**15**:20638–55.

129. Maqueira B, Chatwin H, Evans P. Identification and characterization of a novel family of Drosophila β-adrenergic-like octopamine G-protein coupled receptors. *J Neurochem* 2005;**94**(2):547–60.

130. Evans PD, Maqueira B. Insect octopamine receptors: a new classification scheme based on studies of cloned Drosophila G-protein coupled receptors. *Invert Neurosci* 2005;**5**:111–8.

131. Han KA, Millar NS, Davis RL. A novel octopamine receptor with preferential expression in *Drosophilla* mushroom bodies. *J Neurosci* 1998;**18**:3650–8.

132. Grohmann L, Blenau W, Erber J, Ebert PR, Strünker T, et al. Molecular and functional characterization of an octopamine receptor from honeybee (*Apis mellifera*) brain. *J Neurochem* 2003;**86**:725–35.

133. Duportets L, Barrozo RB, Bozzolan F, Gaertner C, Anton S, et al. Cloning of an octopamine/tyramine receptor and plasticity of its expression as a function of adult sexual maturation in the male moth *Agrotis ipsilon*. *Insect Mol Biol* 2010;**19**:489–99.

134. Balfanz S, Strünker T, Frings S, Baumann A. A family of octopamine receptors that specifically induce cyclic AMP production or Ca^{2+} release in *Drosophila melanogaster*. *J Neurochem* 2005;**93**:440–51.

135. Ohtani A, Arai Y, Ozoe F, Ohta H, Narusuye K, et al. Molecular cloning and heterologous expression of an alpha adrenergic-like octopamine receptor from the silkworm *Bombyx mori*. *Insect Mol Biol* 2006;**15**:763–72.

136. Saudou F, Amlaiky N, Plassat JL, Borrelli E, Hen R. Cloning and characterization of a *Drosophila* tyramine receptor. *EMBO J* 1990;**9**:3611–7.

137. Vanden Broeck J, Vulsteke V, Huybrechts R, De Loof A. Characterization of a cloned locust tyramine receptor cDNA by functional expression in permanently transformed Drosophila S2 cells. *J Neurochem* 1995;**64**:2387–95.
138. Poels J, Suner MM, Needham M, Torfs H, De Rlick J, et al. Functional expression of a locust tyramine receptor in murine erythroleukemia cells. *Insect Mol Biol* 2001;**10**:541–8.
139. Ohta H, Utsumi T, Ozoe Y. B96Bom encodes a *Bombyx mori* tyramine receptor negatively coupled to adenylate cyclase. *Insect Mol Biol* 2003;**12**:217–23.
140. Blenau W, Balfanz S, Baumann A. Amtyr1: characterization of a gene from honeybee (*Apis mellifera*) brain encoding a functional tyramine receptor. *J Neurochem* 2000;**74**:900–8.
141. Robb S, Cheek TR, Hannan FL, Hall LM, Midgley JM, Evans PD. Agonist-specific coupling of a cloned *Drosophila* octopamine/tyramine receptor to multiple second messenger systems. *EMBO J* 1994;**13**:1325–30.
142. Reale V, Hannan F, Midgley JM. The expression of a cloned Drosophila octopamine/tyramine receptor in *Xenopus* oocytes. *Brain Res* 1997;**769**:309–20.
143. Cazzamali G, Klaerke DA, Grimmelikhuijzen CJP. A new family of insect tyramine receptors. *Biochem Biophys Res Commun* 2005;**338**:1189–96.
144. Huang J, Ohta H, Inoue N, Takao H, Kita T, et al. Molecular cloning and pharmacological characterization of a *Bombyx mori* tyramine receptor selectively coupled to intracellular calcium mobilization. *Insect Biochem Mol Biol* 2009;**39**:842–9.
145. Bayliss A, Roselli G, Eans PD. A comparison of the signalling properties of two tyramine receptors from Drosophila. *J Neurochem* 2013;**12**(1):37–48.
146. El-Kholy S, Stephano F, Li Y, Bhandari A, Flink C, Roeder T. Expression analysis of octopamine and tyramine receptors in Drosophila. *Cell Tissue Res* 2015;**361**(3):669–84.
147. Charles S, Luetje CW. Trace amines inhibit insect odorant receptor function through antagonism of the co-receptor subunit. *F1000Res* 2014;**3**:84.

Octopaminergic and Tyraminergic Signaling in the Honeybee (*Apis mellifera*) Brain: Behavioral, Pharmacological, and Molecular Aspects

W. Blenau[1] and A. Baumann[2]

[1]Institute of Zoology, University of Cologne, Cologne, Germany

[2]Forschungszentrum Juelich, Institute of Complex Systems – Cellular Biophysics (ICS-4), Juelich, Germany

OUTLINE

INTRODUCTION

The exchange of information between neurons or between neurons and target cells depends on the coordinated interplay of chemical and electrical signals. Neurons are exquisitely tuned to generate action potentials for fast information fluxes over long distances. In addition, a transient change of a neuron's physiological and biochemical properties is equally important for proper neuronal function. One group of neurotransmitters that induces transient changes of cellular signaling properties is the biogenic amines. The substances can act either locally as neurotransmitters and neuromodulators or as neurohormones at sites further away from their release site. Biogenic amines convey their biological activity by binding to membrane proteins that belong to the superfamily of G-protein-coupled receptors (GPCRs). These membrane proteins are characterized by a common structural motif of seven transmembrane (TM) segments.[1,2] Agonist binding to GPCRs leads to transient changes in intracellular concentrations of second messengers like 3′,5′-cyclic adenosine monophosphate (cAMP), 3′,5′-cyclic guanosine monophosphate (cGMP), inositol-1,4,5-trisphosphate (IP_3), and Ca^{2+}. Since second messengers control and modulate the activity of various proteins, for example, kinases, phosphatases, transcription factors, it is the activation of these proteins that causes dynamic and spatially restricted modifications of effector proteins and thereby modulates a cell's signaling properties.

The group of biogenic amines consists of five main members in both deuterostomes and protostomes. Three amines are shared across phyla, that is, dopamine, histamine, and serotonin (5-hydroxytryptamine). Some compounds seem to be preferentially utilized in either deuterostomes (norepinephrine, epinephrine) or protostomes (tyramine, octopamine). Nevertheless, low amounts of tyramine and octopamine have been identified in vertebrates where they are classified as trace amines due to their low abundance.[3] In protostomes, tyramine, and octopamine bind to specific members of the GPCR superfamily. Tyramine receptors have been molecularly cloned and pharmacologically characterized from insects, for example, honeybees (Apis mellifera) and common fruit flies (Drosophila melanogaster), and were shown to cause inhibition of cAMP production or to induce Ca^{2+} release from intracellular stores. In contrast, octopamine receptors either activate cAMP production or mediate Ca^{2+} release from intracellular stores. Depending on the cell type in which the receptor genes have been heterologously expressed, additional signaling pathways may also be activated.[4–9]

Biogenic amines exert a multitude of effects in insects which are not restricted to the central nervous system but also have been described and examined at the sensory periphery as well as at the level of motor output on muscles and glands.[10–13] In the honeybee brain, the quantities of dopamine, octopamine, serotonin, and tyramine are quite different.

Octopamine is present in relatively small amounts but it modulates numerous behaviors. It has been demonstrated that octopamine is involved in the initiation of foraging behavior, modulates the sensitivity for stimulus modalities, and can increase both acquisition and retrieval of information and thus controls learning and memory formation (for details, see below in "Physiological and behavioral aspects of octopamine and tyramine signaling"). Both dopamine and serotonin are present in rather high amounts in the honeybee brain and often have inhibitory effects on behavior. The role of tyramine is less clear, mainly because tyramine has traditionally been considered as the biochemical precursor of octopamine rather than as being a neuroactive substance on its own.[14]

In the following sections we will concentrate on tyramine and octopamine and their receptors, with special emphasis on their functional role in honeybees.

BIOSYNTHESIS OF TYRAMINE AND OCTOPAMINE

As all biogenic amines, tyramine, and octopamine are derived from a small organic molecule (Fig. 14.1). The proteinogenic amino acid L-tryrosine is first decarboxylated by the enzyme tyrosine decarboxylase (TDC),[15] resulting in the production of tyramine. This phenolamine can be further modified. The β-carbon of the side chain can be hydroxylated. This reaction is catalyzed by the enzyme tyramine β-hydroxylase (TβH),[16,17] and leads to the synthesis of the biogenic amine octopamine. The biochemical pathways generating tyramine and octopamine have been discussed in several comprehensive reviews to which we refer the reader for more details.[13,18,19] Since the chemical structures of octopamine and

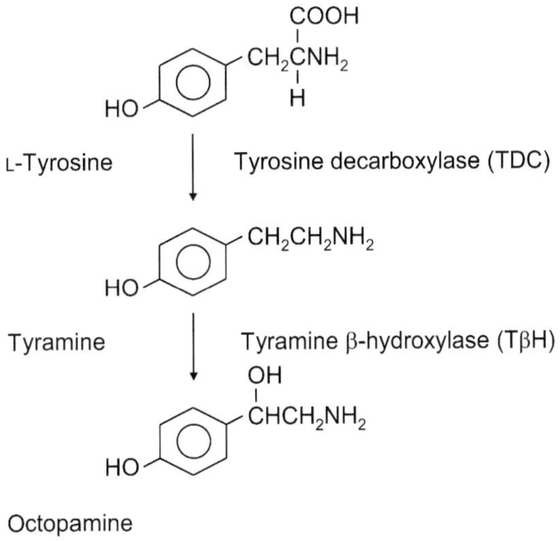

FIGURE 14.1 Biosynthesis of tyramine and octopamine. The aromatic amino acid L-tyrosine is decarboxylated by tyrosine decarboxylase (TDC) to tyramine. This biogenic amine is hydroxylated on the $C_β$ position by tyramine-β-hydroxylase (TβH) resulting in biosynthesis of octopamine.

norepinephrine are very similar, it is generally assumed that the noradrenergic/adrenergic system of deuterostomes is functionally substituted by the tyraminergic/octopaminergic system in protostomes.[14,19]

DISTRIBUTION OF TYRAMINE AND OCTOPAMINE IN THE BRAIN OF ADULT HONEYBEES

Staining tissue sections with antisera raised against biogenic amines revealed that only a relatively small number of interneurons in the insect brain synthesize these bioactive molecules.[20] In comparison to dopamine and serotonin, which are present in the bee brain at concentrations of 12–40 and 6–21 pmol/brain, respectively, the amount of octopamine is lower, ranging between 2 and 10 pmol/brain.[10] Even lower amounts have been measured for tyramine (0.3–2 pmol/brain).[21]

The distribution of tyramine in the honeybee brain has not yet been studied at the cellular level. However, because tyramine is the precursor of octopamine during biosynthesis, tyramine must be present at least in all octopamine-containing cells.

Five cell clusters containing octopamine-immunoreactive somata have been identified in the honeybee brain. These comprise neurosecretory cells, cells located mediodorsal to the antennal lobe, cells distributed on both sides of the protocerebral midline, a group of cells between the lateral protocerebral lobes and the dorsal lobes, as well as single cells on either side of the central body.[22,23] With the exception of the pedunculi of the mushroom bodies and large parts of the α- and β-lobes lobes, varicose octopamine-immunoreactive fibers innervate most neuropils of the honeybee brain.[22,24,25] An interesting pattern has been detected in the central body with a much higher density of octopamine-immunoreactive fibers in the lower than in the upper division.[23] Using in situ hybridization to identify TβH-expressing neurons in honeybee brain revealed very similar locations and numbers of cells as found for octopamine-immunoreactive neurons.[17]

MOLECULAR AND PHARMACOLOGICAL PROPERTIES OF HONEYBEE TYRAMINE AND OCTOPAMINE RECEPTORS

As already mentioned, biogenic amines bind to and activate receptors belonging to the superfamily of GPCRs. Within this gene family, rhodopsin-like receptors, which also include tyramine and octopamine receptors, form the largest subfamily. Structural data for some GPCRs became available more recently,[26–28] and confirmed the predicted membrane organization with seven TM segments (Fig. 14.2). The receptors are classical type II membrane proteins with the N-terminus located extracellularly and the C-terminus facing the cytosol. The N-termini of these receptors are often modified post-translationally by N-linked glycosylation. As depicted in Fig. 14.2, the membrane-spanning regions are linked by three extracellular loops that alternate with three intracellular loops (IL). Cysteine residues in the cytoplasmic tail of the polypeptides are a target of post-translational palmitoylation. Ligand binding to the receptor takes place in a binding pocket formed by the TM regions in the plane of the membrane. An aspartic acid residue (D) in TM3, serine residues (S) in TM5, and

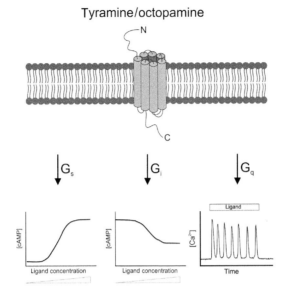

FIGURE 14.2 Signaling pathways mediated by GPCR activation. Binding of tyramine and/or octopamine to specific receptor subtypes can activate either trimeric G_s-, G_i-, or G_q-type GTP-binding proteins which couple to downstream effectors and mediate either cAMP production, inhibition of cAMP production, or IP_3-dependent Ca^{2+} release from intracellular stores, respectively.

a phenylalanine residue (F) in TM6 were shown to determine the ligand-binding properties of biogenic amine receptors.[1,2] The physical interaction between the ligand and its receptor induces a conformational change in the protein which is transferred to trimeric G-proteins interacting with the receptor. Residues in close vicinity to the plasma membrane of IL2, 3, and 4 determine the specificity and efficacy of G-protein activation and, thus, the intracellular signaling pathway that is addressed.

The availability of the completely sequenced honeybee genome[29] has largely facilitated identification and cloning of genes from this insect. Several groups have put much effort into annotation projects providing researchers with detailed information on, for example, gene structures and potential splicing variants.[30,31] Molecular cloning of cDNAs coding for some tyramine and octopamine receptors, however, was achieved even before the honeybee genome sequence was available. Up to now, two cDNAs encoding tyramine receptors (AmTyr1[6]; AmTyr2[8]) and five cDNAs encoding octopamine receptors (AmOctα1R[7,32]; AmOctβR1–AmOctβR4[9]) have been cloned from honeybee brain. Except for the AmTyr2 receptor, the other six receptors were pharmacologically and functionally examined after heterologous expression in mammalian or insect cell lines.[6,7,9,33,34]

The AmTyr1 Receptor

The relative molecular mass of AmTyr1 as deduced from the cDNA is 44.7 kDa. With 80.2% and 83.2% high degrees of homology exist between AmTyr1 and tyramine receptors from, for example, *D. melanogaster* (DmTyr[35,36]) and *Locusta migratoria* (LocTYR[37]), respectively. The AmTyr1 receptor contains a cognate motif for post-translational N-linked glycosylation (N$_2$SS) in the N-terminus and several consensus motifs for phosphorylation by cAMP/cGMP-dependent protein kinases (PKA/PKG) as well as protein kinase C (PKC) in the IL. Notably, the C-terminus of AmTyr1 consists of only 16 amino acid residues.[6]

Pharmacological properties and functional coupling of the receptor were examined in stably transfected HEK 293 cells,[6,34] as well as in insect cells.[33] Activation of the receptor with tyramine inhibited forskolin-stimulated cAMP synthesis with an EC$_{50}$ of ~130 nM. Octopamine was both less potent (EC$_{50}$~3 μM) and less efficacious than tyramine. Mianserin has been shown to block AmTyr1 activity at high concentrations (IC$_{50}$ 73 μM[34]).

In the course of characterizing the AmTyr1 receptor, the effects of tyramine and octopamine on adenylyl cyclase activity in membrane homogenates of honeybee brain were also examined.[6] Adenylyl cyclase activity was stimulated with forskolin. When samples were coincubated with 0.1–1 μM tyramine, cAMP synthesis was attenuated by ~25%, a similar value to that determined on heterologously expressed AmTyr1. At higher concentrations of tyramine, the inhibitory effect was less pronounced, which could be due to binding of tyramine to, for example, octopamine receptors that rather activate than inhibit adenylyl cyclase and which were present in the membrane sample as well,[6] or to another, hitherto not-identified third tyramine-receptor-type that stimulates cAMP biosynthesis.

The developmental and tissue distribution of AmTyr1 expression was studied by in situ hybridization to cryosections of the brain and by Northern blotting.[6,33] In situ hybridization showed that the mRNA encoding AmTyr1 was abundantly expressed in the brain of adult worker honeybees. Labeling of cell somata was observed in many brain neurons, including defined clusters of cells associated with mushroom bodies, optic lobes, and antennal lobes.[6] Intense staining was detected within the three major divisions of mushroom body intrinsic neurons, that is, outer compact, noncompact, and inner compact cells. Age-related increases in receptor transcript levels were observed not only during metamorphosis, but also in the brain of the adult worker bee.[33]

The AmTyr2 Receptor

A second gene coding for a potential tyramine receptor has been identified and annotated from the honeybee genomic sequence.[8] The Am*tyr*2 gene was identified by screening the database with a *Drosophila* gene (DmTyr2, CG7431) that was studied in more detail.[8] When heterologously expressed, the DmTyr2 receptor induced Ca^{2+} release from intracellular stores. In contrast to previously characterized tyramine receptors, including AmTyr1, that could be stimulated both by tyramine and octopamine, the DmTyr2 receptor seemed to be almost exclusively activated by tyramine. Since the honeybee AmTyr2 sequence assembled in the same clade as the DmTyr2 receptor and is clearly set apart from the AmTyr1 receptor-containing clade,[8,30] it is reasonable to speculate that the honeybee receptor couples to Ca^{2+} signaling, as well. Nevertheless, this statement necessitates experimental proof after physical cloning and functional expression of the honeybee cDNA. It would be similarly interesting to examine the developmental and tissue-specific expression pattern of the Am*tyr*2 gene to gain insight into its potential role in honeybee physiology and behavior.

The AmOctα1 Receptor

At the cellular level, activation of specific octopamine receptor subtypes induces either Ca^{2+} signals or cAMP synthesis.[10,13,19] The receptors are classified as α-adrenergic-like or β-adrenergic-like octopamine receptors,[38] due to structural and functional similarities to vertebrate adrenoceptors.

The amino acid sequence of AmOctα1R (previously called AmOA1[7,39]) deduced from the open reading frame of cloned cDNA consists of 587 residues with a calculated molecular weight of 66.5 kD. The primary structure contains several conserved residues that are hallmarks of biogenic amine receptors. The extracellular domains harbor consensus motifs for N-linked glycosylation. Phosphorylation by PKA or PKC might occur at consensus sites found in IL as well as in the C-terminus.[7]

Activation of heterologously expressed AmOctα1R by nanomolar concentrations of octopamine led to oscillations of the intracellular Ca^{2+} concentration ($[Ca^{2+}]_i$).[7] This property is very similar to that described for *Drosophila* α-adrenergic-like octopamine receptors (DmOctαR1A, DmOctαR1B[40,41]). Whether the Ca^{2+} signals induced by AmOctα1R depend on cycling desensitization/resensitization mechanisms mediated by PKC and phosphatase activities, respectively, as has been recently identified for DmOctαR1B,[42] remains to be experimentally addressed for the honeybee receptor. Application of tyramine to AmOctα1R-expressing cells in the same concentration range as octopamine activated Ca^{2+} signals only at high ligand concentrations ($\geq 1 \mu M$). Compared to the effect of octopamine at the same concentration, the tyramine effect was always delayed.[7] The Ca^{2+} responses were efficaciously suppressed, when cells expressing AmOctα1R were treated with $10 \mu M$ concentrations of either cis-(Z)-flupentixol, spiperone, mianserin, or epinastine in the presence of $1 \mu M$ octopamine.[34]

Several groups have addressed the distribution of AmOctα1R by different and independent methods. In situ hybridizations to cryosections of adult worker bee brains identified prominently labeled soma clusters representing mushroom body intrinsic neurons, cells belonging to the optic lobes, antennal lobes, and the deutocerebrum.[7] Antibodies were raised against peptide sequences located in the C-terminal domain of the protein. These antibodies stained similar cell body clusters as detected by in situ hybridization, including the antennal lobes, the calyces, pedunculus, vertical and medial lobes of the mushroom body, optic lobes, subesophageal ganglion, and the central complex.[43] Notably, populations of GABAergic local interneurons in the antennal lobes as well as a subpopulation of GABAergic feedback neurons that project into the mushroom body were identified to express the receptor protein. The data suggest a prominent role for AmOctα1R to modulate the activity of different types of inhibitory neurons in the antennal lobe, yet, expression of the receptor protein is not limited to inhibitory neurons.[43,44] Using RNA interference to impair AmOctα1R expression followed by Western blotting resulted in reduced staining of protein bands obtained from antennal lobes as well as from honeybee brains excluding the antennal lobes.[39] Finally, applying qPCR analyses revealed both age-related as well as task-related changes in AmOctα1R expression in the whole brain as well as certain substructures.[45,46]

A certain aspect of octopamine receptor signaling is worth mentioning. Once examining the cellular responses of α-adrenergic-like octopamine receptors it was observed that in addition to Ca^{2+} signals, cells expressing these receptors also showed an increase in intracellular cAMP.[7,34,40,41] However, cAMP amounts increased linearly with increasing ligand concentrations, did not saturate and, compared to cAMP responses mediated by β-adrenergic-like octopamine receptors, were very small.[7,41] Such responses, however, could be secondary to the AmOctα1R-induced rise in $[Ca^{2+}]_i$ and caused by stimulation of Ca^{2+}-dependent, cell endogenous membrane-bound adenylyl cyclases rather than via interaction between AmOctα1R with stimulating G-proteins, further arguing for specific cellular signaling pathways activated by either α- or β-adrenergic-like octopamine receptors (see below).

The AmOctβ Receptors

The release of the complete genome sequence of the honeybee has paved the way to access hitherto missing members from different gene families. Four gene candidates coding for potential β-adrenergic-like octopamine receptors were identified in a GPCR annotation project.[30] Based on annotation data, physical cloning and functional expression finally led to the successful characterization of receptors that unequivocally stimulated adenylyl cyclase activity in an octopamine-dependent fashion.[9] The deduced amino acid sequences for AmOctβR1 to AmOctβR4 consist of 428, 413, 413, and 401 residues with calculated molecular weights of 48.8, 46.9, 46.2, and 44.6 kDa, respectively. The primary structures of these receptors share a high degree of similarity. The closest relationship exists between AmOctβR3 and AmOctβR4 which share 76.8% of identical or conservatively substituted residues. Notably, a stretch of 107 residues in the C-terminal third of both proteins is completely identical.[9] For the *Drosophila* DmOctαR1A and DmOctαR1B receptors it had been previously found that they share identical N-terminal halves.[41] The proteins originate from a single gene by alternative splicing. The honeybee AmOctβR3 and AmOctβR4 receptors most likely also originate by alternative splicing from one gene where the N-terminal two-thirds of both receptors are encoded by physically separated exons and the identical C-terminal parts of the proteins are encoded by two shared exons located further downstream.[9] As already mentioned for the tyramine and α-adrenergic-like octopamine receptors, the four AmOctβ receptors possess the characteristic features of GPCRs,[1,2] with seven transmembrane regions, potential N-glycosylation sites, and consensus motives for phosphorylation by different kinases.

Pharmacologically, all four receptors could be activated by octopamine as well as tyramine. Half-maximal activation of the receptors with octopamine was in the low nanomolar range (~2–~44 nM) with AmOctβR4 being the most sensitive and AmOctβR1 being the least sensitive receptor. Furthermore, octopamine was at least one order of magnitude more efficient than tyramine at each receptor.[9] The ability of all four receptors to raise Ca^{2+} signals in stably transfected cell lines was tested with octopamine concentrations ranging from 10^{-9} to 10^{-6} M. None of the receptor-expressing cell lines showed a Ca^{2+} response.

A series of potential antagonists was examined for their ability to impair octopamine- and tyramine-stimulated AmOctβR activity. The most efficacious antagonist on octopamine-stimulated AmOctβRs was mianserin with IC_{50s} in the low nanomolar range (~5 nM AmOctβR3; ~23 nM AmOctβR2). Similarly low IC_{50s} for mianserin were determined on tyramine-activated AmOctβRs.[9] The activity of all four receptors was also inhibited by cyproheptadine, metoclopramide, chlorpromazine, and phentolamine. Half-maximal inhibition, however, required sub- to micromolar ligand concentrations.[9] Yohimbine which is considered a tyramine-receptor specific antagonist,[19] did not impair octopamine-stimulated AmOctβR activity. In summary, the same rank order of potency was observed for the different antagonists on either octopamine- or tyramine-stimulated AmOctβRs with mianserin being the most efficient inhibitor.

The expression pattern of all four AmOctβ receptors has not yet been addressed at the cellular level. However, a recent study examining age- and task-related changes of octopamine receptor-gene expression, provided the first data for AmOctβ3 and AmOctβ4.[46] Due to primer design in the conserved region of this alternatively spliced gene (s.a.) transcripts for both receptors were detected simultaneously. Expression was found in the subesophageal ganglion, the antennal lobes, the mushroom bodies, and the optic lobes. The authors did not detect differences in expression levels in these tissues between nurse bees and foragers, except for the mushroom bodies in which nurse bees had higher transcript levels than foragers. Notably, this difference disappeared, when precocious foragers and age-matched nurse bees were examined,[46] suggesting that these receptors rather than being involved in regulating social behavior play a role in aging.

PHYSIOLOGICAL AND BEHAVIORAL ASPECTS OF OCTOPAMINE AND TYRAMINE SIGNALING

In the honeybee (*A. mellifera* L.) a number of studies have shown that octopamine modulates the responsiveness of sensory receptors, interneurons, as well as motoneurons and thus affects complex behavioral responses originating from olfactory or visually perceived inputs. In comparison to octopamine, current knowledge on the functional contribution of tyramine to behavior is less pronounced. In the following paragraph, we summarize some effects that both biogenic amines evoke in honeybee's neuronal function.

Locomotor Activity

Honeybees use active scanning movements of their antennae to analyze the surface of objects, for example, flowers. The frequency of antennal contacts with an object can be quantified and used to characterize antennal scanning behavior.[47] When octopamine was injected into the dorsal lobe of the brain, the sensory motor-center of the antenna, antennal scanning activity was stimulated.[47] The impact of octopamine and tyramine on locomotor activity was further investigated by injecting worker bees with various concentrations of both amines as well as the octopamine and tyramine receptor antagonists, mianserin, and yohimbine, respectively. Effects of the individual treatments were monitored in freely moving bees.[48] Each manipulation evoked complex changes in behavior with alterations correlating with the drug applied, its concentration, as well as the time passed since injection.[48] Flying behavior was differently affected by octopamine and tyramine. Whereas octopamine generally increased flying behavior, tyramine or high concentrations of mianserin led to a decrease in time spent flying.[48] From these results it was concluded that octopamine and tyramine modulate motor function either via interaction with central pattern generators or via effects on sensory perception. In order to test for possible effects of octopamine on plasticity in circadian and diurnal rhythms, young bees were fed with octopamine.[49] This treatment influenced aspects of locomotor behavior that most likely do not originate from the endogenous circadian pacemaker. Octopamine-treated animals showed an earlier daily onset of activity, shorter alpha phases, defined as the period between the daily onset and offset of activity, and a trend toward more bees showing diurnal rhythms in light–dark illumination.[49] Therefore, octopamine might have modulated neuronal centers mediating responses to light, locomotion, or both.[49]

Visual Sensitivity and Phototaxis

Honeybees rely on their visual senses, particularly at foraging age. The direction-specific visual antennal reflex is a useful tool to experimentally study neuromodulation in the bee's visual system. A stripe pattern moving upwards in front of a fixed bee induces downward antennal movements, whereas a downward-moving pattern induces antennal movements in the opposite direction. The difference in the antennal angles for the two directions is used as a measure of the direction specificity of the response.[50,51] Octopamine injections into the lobula enhance direction specificity,[50,51] suggesting that octopamine modulates motion-sensitive neurons in this part of the visual system. The hypothesis was proven by recording field potentials in the lobula. Moving stripe patterns evoked an increase in the initial slope and amplitude of field potentials in octopamine-injected animals.[52]

Task-defined groups of bees also differ naturally in their phototaxis behavior in such a way that pollen foragers display lower light responsiveness than nectar foragers.[53] The lower phototaxis of pollen foragers coincides with higher octopamine titers in the optic lobes while the octopamine content in mushroom bodies of pollen and nectar foragers is very similar. When feeding bees with octopamine to increase the brain octopamine content, their responses to light were reduced but locomotor activity was not affected. In contrast to octopamine, tyramine-fed bees showed increased phototaxis which, at least in part, might have arisen by tyramine effects on general locomotor activities in these animals.[53]

Gustatory and Olfactory Sensitivity

An experimental setup to test for a bee's sensitivity towards gustatory or olfactory stimuli is their proboscis extension response (PER). When an antenna is touched with a droplet of sucrose solution of appropriate concentration, the bee extends its proboscis. The responsiveness or sensitivity can be measured by applying solutions containing different sucrose concentrations.[54] Both octopamine and tyramine decrease a bee's response thresholds to such gustatory stimuli.[55,56] The spontaneous PER to antennal stimulation with an odor originally has been introduced as an indicator of olfactory sensitivity,[57] and it has been shown that octopamine injection into the brain decreases olfactory response thresholds.[58] Furthermore, the response threshold for brood pheromone also decreased in hive bees, when the animals were fed with octopamine for several days.[59]

Honeybee strains have been discovered that naturally differ in their hygienic behavior.[24] Bees of "hygienic" strains detect diseased brood quickly and remove it from the hive. In contrast, animals of "nonhygienic" strains need longer and higher stimulus intensities to perform the task. Octopamine treatment of bees from the "nonhygienic" strain resulted in increased electroantennogram (EAG) responses to odors originating from diseased brood. In contrast, no effect of octopamine was observed in bees of the "hygienic" strain, suggesting that these animals had already reached maximal sensitivity. In line with this interpretation was the finding that an octopamine receptor antagonist, epinastine, led to a reduction of EAG amplitudes in these bees but had no effect on "nonhygienic" bees.[24] Furthermore, bees from the "hygienic" strain do not have lower sucrose response thresholds and do not forage at an earlier age compared to animals from the "nonhygienic" strain.[60] Thus, the effects of octopamine on brood odor discrimination and the expression of hygienic behavior[24] are not caused by an increase in sucrose sensitivity or foraging ontogeny, even though both of these activities are altered by octopamine.[55,56,61]

Recently, the effect of octopamine was directly tested in the antennal lobe, the first olfactory center in the brain.[62] Calcium imaging was applied to monitor background activity and odor-evoked neuronal activity, while superfusing the brain with test solutions either containing octopamine or not. Interestingly, octopamine increased background activity in olfactory output neurons but reduced average calcium levels. The network effect of octopamine was reproducible for a given stimulus within a given animal, yet, it varied across glomeruli, odorants, odorant concentrations, and animals. This finding suggests that octopamine acts on synaptic contacts at the network level rather than on the level of individual neurons. Network properties are further shaped by plasticity depending on an individual animal's history but responses are hardly predictable for different odors or specific animals.[62]

Nonassociative Learning

Learning and memory can be considered as general forms of plasticity in the nervous system and underlying mechanisms are rather conserved across phylogenetic distant species. In comparison to studies in vertebrates, honeybees serve a promising alternative for studying higher brain functions at the cellular and molecular level, including learning and memory tests.

Learning is an essential part of a honeybee's daily life, for example, when foraging for food. Once having identified a food source, they have to remember how to gain nectar or pollen as well as how to return to their hive. For experimental testing nonassociative and associative forms of learning in honeybees, several paradigms have been established that frequently use the PER. As mentioned previously, sensitization can be tested as PER to antennal stimulation with odors,[63] water vapor,[64] or gustatory stimuli.[58] Normally, bees do not show PER in response to antennal stimulation with an odor or water vapor. When bees are stimulated with a high-sucrose concentration, for example, 30% sucrose, shortly before the odor is presented to their antennae, they can be sensitized to the odor. Successful sensitization leads to a PER after subsequent antennal stimulation with that odor.[64] Similarly, bees can be sensitized to antennal stimulation with water.[58] Injection of octopamine either into the hemolymph or into the dorsal lobe increased responsiveness to water and, thus, mimicked the effect of sucrose stimulation.[65] In contrast, sensitization to odors was not affected by octopamine in honeybees whose nervous system had been depleted of biogenic amines by reserpine.[63]

In contrast to sensitization, repeated stimulation of honeybee antennae with sucrose solution leads to habituation of the PER.[66] Tyramine appears to increase the rate of habituation when fed 12h before the experiment because animals fed with tyramine needed fewer trials than control animals to achieve complete habituation of the PER.[66] Interestingly, chlordimeform, an octopamine receptor agonist, had the same effect as tyramine.[66] Originally, it was assumed that the effects of tyramine were mediated by the activation of octopamine receptors either after enzymatic conversion of tyramine to octopamine or by direct binding of tyramine to octopamine receptors.[66] However, now that specific tyramine and octopamine receptors have been molecularly identified and pharmacologically characterized, some experiments may need to be revisited and repeated using ligands that specifically bind to certain receptor subtypes.

Associative Learning

To examine associative learning under laboratory conditions, appetitive olfactory conditioning of the PER is usually employed. Other learning paradigms are established as well (eg, tactile conditioning[67]; avoidance learning[68]). In the olfactory learning paradigm, the conditioned stimulus (CS, odor) is paired with an unconditioned stimulus (US, sucrose reward). The spontaneous responsiveness to the CS is measured first by applying an odor to the antennae. The conditioning trial begins when the bee can smell the odor. While the bee experiences the CS, the PER is elicited by applying a small droplet of sucrose solution to its antennae (US). When the bee extends its proboscis, it is allowed to drink a small volume of sucrose solution as reward. Usually, a few pairings of CS and US suffice for the formation of a memory, which lasts for days.[54,69–72]

Using such tests, it has been demonstrated that octopamine plays a decisive role in acquisition and memory formation in bees. Application of octopamine improves olfactory acquisition, memory formation, and retrieval. Satiated bees normally do not display any associative PER learning. Octopamine injections into the hemolymph of satiated bees can restore the "motivation" to learn. In bees whose nervous system had been depleted of biogenic amines by reserpine, octopamine significantly improved the acquisition performance but not retention of the memory.[63] Injection of mianserin into the antennal lobe strongly reduced acquisition and retrieval performance of the animals.[32] A similar but irreversible effect was observed when the expression of the AmOctα1R receptor was downregulated by injection of Am*octα*1R dsRNA.[32] An independent line of evidence for the role of octopamine in memory formation at the cellular level originates from the analysis of the VUM_{mx1} neuron (ventral unpaired median neuron 1 of the maxillary neuromere[73]). It belongs to a cluster of octopaminergic neurons, the cell bodies of which are located ventrally in the subesophageal ganglion.[22,25,74] The VUM_{mx1} neuron depolarizes in response to the presentation of a sucrose reward to either antennae or proboscis. Current injection into the VUM_{mx1} neuron can substitute for the sucrose reward during olfactory conditioning.[73,75] It is assumed that upon stimulation VUM neurons release octopamine, which could then mediate the US in PER conditioning.[72,75,76] Whether or not tyramine is coreleased with octopamine from activated VUM neurons is currently not known.

Division of Labor

Division of labor in honeybee colonies is characterized by changes in the tasks performed by individual bees as they age. Octopamine levels have been determined in the brains of bees performing different tasks inside and outside the hive and of bees which differ in age. During development, the amine is present in the brains of larvae, pupae, and adult honeybees,[77–80] with levels increasing during the transition from the larval to pupal stage.[80] In adult honeybees, octopamine levels increase as well, with the highest concentration being found in foragers.[17,78,81–83] Whether the difference in octopamine levels are related to age differences or to the different tasks the bees perform was examined in single-cohort colonies which consisted of same-aged bees.[17,82,84] Age-related differences in octopamine levels were observed in the mushroom bodies of nurse bees and foragers, whereas in the antennal lobe, the differences were related to the task a bee performed.[82] Interestingly, levels of octopamine in the antennal lobes were found to be elevated immediately after the onset of foraging, but they did not change as a consequence of preforaging orientation flight activity, diurnal pauses in foraging, or different amounts of foraging experience.[84] These results suggest that octopamine is important to trigger and maintain the foraging behavioral state, and thus is involved in the regulation of honeybee division of labor. In addition to octopamine levels, significantly higher TβH levels were found in forager brains as compared to nurses.[17] This result was seen in both typical colonies and in single-cohort colonies, in which (precocious) foragers were the same age as nurses.[17] Elevated TβH levels are a result of upregulation of TβH expression in existing octopaminergic cells, and were not due to the appearance of new neurons that begin to synthesize octopamine later in life.[17] Tyramine levels have not yet been measured in single-cohort colonies.

Oral octopamine treatment induced hive bees to forage precociously, whereas tyramine had the opposite effect.[59,61,84] In addition to octopamine, the lipophilic juvenile hormone (JH), is also a decisive factor in regulating the onset of foraging behavior.[85] Hormone titers increase with age in adult bees, with foragers having higher levels of JH than hive bees. Removal of the JH synthesizing *corpora allata* delays the initiation of foraging, an effect that can be reversed by treating animals with the JH analog methoprene.[86] Notably, octopamine and JH regulate each other and modulate the onset of foraging behavior and the initiation of other tasks.[87–89] Foragers have high titers of both JH and octopamine, particularly in the antennal lobes.[82,89] Treating 1-day-old bees with methoprene causes increased levels of octopamine in the antennal lobes 12 d later and leads to precocious foraging.[88] When allatectomized bees were fed octopamine, they became normal foragers. Whereas these experiments suggest that octopamine acts downstream of JH, it was also shown that octopamine increases JH release from the *corpora allata* in vitro in a dose-dependent manner.[87] Thus, a precisely regulated interaction of both neuroactive substances seems to be necessary for adequate initiation of foraging behavior.

CONCLUSION

In our contribution, we have focused on two, that is, octopamine and tyramine, of at least five biogenic amines that play a pivotal role in the behavior and physiology of honeybees. Octopamine has been demonstrated to play a decisive role in the acquisition and retrieval of information, thereby contributing to associative and nonassociative forms of learning. Furthermore, octopamine participates in the initiation of honeybee foraging behavior. Although tyramine has lost its status as merely serving as an octopamine precursor during biosynthesis, its contribution to honeybee behavior and physiology still needs to be determined in detail. A total of five octopamine receptors and two tyramine receptors have been cloned and characterized at the molecular level. For most receptor proteins, pharmacological profiles and cellular signaling pathways have been thoroughly examined. These data may now pave the way to either revisit existing behavioral pharmacological data or to perform new in vivo pharmacological experiments that specifically address individual receptor subtypes. Given that molecular biological techniques prosper, it should also be feasible to interfere with the expression of individual receptor genes and to study the effects of reduction or loss of these receptors on honeybee behavior and physiology.

References

1. Lefkowitz RJ. Historical review: a brief history and personal retrospective of seven-transmembrane receptors. *Trends Pharmacol Sci* 2004;**25**:413–22.
2. Kobilka BK. Structural insights into adrenergic receptor function and pharmacology. *Trends Pharmacol Sci* 2011;**32**:213–8.
3. Burchett SA, Hicks TP. The mysterious trace amines: protean neuromodulators of synaptic transmission in mammalian brain. *Progr Neurobiol* 2006;**79**:223–46.
4. Robb S, Cheek TR, Hannan FL, Hall LM, Midgley JM, Evans PD. Agonist-specific coupling of a cloned *Drosophila* octopamine/tyramine receptor to multiple second messenger systems. *EMBO J* 1994;**13**:1325–30.
5. Reale V, Hannan F, Midgley JM, Evans PD. The expression of a cloned *Drosophila* octopamine/tyramine receptor in *Xenopus* oocytes. *Brain Res* 1997;**769**:309–20.
6. Blenau W, Balfanz S, Baumann A. Amtyr1: characterization of a gene from honeybee (*Apis mellifera*) brain encoding a functional tyramine receptor. *J Neurochem* 2000;**74**:900–8.
7. Grohmann L, Blenau W, Erber J, Ebert PR, Strünker T, Baumann A. Molecular and functional characterization of an octopamine receptor from honeybee (*Apis mellifera*) brain. *J Neurochem* 2003;**86**:725–35.
8. Cazzamali G, Klaerke DA, Grimmelikhuijzen CJ. A new family of insect tyramine receptors. *Biochem Biophys Res Commun* 2005;**338**:1189–96.
9. Balfanz S, Jordan N, Langenstück T, Breuer J, Bergmeier V, Baumann A. Molecular, pharmacological, and signaling properties of octopamine receptors from honeybee (*A. mellifera*) brain. *J Neurochem* 2014;**129**:284–96.
10. Scheiner R, Baumann A, Blenau W. Aminergic control and modulation of honeybee behavior. *Curr Neuropharmacol* 2006;**4**:259–76.
11. Walz B, Baumann O, Krach C, Baumann A, Blenau W. The aminergic control of cockroach salivary glands. *Arch Insect Biochem Physiol* 2006;**62**:141–52.
12. Farooqui T. Octopamine-mediated neuromodulation of insect senses. *Neurochem Res* 2007;**32**:1511–29.
13. Verlinden H, Vleugels R, Marchal E, Badisco L, Pflüger HJ, Blenau W, et al. The role of octopamine in locusts and other arthropods. *J Insect Physiol* 2010;**56**:854–67.
14. Lange AB. Tyramine: from octopamine precursor to neuroactive chemical in insects. *Gen Comp Endocrinol* 2009;**162**:18–26.
15. Livingstone MS, Tempel BL. Genetic dissection of monoamine neurotransmitter synthesis in *Drosophila*. *Nature* 1983;**303**:67–70.
16. Monastirioti M, Linn Jr. CE, White K. Characterization of *Drosophila* tyramine β-hydroxylase gene and isolation of mutant flies lacking octopamine. *J Neurosci* 1996;**16**:3900–11.
17. Lehman HK, Schulz DJ, Barron AB, Wraight L, Hardison C, Whitney S, et al. Division of labor in the honey bee (*Apis mellifera*): the role of tyramine β-hydroxylase. *J Exp Biol* 2006;**209**:2774–84.
18. Blenau W, Baumann A. Molecular and pharmacological properties of insect biogenic amine receptors: lessons from *Drosophila melanogaster* and *Apis mellifera*. *Arch Insect Biochem Physiol* 2001;**48**:13–38.
19. Roeder T. Tyramine and octopamine: ruling behavior and metabolism. *Annu Rev Entomol* 2005;**50**:447–77.
20. Homberg U. Distribution of neurotransmitters in the insect brain. *Progr Zool* 1994;**40**:1–88. Fischer Verlag, Stuttgart, Germany.
21. Sasaki K, Nagao T. Brain tyramine and reproductive states of workers in honeybees. *J Insect Physiol* 2002;**48**:1075–85.

22. Kreissl S, Eichmüller S, Bicker G, Rapus J, Eckert M. Octopamine-like immunoreactivity in the brain and subesophageal ganglion of the honeybee. *J Comp Neurol* 1994;**348**:583–95.
23. Bicker G. Biogenic amines in the brain of the honeybee: cellular distribution, development, and behavioral functions. *Microsc Res Tech* 1999;**44**:166–78.
24. Spivak M, Masterman R, Ross R, Mesce KA. Hygienic behavior in the honey bee (*Apis mellifera* L.) and the modulatory role of octopamine. *J Neurobiol* 2003;**55**:341–54.
25. Sinakevitch I, Niwa M, Strausfeld NJ. Octopamine-like immunoreactivity in the honey bee and cockroach: comparable organization in the brain and subesophageal ganglion. *J Comp Neurol* 2005;**488**:233–54.
26. Palczewski K, Kumasaka T, Hori T, Behnke CA, Motoshima H, Fox BA, et al. Crystal structure of rhodopsin: a G protein-coupled receptor. *Science* 2000;**289**:739–45.
27. Cherezov V, Rosenbaum DM, Hanson MA, Rasmussen SGF, Thian FS, Kobilka TS, et al. High-resolution crystal structure of an engineered human β2-adrenergic G protein-coupled receptor. *Science* 2007;**318**:1258–65.
28. Rasmussen SGF, Choi H-J, Rosenbaum DM, Kobilka TS, Thian FS, Edwards PC, et al. Crystal structure of the human β₂ adrenergic G-protein-coupled receptor. *Nature* 2007;**450**:383–7.
29. Honeybee Genome Sequencing Consortium Insights into social insects from the genome of the honeybee *Apis mellifera*. *Nature* 2006;**443**:931–49.
30. Hauser F, Cazzamali G, Williamson M, Blenau W, Grimmelikhuijzen CJP. A review of neurohormone GPCRs present in the fruit fly *Drosophila melanogaster* and the honey bee *Apis mellifera*. *Progr Neurobiol* 2006;**80**:1–19.
31. Hummon AB, Richmond TA, Verleyen P, Baggerman G, Huybrechts J, Ewing MA, et al. From the genome to the proteome: uncovering peptides in the Apis brain. *Science* 2006;**314**:647–9.
32. Farooqui T, Robinson K, Vaessin H, Smith BH. Modulation of early olfactory processing by an octopaminergic reinforcement pathway in the honeybee. *J Neurosci* 2003;**23**:5370–80.
33. Mustard JA, Kurshan PT, Hamilton IS, Blenau W, Mercer AR. Developmental expression of a tyramine receptor gene in the brain of the honey bee, *Apis mellifera*. *J Comp Neurol* 2005;**483**:66–75.
34. Beggs KT, Tyndall JDA, Mercer AR. Honey bee dopamine and octopamine receptors linked to intracellular calcium signaling have a close phylogenetic and pharmacological relationship. *PLoS One* 2011;**6**:e26809.
35. Arakawa S, Gocayne JD, McCombie WR, Urquhart DA, Hall LM, Fraser CM, et al. Cloning, localization, and permanent expression of a *Drosophila* octopamine receptor. *Neuron* 1990;**2**:343–54.
36. Saudou F, Amlaiky N, Plassat J-L, Borelli E, Hen R. Cloning and characterization of a *Drosophila* tyramine receptor. *EMBO J* 1990;**9**:3611–7.
37. Vanden Broeck J, Vulsteke V, Huybrechts R, De Loof A. Characterization of a cloned locust tyramine receptor cDNA by functional expression in permanently transformed *Drosophila* S2 cells. *J Neurochem* 1995;**64**:2387–95.
38. Evans PD, Maqueira B. Insect octopamine receptors: a new classification scheme based on studies of cloned *Drosophila* G-protein coupled receptors. *Invert Neurosci* 2005;**5**:111–8.
39. Farooqui T, Vaessin H, Smith BH. Octopamine receptors in the honeybee (*Apis mellifera*) brain and their disruption by RNA-mediated interference. *J Insect Physiol* 2004;**50**:701–13.
40. Han K-A, Millar NS, Davis RL. A novel octopamine receptor with preferential expression in Drosophila mushroom bodies. *Neuron* 1998;**18**:1127–35.
41. Balfanz S, Strünker T, Frings S, Baumann A. A family of octopamine receptors that specifically induce cyclic AMP production or Ca²⁺ release in *Drosophila melanogaster*. *J Neurochem* 2005;**93**:440–51.
42. Hoff M, Balfanz S, Ehling P, Gensch T, Baumann A. A single amino acid residue controls Ca²⁺ signaling by an octopamine receptor from *Drosophila melanogaster*. *FASEB J* 2011;**25**:2484–91.
43. Sinakevitch I, Mustard JA, Smith BH. Distribution of the octopamine receptor AmOA1 in the honey bee brain. *PLoS One* 2011;**6**:e14536.
44. Sinakevitch IT, Smith AN, Locatelli F, Huerta R, Bazhenov M, Smith BH. *Apis mellifera* octopamine receptor 1 (AmOA1) expression in antennal lobe networks of the honey bee (*Apis mellifera*) and fruit fly (*Drosophila melanogaster*). *Front Syst Neurosci* 2013;**7**:70.
45. McQuillan HJ, Nakagawa S, Mercer AR. Mushroom bodies of the honeybee brain show cell population-specific plasticity in expression of amine-receptor genes. *Learn Mem* 2012;**19**:151–8.
46. Reim T, Scheiner R. Division of labour in honey bees: age- and task-related changes in the expression of octopamine receptor genes. *Insect Mol Biol* 2014;**23**:833–41.
47. Pribbenow B, Erber J. Modulation of antennal scanning in the honeybee by sucrose stimuli, serotonin, and octopamine: behavior and electrophysiology. *Neurobiol Learn Mem* 1996;**66**:109–20.

48. Fussnecker BL, Smith BH, Mustard JA. Octopamine and tyramine influence the behavioral profile of locomotor activity in the honey bee (*Apis mellifera*). *J Insect Physiol* 2006;**52**:1083–92.
49. Bloch G, Meshi A. Influences of octopamine and juvenile hormone on locomotor behavior and period gene expression in the honeybee, *Apis mellifera*. *J Comp Physiol A* 2007;**193**:181–99.
50. Erber J, Kloppenburg P. The modulatory effects of serotonin and octopamine in the visual system of the honey bee (*Apis mellifera* L.): I. Behavioral analysis of the motion-sensitive antennal reflex. *J Comp Physiol A* 1995;**176**:111–8.
51. Erber J, Pribbenow B, Bauer A, Kloppenburg P. Antennal reflexes in the honeybee: tools for studying the nervous system. *Apidologie* 1993;**24**:283–96.
52. Kloppenburg P, Erber J. The modulatory effects of serotonin and octopamine in the visual system of the honey bee (*Apis mellifera* L.). 2. Electrophysiological analysis of motion-sensitive neurons in the lobula. *J Comp Physiol A* 1995;**176**:119–29.
53. Scheiner R, Toteva A, Reim T, Sovik E, Barron AB. Differences in the phototaxis of pollen and nectar foraging honey bees are related to their octopamine brain titers. *Front Physiol* 2014;**5**:116.
54. Matsumoto Y, Menzel R, Sandoz J-P, Giurfa M. Revisiting olfactory classical conditioning of the proboscis extension response in honey bees: a step toward standardized procedures. *J Neurosci Methods* 2012;**211**:159–67.
55. Scheiner R, Plückhahn S, Öney B, Blenau W, Erber J. Behavioural pharmacology of octopamine, tyramine and dopamine in honey bees. *Behav Brain Res* 2002;**136**:545–53.
56. Pankiw T, Page Jr. RE. Effect of pheromones, hormones, and handling on sucrose response thresholds of honey bees (*Apis mellifera* L.). *J Comp Physiol A* 2003;**189**:675–84.
57. Takeda K. Classical conditioned response in the honey bee. *J Insect Physiol* 1961;**6**:168–79.
58. Mercer AR, Menzel R. The effects of biogenic amines on conditioned and unconditioned responses to olfactory stimuli in the honeybee *Apis mellifera*. *J Comp Physiol* 1982;**145**:363–8.
59. Barron AB, Schulz DJ, Robinson GE. Octopamine modulates responsiveness to foraging-related stimuli in honey bees (*Apis mellifera*). *J Comp Physiol A* 2002;**188**:603–10.
60. Goode K, Huber Z, Mesce KA, Spivak M. Hygienic behavior of the honey bee (*Apis mellifera*) is independent of sucrose responsiveness and foraging ontogeny. *Horm Behav* 2006;**49**:391–7.
61. Schulz DJ, Robinson GE. Octopamine influences division of labor in honey bee colonies. *J Comp Physiol A* 2001;**187**:53–61.
62. Rein J, Mustard JA, Strauch M, Smith BH, Galizia CG. Octopamine modulates activity of neural networks in the honey bee antennal lobe. *J Comp Physiol A* 2013;**199**:947–62.
63. Menzel R, Heyne A, Kinzel C, Gerber B, Fiala A. Pharmacological dissociation between the reinforcing, sensitizing, and response-releasing functions of reward in honeybee classical conditioning. *Behav Neurosci* 1999;**113**:744–54.
64. Blenau W, Erber J. Behavioural pharmacology of dopamine, serotonin and putative aminergic ligands in the mushroom bodies of the honeybee (*Apis mellifera*). *Behav Brain Res* 1998;**96**:115–24.
65. Menzel R, Michelsen B, Rüffer P, Sugawa M. Neuropharmacology of learning and memory in honey beesHertting G.Spatz H-C, editors. *Modulation of synaptic transmission and plasticity in nervous systems*, **H19**. Berlin, Heidelberg: Springer-Verlag, NATO ASI Series; 1988. p. 333–50.
66. Braun G, Bicker G. Habituation of an appetitive reflex in the honeybee. *J Neurophysiol* 1992;**67**:588–98.
67. Erber J, Kierzek S, Sander E, Grandy K. Tactile learning in the honeybee. *J Comp Physiol A* 1998;**183**:737–44.
68. Agarwal M, Giannoni Guzmán M, Morales-Matos C, Del Valle Díaz RA, Abramson CI, Giray T. Dopamine and octopamine influence avoidance learning of honey bees in a place preference assay. *PLoS One* 2011;**6**:e25371.
69. Bitterman ME, Menzel R, Fietz A, Schäfer S. Classical conditioning of proboscis extension in honeybees (*Apis mellifera*). *J Comp Psychol* 1983;**97**:107–19.
70. Felsenberg J, Gehring KB, Antemann V, Eisenhardt D. Behavioural pharmacology in classical conditioning of the proboscis extension response in honeybees (*Apis mellifera*). *J Vis Exp* 2011;**24**:2282.
71. Menzel R, Müller U. Learning and memory in honeybees: from behavior to neural substrates. *Annu Rev Neurosci* 1996;**19**:379–404.
72. Giurfa M, Sandoz JC. Invertebrate learning and memory: fifty years of olfactory conditioning of the proboscis extension response in honeybees. *Learn Mem* 2012;**19**:54–66.
73. Hammer M. An identified neuron mediates the unconditioned stimulus in associative olfactory learning in honeybees. *Nature* 1993;**366**:59–63.

74. Schröter U, Malun D, Menzel R. Innervation pattern of suboesophageal ventral unpaired median neurones in the honeybee brain. *Cell Tissue Res* 2007;**327**:647–67.
75. Hammer M, Menzel R. Multiple sites of associative odor learning as revealed by local brain microinjections of octopamine in honeybees. *Learn Mem* 1998;**5**:146–56.
76. Hammer M. The neural basis of associative reward learning in honeybees. *Trends Neurosci* 1997;**20**:245–52.
77. Fuchs E, Dustmann J-H, Stadler H, Schürmann F-W. Neuroactive compounds in the brain of the honeybee during imaginal life. *Comp Biochem Physiol C* 1989;**92**:337–42.
78. Harris JW, Woodring J. Effects of stress, age, season, and source colony on levels of octopamine, dopamine and serotonin in the honey bee (*Apis mellifera*) brain. *J Insect Physiol* 1992;**38**:29–35.
79. Mercer AR, Mobbs PG, Davenport AP, Evans PD. Biogenic amines in the brain of the honey bee, *Apis mellifera*. *Cell Tissue Res* 1983;**234**:655–77.
80. Taylor DJ, Robinson GE, Logan BJ, Laverty R, Mercer AR. Changes in brain amine levels associated with the morphological and behavioural development of the worker honeybee. *J Comp Physiol A* 1992;**170**:715–21.
81. Wagener-Hulme C, Kuehn JC, Schulz DJ, Robinson GE. Biogenic amines and division of labor in honey bee colonies. *J Comp Physiol A* 1999;**184**:471–9.
82. Schulz DJ, Robinson GE. Biogenic amines and division of labor in honey bee colonies: behaviorally related changes in the antennal lobes and age-related changes in the mushroom bodies. *J Comp Physiol A* 1999;**184**:481–8.
83. Schulz DJ, Pankiw T, Fondrk MK, Robinson GE, Page RE. Comparison of juvenile hormone hemolymph and octopamine brain titers in honey bees (Hymenoptera: Apidae) selected for high and low pollen hoarding. *Ann Entomol Soc Am* 2004;**97**:1313–9.
84. Schulz DJ, Elekonich MM, Robinson GE. Biogenic amines in the antennal lobes and the initiation and maintenance of foraging behavior in honey bees. *J Neurobiol* 2003;**54**:406–16.
85. Robinson GE. Genomics and integrative analyses of division of labor in honey bee colonies. *Am Nat* 2002;**160**:S160–72.
86. Sullivan JP, Fahrbach SE, Robinson GE. Juvenile hormone paces behavioral development in the adult worker honey bee. *Horm Behav* 2000;**37**:1–14.
87. Kaatz H, Eichmüller S, Kreissl S. Stimulatory effect of octopamine on juvenile hormone biosynthesis in honey bees (*Apis mellifera*): physiological and immunocytochemical evidence. *J Insect Physiol* 1994;**40**:865–72.
88. Schulz DJ, Barron AB, Robinson GE. A role for octopamine in honey bee division of labor. *Brain Behav Evol* 2002;**60**:350–9.
89. Schulz DJ, Sullivan JP, Robinson GE. Juvenile hormone and octopamine in the regulation of division of labor in honey bee colonies. *Horm Behav* 2002;**42**:222–31.

Octopamine and Tyramine Signaling in Locusts: Relevance to Olfactory Decision-Making

Z. Ma[1,2], X. Guo[1,2] and L. Kang[1,2]

[1]Beijing Institutes of Life Sciences, Chinese Academy of Sciences, Beijing, China [2]State Key Laboratory of Integrated Management of Pest Insects and Rodents, Institute of Zoology, Chinese Academy of Sciences, Beijing, China

O U T L I N E

INTRODUCTION

In insect neurobiology, a limited number of insect types are commonly employed, due to convenience and availability of experimental resources. The swarm-forming locusts in acridoid have been among the most favored subjects in the study of insect neurobiology and behavior.[1] The reasons are their relatively large body size, their ease of culture in the laboratory, and most of all, their importance for agricultural economy. With the development of techniques and concepts in neurobiology and functional genomics, an array of innovative locust (*Locusta migratoria*) studies has been performed and reported.[2–12]

Olfactory preference for volatiles from gregarious locusts (living with high population density) potentially promotes locust swarming.[2,7] Octopamine and tyramine were identified in locusts and confirmed to be relevant in various behaviors and physiology. The roles of these two chemicals in modulating olfactory perception have been confirmed in certain insect species, including flies, bees, and moths.[13–15] However, little is known about the neurobiological and neurogenetic mechanisms that octopamine and tyramine mediate on olfactory preferences in locusts. The present study recognizes the remarkable progress of research in olfactory anatomy; distribution of octopamine, tyramine, and their receptors; and the roles of these two neurochemicals in processing of olfactory stimuli in locusts.

OLFACTORY DECISION-MAKING IN *LOCUSTA*

Decision-Making in Animals

Decision-making for adaptation to different environments is a primary function of the nervous system. The simpler level of decision-making, which may be called behavioral choice, mainly concerns the proximal causes of selecting among alternative sensory cues or behavior.[16] Historically, the neuronal analysis of decision-making has been emphasized on humans and nonhuman primates. Such processes commonly rely on the prediction of environmental changes and the recognition of patterns to discriminate situations.[17] Similarly, invertebrates with very little cognition constantly make choices in handling external stimuli and internal drives, including habitat selection,[18,19] mate selection,[20] and selection between fight or flight.[21,22]

Locust Phase Polyphenism

Phase polyphenism is a universal biological phenomena exhibiting phenotypic plasticity in response to the changes in environmental stimuli.[23a] Locusts show phase polyphenism, gregarious and solitary phases, in response to the fluctuation of population density. In high population density, gregarious locusts are active and exhibit an attractive response to their gregarious conspecifics, whereas in low population density, solitary locusts are sedative and show a repulsive response to their gregarious conspecifics.[23b,24]

Olfactory Decision-Making in the Phase Change of *Locusta*

The migratory locust show density-dependent olfactory choices during phase change.[2] The same volatile information can elicit two different behavioral responses in gregarious and solitary locusts, namely attraction and repulsion (Fig. 15.1A). Solitary insects show an attractive response after 4h of crowding, whereas gregarious ones exhibit repulsive response after 1h of isolation.[2] The choice between different forms of behavior (such as attractive olfaction versus repulsive olfaction) belongs to the neuroethological/behavioral category among the two major traditions in decision-making.[21] Moreover, making a transitional choice between attraction and repulsion depends on the fluctuation of population density (Fig. 15.1B). Thus, decision-making (attraction versus repulsion) for the same olfactory information is population-density-dependent in *Locusta*. Before attaining the preference for gregarious volatiles, multiple decision-making actions are required, including the processing of chemosensory information collected by a peripheral system, and the decision by the brain to approach or avert their conspecifics. Examination of the choices made between attraction and repulsion will provide a powerful model to study the neurogenetic and neurophysiological mechanisms underlying olfactory decision.

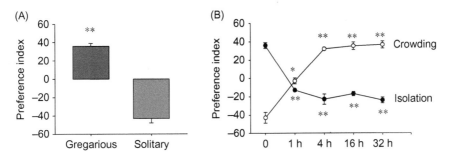

FIGURE 15.1 Migratory locusts make transitional choices (attraction versus repulsion) depending on population density. (A) Olfactory behavioral choice of solitary and gregarious locusts in response to gregarious volatiles. (B) Olfactory behavioral choice of solitary and gregarious locusts during crowding and isolation of *Locusta*. Preference index >0 denotes an attractive response, while preference index <0 denotes a repulsive response. *, $P<0.05$; **, $P<0.01$.

TYROSINE METABOLISM IN GREGARIOUS AND SOLITARY LOCUSTS

The same olfactory information elicits two different responses (attraction versus repulsion) in *Locusta*. A previous study reported that the tyrosine level is higher in gregarious locusts than that in solitary ones.[8] Tyrosine metabolism is more active in gregarious locusts,[3,8] suggesting a significant correlation with population-dependent olfactory decision-making. Tyrosine is the precursor of invertebrate-specific monoamine octopamine and tyramine. These two chemicals respectively modulate attractive and repulsive behavior in *Locusta*.[2] Little is known about the neurogenetic and neurobiological mechanisms underlying this decision-making in *Locusta*.

OCTOPAMINE AND TYRAMINE SIGNALING

Trace amines in mammals, including octopamine and tyramine, have been associated with the etiology of a variety of neurological and neuropsychiatric disorders.[25] The roles of trace amines as neurotransmitters in invertebrates are well-established. As the sole nonpeptide transmitters/hormones that are restricted to invertebrates, octopamine and tyramine take an exceptional position. The roles of octopamine and tyramine as neurotransmitters acting via stimulation of G-protein-coupled receptors (GPCR) in invertebrates systems,[26–29] and the GPCRs for tyramine and octopamine, have been reported in insects.[2,30,31]

Octopamine and Tyramine Signaling in Insects

As a structural analogue of norepinephrine, octopamine is an extensively studied biogenic amine in the invertebrate nervous system of invertebrates, and it modulates a diverse range of behavior from locomotion and flight to learning and memory. Moreover, numerous functions have been assigned to octopamine including the regulation of lipid metabolism and modulation of sensory perception.[32,33] Although tyramine has long been considered as the precursor of octopamine, the identification of this chemical in flies, locusts, honeybees, silk moths, and worms and GPCRs that respond to tyramine has led to the conclusion that tyramine may itself act as a neurotransmitter.[29–33]

Distribution of Octopamine in the Locust Central Nervous System

The distribution of octopamine is closely correlated with its function in specific tissues. Its localization in the nervous system has become realistic with the development of specific antibodies. In *Schistocerca* and *Locusta*, octopamine-like immunoreactivity occurs in the protocerebral bridge, mushroom body, central complex, lateral protocerebrum, and lateral border of tritocerebrum.[34–38] The antennal lobe is the center for olfactory perception modulation. In *Schistocerca*, a group of four pairs of cells with large soma diameter is located laterally and anteriorly in relation to the antennal lobe region. The cells of this cluster became immunopositive for octopamine only in the group of animals that were stressed (handled) before preparation.[39] Interestingly, the neuritis of these cells supplies a variety of neuropiles in the lateral and medial protocerebrum, as well as in the deutocerebrum posterior to the antennal lobes and above the central complex. Moreover, the mushroom body is more restricted to the immunity of octopamine.[39] In *Locusta*, octopamine is highly enriched in the protocerebrum and cerebral ganglion.[40] The distribution of octopamine in brains implicates its roles associated with olfactory behavior in locusts.

Distribution of Tyramine in the Locust Central Nervous System

The tissue-specific distribution of tyramine suggests its roles in modulating specific physiology and behavior. Distinct tyraminergic patterns of expression are believed to exist in locusts.[41,42] In *Schistocerca*, tyramine-like immunoreactivity is expressed by six bilaterally symmetrical sets of neurons, as well as four pairs of tyramine-like immunoreactive single cells in the brain. Numerous tyramine-like immunoreactive fibers invade the antennal lobes. Surprisingly, the immunoreactivity analysis found that tyramine does not exist in the mushroom body.[39] The universal distribution of tyramine in the brain suggests its ubiquitous roles in mediating behavior and physiology in locusts.

Distribution of Octopamine and Tyramine Receptors in the Central Nervous System

Although the distribution of octopamine and tyramine in the brains of *Locusta* and *Schistocerca* implicates their common roles in behavior, the fact that tyramine is the precursor of octopamine may lead to the confusion that tyramine solely acts as a precursor but not as an independent neurotransmitter. The expression of tyramine and octopamine receptors in the brains of *Locusta* and *Schistocerca* may help to clearly distinguish between the functions of the two chemicals.

Thus, to understand the specificity of the action of octopamine in different types of behavior, it will be necessary to describe not only the site of octopamine release but also the distribution of different octopamine receptor subtypes in the brain. Octopamine receptor α is similar to the α-adrenergic-like receptor (OCTα-R), whereas octopamine receptor β is an analogue of the β-adrenergic-like receptor (OCTβ-R).[33] The function of the octopamine receptor has been explored from different aspects and using different methods including electrophysiology, in vitro expression, and molecular biology. However, little is known about the expression patterns of these receptor subtypes and how each subtype contributes to specific behavioral patterns. The function of octopamine receptors in the brain is reflected by the density of the high receptor concentration in corresponding brain areas. Two classes of octopamine receptors have been identified recently in *Schistocerca*[43,44] and *Locusta*.[2] In *Schistocerca*, qPCR analysis showed that SgOctαR shows a much higher transcript level in the brain than that in peripheral tissues. Their receptor subtypes are relatively abundant in mushroom bodies.[44] In *Locusta*, LmOA1 is expressed in the antennal lobe, mushroom body, and lateral protocerebrum.[2] The transcript of OARα is highly expressed in the brain.[45] Structure–activity studies of various types of OA agonists and antagonists were reported using the nervous tissue of *Locusta*.[46] The mushroom bodies are the areas of highest receptor density in locusts, and the receptors in other parts of the brain are also expressed at high levels.[36,47]

The tyramine receptor (TAR) in *Locusta* is an invertebrate equivalent of the mammalian α2-adrenergic receptors.[2] This receptor is widely distributed in the brain and the ventral nerve cord.[30,45] In situ hybridization showed that TAR distributed in antennal lobe and lateral horn, suggesting its correlation with olfaction of locusts.[2]

Convergence and Divergence in Distribution of Octopamine and Tyramine Signaling in the Central Nervous System

The distribution of octopamine and tyramine in brains implicates the convergent and divergent roles of these two neurochemicals in modulating locust behaviors. Octopamine and tyramine are present in small numbers of interneurons, often with large branching patterns, suggesting their neuromodulatory roles in *Schistocerca*.[48] The prominent architecture of the brain, the protocerebral bridge, the central body, and the associated neuropiles, all contain octopamine and tyramine-immunoreactive fibers in *Schistocerca*. Tyramine and octopamine all show immunoreactivity in antennal lobes, but tyramine does not exist in mushroom body.[39] Their immunohistochemical distributions suggested that octopamine and tyramine modulate similar behavioral paradigms. The nonexistence of tyramine in the mushroom body implies that tyramine does not mediate behavior modulated by the mushroom body (Table 15.1). In *Locusta*, OARα and TAR were found expressed in the antennal lobe and lateral horn of whole-mount brains of *Locusta*.[2] OARα is expressed in mushroom bodies, whereas the mRNA of TAR is not expressed in this region. The comparison in localization between these two signaling pathways will provide vital clues to the functional convergences and divergences in *Locusta*.

TABLE 15.1   Distribution of Octopamine and Tyramine in the locust brain

Brain Region	Octopamine	Tyramine
Tritocerebrum		✓
Protocerebrum	✓	
Protocerebral bridge	✓	✓
Subesophageal ganglion		✓
Lateral dorsal		✓
Cerebral ganglion	✓	
Central complex	✓	
Mushroom body	✓	
Lateral protocerebrum	✓	
Antenna lobe	✓	✓
AMMC	✓	
Medulla	✓	✓

AMMC, antennal mechanosensory and motor center.

DISTRIBUTION OF OCTOPAMINE AND TYRAMINE SIGNALING SUGGESTS THEIR RELEVANCE TO OLFACTORY SENSATION

Olfactory perception plays a central role in the behavior of insects to humans. The shared features of organization and function of the olfactory system in invertebrates and vertebrates rendered the discoveries in insects of general interest to the study of other animals. Locusts have been considered as a representative species spanning *holometabola* and *hemimetabola* (respectively) for research on olfactory neurobiology of insects. They instantiate a unique code in a unique olfactory network, distinct from those of other insects such as the sphinx moths, fruit flies, and honey bees. In locusts, the architecture of the major centers for olfaction in the brain includes the antennal lobe, mushroom body, and the lateral horn of the protocerebrum. The locust antenna lobes show the microglomeruli structure and input circuits of odorant receptors cells. The antenna–protocerebral tract is followed by the microglomerular projection neurons of the antennal lobe that are distributed to mushroombody calyx in locusts.[49,50] In *S. americana*, the lateral horn is the source of a strong, feedforward inhibition to Kenyon cells in the mushroom body.[51] Despite these advances, the roles of octopamine and tyramine signaling are not fully explored in locusts.

Octopamine and its receptor OARα distribute in the antennal lobe, lateral horn, and mushroom bodies, whereas tyamine and its receptor distribute in antennal lobes and lateral horn.[2] The distribution regions of these two signaling pathways in locust brains are closely related to the anatomical region for olfactory sensation in *Locusta*. Until now, we found only one report stating that octopamine mediates associated olfactory learning through measuring spike-timing-dependent plasticity (STDP) in *Schistocerca*.[52] The function of octopamine and tyramine in modulating locust olfactory sensation are little reported.

OCTOPAMINE AND TYRAMINE SIGNALING MEDIATE OLFACTORY DECISION-MAKING IN *LOCUSTA*

In *Locusta*, the localization and distribution of octopamine, tyramine, and related receptors lies in the lateral horn and antennal lobes of the brain.[2] Octopamine levels are higher in the brains of gregarious locusts than in those of solitary locusts, whereas tyramine levels are higher in the brains of solitary locusts than in those of gregarious locusts. Moreover, linear regression analysis showed that octopamine–OARα and tyramine–TAR signaling pathways are significantly correlated with attractive and repulsive behavior, respectively.[2] Pharmacological intervention and RNAi knockdown confirmed that octopamine mediates attractive sensation through OARα signaling, whereas tyramine induces the olfactory repulsion response to gregarious volatiles. Moreover, after fixing the concentration of injected octopamine and gradually increasing the concentration of tyramine, gregarious locusts gradually increased their repulsive response to gregarious volatiles. In solitary locusts, after fixing the concentration of injected tyramine and gradually increasing the concentration of octopamine, olfactory response gradually changed from repulsion to attraction. These results suggested that octopamine and tyramine antagonistically mediate olfactory decision-making during the phase change of *Locusta*.[2]

POTENTIAL MECHANISM BY WHICH OCTOPAMINE–OARα AND TYRAMINE–TAR SIGNALING MEDIATE INDIVIDUAL OLFACTORY DECISION IN *LOCUSTA*

Octopamine–OARα and Tyramine–TAR Signaling in the Brain Mediate Individual Olfactory Decision-Making at the Neurochemical Level

The olfactory regulatory circuits from the antennal lobe to the lateral horn and mushroom body have been identified in locusts.[53] Octopaminergic and tyraminergic circuits may regulate the olfactory sensory signals from peripheral chemosensory circuits. Although the differences in anatomical circuits between gregarious and solitary locusts are not known, the differential concentration of octopamine and tyramine between the two phases in *Locusta* may render differential effects on olfactory preferences through octopaminergic and tyraminergic circuits in the brain.[2] Octopamine increases the sensitivity of odorant receptor neurons and enhances the olfactory response to pheromones.[54–56] The olfactory transduction pathway is the same as the G-protein/PLC-dependent cascade in pheromone-sensitive neurons. Octopamine may modulate pheromone-sensitive receptor neurons through a cyclic mononucleotide-dependent pathway.[57–59] Recently, octopamine was found to induce pheromone responses in a more phasic and sensitive manner and raise the spontaneous action potential frequency.[52] Therefore, octopamine is expected to increase the response magnitude of the olfactory neural circuits to external stimuli. Octopamine in the antenna is obligatory for the detection of intermittent pheromone pulses.[60] Octopamine plays a role in modulating the octopaminergic reward system in olfactory learning and memory.[61] Hence, this chemical may also mediate olfactory learning and memory in the phase change of *Locusta*.

The neurocodings at the periphery and central nervous system are possibly differential between the two phases. The tyraminergic process is identified in antennal lobes of locusts.[2,39] Tyramine may thereby increase the effect of repulsive odorants in volatiles and induce repulsion during isolation. Tyramine–TAR signaling in solitary locusts may involve the neural and cellular signals in the peripheral nervous system after binding with repulsive odorants in volatiles. Thus, octopaminergic and tyraminergic neurons may mediate differential responses of gregarious and solitary locusts elicited by the volatiles released from gregarious locusts.

Octopamine–OARα and Tyramine–TAR Signaling in the Central Nervous System Mediate Olfactory Decision-Making at the Neurogenetic Level

To fully understand the molecular mechanism underlying the olfactory preferences of *Locusta*, we screened the differentially expressed genes and receptors of monoamines associated with olfactory preferences during the phase change of *Locusta*.[2–12] In *Locusta*, octopamine–OARα has been identified in brain tissues.[2] This receptor belongs to the superfamily of metabotropic GPCRs. The interaction of octopamine with OCTα–R increases the intracellular Ca^{2+} levels as well as the level of intracellular cAMP to a small degree.[62] OCTα–R in *Locusta* may control olfactory perception through the Ca^{2+} and cAMP second-messenger pathway. Octopamine–OARα signaling in gregarious locusts may process the neural and cellular signals in the peripheral nervous system after sensing attractive odorants in volatiles.

Furthermore, TAR was identified in the brains of *Locusta*,[2] but little is known about the mechanism by which tyramine signaling modulates the olfactory sensation during phase change. TAR mediates the repulsive olfactory behavior in *Drosophila*.[14] In *Locusta*, TAR inhibits adenylyl cyclase activity via coupling to inhibitory G-proteins and inducing a decrease in intracellular cAMP levels.[30,31,63] The knockdown of TAR induces the olfactory preference of solitary locusts from repulsion to attraction, and activation of TAR may decrease intracellular cAMP levels and render the change of olfactory behavior from attraction to repulsion in response to gregarious volatiles.

Octopamine–OARα and Tyramine–TAR Signaling May Mediate Olfactory Decision-Making in a Seesaw Manner

The overlapping of distribution in the brain and the similarities in synthesis and molecular structure between octopamine and tyramine suggest the functional correlation between these two chemicals in locust behavior. Octopamine and tyramine play opposite roles in muscle contraction in *Drosophila* larvae.[14,64,65] In *Manduca*, octopamine and tyramine exert opposite effects on the central pattern generator of the thorax.[66] In *Locusta*, octopamine and tyramine have opposite effects on olfactory preferences of *Locusta* through their respective receptors.[2] Thus, octopamine and tyramine may play reverse roles in olfactory sensation of *Locusta* at neurochemical and neurogenetic levels. Moreover, combined injection of octopamine and tyramine is found to not only respectively affect attractive and repulsive decisions in a dose-dependent manner but also confirm their antagonistic effects on olfactory behavioral choice. The modulation of olfactory preference through octopamine–OARα and tyramine–TAR signaling indicates the existence of a bimodal output system for olfactory decision-making.

Several hypotheses have been proposed to explain the mechanism involved in modulation of attractive or repulsive responses. In *Caenorhabditis elegans*, one set of antagonistic command neurons comprises microcircuits that regulate attractive and aversive behavior via a push–pull mechanism.[67] A central flip-flop circuit that integrates two contradictory sensory inputs generates bistable hormone output circuits to regulate the preference for attractants or repellents.[68] By contrast, *Drosophila* executes attraction and repulsion divergently by a motor-related neuronal circuit and olfactory neural pathways.[69,70] The antagonistic effects between octopamine–OARα and tyramine–TAR signaling on olfactory decision-making suggest a seesaw manner modulating the attractive and repulsive response induced by gregarious volatiles (Fig. 15.2). During the crowding of solitary locusts, activation of octopamine–OARα signaling suppresses the effects the tyramine–TAR signaling and induces the behavioral changes for preferring gregarious volatiles. During the isolation of gregarious locusts, activation of tyramine–TAR signaling suppresses the roles of octopamine–OARα signaling and induces the behavioral changes toward olfactory repulsive behavior. The action mode of octopamine–OARα and tyramine–TAR signaling in modulating olfactory decision-making is similar to the manner that a seesaw changes (Fig. 15.2). Thus, octopamine–OARα and tyramine–TAR signaling may mediate olfactory preferences of *Locusta* in a seesaw manner. The seesaw model establishes a causal association between the two signaling pathways with the transitional choice between attraction and repulsion in the phase change of *Locusta*.

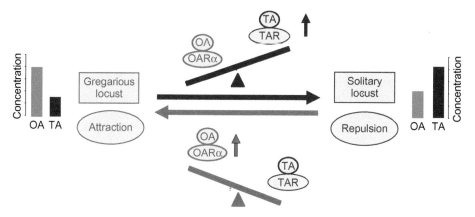

FIGURE 15.2   Octopamine–OARα and tyramine–TAR signaling mediate olfactory decision-making in a seesaw manner. The level of octopamine in the brain of gregarious locusts is higher than that in solitary locusts, and injection of tyramine (100 µg, 25 µg/µL) in the thoracic cavity of gregarious locusts induces a change of olfactory behavior from attraction to repulsion. By contrast, the level of tyramine in the brain of solitary locusts is higher than that in gregarious locusts, and injection of octopamine (100 µg, 25 µg/µL) in thoracic cavity of solitary locusts induces a change of olfactory behavior from repulsion to attraction. Red arrow indicates an increase in attraction whereas blue arrow indicates an increase in repulsion. *OA*, octopamine; *TA*, tyramine; *TAR*, tyramine receptor; *OARα*, octopamine receptor α.

CONCLUSION

Octopamine–OARα and tyramine–TAR signaling mediate a population-density-dependent transitional decision-making in *Locusta*. This study deepens the understanding of the conservative roles of octopamine and tyramine in regulating olfactory decision-making across phyla. The examination of the roles of octopamine and tyramine in animal nervous systems is likely to shed light on the conservation of the signal-transduction mechanism. This study also elicits a series of issues, particularly on how octopamine and tyramine mediate the olfactory transitional choice at the neurobiological level and how these two neurochemicals mediate olfactory decision-making through octopaminergic and tyraminergic neurites? Moreover, despite the knowledge that octopamine and tyramine signaling facilitate olfactory preferences through their specific receptors, the underlying molecular mechanism by which octopamine and tyramine arbitrate the population-dependent transitional choice should be explored. The development of new-generation sequencing and CRISPRA/Cas 9 system may help elucidate the mechanisms by which these two chemicals regulate olfactory decision-making in locusts.

Acknowledgments

This research was supported by National Natural Science Foundation of China (Nos. 31272361, 31472040 and 31522054), Strategic Priority Research Program of the Chinese Academy of Sciences (No. XDB11010000) and the National Basic Research Program of China (No. 2012CB114102).

References

1. Rowell CHF. *Locust neurobiology: a bibliography, 1871–1991.* Springer; 1992.
2. Ma ZY, Guo XJ, Lei H, Li T, Hao SG, Kang L. Octopamine and tyramine respectively regulate attractive and repulsive behavior in locust phase changes. *Sci Rep* 2015;**5**:8036.
3. Ma ZY, Guo W, Guo XJ, Wang XH, Kang L. Modulation of behavioral phase changes of the migratory locust by the catecholamine metabolic pathway. *Proc Natl Acad Sci USA* 2011;**108**:3882–7.
4. Ma ZY, Yu J, Kang L. Locust DB: a relational database for the transcriptome and biology of the migratory locust (*Locusta migratoria*). *BMC Genomics* 2006;**7**:e11.
5. Kang L, Chen XY, Zhou Y, Liu B, Zheng W, Li RQ, et al. The analysis of large-scale gene expression correlated to the phase changes of the migratory locust. *Proc Natl Acad Sci USA* 2004;**101**:17611–5.
6. Wang XH, Fang XD, Yang PC, Jiang XT, Jiang F, Zhao DJ, et al. The locust genome provides insight into swarm formation and long-distance flight. *Nat Commun* 2004;**5**:2957.
7. Guo W, Wang XH, Ma ZY, Xue L, Han JY, Yu D, et al. CSP and takeout genes modulate the switch between attraction and repulsion during behavioral phase change in the migratory locust. *PLoS Genet* 2011;**7**:e1001291.
8. Wu R, Wu ZM, Wang XH, Yang PC, Yu D, Zhao CX, et al. Metabolomic analysis reveals that carnitines are key regulatory metabolites in phase transition of the locusts. *Proc Natl Acad Sci USA* 2012;**109**:3259–63.
9. Yang ML, Wei YY, Jiang F, Wang YL, Guo XJ, He J, et al. MicroRNA-133 inhibits behavioral aggregation by controlling dopamine synthesis in locusts. *PLoS Genet* 2014;**10**:e1004206.
10. Chen S, Yang PC, Jiang F, Wei YY, Ma ZY, Kang L. De novo analysis of transcriptome dynamics in the migratory locust during the development of phase traits. *PLoS One* 2010;**5**:e15633.
11. Guo XJ, Ma ZY, Kang L. Serotonin enhances solitariness in phase transition of the migratory locust. *Front Behav Neurosci* 2013;**7**:129.
12. Guo XJ, Ma ZY, Kang L. Two dopamine receptors play differential roles in phase change of the migratory locust. *Front Behav Neurosci* 2015;**1**:9.
13. Farooqui T. Octopamine-mediated neuromodulation of insect senses. *Neurochem Res* 2007;**32**:1511–29.
14. Kutsukake M, Komatsu A, Yamamoto D, Ishiwa-Chigusa S. A tyramine receptor gene mutation causes a defective olfactory behavior in *Drosophila melanogaster*. *Gene* 2000;**245**:31–42.
15. Anton S, Dufour M, Gadenne C. Plasticity of olfactory-guided behaviour and its neurobiological basis: lessons from moths and locusts. *Entomol Exp Appl* 2007;**123**:1–11.
16. Palmer CR, Kristan WB. Contextual modulation of behavioral choice. *Curr Opin Neurobiol* 2011;**21**:520–6.
17. Menzel R. The honeybee as a model for understanding the basis of cognition. *Nat Rev Neurosci* 2012;**13**:758–68.
18. Bargmann CI. Chemosensation in *C. elegans*. *WormBook* 2006:1–29.
19. Britton NF, Franks NR, Pratt SC, Seeley TD. Deciding on a new home: how do honeybees agree? *P Roy Soc B-Biol Sci* 2002;**269**:1383–8.
20. Richardson RH, Vasco DA. Habitat and mate selection in Hawaiian *Drosophila*. *Behav Genet* 1987;**17**:571–96.
21. Kristan WB. Neuronal decision-making circuits. *Curr Biol* 2008;**18**:R928–932.
22. Stevenson PA, Rillich J. The decision to fight or flee—insights into underlying mechanism in crickets. *Front Neurosci* 2012;**6**:118.
23. (a)West-Eberhard MJ. *Developmental plasticity and evolution.* Oxford: Oxford University Press; 2003.(b)Pener MP, Simpson SJ. Locust phase polyphenism: an update. *Adv Insect Physiol* 2009;**36**:1–272.
24. Wang XH, Kang L. Molecular mechanisms of phase change in locusts. *Annu Rev Entomol* 2014;**59**:225–44.
25. Branchek TA, Blackburn TP. Trace amine receptors as targets for novel therapeutics: legend, myth and fact. *Curr Opin Pharmacol* 2003;**3**:90–7.
26. Axelrod J, Saavedra JM. Octopamine. *Nature (London)* 1977;**265**:501–4.
27. David JC, Coulon JF. Octopamine in invertebrates and vertebrates, a review. *Prog Neurobiol* 1985;**24**:141–85.
28. Evans PD, Robb S. Octopamine receptor subtypes and their mode of action. *Neurochem Res* 1993;**18**:869–74.
29. Roeder T. Octopamine in invertebrates. *Prog Neurobiol* 1999;**59**:533–61.
30. Vanden Broeck J, Vulsteke V, Huybrechts R, De Loof A. Characterization of a cloned locust tyramine receptor cDNA by functional expression in permanently transformed *Drosophila* S2 cells. *J Neurochem* 1995;**64**:2387–95.
31. Verlinden H, Vleugels R, Marchal E, Badisco L, Pfluger HJ, Blenau W, et al. The role of octopamine in locusts and other arthropods. *J Insect Physiol* 2010;**56**:854–67.
32. Roeder T. Tyramine and octopamine: ruling behavior and metabolism. *Annu Rev Entomol* 2005;**50**:447–77.
33. Farooqui T. Review of octopamine in insect nervous systems. *Open Access Insect Physiol* 2012;**4**:1–17.

34. Roeder T. High-affinity antagonists of the locust neuronal octopamine receptor. *Eur J Pharmacol* 1990;**191**:221–4.
35. Roeder T, Gewecke M. Octopamine receptors in locust nervous tissue. *Biochem Pharmacol* 1990;**39**:1793–7.
36. Degen J, Gewecke M, Roeder T. Octopamine receptors in the honey bee and locust nervous system: pharmacological similarities between homologous receptors of distantly related species. *Br J Pharmacol* 2000;**130**:587–94.
37. Konings PNM, Vullings HGB, Geffard M, Buijs RM, Diederen JHB, Jansen WF. Immunohistochemical demonstration of octopamine-immunoreactive cells in the nervous system of *Locusta migratoria* and *Schistocerca gregaria*. *Cell Tissue Res* 1988;**251**:371–9.
38. Stern M. Octopamine in the locust brain: cellular distribution and functional significance in an arousal mechanism. *Microsc Res Tech* 1999;**45**:135–41.
39. Kononenko NL, Wolfenberg H, Pflüger HJ. Tyramine as an independent transmitter and a precursor of octopamine in the locust central nervous system: an immunocytochemical study. *J Comp Neurol* 2009;**512**:433–52.
40. Matheson T. Octopamine modulates the responses and presynaptic inhibition of proprioceptive sensory neurones in the locust *Schistocerca gregaria*. *J Exp Biol* 1997;**200**:1317–25.
41. Downer RG, Hiripi L, Juhos S. Characterization of the tyraminergic system in the central nervous system of the locust, *Locusta migratoria migratoides*. *Neurochem Res* 1993;**18**:1245–8.
42. da Silva R, Lange AB. Tyramine as a possible neurotransmitter/neuromodulator at the sperm theca of the African migratory locust, *Locusta migratoria*. *J Insect Physiol* 2008;**54**:1306–13.
43. Roeder T, Nathanson JA. Characterization of insect neuronal octopamine receptors (OA3 receptors). *Neurochem Res* 1993;**18**:921–5.
44. Pflüger H-J, Blenau W, Van den Broeck J. The cloning, phylogenetic relationship and distribution pattern of two new putative GPCR-type octopamine receptors. *J Insect Physiol* 2010;**56**:868–75.
45. Hiripi L, Juhos S, Downer RGH. Characterization of tyramine and octopamine receptors in the locust (*Locusta migratoria migratorioides*) brain. *Brain Res* 1994;**633**:119–26.
46. Roder T. Biochemistry and molecular biology of receptors for biogenic amines in locusts. *Microsc Res Tech* 2002;**56**:237–47.
47. Howell KRM, Evans PD. The characterization of presynaptic octopamine receptors modulating octopamine release from an identified neuron in the locust. *J Exp Biol* 1998;**201**:2053–60.
48. Homberg U. Neurotransmitters and neuropeptides in the brain of the locust. *Microsc Res Tech* 2002;**56**:189–209.
49. Ignell R. Monoamines and neuropeptides in antennal lobe interneurons of the desert locust, *Schistocerca gregaria*: an immunocytochemical study. *Cell Tissue Res* 2001;**306**:143–56.
50. Laurent G, Naraghi M. Odorant-induced oscillations in the mushroom bodies of the locust. *J Neurosci* 1994;**14**:2993–3004.
51. Perez-Orive J, Bazhenov M, Laurent G. Intrinsic and circuit properties favor coincidence detection for decoding oscillatory input. *J Neurosci* 2004;**24**:6037–47.
52. Cassenaer S, Laurent G. Conditional modulation of spike-timing-dependent plasticity for olfactory learning. *Nature* 2012;**482**:47–52.
53. Martin JP, Beyerlein A, Dacks AM, Reisenman CE, Riffell JA, Lei H, et al. The neurobiology of insect olfaction: sensory processing in a comparative context. *Prog Neurobiol* 2011;**95**:427–47.
54. Greenwood M, Chapman RF. Differences in numbers of sensilla on the antennae of solitarious and gregarious *Locusta migratoria* L. (Orthoptera: Acrididae). *Int J Insect Morph Embyol* 1984;**13**:295–301.
55. Ott SR, Rogers SM. Gregarious desert locusts have substantially larger brains with altered proportions compared with the solitarious phase. *Proc Roy Soc B* 2010;**277**:3087–96.
56. Sombati S, Hoyle G. Central nervous sensitization and dishabituation of reflex action in an insect by the neuromodulator octopamine. *J Neurobiol* 1984;**15**:455–80.
57. Pophof B. Octopamine modulates the sensitivity of silk moth pheromone receptor neurons. *J Comp Physiol A* 2000;**186**:307–13.
58. Pophof B, Van der Goes van Naters W. Activation and inhibition of the transduction process in silk moth olfactory receptor neurons. *Chem Senses* 2002;**27**:435–43.
59. Pophof B. Octopamine enhances moth olfactory responses to pheromones but not those to general odorants. *J Comp Physiol A* 2002;**188**:659–62.
60. Flecke C, Stengl M. Octopamine and tyramine modulate pheromone-sensitive olfactory sensilla of the hawk moth *Manduca sexta* in a time-dependent manner. *J Comp Physiol A* 2009;**195**:529–45.

61. Unoki S, Matsumoto Y, Mizunami M. Participation of octopaminergic reward system and dopaminergic punishment system in insect olfactory learning revealed by pharmacological study. *Eur J Neurosci* 2005;**22**:1409–16.

62. Zeng H, Loughton BG, Jennings KR. Tissue specific transduction systems for octopamine in the locust (*Locusta migratoria*). *J Insect Physiol* 1996;**42**:765–9.

63. Poels J, Suner MM, Needham M, Torfs H, De Rijck J, De Loof A, et al. Functional expression of a locust tyramine receptor in murine erythroleukaemia cells. *J Insect Mol Biol* 2001;**10**:541–8.

64. Nagaya Y, Kutsukake M, Chigusa SI, Komatsu A. A trace amine, tyramine, functions as a neuromodulator in *Drosophila melanogaster*. *Neurosci Lett* 2002;**329**:324–8.

65. Saraswati S, Fox LE, Soll DR, Wu CF. Tyramine and octopamine have opposite effects on the locomotion of *Drosophila* larvae. *J Neurobiol* 2004;**58**:425–41.

66. Vierk R, Pflueger HJ, Duch C. Differential effects of octopamine and tyramine on central pattern generator for *Manduca* flight. *J Comp Physiol A* 2009;**195**:265–77.

67. Faumont S, Lindsay TH, Lockery SR. Neuronal microcircuits for decision making in *C. elegans*. *Curr Opin Neurobiol* 2012;**22**:580–91.

68. Li ZY, Li YD, Yi YL, Huang WM, Yang S, Niu WP, et al. Dissecting a central flip-flop circuit that integrates contradictory sensory cues in *C. elegans* feeding regulation. *Nat Commun* 2012;**3**:776.

69. Gao XJ, Potter CJ, Gohl DM, Silies M, Katsov AY, Clandinin TR, et al. Specific kinematics and motor-related neurons for aversive chemotaxis in *Drosophila*. *Curr Biol* 2013;**23**:1163–72.

70. Chu LA, Fu TF, Dickson BJ, Chiang AS. Parallel neural pathways mediate CO_2 avoidance responses in *Drosophila*. *Science* 2013;**340**:1338–41.

TRACE AMINES AND NEUROLOGICAL DISORDERS

Neurochemical Aspects of Neurological Disorders

A.A. Farooqui

Department of Molecular and Cellular Biochemistry, The Ohio State University, Columbus, OH, United States

INTRODUCTION

It is well known that brain has a very high metabolic rate. It accounts for 2% of body weight, but it receives about 15% of the cardiac output, consumes approximately 25% of glucose, and 20% of all inhaled oxygen at rest. This enormous metabolic demand of glucose and oxygen is due to the fact that neurons are highly differentiated cells requiring large amounts of Adenosine triphosphate (ATP) in order to maintain ionic gradients across cell membranes and maintain physiological neurotransmission. Diseases associated with metabolic dysfunction of brain, spinal cord, and nerves are called neurological disorders. More than 600 neurological disorders have been described in the literature. Neurological disorders may cause structural,

Trace Amines and Neurological Disorders
DOI: http://dx.doi.org/10.1016/B978-0-12-803603-7.00016-1

neurochemical, and electrophysiological abnormalities in the brain, spinal cord, and nerves, leading to neurodegeneration, which is accompanied by paralysis, muscle weakness, poor coordination, seizures, confusion, and pain.[1,2] Neurodegeneration in neurological disorders is a complex multifactorial process that causes neuronal death and brain dysfunction. The molecular mechanisms contributing to neurodegeneration include oxidative stress, axonal transport deficits, protein oligomerization, and aggregation, calcium deregulation, mitochondrial dysfunction, neuron–glial interactions, neuroinflammation, DNA damage, and aberrant RNA processing.[1]

Neurodegeneration is regulated by many factors, including genetic abnormalities and immune system problems.[1] For the sake of simplicity, I will classify neurological disorders into three groups: neurotraumatic diseases, neurodegenerative diseases, and neuropsychiatric diseases. Common neurotraumatic diseases are strokes, spinal cord injury (SCI), traumatic brain injury (TBI), and epilepsy.[1] Common neurodegenerative diseases include Alzheimer disease (AD), Parkinson disease (PD), Huntington disease (HD), amyotrophic lateral sclerosis (ALS), multiple sclerosis, and prion diseases.[1] Neuropsychiatric diseases include both neurodevelopmental disorders and behavioral or psychological difficulties associated with some neurological disorders. Examples of neuropsychiatric disorders are depression, schizophrenia, some forms of bipolar affective disorders, autism, mood disorders, attention-deficit disorder, dementia, tardive dyskinesia, and chronic fatigue syndrome. Neuropsychiatric diseases involve the abnormalities in cerebral cortex and limbic system (thalamus, hypothalamus, hippocampus, and amygdala). In addition, various types of brain tumors also fall under neurological disorders. Neuropsychiatric diseases not only involve alterations in serotonergic, dopaminergic, noradrenergic, cholinergic, glutamatergic, and γ-aminobutyric acid (GABA)-ergic signaling within the visceromotor network,[3] but are also associated with alterations in synaptogenic growth factors (brain-derived neurotrophic factor, BDNF), fibroblast growth factor, and insulin-like growth factors.[3,4]

NEUROTRAUMATIC DISEASES

Among neurotraumatic diseases, stroke is a metabolic insult, which is caused by severe reduction or blockade in cerebral blood flow, leading not only to a deficiency of oxygen and reduction in glucose metabolism, but also decrease in ATP production and accumulation of toxic products. TBI and SCI are caused by mechanical trauma to brain and spinal cord, which occur following falls and motorcycle and car accidents.[1] Neurotraumatic diseases also cause muscle dystrophy, leading to a decline in neuronal and muscular functions, which often limit quality of life as well as lifespan.

Neurotraumatic disorders share oxidative stress and neuroinflammation as common mechanisms of brain injury and neural cell death.[1] The onset of neurotraumatic diseases is often subtle and accompanied by a reduction in ATP, disturbance in transmembrane potential, and sudden collapse of ion gradients at a very early stage along with breakdown of the blood–brain barrier (BBB) (Fig. 16.1). This not only results in transmigration of numerous immune system cells, including monocytes and lymphocytes, but also leads to hyperpermeability induced by enhanced transcytosis and gap junction abnormalities between neural and endothelial cells. In addition to oxidative stress and neuroinflammation, neurochemical changes in neurotraumatic diseases also include release of glutamate, overstimulation

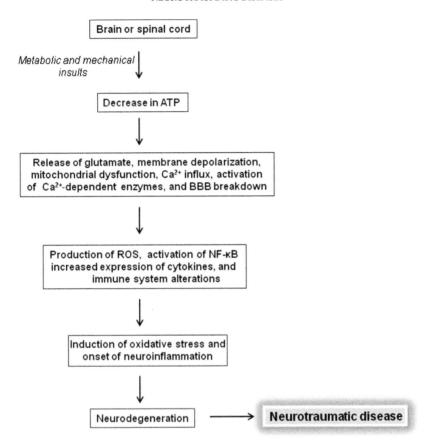

FIGURE 16.1 Factors contributing to the pathogenesis of neurotraumatic diseases.

of glutamate receptors, elevation in intracellular Ca^{2+}, and marked reduction in ATP that is needed not only for maintaining the appropriate ionic gradients across the neural membranes but also for creating the proper cellular redox potentials. Neurotraumatic injury to the brain and spinal cord results in rapid neurodegeneration (in days) because of sudden lack of oxygen and a quick drop in ATP along with alterations in ion homeostasis (Fig. 16.1). The initial response to a decrease in ATP is neural membrane depolarization resulting in Na^+ influx into axons. Prolonged decrease in ATP results in a massive influx of Ca^{2+} with neurodegeneration resulting in irreversible loss of neurologic function.[5] Ca^{2+}-mediated activation of phospholipases, kinases, and proteases not only generates high levels of proinflammatory lipid mediators (eicosanoids and platelet-activating factors) and protein-derived proinflammatory cytokines and chemokine (TNF-α, IL-1β, IL-6, MCP1, and CXCL3), but also produce other proapoptotic metabolites, such as nitric oxide and peroxynitrite. Mitochondria and endoplasmic reticulum (ER) play a central role in apoptotic neural cell death. The release of cytochrome c from mitochondria and abnormal protein processing in ER are key processes that contribute to apoptotic cell death in neurotraumatic diseases.[1,5]

Molecular Aspects of Stroke

As stated above, onset of stroke involves the reduction or blockade of blood flow to the brain due to formation of a clot leading not only to the deficiency of oxygen and reduction in glucose metabolism, but also a decrease in ATP production and breakdown of the BBB along with accumulation of toxic products.[1] Two types of stroke have been shown to occur in humans: ischemic and hemorrhagic. Ischemic strokes are caused by a critical decrease in blood flow to various brain regions causing neuronal cell death. Ischemic stroke is the most common type of stroke, constituting around 80% of all strokes, of which 60% are attributable to large-artery ischemia. Hemorrhagic strokes are caused by a break in the wall of the artery resulting in spillage of blood inside the brain or around the brain. Age is a prominent risk factor for stroke. Loss of synaptic spine seems to be the earliest event of cerebral ischemia and generally contributes to the subsequent brain damage.[1] Stroke mediates neuronal damage through the overstimulation of NMDA type of glutamate receptors, rapid Ca^{2+}-influx, and stimulation of phospholipases A_2, C, and D (PLA_2, PLC, and PLD), calcium/calmodulin-dependent kinases (CaMKs), mitogen-activated protein kinases such as extracellular signal-regulated kinase, p38, and c-Jun N-terminal kinase, nitric oxide synthases (NOS), calpains, calcineurin, and endonucleases leading to phospholipid hydrolysis, proteolysis, and a disturbed docking of glutamate-containing vesicles resulting from impaired phosphorylation.[1] Many of above-mentioned enzymes contribute to neuronal damage by increasing the production of ROS via cell membrane-bound NADPH oxidase, uncontrolled arachidonic acid cascade (via the activation of PLA_2 and cyclooxygenase and lipoxygenase) and mitochondrial dysfunction. Stroke also triggers a robust inflammatory reaction characterized by peripheral leukocyte influx into the cerebral parenchyma and activation of endogenous microglia.[1,5–7] Following stroke-mediated brain damage, neurons secrete inflammatory cytokines and chemokines that cause, among other things, adhesion molecule upregulation in the cerebral vasculature which leads to peripheral leukocyte recruitment. In addition, oxidation of biogenic amines by monoamine oxidases generates hydrogen peroxide (H_2O_2), which in the presence of copper generates hydroxyl radicals ($^{\bullet}OH$). In addition to oxidative stress and neuroinflammation, immunological changes (changes in metabolism of neutrophils and macrophages) are key elements of the pathobiology of stroke. While the immune system participates in the brain damage caused by stroke, the damaged brain, in turn, exerts a powerful immunosuppressive effect that facilitates fatal intercurrent infections and threatens the survival of stroke patients. Thus, oxidative stress, neuroinflammation, and immunological alterations contribute to the ischemic cascade from the early damaging events triggered by arterial occlusion, to the late regenerative processes underlying postischemic tissue repair.[7] Converging evidence suggests that multiple mechanisms contribute to neuronal injury and neural cell death following stroke-mediated brain injury.[1,8]

Molecular Aspects of SCI

SCI is a devastating neurological disorder that may result in the loss of sensory and motor function and, depending on the extent of injury, may lead to paralysis and death.[1,9] SCI is accompanied two broadly defined events: the first of which is a primary event, is caused by the mechanical insult. This event is instantaneous, causing neuronal fiber damage and neural cell necrosis. Primary mechanical insult is beyond therapeutic management. In contrast, the secondary event involves a series of systemic and local neurochemical changes that occur in spinal cord tissue after the primary injury. Neurochemical changes in secondary event develop slowly (hours to days) after SCI. At the core of primary injury site, SCI causes a rapid deformation of spinal cord tissue due to compression, contusion, and laceration due to penetrating injury along with acute stretching of the spinal cord as a result of iatrogenic vertebral distraction, rupturing of neural cell membranes, resulting in the release of neuronal intracellular contents. These morphological changes result in behavioral and functional impairments, due to the release of glutamate, induction excitotoxicity, influx of calcium ions, activation of calcium-dependent enzymes (phospholipase A_2, NOS, proteases, endonucleases, and matrix metalloproteinase), release of proinflammatory cytokines and chemokines, and generation of proinflammatory lipid mediators (eicosanoids).[10,11] These neurochemical processes are supported by the activation of microglial cells, recruitment of neutrophils, and activation of macrophages and vascular endothelial cells and T cells leading to the onset of acute neuroinflammation, and oxidative stress.[1] Production of ROS directly downregulates proteins of tight junctions and indirectly activates matrix metalloproteinases (MMPs) that contribute to opening the BBB[1]. Loosening of the vasculature and perivascular unit by oxidative stress-induced activation of MMPs and fluid channel aquaporins promotes vascular or cellular fluid edema, and enhances leakiness of the BBB.[9] These processes contribute to a failure in normal neural function and spinal shock, and represent a generalized failure of circuitry in the spinal neural network. Hemorrhage occurs, with localized edema, loss of microcirculation by thrombosis, vasospasm, and mechanical damage, and loss of vasculature autoregulation, all of which further exacerbate the neural injury. Inhibitory elements (neurite outgrowth inhibitor, myelin-associated glycoprotein, oligodendrocyte-myelin glycoprotein, and chondroitin sulfate proteoglycan) in the spinal cord tissue inhibit damaged nerve fibers to exhibit regenerative sprouting.[1] Converging evidence suggests that SCI is an irreversible condition that causes damage to myelinated fiber tracts that carry sensation and motor signals to and from the brain. It involves primary and secondary damage to the spinal cord. Primary damage to the spinal cord is caused by the mechanical damage leading to deformation of the spinal cord resulting in necrotic cell death. The secondary damage in SCI involves a cascade of biochemical and cellular processes, such as release of glutamate; overstimulation of glutamate receptors and calcium influx; stimulation of PLA_2, COX-2, NOS, calpains, caspases, and MMP; formation of free radicals, oxidative stress, vascular ischemia, edema, mitochondrial dysfunction, activation of transcription factors; induction of cytokines and chemokines, posttraumatic inflammatory reaction, activation of the complement system; and apoptotic cell death. SCI increases the risk of cardiovascular complications, deep vein thrombosis, osteoporosis, pressure ulcers, autonomic dysreflexia, and neuropathic pain (Fig. 16.2).[1,12] Therefore, it is important to be aware of chronic complications of SCI and learn how to manage these complications for the recovery and rehabilitation process.

Molecular Aspects of TBI

TBI, which is caused by falls or motorcycle or car accidents is accompanied by mechanical trauma to head (primary injury), which involves rapid deformation of brain tissue and rupture of neural cell membranes leading to the release of intracellular contents, disruption

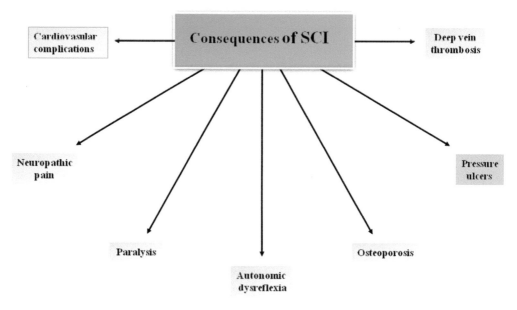

FIGURE 16.2   Consequences of spinal cord injury.

of blood flow, breakdown of the BBB, intracranial hemorrhage, brain edema, and axonal shearing, in which the axons of neurons are stretched and torn.[1] The primary injury is followed by secondary injury, which at the regional level involves changes in hippocampal, prefrontal cortical, and limbic region along with alterations in synaptogenesis, dendritic remodeling, and neurogenesis. At the cellular level, microglial cells, astrocytes, and oligodendroglial cells contribute to inflammation, gliosis, and demyelination in TBI. At the molecular level TBI involves a complex cascade of signal transduction processes associated with the onset of oxidative stress, excitotoxicity, ischemia, edema, and neuroinflammation.[1,13] Mitochondrial dysfunction at the neuronal/astrocytic level is another characteristic feature of TBI pathophysiology.[14] In addition, adult brain TBI is accompanied by induction of reactive gliosis and reduction in levels of BDNF leading to cognitive impairment. Processes that mediate induction of BDNF and activation of its intracellular receptors can produce neural regeneration, reconnection, and dendritic sprouting, and can improve synaptic efficacy.[15] It is important to note that moderate-to-severe TBI is accompanied by progressive atrophy of gray and white matter structures that may persist months to years after injury.[1,16] In addition, multiple studies support a link between single moderate–severe TBI and AD,[17] PD,[18] and ALS.[19] In a meta-analysis of 15 case-control studies, males who had a single head injury associated with LOC had a 50% increased risk of AD dementia.[17]

NEURODEGENERATIVE DISEASES

Neurodegenerative diseases also share oxidative stress and neuroinflammation as common mechanisms of brain injury and neural cell death. The molecular mechanism

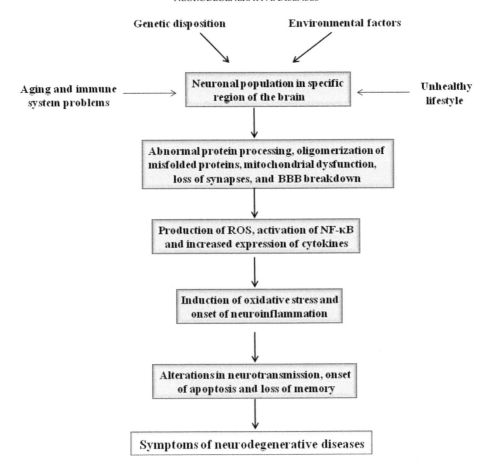

Genetic disposition

Environmental factors

Aging and immune system problems

Neuronal population in specific region of the brain

Unhealthy lifestyle

Abnormal protein processing, oligomerization of misfolded proteins, mitochondrial dysfunction, loss of synapses, and BBB breakdown

Production of ROS, activation of NF-κB and increased expression of cytokines

Induction of oxidative stress and onset of neuroinflammation

Alterations in neurotransmission, onset of apoptosis and loss of memory

Symptoms of neurodegenerative diseases

FIGURE 16.3   Factors contributing to the pathogenesis of neurodegenerative diseases.

associated with the pathogenesis of neurodegenerative diseases is not fully understood. However, it is becoming increasingly evident that neurodegenerative diseases are accompanied by the accumulation of misfolded proteins, mitochondrial and proteasomal dysfunction, loss of synapses, and progressive premature and selective slow death of specific neuronal populations in a specific region of the brain (Fig. 16.3).[1,20] Mitochondrial malfunction leads to a reduction in ATP production, impairment in Ca^{2+} buffering, and generation of ROS in both aging and neurodegenerative disease.[1] The innermembrane structural alterations, in particular numerous dilated or swollen cristae, are closely and consistently associated with induction of oxidative stress and apoptosis in various neurodegenerative diseases.[21] It is speculated that interplay among pathological factors (aging, environmental factors, genetic predisposition, and cellular redox status)[22] along with increase in metal ion like iron and expression of cytokines and chemokines play a central role in the pathogenesis of neurodegenerative diseases.[1,22–24] The majority of cases of neurodegenerative diseases (>93–95%) are of sporadic and only 5–7% cases appear to be primarily of genetic origin.

Unlike neurotraumatic disease, where neurodegeneration occurs (in hours to days), in neurodegenerative diseases oxygen and nutrients are available to neural cells to some extent facilitating generation of some ATP for maintaining ion homeostasis, so onset of neurodegeneration occurs progressively. This type of neuronal damage takes a longer time (years) with slow appearance of symptoms due to lingering chronic oxidative stress and neuroinflammation.[1] In AD, neurodegeneration occurs in the nucleus basalis and hippocampus, whereas in PD, neurons die in the substantia nigra. The most severely affected neurons in HD are striatal medium spiny neurons.[1] Another feature of neurodegenerative diseases is the accumulation of disease-specific proteins, such as accumulation of Aβ and its aggregates in the cerebral cortex and hippocampal region in AD, α-synuclein and its aggregates in the brain stem in PD, huntingtin and its aggregates in striatal medium spiny neurons in HD, abnormalities in Cu/Zn-superoxide dismutase in ALS, and misfolded PrP^{Sc} polymerized amyloid fibril is involved in neurodegeneration in prion diseases.[1,25,26] Converging evidence suggests that each neurodegenerative disease involves selective vulnerability of the neuronal population, which undergoes neurodegeneration as a consequence of potentially interrelated processes, involving but not limited to oxidative-related damage and impairments in the ubiquitin-proteasome system, and RNA metabolism, mitochondrial dysfunction, and protein aggregation and propagation. The causative link between protein aggregate formation and neurodegenerative diseases has not yet been clearly established. However, it is becoming increasingly evident that the toxic action of soluble oligomers and protofibrillar derivatives of misfolded proteins may contribute to neurodegeneration in neurodegenerative diseases.[27] This suggestion is supported by the observation that a single-domain antibody can recognize a common conformational epitope, which is displayed by several disease-associated proteins, including Aβ, α-synuclein, τ-protein, prions, and polyglutamine (polyQ)-containing peptides.[27] Furthermore, most neurodegenerative diseases are accompanied by the progressive synaptic and cognitive dysfunctions and motor disabilities with devastating consequences to patients. Despite these developments and ongoing refinements the basic insights into the pathogenesis and understanding of neurodegenerative disease remain fragmentary and poorly defined. There are major problems associated with reliable clinical tools for risk stratification, early diagnosis and prognostication, and monitoring disease progression and there are few, if any, therapies currently available to modify the natural history of these diseases.

Risk factors for neurodegenerative diseases (Fig. 16.3) include old age, race/ethnicity, a positive family history, exposure to metal ions (iron, copper, zinc, and mercury), and unhealthy lifestyle,[1] which includes diet, exercise, and sleep. Long-term overconsumption of a Western diet, which is enriched in refined grains, saturated and omega-6 fats, proteins of animal origins, high salt, and low in fiber, increases obesity, hyperglycemia, insulin resistance, dyslipidemia, and hypertension. Prolonged hyperglycemia, insulin resistance, dyslipidemia, hypertension, and accumulation of ROS along with endothelial dysfunction are closely associated with the pathogenesis of metabolic syndrome (Fig. 16.4). This pathological condition is an important risk factor for neurotraumatic, neurodegenerative, and neuropsychiatric diseases.[28,29] The onset of neurodegenerative diseases is often subtle and usually occurs in mid to late life and their progression depends not only on genetic, but also on environmental factors.[20] The onset of neurodegenerative diseases occurs when neurons fail to respond adaptively to age- and lifestyle-related increases in oxidative and

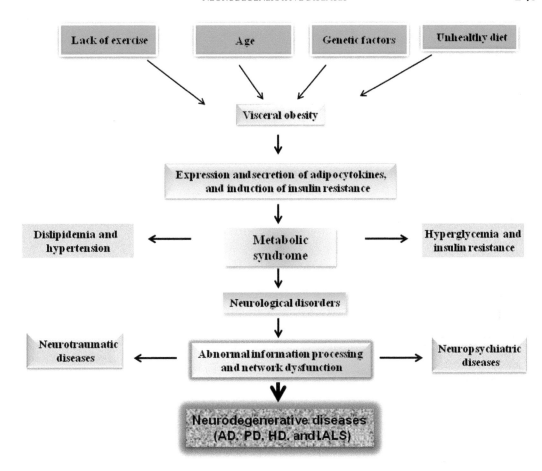

FIGURE 16.4 Factors contributing to the pathogenesis of metabolic syndrome.

nitrosative stress and neuroinflammation. Persistence and progressive increase in oxidative stress and neuroinflammation cause the accumulation of oxidative-inducing lipid mediators (4-hydroxynonenal, malondialdehyde, acrolein, and isoprostanes), oxidized proteins and DNA products (such as 8-hydroxy-2-deoxyguanosine), and membrane fragments leading to synaptic loss and neurodegeneration (Table 16.1).[1] Persistant presence of oxidative stress and neuroinflammation in neurodegenerative diseases are accompanied by a significant decline in glutathione, glutathione peroxidase, glutathione-*S*-transferase, and superoxide dismutase, supporting the view that production of high levels of oxidative stress and neuroinflammation in brains of patients with neurodegenerative diseases may contribute to neurodegeneration.[1]

Molecular Aspects of AD

AD is an age-associated progressive neurodegenerative disorder characterized by neuro-degeneration in the nucleus basalis, hippocampus, and cerebral cortex along the combined

TABLE 16.1	Neurochemical Alterations in Neurodegenerative Diseases

Parameter	AD	PD	HD	ALS	Prion Diseases
Phospholipid metabolism	Abnormal	Abnormal	Abnormal	Abnormal	Abnormal
PLA$_2$ activity	Increased	Increased	–	–	Increased
Eicosanoids	Increased	Increased	Increased	–	Increased
Lipid peroxidation	Increased	Increased	Increased	Increased	Increased
4-HNE	Increased	Increased	–	Increased	–
Aggregation of accumulated protein	Abnormal APP processing	Increase in α-synuclein generation	Increase in huntingtin	Increase in mutant SOD	Abnormal PrP processing
Oxidative stress	Increased	Increased	Increased	Increased	Increased
Neuroinflammation	Increased	Increased	Increased	Increased	Increased
Neurodegradation	Increased	Increased	Increased	Increased	Increased

Summarized from Refs. Farooqui AA. Neurochemical aspects of neurotraumatic and neurodegenerative diseases. New York, NY:Springer; 2010; Farooqui AA, Horrocks LA. Glycerophospholipids in the brain: phospholipases A2 in neurological disorders. New York, NY; Springer; 2007; Rao SD, Weiss JH. Excitotoxic and oxidative cross-talk between motor neurons and glia in ALS pathogenesis. Trends Neurosci 2004; 27: 17–23; Yoshinaga N, Yasuda Y, Murayama T, Nomura Y. Possible involvement of cytosolic phospholipase A(2) in cell death induced by 1-methyl-4-phenylpyridinium ion, a dopaminergic neurotoxin, in GH3 cells. Brain Res 2000; 855: 244–51; Klivenyi P, Kiaei M, Gardian G, Calingasan NY, Beal MF. Additive neuroprotective effects of creatine and cyclooxygenase 2 inhibitors in a transgenic mouse model of amyotrophic lateral sclerosis. J Neurochem 2004; 88: 576–82; Thomas EA. Striatal specificity of gene expression dysregulation in Huntington's disease. J Neurosci Res 2006; 84: 1151–64; Rocha NP, de Miranda AS, Teixeira AL. Insights into neuroinflammation in Parkinson's disease: from biomarkers to anti-inflammatory based therapies. Biomed Res Int 2015; 2015: 628192; Morreale MK. Huntington's disease: looking beyond the movement disorder. Adv Psychosom Med 2015; 34: 135–42 [30–34].

presence of two lesions in the brain: extracellular amyloid-beta (Aβ) plaques (senile plaques, SP) and intracellular neurofibrillary tangles (NFTs) with brain atrophy. The extracellular deposits contain aggregated Aβ peptides, while intraneuronal NFTs are aggregates of hyperphosphorylated forms of the neurofilament-associated protein tau.[1,25] Approximately 5% of patients with AD have familial form of AD—that is, related to a genetic predisposition, including mutations in the amyloid precursor protein, presenilin 1 and presenilin 2 genes, and 95% of AD cases are sporadic. The initiating event in Aβ production is the cleavage of the amyloid precursor protein (APP) at the β site APP cleaving enzyme 1 (BACE-1), a neuronal specific aspartyl protease.[1] This results in generation and release of a soluble N-terminus exodomain (soluble APPβ) into the lumen and a membrane-bound β-C-terminus fragment (β-CTF). Gamma secretase-mediated cleavage of the membrane-anchored β-CTF releases Aβ peptides of different lengths, including Aβ38, Aβ40, and Aβ42.[25] Aβ42 readily aggregates into neurotoxic oligomers and eventually forms mature fibrils and plaques. The generation of amyloid plaques in humans and animal models is invariably accompanied by activation of astrocytes and microglial cells with elevated levels of proinflammatory cytokines (TNF-α, IL-1β, and IL-6) and chemokines (MCP1 and CXCL3).[1] Mounting evidence suggests that an imbalance between the production and clearance of Aβ in the brain results in the accumulation and aggregation of Aβ. The toxic Aβ aggregates in the form of soluble Aβ oligomers, intraneuronal Aβ, and amyloid plaques injure synapses and ultimately cause neurodegeneration and dementia. The toxicity of Aβ seems to depend on the presence of microtubule-associated protein tau, the hyperphosphorylated forms of which aggregate and deposit in AD brains as neurofibrillary tangles.[35] Amyloid plaque burden and hyperphosphorylation of tau correlate poorly with memory deficits in AD patients; however, synaptic loss is a strong predictor of the clinical symptoms of AD. Clinically, AD is characterized by a gradual decline in cognition along with loss of memory. Other changes in AD include alterations in behavior and personality, difficulty in reasoning, disorientation, and language problems. Neurochemically, AD is characterized by induction of oxidative stress, neuroinflammation, activation of phospholipases A_2 (PLA$_2$), sphingomyelinases (SMase), and cholesterol hydroxylases (CYP46), increase in levels of phospholipid-, sphingolipid-, and cholesterol-derived lipid mediators, mitochondrial dysfunction, activation of caspases, stimulation of protein phosphorylation, and loss of synapses along with onset of dementia along with depletion of neurotransmitter systems in the hippocampus and cerebral cortex.[1,36] The pathogenesis of AD develops over many years before clinical symptoms appear. The severity of AD pathology is associated with protein misfolding and abnormal increase in intracellular calcium along with abnormal protein clearance defects through the ubiquitin–proteasome system-mediated synaptic degeneration, neuronal loss.

Molecular Aspects of PD

PD is a chronic, disabling neurodegenerative proteinopathy characterized by the selective degeneration of dopaminergic neurons of the substantia nigra pars compacta. The three cardinal clinical features of PD are rigidity, resting tremor, and bradykinesia, and these occur when approximately 50% of dopaminergic neurons projecting from the substantia nigra pars compacta (SNc) to the striatum are lost.[37] The degeneration of dopaminergic neurons results in the depletion of dopamine leading to abnormal dopaminergic neurotransmission in the basal ganglia motor circuit. Characteristic features of PD such as rigidity, akinesia, rest tremor, and postural instability are attributed to the degeneration of dopaminergic neurons, while the nonmotor alterations, such as hyposmia, autonomic, and other dysfunctions are linked to widespread distribution of α-synuclein, a highly conserved 140-amino-acid protein that is predominantly expressed in the central, autonomic and peripheral nervous system and multiple organs. It is encoded by a single gene consisting of seven exons located in chromosome 4. Recent advances have indicated pathogenic mechanisms related to deposition of phosphorylated α-synuclein as Lewy bodies and neuritis are hallmark of PD.[38] Very little is known about the role of α-synuclein in the brain. However, it is suggested that this protein plays an important role in the regulation of synaptic vesicle release and trafficking, maintenance of synaptic vesicle pools, fatty acid binding, neurotransmitter release, synaptic plasticity, and neuronal survival.[39] Within cells, α-synuclein normally adopts an α-helical conformation. However, under oxidative stress this protein undergoes a profound conformational transition to a β-sheet-rich structure that polymerizes to form toxic oligomers. The transformation of soluble oligomeric and protofibrillar forms of α-synuclein into aggregates in the pathogenesis of PD is not only supported by the consistent detection of α-synuclein deposits in affected brain areas, but also by pathogenic mutations affecting the α-synuclein gene in familial PD. Recent studies on neurodegenerative potency of α-synuclein fibrils have indicated that toxicity of α-synuclein fibrils may be due to its ability to penetrate neural cell membranes.[40] Thus, drugs that inhibit α-synuclein aggregation and fibrillization and stabilize it in a nontoxic state can therefore serve as therapeutic molecules for both prevention of accumulation of aggregated α-synuclein and maintenance of normal physiological concentrations of α-synuclein.[41]

Molecular Aspects of HD

HD is an autosomal, progressive neurodegenerative disorder, which prominently affects the basal ganglia, leading to significant motor dysfunction, cognitive and behavioral decline, and psychiatric symptoms.[1] Symptoms of HD include midlife onset of involuntary movements, cognitive, physical and emotional deterioration, personality changes, and dementia leading to premature death. At the genetic level, HD is caused by a mutation in the IT-15 gene that abnormally expands the number of CAG nucleotide repeats.

Normal HD alleles have 37 or fewer glutamines in this polymorphic tract, more than 37 of these residues cause HD.[42] The length of the CAG tract is directly correlated with the onset of HD, with longer expansions leading to earlier onset of HD. Insoluble aggregates containing Huntingtin occur in cytosol and nuclei of HD patients, transgenic animals, and cell culture models of HD. The molecular mechanism involved in aggregation of Huntingtin is not fully understood. Many possible mechanisms are being explored. In particular, factors promoting apoptosis, phenomena causing the toxic aggregation of proteins, the blockage of trophic factors, mitochondrial dysfunction, and excitotoxicity have been studied.[1] Wild-type huntingtin reduces the cellular toxicity of mutant Huntingtin in vitro and in vivo conditions and mediates neuroprotection by a mechanism through the involvement of inhibition of procaspase-9 processing or caspase.[43] Futhermore, wildtype Huntingtin may also prevent PAK2 cleavage by caspase-3 and caspase-8, which activates PAK2 by releasing a constitutively active C-terminal kinase domain that mediates cell death.[44] Based on these results, it is suggested that loss-of-function of huntingtin may mediate neuronal toxicity resulting from the polyQ expansion. In contrast, other studies support the view that the direct aberrant interactions between mutant Huntingtin and myriad specific effector proteins produce neurotoxicity. Mutant Huntingtin can also produce neurodegeneration indirectly by stressing the protein homeostasis system in such a way that other metastable proteins fail to fold properly, leading to widespread dysfunction of the proteome.[45] Mutant Huntingtin induces neurodegeneration predominantly through gain-of-function mechanisms, although Huntingtin loss-of-function may have a role. Converging evidence suggests that Huntingtin aggregates promote neurodegeneration not only by disrupting normal synaptic transmission and modulating gene transcription, protein interactions, protein transport inside the nucleus and cytoplasm, but also by regulating vesicular transport along with mitochondrial and proteosomal functions, axonal transport deficit, apoptosis, and excitotoxicity in the striatum.[46] It has been reported that the accumulation of mutated Huntingtin inclusions is not a consequence of direct proteasomal inhibition but rather results from the gross failure of protein quality control systems in association with the sequestration of molecular chaperones.[47] Furthermore, expression of the polyQ-expanded form of Huntingtin not only leads to mitochondrial dysfunction, but also in mutant Huntingtin-mediated alterations in activity of the NMDA-type glutamate receptor, especially in the striatum. Another important finding related to HD is a decrease in expression of hippocalcin, a neuronal calcium sensor protein, which is expressed in the medium spiny striatal output neurons that degenerate selectively in HD. The role of hippocalcin in HD is not fully understood. However, a decrease in hippocalcin expression is known to occur in parallel with the onset of disease in mouse models of HD. In situ hybridization histochemistry studies have indicated that hippocalcin RNA is diminished by 63% in human HD brain, suggesting that hippocalcin may be associated with neuronal viability and plasticity.[48]

Molecular Aspects of ALS

ALS is a neurodegenerative disease characterized by progressive loss of upper and lower motor neurons. Although the etiology remains unclear, disturbances in calcium homeostasis, protein folding, are essential features of neurodegeneration in this disorder. The progressive loss of upper and lower motor neurons leads to muscle loss, paralysis, and death from respiratory failure. Approximately 95–90% of ALS cases are sporadic (sALS) and less than 5% cases are familial diseases (fALS), where mutations in superoxide dismutase 1 (SOD-1) have been reported.[49] Other possible mechanisms of neurodegeneration in ALS include excitotoxicity,[50] excessive production of ROS and RNS,[51] mitochondrial dysfunction,[49] induction of ER stress, axonal deterioration, and deposition of toxic ubiquitinated neuronal inclusions, where transactive response DNA-binding protein 43 kDa (TDP-43), and fused in sarcoma are major protein components.[49] Most of above-mentioned mechanisms are interconnected and interactions among excitotoxicity, oxidative stress, and neuroinflammation may play a major role in the pathogenesis of ALS.[51,52] In addition, there is evidence not only for the involvement of the immune system in the ALS, but also for the activation of components of the classical complement pathway in the serum, cerebrospinal fluid, and neuronal tissue of individuals with ALS.[49,50]

Molecular Aspects of Prion Diseases

Prion diseases are fatal disorders characterized by not only progressive loss of neurons, lack of classical inflammation, and appearance of vacuolation in the neuropil (spongiform encephalopathy), but also deposition of abnormal conformers of prion protein (PrP^{Sc}); and transmissibility in most forms of the disease (Fig. 16.5). The abnormal conformer of prion protein (PrP^{Sc}) accumulates in the brain parenchyma. Both PrP^C and PrP^{Sc} are glycoproteins, which are encoded by the *Prnp* gene. There are no differences in the primary structure of PrP^C and PrP^{Sc}, suggesting that PrP^C differs from PrP^{Sc} in its conformation.[53] Deposition of aggregated and misfolded protein into large amyloid plaques and fibrous structure is a fundamental mechanistic event in prion diseases. The propensity of the prion protein to oligomerize or fibrillize is correlated with acidic pH.[54] The misfolding and oligomerization of PrP^C in the cell originates in the endocytic pathway in late endosomes or in lysosomes, which have a low internal pH. In prion diseases synaptic loss precedes neuronal degeneration, in particular since both PrP^C and PrP^{Sc} are located at synapses. It is proposed that Notch-1 signaling may contribute to the progressive loss of dendritic spines.[55] In humans prion protein causes Creutzfeldt–Jakob and kuru diseases (CJD), in cows it promotes bovine spongiform encephalopathy and in sheep it produces scrapie. The function of PrP^C remains elusive. However, it is proposed that PrP^C plays an important role in cell adhesion, neuroprotection, basic biology of embryonic and tissue-specific stem cells, T-cell regulation and immune function, oxidative stress homeostasis, and synaptic function.[56]

Brain iron dyshomeostasis is a prominent feature of human and animal prion disorders (Fig. 16.5). Biochemical analysis of brain tissue from sporadic-(CJD) and scrapie-infected mouse and hamster models shows increased reactivity for redox-active iron and, paradoxically, a phenotype of neuronal iron deficiency.[57] It is demonstrated that PrP^C is a ferrireductase (FR), and its absence causes systemic iron deficiency in PrP knock-out mice ($PrP^-/^-$).[57] Chronic exposure to excess dietary iron corrects this deficiency. Unlike wildtype ($PrP^+/^+$) controls, $PrP^-/^-$ mice revert back to the iron-deficient phenotype after 5 months of chase on normal diet. Detailed investigations have indicated that there is a correlation between PrP^C expression and cellular iron levels.[57] Collective evidence suggests the imbalance of brain metal homeostasis as a common cause of neuronal death in several age-dependent neurodegenerative diseases.[58] Whether the accumulation of these metals is a cause or consequence of the disease process is a subject of much dispute. It is proposed that a redox-active metal interacts with a specific protein and is reduced in its presence, leading to the generation of ROS, H_2O_2, and $OH^\bullet$ that cause aggregation of the involved protein.[59]

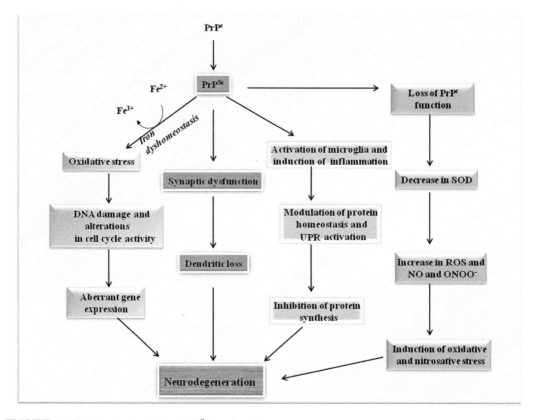

FIGURE 16.5   Neurotoxic effects of PrP^{Sc} in prion diseases.

NEUROPSYCHIATRIC DISEASES

Neuropsychiatric diseases involve mild oxidative and neuroinflammation.[60] At the nuclear level, abnormalities in neuropsychiatric diseases may be caused by abnormal formation of neuronal networks, disbalanced neurotransmission, which are regulated by overexpression or underexpression of genes and alterations in neurotransmitters that modulate behavioral symptoms, such as thoughts or actions, delusions, delirium, and hallucinations. In addition, environmental factors (exposure to heavy metals or other toxins), and hormonal impairments can also contribute to the pathogenesis of neuropsychiatric diseases (Fig. 16.6). These behavioral abnormalities are the hallmarks of many neuropsychiatric diseases. In addition to signal transduction processes associated with dopamine, glutamate, and GABA receptor-mediated behavioral abnormalities,[61] neuropsychiatric disorders also involve gray

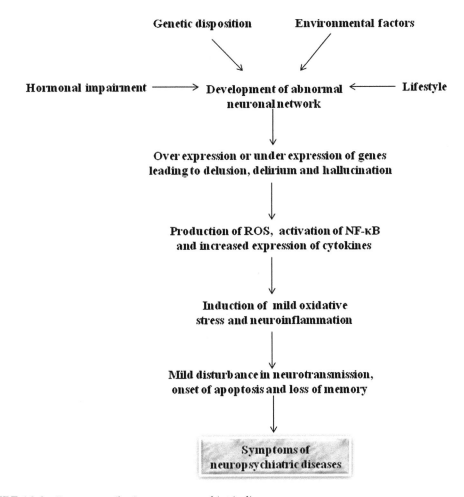

FIGURE 16.6 Factors contributing to neuropsychiatric diseases.

matter atrophy caused by reduction in neuronal and glial size, increase in cellular packing density, disruption in neuronal connectivity, particularly in the dorsolateral prefrontal cortex, and distortions in neuronal orientation.[62] These neurochemical and morphological changes may simultaneously mediate alterations within a single microcircuit in more than one region. Changes in microcircuits and neurotransmitters (synthesis and transport) may not only vary on a region-by-region basis but also from one neuropsychiatric disease to another. Both macro- and microcircuitry within the specific brain system (such as limbic system) may serve as "triggers" for the onset of neuropsychiatric condition.[60] Neurochemical and neuroimaging studies have also indicated alterations in cerebral blood flow and glucose utilization in the limbic system and prefrontal cortex of patients with major depression and other neuropsychiatric diseases.[60] Converging evidence suggests that neuropsychiatric diseases are mediated by genetic factors, alterations in blood flow, disruption of cellular connectivity, decrease in neurogenesis, alterations in microcircuitry, decrease in neuroplasticity along with mild oxidative stress, and mild neuroinflammation.

Molecular Aspects of Depression

Depression is a multisystem and multifactorial mental disorder characterized by behavioral changes such as sleep disturbances (insomnia or hypersomnia), psychomotor retardation or agitation, fatigue, feelings of worthlessness or guilt, and psychomotor changes leading to diminished cognitive functioning, loss of energy, concentration difficulties/indecisiveness, irritability, and low self-esteem.[63] Not all individuals show all of the symptoms of depression.[64] Clinical features of depression include elevated blood levels of IL-6 and TNF-α,[65] low levels of magnesium, overactivity of hypothalamic–pituitary–adrenal (HPA) axis, alterations in cerebral structures such as an increased ventricle/brain ratio and localized atrophy of the prefrontal cortex, cingulated gyrus, ventral striatum, amygdale, cerebellum, and hippocampus. The molecular mechanism of depression is not fully understood. However, it is proposed that the pathogenesis of depression involves the disturbance in neurotransmitters (dopamine, norepinephrine, and serotonin), increase in inflammatory processes, defects in neurogenesis, decrease in synaptic plasticity, mitochondrial dysfunction, and redox imbalance. In addition, changes in neuropeptides (vasopressin), cytokines, and gene–environmental interactions may contribute to the pathogenesis of depression.[64,65] Meta-analysis has indicated that genetic factors can predispose and contribute to the pathogenesis of depression.[66] For example, aberrant genes can predispose one to depression by decreasing the production of growth factors that act and play important roles during brain development. Aberrant genes have also been reported to modulate the release of neurotransmitter at the synapse.[67]

Molecular Aspects of Autism

Autism is a complex neurodevelopmental disorder of unknown etiology. It is characterized by qualitative impairments in social interaction, deficits in verbal and nonverbal communication, and restricted repetitive and stereotyped patterns of behavior and interests. There is growing evidence that an abnormal immune response may exert a negative influence on neurodevelopment, potentially contributing to the etiology of some cases of autism. Alterations in appropriate regulation of the immune response may result in chronic inflammation, autoimmunity, or an inappropriate response to immune challenge in children with autism.[68] Furthermore, abnormally regulated immune responses may also contribute to neuroinflammation of the CNS or brain leading to altered neurodevelopment. Structural abnormalities have also been described in the cerebellum, hippocampus, amygdala, and insular cortex of autistic patients.[69] Animal studies indicate that stress reduces BDNF expression or activity in the hippocampus and that this reduction can be prevented by treatment with antidepressant drugs.[70] Recent evidence supports the view that pathogenesis of autism may involve not only genetic factors, but supported by strong environmental components along with persistent neuroinflammation.[71,72] Studies on CSF of autistic subjects have indicated that autism is characterized by a neuroinflammatory response, regardless of age (in patients between 5 and 46 years of age), involving excess microglial activation and increased proinflammatory cytokine profiles. Determination of cytokines in brain tissue of autistic subjects indicates high levels of cytokines in the brain.[73,74] Other immune abnormalities such as reduction in TGF-β1 in autistic subjects versus controls along with high cytokines levels are also observed. Alterations in cytokines and reduction in TGF-β1 may contribute to irritability, lethargy, stereotypy, and hyperactivity as well as with levels of social adaptability.[75] Collective evidence suggests that autism is a neuropsychiatric disorder of unknown pathogenesis. Autism is characterized by neuroinflammation, peripheral immune abnormalities, and environmental factors. Interactions among these factors may explain the symptomatology of autism.

Trace amines and their receptors (TAARs) have been reported to contribute to some aspects of neuropsychiatric diseases (depression, autism, and attention deficit hyperactivity disorder).[76] Nothing is known about the contribution of TAAR in neurotraumatic and neurodegenerative diseases. However, depression usually develops following stroke, SCI, and TBI, as well as neurodegenerative diseases. It remains to be seen whether TAARs contribute to the depressive behavior following neurotraumatic insults and neurodegenerative diseases.

CONCLUSION

Neurological disorders constitute a group of brain and spinal cord diseases characterized by a progressive deterioration of structure and/or function of neuronal cells. Neurological disorders not only show different symptoms, but may be caused by a multitude of unknown causes and factors. Most neurological disorders are accompanied by induction of oxidative stress and onset of neuroinflammation along with mitochondrial and proteasome system dysfunctions. Neurological disorders are classified into three groups namely neurotraumatic diseases, neurodegenerative diseases, and neuropsychiatric diseases. The major basic mechanisms leading to neurodegeneration are multifactorial, caused by genetic, environmental and endogenous factors related to aging. Aging is the most important nonmodifiable risk factor for stroke, AD, PD, and HD.

References

1. Farooqui AA. *Neurochemical aspects of neurotraumatic and neurodegenerative diseases*. New York, NY: Springer; 2010.
2. Deleidi M, Jäggle M, Rubino G. Immune aging, dysmetabolism, and inflammation in neurological diseases. *Front Neurosci* 2015;9:172.

3. Corvin AP, Molinos I, Little G, Donohoe G, Gill M, Morris DW, et al. Insulin-like growth factor 1 (IGF1) and its active peptide (1-3)IGF1 enhance the expression of synaptic markers in neuronal circuits through different cellular mechanisms. *Neurosci Lett* 2012;**520**:51–6.

4. Williams AJ, Umemori H. The best-laid plans go of awry: synaptogenic growth factor signaling in neuropsychiatric disease. *Front Synaptic Neurosci* 2014;**6**:4.

5. Farooqui AA. *Inflammation and oxidative stress in neurological disorders*. New York, NY: Springer; 2014.

6. Allan SM, Tyrrell PJ, Rothwell NJ. Interleukin-1 and neuronal injury. *Nat Rev Immunol* 2005;**5**:629–40.

7. Iadecola C, Anrather J. The immunology of stroke: from mechanisms to translation. *Nat Med* 2011;**17**:796–808.

8. Farooqui AA, Horrocks LA. *Glycerophospholipids in the brain: phospholipases A_2 in neurological disorders*. New York, NY: Springer; 2007.

9. Fehlings MG, Tighe A. Spinal cord injury: the promise of translational research. *Neurosurg Focus* 2008;**25**:E1.

10. Witiw CD, Fehlings MG. Acute spinal cord injury. *J Spinal Disord Tech* 2015;**28**:202–10.

11. Klussmann S, Martin-Villalba A. Molecular targets in spinal cord injury. *J Mol Med (Berl)* 2005;**83**:657–71.

12. Sezer N, Akkuş S, Uğurlu FG. Chronic complications of spinal cord injury. *World J Orthop* 2015;**6**:24–33.

13. Maas AI, Stocchetti N, Bullock R. Moderate and severe traumatic brain injury in adults. *Lancet Neurol* 2008;**7**:728–41.

14. Motori E, Puyal J, Toni N, Ghanem A, Angeloni C, et al. Inflammation-induced alteration of astrocyte mitochondrial dynamics requires autophagy for mitochondrial network maintenance. *Cell Metabolism* 2013;**18**:844–59.

15. Rostami E, Krueger F, Plantman S, Davidsson J, Agoston D, et al. Alteration in BDNF and its receptors, full-length and truncated TrkB and p75(NTR) following penetrating traumatic brain injury. *Brain Res* 2014;**1542**:195–205.

16. Farkas O, Povlishock JT. Cellular and subcellular change evoked by diffuse traumatic brain injury: a complex web of change extending far beyond focal damage. *Prog Brain Res* 2007;**2007**(161):43–59.

17. Fleminger S, Oliver DL, Lovestone S, Rabe-Hesketh S, Giora A. Head injury as a risk factor for Alzheimer's disease: the evidence 10 years on; a partial replication. *J Neurol Neurosurg Psychiatry* 2003;**74**:857–62.

18. Bower JH, Maraganore DM, Peterson BJ, McDonnell SK, Ahlskog JE, Rocca WA. Head trauma preceding PD: a case-control study. *Neurology* 2003;**60**:1610–5.

19. Chen H, Richard M, Sandler DP, Umbach DM, Kamel F. Head injury and amyotrophic lateral sclerosis. *Am J Epidemiol* 2007;**166**:810–6.

20. Graeber MB, Moran LB. Mechanisms of cell death in neurodegenerative diseases: fashion, fiction, and facts. *Brain Pathol* 2002;**12**:385–90.

21. Mannella CA. Structural diversity of mitochondria: functional implications. *Ann NY Acad Sci* 2008;**1147**:171–9.

22. Meadowcroft MD, Connor JR, Yang QX. Cortical iron regulation and inflammatory response in Alzheimer's disease and $APP_{SWE}/PS1_{\Delta E9}$ mice: a histological perspective. *Front Neurosci* 2015;**9**:255.

23. Andersen HH, Johnsen KB, Moos T. Iron deposits in the chronically inflamed central nervous system and contributes to neurodegeneration. *Cell Mol Life Sci* 2014;**71**:1607–22.

24. Farooqui AA. *Therapeutic potentials of Curcumin for Alzheimer's disease*. Springer, New York, NY, 2016.

25. Selkoe DJ. Folding proteins in fatal ways. *Nature* 2003;**426**:900–4.

26. Bates G. Huntingtin aggregation and toxicity in Huntington's disease. *Lancet* 2003;**361**:1642–4.

27. Jellinger KA. Recent advances in our understanding of neurodegeneration. *J Neural Transm* 2009;**116**:1111–62.

28. Farooqui AA. *Metabolic syndrome: an important risk factors for stroke, Alzheimer's disease, and depression*. New York, NY: Springer; 2013.

29. Farooqui AA. *High calorie diet and the human brain*. Switzerland: Springer, International Publishing; 2015.

30. Yoshinaga N, Yasuda Y, Murayama T, Nomura Y. Possible involvement of cytosolic phospholipase A(2) in cell death induced by 1-methyl-4-phenylpyridinium ion, a dopaminergic neurotoxin, in GH3 cells. *Brain Res* 2000;**855**:244–51.

31. Klivenyi P, Kiaei M, Gardian G, Calingasan NY, Beal MF. Additive neuroprotective effects of creatine and cyclooxygenase 2 inhibitors in a transgenic mouse model of amyotrophic lateral sclerosis. *J Neurochem* 2004;**88**:576–82.

32. Thomas EA. Striatal specificity of gene expression dysregulation in Huntington's disease. *J Neurosci Res* 2006;**84**:1151–64.

33. Rocha NP, de Miranda AS, Teixeira AL. Insights into neuroinflammation in Parkinson's disease: from biomarkers to anti-inflammatory based therapies. *Biomed Res Int* 2015;**2015**:628192.

34. Morreale MK. Huntington's disease: looking beyond the movement disorder. *Adv Psychosom Med* 2015;**34**:135–42.

35. Roberson ED, Scearce-Levie K, Palop JJ, Yan F, Cheng IH, et al. Reducing endogenous tau ameliorates amyloid β-induced deficits in an Alzheimer's disease mouse model. *Science* 2007;**316**:750–4.

36. Farooqui AA, Ong WY, Farooqui T. Lipid mediators in the nucleus: their potential contribution to Alzheimer's disease. *Biochim Biophys Acta* 2010;**1801**:906–16.

37. Samii A, Nutt JG, Ransom BR. Parkinson's disease. *Lancet* 2004;**363**:1783–93.

38. Jellinger KA. The pathomechanisms underlying Parkinson's disease. *Expert Rev Neurother* 2014;**14**:199–215.

39. Uversky VN, Eliezer D. Biophysics of Parkinson's disease: structure and aggregation of alpha-synuclein. *Curr Protein Pept Sci* 2009;**10**:483–99.

40. Volles MJ, Lee SJ, Rochet JC, Shtilerman MD, Ding TT, et al. Vesicle permeabilization by protofibrillar alpha-synuclein: implications for the pathogenesis and treatment of Parkinson's disease. *Biochemistry* 2001;**40**:7812–9.

41. Li J, Zhu M, Rajamani S, Uversky VN, Fink AL. Rifampicin inhibits alpha-synuclein fibrillation and disaggregates fibrils. *Chem Biol* 2004;**11**:1513–21.

42. Rubinsztein DC, Leggo J, Coles R, Almqvist E, Biancalana V, et al. Phenotypic characterization of individuals with 30–40 CAG repeats in the Huntington's disease (HD) gene reveals HD cases with 36 repeats and apparently normal elderly individuals with 36–39 repeats. *Am J Hum Genet* 1996;**59**:16–22.

43. Rigamonti D, Sipione S, Goffredo D, Zuccato C, Fossale E, Cattaneo E. Huntingtin's neuroprotective activity occurs via inhibition of procaspase-9 processing. *J Biol Chem* 2001;**276**:14545–8.

44. Luo S, Rubinsztein DC. Huntingtin promotes cell survival by preventing Pak2 cleavage. *J Cell Sci* 2009;**122**:875–85.

45. Finkbeiner S. Huntington's disease. *Cold Spring Harb Perspect Biol* 2011;**3**:a007476.

46. Gil JM, Rego AC. Mechanisms of neurodegeneration in Huntington's disease. *Eur J Neurosci* 2008;**27**:2803–20.

47. Hipp MS, Patel CN, Bersuker K, Riley BE, Kaiser SE, et al. Indirect inhibition of 26S proteasome activity in a cellular model of Huntington's disease. *J Cell* 2012;**196**:573–87.

48. Rudinskiy N, Kaneko YA, Beesen AA, Gokce O, Regulier E, et al. Diminished hippocalcin expression in Huntington's disease brain does not account for increased striatal neuron vulnerability as assessed in primary neurons. *J Neurochem* 2009;**111**:460–72.

49. Sreedharan J, Brown Jr. RH. Amyotrophic lateral sclerosis: problems and prospects. *Ann Neurol* 2013;**74**:309–16.

50. Rothstein JD. Current hypotheses for the underlying biology of amyotrophic lateral sclerosis. *Ann. Neurol* 2009;**65**(Suppl. 1):S3–S9.

51. Barber SC, Shaw PJ. Oxidative stress in ALS: key role in motor neuron injury and therapeutic target. *Free Radic Biol Med* 2010;**48**:629–41.

52. Rao SD, Weiss JH. Excitotoxic and oxidative cross-talk between motor neurons and glia in ALS pathogenesis. *Trends Neurosci* 2004;**27**:17–23.

53. Supattapone S. Prion protein conversion in vitro. *J Mol Med* 2004;**82**:348–56.

54. Jain S, Udgaonkar JB. Evidence for stepwise formation of amyloid fibrils by the mouse prion protein. *J Mol Biol* 2008;**382**:1228–41.

55. Ishikura N, Clever JL, Bouzamondo-Bernstein E, Samayoa E, Prusiner SB, et al. Notch-1 activation and dendritic atrophy in prion disease. *Proc Natl Acad Sci USA* 2005;**102**:886–91.

56. Resenberger UK, Harmeier A, Woerner AC, Goodman JL, Muller V, et al. The cellular prion protein mediates neurotoxic signalling of beta-sheet-rich conformers independent of prion replication. *EMBO J* 2011;**30**:2057–70.

57. Singh A, Qing L, Kong Q, Singh N. Change in the characteristics of ferritin induces iron imbalance in prion disease affected brains. *Neurobiol Dis* 2012;**45**:930–8.

58. Lovell MA, Robertson JD, Teesdale WJ, Campbell JL, Markesbery WR. Copper, iron and zinc in Alzheimer's disease senile plaques. *J Neurol Sci* 1998;**158**:47–52.

59. Turnbull S, Tabner BJ, Brown DR, Allsop D. Copper-dependent generation of hydrogen peroxide from the toxic prion protein fragment PrP106-126. *Neurosci Lett* 2003;**336**:159–62.

60. Morris G, Berk M. The many roads to mitochondrial dysfunction in neuroimmune and neuropsychiatric disorders. *BMC Med* 2015;**13**:68.

61. Luscher B, Shen Q, Sahir N. The GABAergic deficit hypothesis of major depressive disorder. *Mol Psychiatry* 2011;**16**:383–406.

62. Blitzer RD, Iyengar R, Landau EM. Postsynaptic signaling networks: cellular cogwheels underlying long-term plasticity. *Biol Psychiatry* 2005;**57**:113–9.

63. Davidson RJ, Pizzagalli D, Nitschike JB, Putnam K. Depression: perspectives from affective neuroscience. *Annu Rev Psychol* 2002;**53**:545–74.
64. Anisman II, Matheson K. Stress, depression, and anhedonia: caveats concerning animal models. *Neurosci. Biobehav Rev* 2005;**29**:525–46.
65. Dowlati Y, Herrmann N, Swardfager W, Liu H, Sham L, et al. A meta-analysis of cytokines in major depression. *Biol Psychiatry* 2010;**67**:446–57.
66. Sullivan PF, Neale MC, Kendler KS. Genetic epidemiology of major depression: review and meta-analysis. *Am J Psychiatry* 2000;**157**:1552–62.
67. Wurtman RJ. Genes, stress, and depression. *Metabolism* 2005;**54**(Suppl. 1):16–19.
68. Ashwood P, Wills S, Van de Water J. The immune response in autism: a new frontier for autism research. *J Leukoc Biol* 2006;**80**:1–15.
69. Pardo CA, Eberhart CG. The neurobiology of autism. *Brain Pathol* 2007;**17**:434–47.
70. Castrén E, Võikar V, Rantamäki T. Role of neurotrophic factors in depression. *Curr Opin Pharmacol* 2007;**7**:18–21.
71. Hallmayer J, Cleveland S, Torres A, Phillips J, Cohen B, et al. Genetic heritability and shared environmental factors among twin pairs with autism. *Arch Gen Psychiatry* 2011;**76**:v1.
72. Vargas DL, Nascimbene C, Krishnan C, Zimmerman AW, Pardo CA. Neuroglial activation and neuroinflammation in the brain of patients with autism. *Ann Neurol* 2005;**57**:67–81.
73. Tetreault NA, Hakeem AY, Jiang S, Williams BA, Allman E, Wold BJ, et al. Microglia in the cerebral cortex in autism. *J Autism Dev Disord* 2012;**42**:2569–84.
74. Pardo CA, Vargas DL, Zimmerman AW. Immunity, neuroglia and neuroinflammation in autism. *Int Rev Psychiatry* 2005;**17**:485–951.
75. Ashwood P, Enstrom A, Krakowiak P, Hertz-Picciotto I, Hansen RL, et al. Decreased transforming growth factor beta1 in autism: a potential link between immune dysregulation and impairment in clinical behavioral outcomes. *J Neuroimmunol* 2008;**204**:149–53.
76. Cichero E, Espinoza S, Gainetdinov RR, Brasili L, Fossa P. Insights into the structure and pharmacology of the human trace amine-associated receptor 1 (hTAAR1): homology modelling and docking studies. *Chem Biol Drug Des* 2013;**81**:509–16.

17

Trace Amines and Their Relevance to Neurological Disorders: A Commentary

T. Farooqui[1] and A.A. Farooqui[2]

[1]Department of Entomology, The Ohio State University, Columbus, OH, United States [2]Department of Molecular and Cellular Biochemistry, The Ohio State University, Columbus, OH, United States

INTRODUCTION

Neurological disorders are diseases of the central and peripheral nervous systems, which occur due to structural, biochemical or electrical abnormalities in the brain, spinal cord, cranial and peripheral nerves, and autonomic nervous system, resulting in a range of symptoms, such as loss of sensation, confusion, poor coordination, altered consciousness, and seizures. Neurological disorders include neurotraumatic (stroke, spinal cord injury, traumatic brain injury, and epilepsy); neurodegenerative (Alzheimer disease (AD), Parkinson disease (PD), Huntington disease (HD), amyotrophic lateral sclerosis (ALS), and multiple sclerosis (MS)); and neuropsychological (depression, dementia, schizophrenia, and bipolar) diseases (Table 17.1). Stroke, spinal cord trauma, and traumatic brain injury are caused by

TABLE 17.1   Classification of Neurological Disorders

Neurotraumatic Diseases	Neurodegenerative Diseases	Neuropsychiatric Diseases
Stroke	Alzheimer disease	Depression
Spinal cord injury	Parkinson disease	Schizophrenia
Traumatic brain injury	Amyotrophic lateral sclerosis	Bipolar disorders
Epilepsy	Prion diseases	Autism
	Multiple sclerosis	Mood disorders
		Attention deficit disorders
		Dementia

metabolic and mechanical trauma to brain and spinal cord. Neurodegenerative diseases are triggered by the accumulation of protein aggregates formed from abnormally modified proteins (excessive misfolding), induction of oxidative stress, and onset of neuroinflammation. These processes contribute to a progressive loss of neurons in an age-dependent manner. The exact mechanisms of abnormal folding are not fully understood, however, speculation leads to the presumption that metabolic, genetic and environmental factors are closely associated with the pathogenesis of neurodegenerative diseases.[1] Some risk factors for neurodegenerative diseases are modifiable while others are and nonmodifiable. The modifiable risk factors include unhealthy lifestyle (long-term consumption of Western diet, lack of exercise, and lack of sleep) whereas, nonmodifiable risk factors include factors, such as old age, gender, race/ethnicity, and heredity/genetic (family history) for neurological disorders.[2,3]

Neuropsychiatric diseases are neurodevelopmental disorders, which are accompanied by dysregulation in the neuronal network, leading to attentional biases, alterations in learning (aberrant learning), and an absence of top-down cognitive control by the prefrontal cortex along with onset of mild oxidative stress and neuroinflammation.[4] Neuroimaging studies have indicated alterations in cerebral blood flow and glucose utilization in the limbic system and prefrontal cortex of patients, contributing to the pathogenesis of neuropsychiatric diseases.[5] Molecular mechanisms associated with the pathogenesis of neurological diseases remain elusive. The present commentary focuses on the possible link between trace amines and neurological disorders.

ASSOCIATION BETWEEN TRACE AMINES AND NEUROLOGICAL DISORDERS

Trace amines have been established to exist in the mammalian brain for more than 40 years.[6] They belong to a family of endogenous amines, primarily consisting of β-phenylethylamine (β-PEA), m- and p-tyramine (TA), m- and p-octopamine (OA), synephrine (SYN), and tryptamine (TRY), which are structurally and metabolically related to classical monoamine neurotransmitters, such as dopamine (DA), norepinephrine (NE), and serotonin or 5-hydroxytryptamine (5-HT), and share common metabolic pathways with

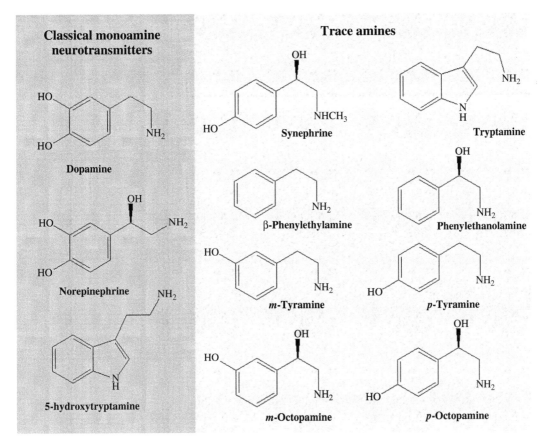

FIGURE 17.1 Chemical structures of classical monoamines, neurotransmitters, and trace amines.

these neurotransmitters (Fig. 17.1).[7–11] Trace amines can be synthesized within parent mono-amine neurotransmitter systems, involving decarboxylation of precursor amino acids by the enzyme aromatic L-amino acid decarboxylase (AAAD).[9] However, an additional member of this family, in particular 3-iodothyronamine (T_1AM), is an endogenous compound with chemical features similar to thyroid hormone (T_4). T_1AM has a carbon skeleton identical to T_4 and contains a single carbon–iodine bond. T_1AM is produced by enzymatic decarboxylation and deiodination (Fig. 17.2). T_1AM exerts distinct biological effects, including decrease in heart rate, cardiac output, body temperature, and metabolic rate, which are largely opposite to those produced on a longer timescale by thyroid hormone.[12] There are structural similarities among thyroxine, T_1AM, and monoamine neurotransmitters, implicating an intriguing role for T_1AM as both a neuromodulator and a hormone-like molecule constituting a part of thyroid hormone action.[12] T_1AM is not a ligand for nuclear thyroid hormone receptors, but stimulates with nanomolar affinity trace amine-associated receptor 1 (TAAR1), and other molecular targets.[12] It produces significant metabolic and neurological

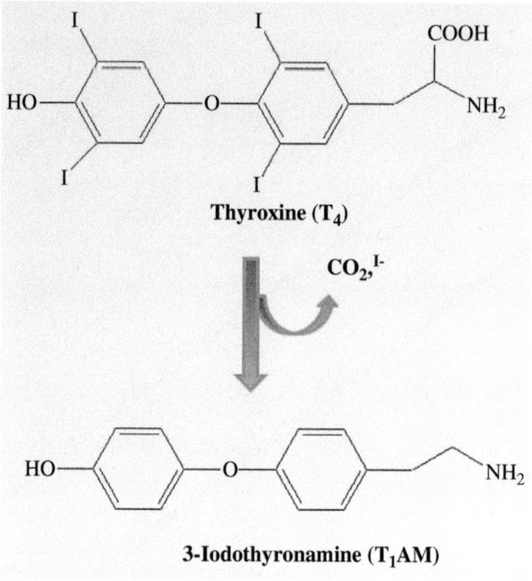

FIGURE 17.2   Biosynthesis of 3-iodothyyronamine from thyroxine.

effects, affecting adrenergic and/or histaminergic neurons.[13] Moreover, intracerebral T_1AM administration has been reported to improve learning ability, modulate sleep and feeding, and reduce pain threshold to hot stimuli in mice, implicating its potential use for treating endocrine and neurodegenerative-induced memory disorders.[14]

Under physiological conditions, endogenous levels of trace amines are several hundred-fold below those of monoamine neurotransmitters (DA, NE, and 5-HT) in the central and peripheral nervous systems of vertebrates due to high rates of metabolism. At low concentrations trace amines maintain the neuronal activity of monoamine neurotransmitters within defined physiological limits and play a significant role in biogenic monoamine-based synaptic physiology. However, altered levels of these amines have been reported in plasma or urine excretion of many patients suffering from various human brain disorders, such as schizophrenia, depression, PD, attention deficit hyperactivity disorder (ADHD), and migraine.[7,11,15] Contrary to the situation in vertebrates, biogenic amine neurotransmitters, such as NE and epinephrine (EPI), are essentially lacking in invertebrates and are replaced by p-TA and p-OA that act as bonafide neurotransmitters/neuromodulators/neurohormones, mediating diverse complex behaviors and many vital function.[16–20]

Trace amines are endogenous amines that are implicated in several physiological processes including modulation of aminergic neurotransmission. Trace amines mediate their effect by binding and activating a class of G-protein-coupled receptors called Trace amine-associated receptors (TAARs). The binding of trace amines with TAAR1 in the brain elevates intracellular cyclic adenosine monophosphate (cAMP) levels.[21] This second messenger is closely associated with cAMP-dependent protein kinases A (PKA), and cAMP response element-binding protein (CREB) signaling pathways.[22] Most of the studies have focused on studying the role of TAAR1 on modulation of the DA transmission.[23] DA supersensitivity is a common feature of schizophrenia, which also occurs in genetically altered, drug-altered, and lesion-altered animals. It is correlated with high-affinity D_2 receptors, suggesting that psychosis pathways converge via high-affinity D_2 receptors.[24] Genetic studies using TAAR1-knockout (TAAR1-KO) mice display a deficit in prepulse inhibition, lack of agonistic effect of amphetamine, increased sensitivity to the psychomotor stimulatory effects of amphetamine temporally correlated with significantly larger increase in the release of DA and NE in the dorsal striatum compared with wildtype, and significant increase in striatal high-affinity D_2 receptor expression, supporting the role of TAAR1 in the modulation of dopaminergic (DA-ergic) activity.[25,26] Thus TAAR1-KO mice may be used as a potential model with relevance to schizophrenia. Moreover, TAAR1-KO mice show acute DA deficiency, exploring the role of TAAR1 in DA-dependent movement control, implicating the potential treatment of PD with TAAR1 antagonists.[27] Double knockout mice, lacking TAAR1 as well as DA transporter, display a significantly enhanced level of spontaneous activity compared to DA transporter knockout mice, suggesting that specific TAAR1-targeted drugs also influence DA-induced movement or hyperactivity.[27,28] In addition to DA, TAAR1 also regulates noradrenergic and serotonergic neurotransmission in brain.[29,30] Thus multiple risk factors (environmental and genetic) as well as multiple neurotransmitter and neuromodulator signaling pathways may be involved with schizophrenia.[31]

Aromatic amino acids, such as phenylalanine, tyrosine, and tryptophan, are precursors of several neurotransmitters, trace amines, and hormones. Many inherited disorders affecting aromatic amino acid metabolism may lead to deficiencies of neurotransmitters, including DA, NE, and 5-HT, in the central nervous system, resulting in most neurological disorders, including depression (Fig. 17.3).[32] Diminished monoamine function has been previously shown to be significantly associated with clinical depression. However, catecholamine (DA, NE) or indoleamine (5-HT) depletion alone has not been associated with significant mood changes in unmedicated depressed subjects or never-depressed control subjects.[33] This supports the notion that monoamines regulate mood in actively depressed patients via indirect mechanisms. There are multiple physical symptoms of neurological disorders such as cognitive dysfunction, partial or complete paralysis, dyskinesia, muscle weakness, seizures, and partial or complete loss of sensation. Depression (disturbed thoughts, emotions, and behavior) is common in patients with chronic neurological disorders.[33–35] However, the etiology of depression is complex and may be multifactorial due to involvement of genetic, biological, and psychosocial factors in individual patients. The association between most neurological disorders and aromatic amino acid metabolites, showing depression as a common characteristic, has been routinely reported in the literature.[32–35] Depression can either occur as a primary neurological disorder or as an early symptom, as well as a consequence of mental diseases. Moreover, it can also occur as a secondary effect, such as a side effect of certain prescribed medications.

The association between phenylalanine, tyrosine, and tryptophan-derived classical monoamine neurotransmitters, trace amines, and neurological disorders with their clinical relevance is presented in Table 17.2.[2–30,35–56]

Briefly, neural transmission abnormalities in neuropsychiatric disorders, such as depression, schizophrenia, and addiction may be associated with abnormalities of serotonergic, dopaminergic, noradrenergic, cholinergic, glutamatergic, GABA-ergic, glucocorticoid and peptidergic functions.[36] Among the above neuropsychiatric disorders, important mechanisms of depression involve interplay (crosstalk) among serotonergic, noradrenergic, and glutamatergic neurotransmission and tryptophan–kynurenine metabolism along with immune system activation (Fig. 17.3).

Tryptophan is an essential amino acid with an indole ring structure, which is obtained from the dietary source and the reference value of plasma tryptophan ranges from 45–60 µmol/L.[57] The activation of indoleamine 2,3-dioxygenase (IDO) contributes to the symptoms of depression. This enzyme is expressed in non-neural cells (fibroblasts, dendritic cells, monocytes, and macrophages) and neural cells (microglial cells). In brain microglial cells express and secrete a number of cytokines such as IFN-γ, IFN-α, and tumor necrosis factor-α (TNF-α) either alone or in combination with astrocytes.[58] Cytokine-mediated activation of IDO results in transformation of L-tryptophan into L-kynurenine (KYN), thereby reducing the availability of tryptophan, which not only plays an important role in the

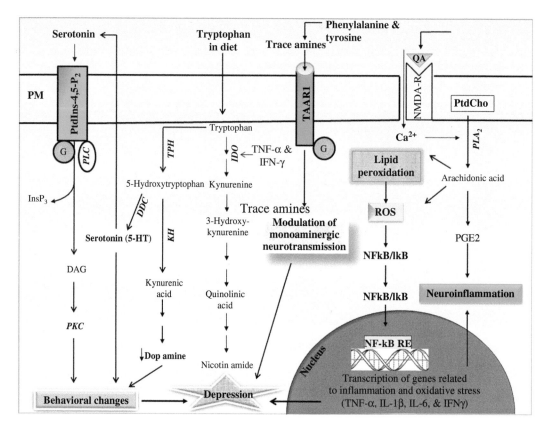

FIGURE 17.3   Interactions among serotonergic, dopaminergic and glutamatergic neurotransmission and tryptophan–kynurenine metabolism in depression. *NMDA-R*, N-methyl-D-aspartate receptor; *PtdCho*, phosphatidylcholine; *cPLA$_2$*, cytosolic phospholipase A$_2$; *PGE2*, prostaglandin E2; *ROS*, reactive oxygen species; *PLC*, phospholipase C; *IDO*, indolamine 2,3-dioxygenase; *TPH*, tryptophan hydroxylase; *KMO*, kynurenine 3-monooxygenase; *TNF-α*, tumor necrosis factor-α; *IL-1β*, interleukin-1beta; *INF-γ*, interferon-gamma; *DAG*, diacylglycerol; and *InsP$_3$*, inositol 1,4,5-P$_3$.

TABLE 17.2 Phenylalanine, Tyrosine, and Tryptophan-Derived Classical Monoamine Neurotransmitters and Trace Amines in Neurological Disorders

Diseases	Classical Monoamine Neurotransmitters and Trace Amines	Clinical Relevance
Parkinson (PD)	Nigrostriatal DA depletion	Progressive loss of dopamine neurons in SNc results in DA deficiency and clinical severity of disease[36]
PD	↓ OA in plasma levels in early PD	↓ OA plasma levels may be used as sign for early and possibly oncoming PD[37]
PD	– Decreased urinary β-PEA excretion in PD patients – Administration of β-PEA to rodents reduces striatal DA content and induces movement disorders	– Drugs that relieve or produce depression and PD resulted in increased or decreased β-PEA levels, respectively[38] – β-PEA potentially contributes to the progressive development of PD[39]
PD	Acute DA deficiency in TAAR1-KO mice	TAAR1 is a regulator of dopaminergic neurotransmission, and TAAR1 antagonists could be beneficial for treating PD[23,26,27]
Alzhimer disease (AD)	Degeneration of the locus coeruleus (main source of NE)	β-Amyloid containing plaques, tau-containing neurofibrillary tangles, and cholinergic neuronal loss[40]
Schizophrenia	– ↑ DA (hyperdopaminergia) – DA dysregulation – Hypoglutamatergic activity	– TAAR1 opens treatment opportunities for psychiatric disorders[30]
Acute schizophrenia	↑ β-PEA plasma levels in patients suffering from acute schizophrenia	β-PEA neuromodulates catechol aminergic transmission.[11,21,41] It is independently associated with various neuropsychiatric disorders
Paranoid schizophrenia	Excess urinary excretion of β-PEA	– β-PEA acts as an endogenous amphetamine[11,21,28,41–43] – β-PEA is a "safe" alternative to drugs (such as amphetamine or methylphenidate) because they are accompanied by side effects
Addiction	TAAR1-KO mice has greater sensitivity to amphetamine, ↑ sensitivity to DA-ergic activation; TAAR 1 agonists ↓ the neurochemical effects of cocaine and amphetamines	TAAR1 agonists have potential for treating psychostimulant addiction[29,44]
ADHD	↓ Urinary β-PEA levels in ADHD patients in comparison to controls	Symptoms are hyperactivity and poor concentration. Amphetamine and methylphenidate increase β-PEA biosynthesis[45]

(*Continued*)

TABLE 17.2   (Continued)

Diseases	Classical Monoamine Neurotransmitters and Trace Amines	Clinical Relevance
ADHD	– ↓ Urine and plasma levels of β-PEA metabolite "phenylacetic acid" – Precursors β-PEA and tyrosine are found along with decrease in plasma TA	Following methylphenidate tratment, ADHD patients show a normalization of urinary β-PEA[45–47]
Primary headaches (migraine and cluster headache)	High level of DA, low level of NE, and elevated levels of OA and SYN in episodic MWoA and CH	Abnormal biogenic amine metabolism is a characteristic biochemical trait found in headache sufferers[48,49]
Migraine headache Chronic migraine	– Imbalance in circulating 5-HT levels and its metabolites during attack – Low tryptamine plasma levels	– Selective $5\text{-HT}_{1B/1D}$ agonists are effective in treating acute migraine headache[50–52] – Common insufficient 5-HT-ergic control of the pain threshold[49]
Stroke	– T_1AM and T_0AM induce transient hypothermia in rodents – 5-HT deficiency	– T_1AM and T_0AM can serve as neuroprotectant against acute stroke[53] – Prestroke SSRI increases stroke severity and mortality in hemorrhagic stroke patients, but was not found in ischemic stroke[54]
Depression	– β-PEA deficiency, and ↓ PAA levels in urine, plasma and CSF	The administration of β-PEA or of its precursor L-Phe improves mood in depressed patients, implicating a relationship between β-PEA-ergic system and depression[55,56]

NT, neurotransmitter; NM, neuromodulator; SNc, substantia nigra compata; NE, norepinephrine; β-PEA, β-phenylethylamine; PAA, phenylacetic acid; DA, dopamine; OA, octopamine; SYN, synephrine; DA-ergic, dopaminergic; TAAR1, trace amine-associated receptor 1; KO, knockout; ADHD, attention deficit hyperactivity disorder; CSF, cerebrospinal fluid; MWoA, migraine without aura; CH, cluster headache; 5-HT, serotonin; thyroxine derivatives: T_0AM, thyronamine and T_1AM, 3-iodothyronamine; L-Phe, L-phenylalanine; SSRI, selective serotonin reuptake inhibitor.

regulation of T cells,[59] but also is a precursor for serotonin (5-HT), a monoamine that is believed to play a prominent role in the neurobiology of mood disorders.[60,61] Increase in levels of proinflammatory cytokines also induces the production of neurotoxic end products of the tryptophan–kynurenine pathway, such as 3-hydroxykynurenine and quinolinic acid. 3-Hydroxy-kynurenine is neurotoxic because of its ability to generate superoxide radicals and quinolinic acid. This metabolite acts as an agonist for NMDA receptor. It is also reported that quinolinic acid inhibits glutamate uptake by astrocytes, leading to an accumulation of glutamate in the microenvironment inducing slow excitotoxicity, a process that may contribute to the pathogenesis of depression.[62]

Phenylalanine is another essential amino acid that is converted into tyrosine. Phenylalanine, like tyrosine, is also a substrate for the enzyme tyrosine hydroxylase, which catalyzes the rate-limiting step in catecholamine synthesis. Tyrosine is the preferred substrate; consequently, unless tyrosine levels are abnormally low, variations in phenylalanine levels do not affect catecholamine synthesis. Unlike tyrosine, phenylalanine does not cause substrate inhibition.[63] Tyrosine in turn is converted into L-DOPA, which is rapidly decarboxylated to DA by AAAD. In neurons that use DA as a neurotransmitter, no further enzymatic modification occurs; however, neurons using NE as a neurotransmitter contain an additional enzyme, DA-β-hydroxylase (DBH), which converts DA to NE, whereas neurons using epinephrine as a neurotransmitter contain the enzyme phenylethanolamine-N-methyl transferase (PNMT), which catalyzes the conversion of NE to EPI.[63] Thus, tyrosine is converted into monoamine-signaling molecules. The deficiency and imbalances mainly in the monoamine neurotransmitters (5-HT, DA, and NE) may be the cause of depression.[64–68] There is also evidence indicating that trace amines, such as β-PEA, may contribute to mood disorders (Table 17.2).[55,56] Thus, depression probably occurs due to a combination of biological, psychological, and social factors; therefore, it cannot be explained on the basis of one single neurotransmitter system, but rather associated with complex interactions among different signaling pathways. The current therapies are only effective in a fraction of patients; therefore, one should extend the focus of depression therapy including nonmonoaminergic systems. Similarly, abnormal chemical imbalance (excess or deficiency in the neurotransmitters, neuromodulators, and hormones levels), absence of corresponding genes, low enzyme activity, disrupted metabolism, and impaired signaling may lead to a variety of neurological disorders having depression as a common characteristic among them (Table 17.2).

Nothing is clearly known about the contribution of trace amines in the pathogenesis of neurotraumatic and neurodegenerative diseases. It is becoming increasingly evident that TAAR1 is a negative regulator of dopamine transmission making TAAR1 a novel target for neuropsychiatric disorders that arises from dopamine dysfunction (schizophrenia, depression, and drug addiction).[22] TAAR1 also modulates NMDA receptor-mediated glutamate transmission in the prefrontal cortex and related functions, suggesting TAAR1-based drugs could provide a novel therapeutic approach for the treatment of disorders related to aberrant cortical functions.[69] TAAR1 expression is concomitant with lymphocyte immune activation, suggesting its role in modulation of immune system.[70] Moreover, methamphetamine and β-PEA (TAAR1 agonists) increase intracellular cAMP in human astrocytes, and modulate glutamate clearance abilities.[71] TAAR1 may play an important role in psychostimulant action, thus use of its antagonists may offer a potential treatment for psychostimulant abuse.

CONCLUSION

Collectively, trace amines are biogenic amines present in very low levels in mammalian tissues. They have defined roles as neurotransmitters in invertebrates, but the extent to which they function as true neurotransmitters in mammals remains unclear. TAAR1 is an amine-activated G protein (Gs and Gq) coupled receptor that is expressed in brain as well as in peripheral tissues, suggesting that it plays a role in neurological as well as non-neurological pathways. TAAR1 regulates DA, NE, and 5-HT neurotransmission in the CNS, and also regulates the immune system in lymphocytes. Trace amines play significant roles in coordinating biogenic monoamine-based synaptic physiology. Dysregulation of trace amines may also contribute to the etiology of depression and mania. There is increasing evidence that neuropsychiatric disorders are accompanied by an increase in oxidative stress, induction of inflammatory signaling, and slow immune responses in the brain tissue. Thus TAAR1 may open an opportunity to treat neurological disorders.

References

1. Farooqui AA. *Neurochemical aspects of neurotraumatic and neurodegenerative diseases.* New York, NY: Springer, International Publishing; 2010.
2. Farooqui AA. *Inflammation and oxidative stress in neurological disorders.* Switzerland: Springer, International Publishing; 2014.
3. Farooqui AA. *High calorie diet and the human brain: metabolic consequences of long term consumption.* Switzerland: Springer, International Publishing; 2015.
4. Sahakian BJ. What do experts think we should do to achieve brain health? *Neurosci Biobehav Rev* 2014;**43**:240–58.
5. Drevets WC, Price JL, Furey ML. Wayne brain structural and functional abnormalities in mood disorders: implications for neurocircuitry models of depression. *Brain Struct Funct* 2008;**213**:93–118.
6. Boulton AA. Identification, distribution, metabolism and function of meta and para tyramine, phenylethylamine and tryptamine in brain. *Adv Biochem Psychopharmacol* 1976;**15**:57–67.
7. Premont RT, Gainetdinov RR, Caron MG. Following the trace of elusive amines. *Proc Natl Acad Sci USA* 2001;**98**(17):9474–5.
8. Branchek TA, Blackburn TP. Trace amine receptors as targets for novel therapeutics: legend, myth and fact. *Curr Opin Pharmacol* 2003;**3**:90–7.
9. Berry MD. Mammalian central nervous system trace amines. Pharmacologic amphetamines, physiologic neuromodulators. *J Neurochem* 2004;**90**:257–71.
10. Burchett SA, Hicks TP. The mysterious trace amines: protean neuromodulators of synaptic transmission in mammalian brain. *Prog Neurobiol* 2006;**79**:223–46.
11. Berry MD. The potential of trace amines and their receptors for treating neurological and psychiatric diseases. *Re Recent Clin Trials* 2007;**2**(1):3–19.
12. Ianculescu AG, Scanlan TS. 3-Iodothyronamine (T(1)AM): a new chapter of thyroid hormone endocrinology? *Mol Biosyst* 2010;**6**(8):1338–44.
13. Zucchi R, Accorroni A, Chiellini G. Update on 3-iodothyronamine and its neurological and metabolic actions. *Front Physiol* 2014;**5**:402.
14. Manni ME, De Siena G, Saba A, Marchini M, Landucci E, et al. Pharmacological effects of 3-iodothyronamine (T1AM) in mice include facilitation of memory acquisition and retention and reduction of pain threshold. *Br J Pharmacol* 2013;**168**(2):354–62.
15. Narang D, Tomlinson S, Holt A, Mousseau DD, Baker GB. Trace amines and their relevance to psychiatry and neurology: a brief overview. *Bull Clin Psychopharmacol* 2011;**21**:73–9.
16. Axelrod J, Saavedra JM. Octopamine. *Nature* 1977;**265**:501–4.
17. David JC, Coulon JF. Octopamine in invertebrates and vertebrates. A review. *Prog Neurobiol* 1985;**24**:141–85.
18. Evans PD, Robb S. Octopamine receptor subtypes and their modes of action. *Neurochem Res* 1993;**18**:869–74.
19. Roeder T. Octopamine in invertebrates. *Prog Neurobiol* 1999;**59**:533–41.
20. Farooqui T. Review of octopamine in insect nervous systems. *Open Access Insect Physiol* 2012;**4**:1–17.
21. Lindemann L, Hoener MC. A renaissance in trace amines inspired by a novel GPCR family. *Trends Pharmacol Sci* 2005;**26**:274–81.
22. Borowsky B, Adham N, Jones KA, Raddatz R, Artymyshyn R, Ogozalek KL, et al. Trace amines: identification of a family of mammalian G protein-coupled receptors. *Proc Natl Acad Sci USA* 2001;**98**:8966–71.
23. Lam VM, Espinoza S, Gerasimov AS, Gainetdinov RR, Salahpour A. In-vivo pharmacology of trace-amine associated receptor 1. *Eur J Pharmacol* 2015;**763**(Pt B):136–42.
24. Seeman P, Schwarz J, Chen JF, Szechtman H, Perreault M, et al. Psychosis pathways converge via D2 high dopamine receptors. *Synapse* 2006;**60**(4):319–46.

25. Wolinsky TD, Swanson CJ, Smith KE, Zhong H, Borowsky B, et al. The trace amine 1 receptor knockout mouse: an animal model with relevance to schizophrenia. *Genes Brain Behav* 2007;**6**(7):628–39.

26. Lindemann L, Meyer CA, Jeanneau K, Bradaia A, Ozmen L, et al. Trace amine-associated receptor 1 modulates dopaminergic activity. *J Pharmacol Exp Ther* 2008;**324**(3):948–56.

27. Sotnikova TD, Zorina OI, Ghisi Caron MG, Gainetdino RR. Trace amine associated receptor 1 and movement control. *Parkinsonism Relat Disord* 2008;**14**(Suppl. 2):S99–S102.

28. Sotnikova TD, Budygin EA, Jones SR, Dykstra LA, Caron MG, et al. Dopamine transporter-dependent and -independent actions of trace amine beta-phenylethylamine. *J Neurochem* 2004;**91**:362–73.

29. Miller GM. The emerging role of trace amine-associated receptor 1 in the functional regulation of monoamine transporters and dopaminergic activity. *J Neurochem* 2011;**116**(2):164–76.

30. Revel FG, Moreau J-L, Gainetdinov RR, Bradaia A, Sotnikova TD, et al. TAAR1 activation modulates monoaminergic neurotransmission, preventing hyperdopaminergic and hypoglutamatergic activity. *Proc Natl Acad Sci USA* 2011;**108**(20):8485–90.

31. Howes OD, Kapur S. The dopamine hypothesis of schizophrenia: version III—the final common pathway. *Schizophrenia Bull* 2009;**35**(3):549–62.

32. Roiser JP, McLean A, Ogilvie AD, Blackwell AD, Bamber DJ, et al. The subjective and cognitive effects of acute phenylalanine and tyrosine depletion in patients recovered from depression. *Neuropsychopharmacology* 2005;**30**(4):775–85.

33. Berman RM, Sanacora G, Anand A, Roach LM, Fasula MK, et al. Monoamine depletion in unmedicated depressed subjects. *Biol Psychiatry* 2002;**51**:469–73.

34. Raskind MA. Diagnosis and treatment of depression comorbid with neurologic disorders. *Am J Med* 2008;**121**(11 Suppl. 2):S28–37.

35. Gilliam FG. Diagnosis and treatment of mood disorders in persons with epilepsy. *Curr Opin Neurol* 2005;**18**:129–33.

36. Alexander GE. Biology of Parkinson's disease: pathogenesis and pathophysiology of a multisystem neurodegenerative disorder. *Dialogues Clin Neurosci* 2004;**6**(3):259–80.

37. D'Andrea G, Nordera G, Pizzolato G, Bolner A, Colavito D, et al. Trace amine metabolism in Parkinson's disease: low circulating levels of octopamine in early disease stages. *Neurosci Lett* 2010;**469**(3):348–51.

38. Wolf ME, Mosnaim AD. Phenylethylamine in neuropsychiatric disorders. *Gen Pharmacol* 1983;**14**(4):385–90.

39. Borah A, Paul R, Mazumder MK, Bhattacharjee N. Contribution of β-phenethylamine, a component of chocolate and wine, to dopaminergic neurodegeneration: implications for the pathogenesis of Parkinson's disease. *Neurosci Bull* 2013;**29**(5):655–60.

40. Weinshenker D. Functional consequences of locus coeruleus degeneration in Alzheimer's disease. *Curr Alzheimer Res* 2008;**5**(3):342–5.

41. Milller GM. The emerging role of trace amine associated receptor 1 in the functional regulation of monoamine transporters and dopaminergic activity. *J Pharmacol Exp Ther* 2008;**325**:617–28.

42. Janssen PA, Levsen JE, Megen AA, Awouters FH. Does phenylethylamine act as an endogenous amphetamine in some patients? *Int J Neuropsychopharm* 1999;**2**:229–40.

43. O'Reilly RL, Dais BA. Phenylethylamine and schizophrenia. *Prog Neuropsychopharmacol Biol Psychiatry* 1994;**18**(1):63–75.

44. Jing L, Li JX. Trace amine-associated receptor 1: a promising target for the treatment of psychostimulant addiction. *Eur J Pharmacol* 2015;**1**:345–52.

45. Kusaga A. Decreased beta-phenylethylamine in urine of children with attention deficit hyperactivity disorder and autistic disorder. *No To Hattatsu* 2002;**34**(3):243–8.

46. Kusanga A, Yamashita Y, Koeda T, Hiratani M, Kaneko M, et al. Increased urinary phenylethylamine after methylphenidate treatment in children with ADHD. *Ann Neurol* 2002;**52**(3):372–4.

47. Baker GB, Bornstein RA, Rouget AC, Ashton SE, van Muyden JC, et al. Phenylethylaminergic mechanisms in attention-deficit disorder. *Biol Psychiatry* 1991;**29**(1):15–22.

48. D'Andrea G, Terrazzino S, Leon A, Fortin D, Perini F, et al. Elevated levels of circulating trace amines in primary headaches. *Neurology* 2004;**62**(10):1701–5.

49. D'Andrea G, Cevoli S, Colavito D, Leon A. Biochemistry of primary headaches: role of tyrosine and tryptophan metabolism. *Neurol Sci* 2015;**36**(Suppl. 1):17–22.

50. Ferrari MD, Goadsby P, Roon KI, Lipton RB. Triptans (serotonin, 5-HT1B/1D agonists) in migraine: detailed results and methods of a meta-analysis of 53 trials. *Cephalalgia* 2002;**22**(8):633–58.

51. Goadsby PJ. Serotonin receptor ligands: treatments of acute migraine and cluster headache. *Handb Exp Pharmacol* 2007;**177**:129–43.

52. Saper JR. The use of rizatriptan in the treatment of acute, multiple migraine attacks. *Neurology* 2000;**55**(9 Suppl. 2):S15–8.

53. Doye KP, Suchland KL, Ciesieski TMP, Lessoy NS, Grandy DK, et al. Novel thyroxine derivatives, thyronamine and 3-iodothyronamine, induce transient hypothermia and marked neuroprotection against stroke injury. *Stroke* 2007;**38**(9):2569–76.

54. Mortensen JK, Larsson H, Johnsen SP, Andersen G. Impact of prestroke selective serotonin reuptake inhibitor treatment on stroke severity and mortality. *Stroke* 2014;**45**:2121–3.

55. Davis BA, Kennedy SH, D'Souza J, Durden DA, Goldbloom DS, et al. Correlations of plasma and urinary phenylacetic acid and phenylethylamine concentrations with eating behavior and mood rating scores in brofaromine-treated women with bulimia nervosa. *J Psychiatry Neurosci* 1994;**19**(4):282–8.

56. Sandler M, Ruthven CR, Goodwin BL, Reynolds GP, Rao VA, et al. Trace amine deficit in depressive illness: the phenylalanine connexion. *Acta Psychiatr Scand Suppl* 1980;**280**:29–39.

57. Eynard N, Flachaire E, Lestra C, Broyer M, Zaidan R, Claustrat B, et al. Platelet serotonin content and free and total plasma tryptophan in healthy volunteers during 24 hours. *Clin Chem* 1993;**39**(11 Pt 1):2337–40.

58. Robinson CM, Hale PT, Carlin JM. NF-kappaB activation contributes to indoleamine dioxygenase transcriptional synergy induced by IFN-gamma and tumor necrosis factor-alpha. *Cytokine* 2006;**35**:53–61.

59. Mellor AL, Munn D, Chandler P, Keskin D, Johnson T, Marshall B, et al. Tryptophan catabolism and T cell responses. *Adv Exp Med Biol* 2003;**527**:27–35.

60. Owens MJ, Nemeroff CB. Role of serotonin in the pathophysiology of depression: focus on the serotonin transporter. *Clin Chem* 1994;**40**:288–95.

61. Flores BH, Musselman DL, DeBattista C, Garlow SJ, Schatzberg AF, et al. In: Schatzberg AF, Nemeroff CB, editors. *Biology of mood disorders in textbook of psychopharmacology* 3rd ed. Washington, DC: America Psychiatric Publishing, Inc.; 2004. p. 717–63.

62. Tavares RG, Tasca CI, Santos CE, et al. Quinolinic acid stimulates synaptosomal glutamate release and inhibits glutamate uptake into astrocytes. *Neurochem Int* 2002;**40**:621–7.

63. Fernstrom JD, Fernstrom MH. Tyrosine, phenylalanine, and catecholamine synthesis and function in the brain. *J Nutr* 2007;**137**(6 Suppl. 1):1539S–47S.

64. Rickards H. Depression in neurological disorders: Parkinson's disease, multiple sclerosis, and stroke. *J Neurol Neurosurg Psychiatry* 2005;**76**(Suppl. I):i48–52.

65. Albert PR, Benkelfat C, Descarries L. The neurobiology of depression – revisiting the serotonin hypothesis. I. Cellular and molecular mechanisms. *Philos Trans R Soc Lond B Biol Sci* 2012;**36**(1601):238–2381.

66. Albert PR, Benkelfat C, Descarries L. The neurobiology of depression—revisiting the serotonin hypothesis. II. Genetic, epigenetic and clinical studies. *Trans R Soc Lond B Biol Sci* 2013;**368**(1615) 20120535.

67. Duman RS, Voleti B. Signaling pathways underlying the pathophysiology and treatment of depression: novel mechanisms for rapid-acting agents. *Trends Neurosci* 2012;**35**(1):47–56.

68. Lee S, Jeong J, Kwak Y, Park SK. Depression research: where are we now? *Mol Brain* 2010;**3**:8.

69. Espinoza S, Lignani G, Caffino L, Maggi S, Sukhano I, et al. TAAR1 modulates cortical glutamate NMDA receptor function. *Neuropsychopharmacology* 2015;**40**(9):2217–27.

70. Panas MW, Xie Z, Panas HN, Hoener MC, Vallender EJ, et al. Trace amine associated receptor 1 signaling in activated lymphocytes. *J Neuroimmune Pharmacol* 2012;**7**(4):866–76.

71. Cisneros IE, Ghorpade A. Methamphetamine and HIV-1-induced neurotoxicity: role of trace amine associated receptor 1 cAMP signaling in astrocytes. *Neuropharmacology* 2014;**85**:499–507.

Trace Amines in Neuropsychiatric Disorders

S.I. Sherwani[1] and H.A. Khan[2]

[1]Department of Internal Medicine, Division of Pulmonary Medicine, Dorothy M. Davis Heart and Lung Research Institute, The Ohio State University College of Medicine, Columbus, OH, United States [2]Department of Biochemistry, College of Science, King Saud University, Riyadh, Saudi Arabia

INTRODUCTION

According to the World Health Organization (WHO), approximately 20% of the children and adolescents in the world are afflicted with mental disorders or related problems. Around half of these mental disorders, across different cultures, commence before the age of 14 years. The neuropsychiatric disorders are the leading cause of disability among young people globally.[1–3] Unfortunately, the world regions with the highest percentage of youth

Trace Amines and Neurological Disorders
DOI: http://dx.doi.org/10.1016/B978-0-12-803603-7.00018-5

population (<19 years) are deprived of the necessary mental health resources. In most low- and middle-income countries only one child psychiatrist is available per 1–4 million individuals.[1,2] Wars and other manmade and natural disasters have an enormous impact on the psychosocial wellbeing and mental health of affected individuals. In the United States (US), one in four adults (about 61.5 million) experiences mental illness in a given year and one in 17 adults (about 13.6 million) live with a serious mental illness (SMI) such as schizophrenia, bipolar disorder, or major depression.[4] Approximately 20% of youth (13–18 years) have experienced severe mental disorders in a given year.[4] The neuropsychiatric disorders are the leading cause of disability, followed by cardiovascular diseases (CVDs) and the diseases of the circulatory system and neoplasms.[4] The disease-related burden of disability is measured in terms of disability-adjusted life years (DALYs) and is a reflection of the years lost due to disability, illness, or premature death in a population.[4] The incidence of neuropsychiatric disorders in the US has been on the rise with mental and behavioral disorders accounting for 13.6% US DALYs as compared to neurological disorders being 5.1% US DALYs.[5] Neuropsychiatric disorders are common in the US but it is the SMI that is a primary concern due to its ever-increasing cost in terms of healthcare expenditure and lost wages due to disability. The National Survey on Drug Use and Health (NSDUH) defines SMI as a mental, behavioral or emotional disorder (excluding developmental and substance use disorders), which is diagnosable currently or within the past year and is of sufficient duration to meet diagnostic criteria specified within the fourth edition of the Diagnostic and Statistical Manual of Mental Disorders (DSM-IV) and results in serious functional impairment, which substantially interferes with or limits one or more major life activities.[4] In 2014, an estimated 9.8 million (4.2%) adults (18 years or older) lived with SMI in the US. According to the National Institute of Mental Health (NIMH), the total costs associated with SMI are approximately $317 billion annually. In the US in 2006, as part of healthcare, 36.2 million people sought mental health services for a total of $57.5 billion. Among the mental health services seekers, there were 4.6 million children alone, whose expenses totaled about $9 billion. The expense for the average mental health services seeker was $1591 per person while it was higher at $1931 per child.[6] So, from an epidemiological perspective, central nervous system (CNS) disorders are a major health problem in developed countries such as the US. Drug abuse and drug addiction, particularly cocaine abuse, on the other hand, remain a major clinical challenge for a plethora of familial, social and environmental problems.[7] Poly-drug abuse presents a variety of addictions and related neuropsychiatric disorders that require pharmacological interventions and evidence-based treatment options with very few effective pharmacotherapeutics available in the market to treat such disorders.[8–10]

DISCOVERY OF TRACE AMINES

Comprehending the enormity of neuropsychiatric disorders is paramount to continued research in the area of neuropsychiatry, pharmacology, biochemistry, and other related fields with the urgent purpose of identifying new biomarkers that can be used as possible drug and therapeutic targets. With the ever-aging population (baby-boomers), the incidence of neuropsychiatric and neurological disorders is expected to continue to rise.[11–13] The aberrant regulation of brain monoamines is a hallmark of the neuropsychiatric disorders.[14–16]

A sustained research in neurology has led to the identification of novel proteins called trace amine-associated receptors (TAAR).[17–20] Both trace amines (TAs) and their associated receptors can be used as promising potential drug targets by manipulating the dopaminergic system and blocking specific signals during the signal transduction pathway.[17,21–24] Trace amines have been implicated in the etiology and pharmacotherapy of an array of neuropsychiatric disorders such as depression, schizophrenia, and attention deficit hyperactivity disorder (ADHD). The role of TAs in depression and other neuropsychiatric disorders is well-established, however their role in the mechanism associated with neurotransmission in mammalian systems is yet to be elucidated.[25,26] Initially, TAs were thought to function indirectly via the release of the traditional amines endogenously. Subsequent studies have suggested that TAs play an independent role in mediating the neurotransmission pathways through various receptors such as the G-protein-coupled receptors (GPCR).[21–24] But the other receptors, such as olfactory receptors, do not have high affinity for TAs.[27] The TAARs are a member of the GPCR family and play an important role in neurotransmission.[22–24] These receptors may present putative targets for monoamines and their metabolites, which could be exploited as viable targets for pharmaceutical drug development for various neuropsychiatric disorders. Genome-wide association studies (GWASs) have presented thousands of such novel proteins.[28,29] Approximately 15–18 mammalian receptors have been identified which may represent TAs.[22–24] Trace amines are structurally closely related to classic monoamines, as well as to some psychotropic molecules, such as amphetamines and their analogs, such as the recreational psychoactive drug, ecstasy, chemically known as 3,4-methylenedioxy-methamphetamine.[30,31] The efficacy of antidepressant drugs and their mode of action are affected by alterations in the metabolism of TAs. Trace amines have been detected in trace concentrations, specifically in the range of 0.1–10 nanomoles and constitute <1% of the different amines prevalent in the mammalian nervous system.[32,33]

DISTRIBUTION OF TRACE AMINES

Trace amines are not as well-characterized as their endogenous counterparts—the amine neurotransmitters such as dopamine (DA), norepinephrine (NE), and serotonin (5-HT). But, like the endogenous amines, the TAs are also the end-products of the metabolic breakdown of amino acids and are found mostly localized in the brain, along with other organs and tissues.[23–25] In humans, the TAs are primarily comprised of tyramine, tryptamine, octopamine, and β-phenylethylamine, however, the mammalian synapses have not been shown to utilize any of these in the traditional pathways.[18,20,33,34] In diseased stages, particularly during various neuropsychiatric disorders, the levels of TAs are found to be elevated significantly, laying the claim for their important role which is yet to be determined completely. If the amino acid metabolism and the signaling pathways are compromised, the amines are not degraded as expected and thus the TAs become accumulated in various compromised states. Trace amines have been implicated in competing with the biologically active amines during neuromodulation with a mechanism similar to that of the amphetamines.[23–25] In mammals, several GPCR have been identified and, prominent among them is the TAAR1 which has a high affinity for, and binds to, specifically, the tyramine and β-phenylethylamine.[14,15,18,35] So far, there is no evidence to suggest that the TAs are utilized by neurons in signal transduction or other signaling pathways but the TAs have been found to have the capacity to compete with the biogenic amines and participate in neuromodulation.[14,15,17,18,20] Among patients with neuropsychiatric disorders such as migraines, a diet with high levels of tyramine (cheese) and β-phenylethylamine (chocolate) may lead to increased bouts of severe migraines and hypertension, particularly in patients on monoamine oxidase inhibitors.[36–38] The invertebrates, on the other hand, utilize octopamine as the primary neuromodulator as they lack the noradrenaline system.[39,40]

The activation of TAAR1 elicits the initiation of a cascade of cell-signaling mechanisms, which include the protein kinase A (PKA)- and protein kinase C (PKC)-mediated phosphorylation of DA release. The TAs, along with amphetamines and methamphetamines, compete with DA to regulate the release and reuptake of neurotransmitters into the neuronal canals and synaptic cleft in the nervous system, in the process, raising the extracellular levels of DA.[15,41,42] These elevated levels of DA are not specifically synaptic in location and may be prevalent throughout the neuronal dendrites and the axon. The PKA- and PKC-mediated phosphorylation of dopamine transporters (DAT), effectively mitigates the reuptake of TAs.[15,41,42] The TAAR1 deficiency inhibits the release of the amphetamine and methamphetamine agonists into the cellular environment. It also leads to an increase in the firing frequency, leading to elevated DA levels in the synaptic region causing the elevated behavioral responses in individuals abusing amphetamine and methamphetamine and other psychostimulant and recreational drugs.[15,41,42]

GENETICS AND LOCATION OF TAAR1

TAAR1 was discovered by Borowsky et al.[22] and Bunzow et al.[23] independently in 2001, about nine decades after Barger and Walpole[43] conducted their first experiments with the trace amine, β-phenylethylamine (PEA). The PEA levels are compromised in neuropsychiatric disorders such as, depression, schizophrenia, Parkinson disease (PD), and attention deficit hyperactivity disorder (ADHD), diseases which involve the monoaminergic systems. TAAR1 agonists such as amphetamine, methamphetamine, methylenedioxy methamphetamine, and 3-iodothyronamine have been shown to stimulate metabolism, physiology, and behavior.[17] It has been reported that increased firing frequency of the DA neurons in midbrain in TAAR1 knockout mice leads to the inhibition of DA neuron activity.[44,45] TAAR1 has been found to be less responsive to the endogenous tryptamine, norepinephrine, serotonin, and histamine.[44,45] The *TAAR* gene has been found to be highly conserved among humans, mice, and fish, suggesting that the TAARs and the chemosensory odorant receptors operate differently.[46]

In humans, the *TAAR* gene is 109 kb long and located on the chromosome 6, while its lengths and locations are different in mice (192 kb, chromosome 10A4) and rats (216 kb, chromosome 1p12). In humans, *TAAR1* is located on chromosome 6q23.2 (cytogenetic location) between base pairs 132,644,983 and 132,646,002 (molecular location). TAAR1 shares 76% and 79% amino acid homology with mouse and rat TAAR1, respectively. The TAAR1 plays an important role in the regulation of neurotransmission and neuromodulation by different endogenous amines, such as dopamine, serotonin, and norepinephrine in the mammalian brain. So far, in humans, six TAARs and three pseudogenes have been identified to be located on chromosome 6 and are coded by intronless genes which are arranged in clusters.[24] Of all the TAARs, TAAR1 has been studied and characterized the most.[17,22,23] The brain mRNA coding for TAAR1 is localized in the limbic system and the regions associated with catecholamines.[15,22-24] As a result, TAAR1 is a highly influential neuromodulator for emotional and motivated behaviors associated with psychotropic drug abuse usually associated with monoaminergics. Recent evidence suggests that TAARs may represent a new class of chemosensory receptors localized in mammalian brains but more research studies will need to corroborate these findings.[46] The human brain harbors and expresses approximately 80% of the functional genes of the human genome. More than 1200 genes are involved with various neuropsychiatric disorders, such as the diseases of the CNS, including dementia.[47,48] Despite the promise that various receptors such as TAAR hold as target biomarkers for developing antidotes for psychotropic drug abuse, the progress has been slow due to the trace amounts of these amines and very short half-life. Given the ever-increasing trajectory of neurodegenerative diseases and the incidence of drug abuse, novel approaches can be designed for disease management by taking into consideration the various agonists and antagonists for TAAR1 and other receptors. In order to study these receptors, inadvertently, they need to be transferred into different cell systems and, oftentimes, they do not mimic the properties as they do in their specific and intact mammalian systems.[14,17,22-24]

MONOAMINES AND THEIR EFFECT

Amphetamine and methamphetamine are addictive psychostimulants whose abuse has significant cognition (memory and attention) and emotional (euphoria, impulsion, enhanced self-esteem, and productivity) consequences which are well-documented.[15,49] These neurotoxic drugs act by elevating the endogenous monoamine (eg, dopamine) levels by compromising their reuptake and releasing them into the synapse, thus causing psychostimulation.[15,49,50] This affects the communication mechanism of the brain by disrupting the normal mechanism in which neurons and other cells transmit, receive and process neurological information. Some antipsychotics express their effect by imitating the brain's own neurotransmitters, while others become effective by overstimulating the brain's reward circuit to release extra amounts of these neurotransmitters. Drugs like heroin and marijuana imitate the endogenous neurotransmitters and fool the brain into activating the receptors, thus transmitting messages prematurely leading to euphoria.[51,52] Amphetamine, methamphetamine, and cocaine initiate the neurons into releasing the excess amounts of endogenous neurotransmitters or block their reuptake, which is necessary for turn-off mechanism.[22-24] The disruption in neuronal communication causes the neurotransmitters to fire with abnormal frequency. The DA neurotransmitter is associated with movement, emotion, motivation, and pleasure, apart from natural instincts of survival (eating and associating with loved ones), and leads to euphoria among those abusing psychostimulants. As the brain learns to use the reward-system, bypassing the natural circuitry, the abusers become addicted to repeat the abusive behavior. The drug abuse is one of the biggest challenges of neuropsychiatry, so the promise of TAs and TAAR1 being possible drug targets is of phenomenal importance.[22-24]

MECHANISM OF ACTION OF TAAR1

At the site of the action (synaptic cleft), the TAAR1 agonists, present inside the neurons, increase the concentration of the associated monoamines in the synaptic cleft, causing the increased postsynaptic receptor binding.[15,53] TAAR1 induces the activation of G protein-coupled inwardly rectifying potassium channels (GIRKs) which attenuates the firing frequency of DA neurons, thus, mitigating the establishment of the hyperdopaminergic state.[15,54] The dopamine transporter (DAT) uptake triggers the competitive inhibition between the TAs and amphetamine for their reuptake to the transporter.[55,56] The presynaptic neuronal cleft harbors the TAs and amphetamine, which cause the activation of TAAR1 via the PKA and PKC signaling pathway leading to DAT phosphorylation.[57,58]

The TA signaling pathway is one of the most recent pathways which have not yet been elucidated comprehensively. It presents an excellent model system to understand the crosstalk between various cell receptors, transporters, channels, neuronal receptors, metabolic or signaling enzymes (PKA and PKC), secreted factors, DAT phosphorylation, signal transduction proteins, and transcription factors.[59] The trace amounts of amphetamines diffuse into the neurons either via the neuronal membrane at the presynaptic site or through the DAT. Upon interaction with TAAR1 inside a synaptic vesicle, the TAs bind to the receptors and, in the process, release the DA into the cytosol and reduce the postsynaptic DA receptor. This triggers the PKA and PKC signal transduction pathway by phosphorylating the DAT. The phosphorylation of PKA and PKC may cause the retreat of DAT into the presynaptic neuron and inhibit transport across the cell and become internalized. Like amphetamines, TAs may also cause the reuptake of calcium ions due to DAT phosphorylation via the calcium/calmodulin-dependent protein kinase type II alpha chain-dependent pathway or the CAMKIIα pathway which produces DA efflux. The binding of TA to the TAAR1 stimulates TA uptake and phosphorylates the TAAR 1 substrate proteins that are associated with the activation of the two main signaling pathways: (1) PKA pathway, which controls the metabolic actions of TA and (2) PKC pathway, which regulates the expression of other receptor genes and crosstalks with the PKA pathway for controlling the neuronal cell growth and differentiation. The interactions between various amines (TAs, DA, amphetamines, etc.) and various enzymes (PKA and PKC) are a function of the significant role that TAAR1 plays in regulating the neurotransmission in the CNS. Once the biomembrane of the presynaptic neuron reaches the threshold frequency, it fires and releases the neurotransmitter which activates the receptors on the adjacent postsynaptic cell, thus transmitting the information next cell.[60]

For the last four decades, TAs have garnered an increased interest among neurologists and psychiatrists about studying their role in the human brain.[61,62] Initially referred to as microamines, as a consensus, these biomarkers are now referred to as TAs.[61–63] Despite their present nomenclature, TAs may be found in trace or microgram amounts in brain and other tissues, respectively. The turnover rate of TAs is invariably high and their half-life is very short, making them difficult to identify and synthesize. The elevated levels of TAs have been found to be associated with depression and other neuropsychiatric disorders.[64] The TAs play a major role in the brain recirciuting mechanism associated with the abuse of psychostimulants and other recreational drugs.[65] The variable electrophysiological responses generated by the TAs, as compared to the classical amines, make them unique biomarker targets for drug development for treating neuropsychiatric disorders and drug addiction.[66,67]

TAAR1, which mediates the vital role of neurotransmission and neuromodulation by controlling synaptic firing, is distributed in trace amounts in the brain and from moderate expression (stomach) to lower expression in organs such as kidney, lung, small intestine, and amygdala.[18,22] Despite the identification of several TAARs, TAAR1 remains the most studied receptor, which plays a critical role in neuronal signaling. The TAAR1 uptake plays a complementary role in the regulation of the neuronal signaling pathway involving the PKA and PKC enzymes.[15,42] The function of the TAAR1 receptor is regulated by their expression levels and post-translational modifications and their dysfunction leads to postsynaptic dysregulation. Understanding the expression of genes involved in neurotransmission and neuromodulation is critical in identifying new sites for drug gene target selection and potential interventions to provide treatment options or therapeutic regimens for preventing neuropsychiatric disorders. This represents an ideal system for studying the convergence of signal transduction and protein trafficking. The TAAR1 translocation process, associated with signal neurotransmissions and protein trafficking is complex and the regulatory steps associated with TAAR1 translocation are yet to be elucidated. Most of the knowledge about the role of TAAR1 in managing and mitigating neuropsychiatric disorders comes from either the in vitro experiments conducted to elucidate the neurological pathways and the signaling mechanisms or the well-established mammalian models of mice and rats representing neurodegenerative diseases or clinical research studies conducted on human subjects with neuropsychiatric disorders.[49,68] The targeted disruption of TAAR1 in the mouse model of PD can be achieved using the homologous recombination approach.[69–71] Likewise, in patients with comorbidities, it is important to evaluate the in vivo and in vitro mechanisms of drug absorption and disposition for long-term neuropsychiatric disorders like PD.[69–71] The management of neuropsychiatric disorders, such as depression, anxiety, bipolar disorder, loneliness, panic attacks, phobias, post-traumatic stress disorder (PTSD), obsessive-compulsive disorder, and thoughts of suicide will require a plethora of approaches which include psychotherapy and neuropsychiatry among other interventions. The incidence of PTSD among war veterans is an issue which must be addressed on a war-footing.[70–72] The neuromodulator and psychostimulant properties of TAAR1 may present new vistas for managing neuropsychiatric indications and related neurological disorders, including drug addiction.[14,15]

PHARMACOGENETIC PERSPECTIVE

The pharmacogenetic studies have been slow to reveal the relationships that may exist between specific candidate genes and the effect (adverse versus positive) that drug treatment may cause among patients with neuropsychiatric disorders.[24,46,73] The treatment responses among individuals vary depending upon the variations in specific candidate genes of the CYP superfamily and the genetic variants linked to dementia. Pharmacogenetic studies will allow the researchers to establish the basis for the design of the drug interaction studies addressing the class effects of oral antineuropsychiatric drugs—interaction of CYP enzymes with neuroleptics, antidepressants, and benzodiazepines and their interactions with neuronal secretion of various amines.[24,46,73]

The biomarkers belonging to the CYP family of genes, as part of the signaling pathway, are targets for the drug gene target selection, studying models of neuropsychiatric disorders and profiling the effect of other epidemiological and environmental factors on gene expression. Since patients with neuropsychiatric disorders usually have associated illnesses and are on multidrug regimens, it is important to address the combined effect of the administered drugs, which are regulated by an array of genes, which could be turned-on and turned-off, based upon various loss-of-function and gain-of-function interactions. Future studies are warranted on understanding the neuropsychiatric effects driven by TAAR1 and other receptors and whether the individuals would require detailed models of the entire neuronal signaling pathway to help us understand the neuropsychiatric disorders better.[24,46,73]

TRACE AMINES IN NEUROPSYCHIATRIC DISORDERS

Age-related neurodegenerative diseases such as, Alzheimer disease (AD), PD, dementia, etc., present differently in different individuals. Biopsychosocial stress has been found to be an important confounding factor associated with early onset of aging and age-related diseases.[74–76] Obesity has been found to be associated with AD, PD, dementia, chronic inflammation, and other diseases such as, cancers, CVDs, diabetes, and arthritis, through the secretion of IL-6 which, in turn, stimulates the secretion of C-reactive protein (CRP), a biomarker of inflammation.[77,78] So, as we continue to advance in an era of personalized medicine, the gene therapy also appears to be promising to pursue for different aspects of neuropsychiatric disorders. The anomalous regulation of brain monoamines is a hallmark of neuropsychiatric disorders and is indicated in the dopaminergic activity. The role of TAs in the pathogenesis of PD is still in the early stages of research but, increasingly, the evidence is pointing to the implication of TAs in PD.[77–79] Several studies have revealed a role for the circulating octopamines in early detection of PD.[79,80] These biomarkers are a product of abnormal enzyme activity during tyrosine decarboxylation, an enzyme involved in TA production. Imbalances in tyramine levels may lead to PD, depression, schizophrenia, migraines, and elevated blood pressure. β-Phenylethylamine, on the other hand, has been found to be associated with depression, schizophrenia, migraines, attention deficit hyperactivity disorder (ADHD), and phenylketonuria.[77–80] The tryptamine levels may cause depression, schizophrenia, and hepatic encephalopathy. The octopamine dysregulation may lead to depression, hepatic encephalopathy, elevated blood pressure, and phenylketonuria.[17,78,79,81] Brain levels of TAs have been reported to be altered by several drugs used to treat neuropsychiatric disorders. Administration of monoamine oxidase (MAO) inhibitor antidepressants, such as phenelzine and tranylcypromine, results in a greater increase in brain levels of TAs than of classical neurotransmitter amines.[82,83] The role of TAs in some prominent neuropsychiatric disorders is summarized in the following text.

Schizophrenia

Schizophrenia is a chronic, severe, and disabling brain disorder. The causes of schizophrenia are not fully known, however it appears that schizophrenia results from a complex interaction between genetic and environmental factors. There is no magic bullet for the treatment of schizophrenia. The current treatments basically manage the positive symptoms, while the prominent features of the disease (negative and cognitive symptoms) remain partially improved. Moreover, antipsychotic treatments often trigger serious side effects resulting in discontinuation of the treatment. The activation of TAAR1, a modulator of monoaminergic neurotransmission, represents a novel therapeutic option for neuropsychiatric diseases. Much of the early interest in the possible role of trace amines in schizophrenic conditions stemmed from the close structural similarity of trace amines and in particular PEA, to amphetamine, which can induce paranoid schizophrenia-like symptoms.[84] Phenylethylamine is structurally similar to amphetamine and has well-documented amphetaminergic effects including induction of similar stereotyped behaviors in animals.[85,86] On this basis, PEA has been suggested to be an endogenous amphetamine.[87,88] This observation has been reinforced with the finding that amphetamine-type hallucinogens activate TAAR family members.[23] Both urinary[89,90] and plasma[91,92] PEA levels have also been reported to be increased during acute phases of the disease, suggesting the hypothesis that an increased PEA activity plays a role in schizophrenia.[93] Similarly, altered metabolism of tyramine and tryptamine in schizophrenia has been suggested.[34] Indeed, urinary tryptamine levels have been shown to correlate with disease severity.[94]

In rodents, activation of TAAR1 by two novel and pharmacologically distinct compounds, the full-agonist RO5256390 and the partial-agonist RO5263397, blocks psychostimulant-induced hyperactivity and produces a brain activation pattern reminiscent of the antipsychotic drug olanzapine, suggesting antipsychotic-like properties of these TAAR1 agonists.[95] These data suggest that TAAR1 agonists may provide a novel and differentiated treatment of schizophrenia as compared with current medication standards. TAAR1 agonists may improve not only the positive symptoms but also the negative symptoms and cognitive deficits, without causing adverse effects such as motor impairments or weight gain.[95] Wolinsky et al.[44] produced knockout mice lacking the TA1 receptor to characterize its contribution to the regulation of behavior. Compared with wildtype littermates, TA1 knockout mice displayed a deficit in prepulse inhibition. The knockout animals, in which the TA1-agonist influence of amphetamine was absent, showed enhanced sensitivity to the psychomotor-stimulating effect of this drug, which was temporally correlated with significantly larger increases in the release of both DA and NE in the dorsal striatum and associated with a 262% increase in the proportion of striatal high-affinity D2 receptors. TA1 therefore appears to play a modulatory role in catecholaminergic function and represents a potentially novel mechanism for the treatment of neuropsychiatric disorders while the TA1 knockout mouse may provide a useful model for the development of treatments for some positive symptoms of schizophrenia.[44]

Migraine Headaches

Migraine is a form of vascular headache caused by a combination of vasodilatation and the release of chemicals from nerve fibers surrounding the blood vessels. Cluster headache (CH) is characterized by unilateral pain and ipsilateral autonomic features resulting in desperation. An association of CH and migraine to TAs has been demonstrated. The levels of TAs in plasma and platelets were found to be elevated in CH sufferers both in active and remission periods as compared to healthy controls.[96] A cluster of gene-encoding GPCRs that bind and were activated by TAs was identified in the long arm of chromosome 6q23.

Aridon et al.[96] have evaluated two families with cluster headaches by linkage analysis to 6q23 region and the mutation scanning of the TAR 1, TAR 3, TAR 4, TAR 5, putative neurotransmitter receptor (PNR), and GPR58 genes. Plasma levels of TAs were found to be significantly higher in CH patients, in both the remission and the active phases, when compared with control subjects or subjects with migraine.[97] In the same study, intraplatelet levels of octopamine, synephrine, and tyramine were higher in CH patients than in control subjects. In migraine patients, plasma levels of octopamine and synephrine were higher compared with controls, although in migraine with aura, the difference was not significant. The elevation of plasma TA levels in both migraine and CH supports the hypothesis that disorders of biogenic amine metabolism may be a characteristic biochemical trait in primary headache sufferers; the observation that such alterations are more prominent in patients with CH than migraine patients suggests that they may reflect sympathetic or hypothalamic dysfunction.[97]

D'Andrea et al.[98] determined the concentrations of TAs in platelets of migraine without aura (MoA) and migraine with aura (MA) patients in headache-free periods, compared with controls. They observed that platelet concentrations of TAs, although elevated in both migraine types, showed a different profile in MoA and MA. Octopamine was significantly higher in MoA sufferers compared with both control subjects and MA patients. Synephrine was significantly higher in MA patients with respect to both controls and MoA sufferers.[98] The same group of investigators measured the plasma levels of dopamine (DA), norepinephrine (NE), and TAs, including tyramine and octopamine, in 73 patients with migraine, 13 patients with chronic tension-type headache, and 37 controls. The plasma levels of DA and NE were several-fold higher in migraine patients compared with control subjects. The plasma levels of tyramine were also extremely elevated and progressively increased with the duration of migraine. These results strengthen the hypothesis that tyrosine metabolism is deranged in migraine and may participate in its pathophysiology. The high plasma levels of tyramine, a potent agonist of the TAAR1, may ultimately downregulate this receptor because of loss of inhibitory presynaptic regulation, resulting in uncontrolled neurotransmitter release that may produce functional metabolic consequences in the synaptic clefts of the pain matrix implicated in migraine.[99] It has also been suggested that conversion of tyramine to octopamine is required for precipitation of migraine attacks.[100]

Depression

Elevated PEA levels may be associated with increased stress and anxiety in laboratory animals and humans.[101,102] Whereas, chronic administration of PEA to rats produces a β-adrenoceptor downregulation similar to that observed with some antidepressants,[103] reserpine depletes central levels of some TAs[83], and the antidepressant effects of exercise have been suggested to be due to an elevation of PEA.[104] Deprenyl (selegiline), a selective inhibitor of MAO-B, is used in the treatment of PD and produces a marked increase in brain levels of PEA relative to other amines.[105,106] In rodents, acute administration of the antipsychotics chlorpromazine, fluphenazine, and haloperidol has been shown to decrease striatal p-TA levels, and similar studies with PEA have shown that these antipsychotics increase the rate of accumulation of this TA in the striatum.[107,108]

Chronic treatment of experimental animals with antidepressants has been reported to decrease the numbers of β-adrenergic receptor,[109] an effect that is also exerted by chronic

PE treatment.[103] Reserpine, an antipsychotic drug, not only precipitates episodes of depression[110] but also decreases central TA levels.[111] It has also been suggested that the anti-depressant effects of exercise are due to an exercise-induced elevation of phenylethylamine levels.[104] Sabelli et al.[112] have reported a decrease in depressive symptomatology in 60% of patients treated with either phenylalanine or phenylethylamine. Other trace amines such as tyramine, tryptamine, and octopamine have also been suggested to play a role in depressive symptomatology.[34,113,114]

Revel et al.[45] showed that a selective TAAR1 agonist, RO5166017, inhibited the firing frequency of dopaminergic and serotonergic neurons in regions where TAAR1 is expressed (ventral tegmental area and dorsal raphe nucleus, respectively). In contrast, RO5166017 did not affect the firing frequency of noradrenergic neurons in the locus coeruleus, an area devoid of TAAR1 expression. Administration of RO5166017 prevented stress-induced hyperthermia and blocked dopamine-dependent hyperlocomotion in cocaine-treated and dopamine transporter knockout mice as well as hyperactivity induced by an NMDA antagonist. These findings associated TAAR1 to the control of monoamine-driven behaviors and suggested anxiolytic- and antipsychotic-like properties for agonists such as RO5166017, opening treatment opportunities for psychiatric disorders.[45] Another selective and potent TAAR1 partial-agonist, RO5203648, showed high affinity and potency at TAAR1, high selectivity versus other targets, and favorable pharmacokinetic properties.[115] In mouse brain slices, RO5203648 increased the firing frequency of dopaminergic and serotonergic neurons in the ventral tegmental area and the dorsal raphe nucleus, respectively. This agonist demonstrated clear antipsychotic- and antidepressant-like activities as well as potential anxiolytic-like properties in rodents and monkeys.[115] It also attenuated drug-taking behavior and was highly effective in promoting attention, cognitive performance, and wakefulness, suggesting that TAAR1 agonists might have therapeutic potential in one or more neuropsychiatric domains.

Attention Deficit Hyperactivity Disorder

Attention deficit hyperactivity disorder (ADHD) is one of the most common childhood disorders and can continue through adolescence and adulthood. Symptoms of ADHD include difficulty staying focused and paying attention, difficulty controlling behavior, and hyperactivity. The administration of PEA in experimental animals has been shown to induce hyperactivity and aggression, two of the cardinal features of ADHD.[116] Amphetamines, which are good ligands at TAAR, also possess clinical utility for the treatment of ADHD.[23] The wake-promoting drug, modafinil, which is beneficial in ADHD patients,[117] has been reported to enhance the activity of PEA at TAAR1.[118]

Several studies have reported decreased urinary PEA levels in ADHD patients in comparison with controls.[119,120] It was also observed that after treatment with methylphenidate, patients who responded positively showed a normalization of urinary PEA whereas nonresponders showed no change in the baseline values.[120] Baker et al.[121] have noticed significant decreases in plasma tyramine as well as in urinary and plasma PEA metabolite (phenylacetic acid) and precursor (phenylalanine and tyrosine) levels in ADHD patients. Although there are limited data on the role of trace amines in ADHD, the findings so far available suggest that there may be a decrease in trace aminergic functioning, which may be normalized by currently available therapeutics.

CONCLUSION

Trace amines play an important role in the pathogenesis of neuropsychiatric disorders. Most of our knowledge comes from the studies on the role that TAAR1 plays as part of the monoaminergic systems. The interaction of TAAR1 with dopamine transporter via the TAAR1-expressing neurons reveals that TAAR1 is associated with the regulation of the dopamine system and also participates in the regulation of the signaling pathways associated with triggered by the use of psychostimulants similar to amphetamine. Future studies in the area of TA, TAAR1, and the development of drugs/therapeutics against neuropsychiatric disorders and to curb the effect of psychostimulants will need to explore further the specific localization of TAAR1 in the brain and intra- and interneuronal crosstalk. In-depth research in the field of monoamine transporters would help pave the way for designing new drugs for providing relief to patients of neuropsychiatric disorders and also combat the abuse and addiction associated with the use of psychostimulants. The TAs and the TAAR1 biomolecules promise to be the panacea biomarkers that neuropsychiatry and related fields have been researching for the last several decades.

References

1. *Ten facts on mental health. WHO.* Retrieved on December 21, 2015 from, <http://www.who.int/features/factfiles/mental_health/mental_health_facts/en/>.
2. *Discussion Paper "Mental health, poverty and development", July 2009.* Retrieved on December 21, 2015 from, <http://www.who.int/nmh/publications/discussion_paper_en.pdf>.
3. Kessler RC, Aguilar-Gaxiola S, Alonso J, Chatterji. The global burden of mental disorders: an update from the WHO World Mental Health (WMH) Surveys. *Epidemiologia E Psichiatria Sociale* 2009;**18**:23–33.
4. *Serious mental illness (SMI) among US adults 2014.* Retrieved on December 21, 2015 from, <http://www.nimh.nih.gov/health/statistics/prevalence/serious-mental-illness-smi-among-us-adults.shtml>.
5. *U.S. DALYs contributed by mental and behavioral disorders 2010.* Retrieved on December 21, 2015 from, <http://www.nimh.nih.gov/health/statistics/disability/us-dalys-contributed-by-mental-and-behavioral-disorders.shtml>.
6. *Annual total direct and indirect costs of serious mental illness 2002.* Retrieved on December 21, 2015 from, <http://www.nimh.nih.gov/health/statistics/cost/index.shtml>.
7. Li JX. Trace amines and cocaine abuse. *ACS Chem Neurosci* 2014;**5**:497–8.
8. Jones JD, Mogali S, Comer SD. Polydrug abuse: a review of opioid and benzodiazepine combination use. *Drug Alcohol Depend* 2012;**125**:8–18.
9. McCabe SE, Cranford JA, Morales M, Young A. Simultaneous and concurrent polydrug use of alcohol and prescription drugs: prevalence, correlates, and consequences. *J Stud Alcohol* 2006;**67**:529–37.
10. Jhanjee S. Evidence based psychosocial interventions in substance use. *Ind J Psychol Med* 2014;**36**:112–8.
11. Kalapatapu RK, Sullivan MA. Prescription use disorders in older adults. *Am J Addict* 2010;**19**:515–22.
12. Goldman DP, Zheng Y, Girosi F, Michaud PC, Olshansky SJ, Cutler D, et al. The benefits of risk factor prevention in Americans aged 51 years and older. *Am J Public Health* 2009;**99**:2096–101.
13. Laks J, Engelhardt E. Peculiarities of geriatric psychiatry: a focus on aging and depression. *CNS Neurosci Ther* 2010;**16**:374–9.
14. Miller GM. The emerging role of trace amine associated receptor 1 in the functional regulation of monoamine transporters and dopaminergic activity. *J Neurochem* 2011;**116**:164–76.
15. Xie Z, Miller GM. Trace amine-associated receptor 1 as a monoaminergic modulator in brain. *Biochem Pharmacol* 2009;**78**:1095–104.
16. Di Benedetto B, Rupprecht R. Targeting glia cells: novel perspectives for the treatment of neuropsychiatric diseases. *Curr Neuropharmacol* 2013;**11**:171–85.

17. Grandy DK. Trace amine-associated receptor 1—family archetype or iconoclast? *Pharmacol Ther* 2007;**116**:355–90.

18. Zucchi R, Chiellini G, Scanlan TS, Grandy DK. Trace amine-associated receptors and their ligands. *Br J Pharmacol* 2006;**149**:967–78.

19. Maguire JJ, Parker WAE, Foord SM, Bonner TI, Neubig RR, Davenport AP. International union of pharmacology. LXXII. Recommendations for trace amine receptor nomenclature. *Pharmacol Rev* 2009;**61**:1–8.

20. Sotnikova TD, Caron MG, Gainetdinov RR. Trace amine-associated receptors as emerging therapeutic targets. *Mol Pharmacol* 2009;**76**:229–35.

21. Ghanemi A. Targeting G protein coupled receptor-related pathways as emerging molecular therapies. *Saudi Pharm J* 2015;**23**:115–29.

22. Borowsky B, Adham N, Jones KA, et al. Trace amines: identification of a family of mammalian G protein-coupled receptors. *Proc Natl Acad Sci USA* 2001;**98**:8966–71.

23. Bunzow JR, Sonders MS, Arttamangkul S, et al. Amphetamine, 3,4-methylenedioxy methamphetamine, lysergic acid diethylamide, and metabolites of the catecholamine neurotransmitters are agonists of a rat trace amine receptor. *Mol Pharmacol* 2001;**60**:1181–8.

24. Lindemann L, Ebeling M, Kratochwil NA, Bunzow JR, Grandy DK, Hoener MC. Trace amine-associated receptors form structurally and functionally distinct subfamilies of novel G protein-coupled receptors. *Genomics* 2005;**85**:372–85.

25. Saltiel PF, Silvershein DI. Major depressive disorder: mechanism-based prescribing for personalized medicine. *Neuropsychiatr Dis Treat* 2015;**11**:875–88.

26. Autry AE, Monteggia LM. Brain-derived neurotrophic factor and neuropsychiatric disorders. *Pharmacol Rev* 2012;**64**:238–58.

27. Hussain A, Saraiva LR, Ferrero DM, Ahuja G, Krishna VS, Liberles SD, et al. High-affinity olfactory receptor for the death-associated odor cadaverine. *Proc Natl Acad Sci USA* 2013;**110**:19579–84.

28. Visscher PM, Brown MA, McCarthy MI, Yang J. Five years of GWAS discovery. *Am J Hum Genet* 2012;**90**:7–24.

29. Bergen SE, Petryshen TL. Genome-wide association studies (GWAS) of schizophrenia: does bigger lead to better results? *Curr Opin Psychiatr* 2012;**25**:76–82.

30. Kalant H. The pharmacology and toxicology of "ecstasy" (MDMA) and related drugs. *Can Med Assoc J* 2001;**165**:917–28.

31. Sáez-Briones P, Hernández A. MDMA (3,4-Methylenedioxymethamphetamine) analogues as tools to characterize MDMA-like effects: an approach to understand entactogen pharmacology. *Curr Neuropharmacol* 2013;**11**:521–34.

32. Wee XK, Ng KS, Leung HW, et al. Mapping the high-affinity binding domain of 5-substituted benzimidazoles to the proximal N-terminus of the GluN2B subunit of the NMDA receptor. *Br J Pharmacol* 2010;**159**:449–61.

33. Berry MD. Mammalian central nervous system trace amines. Pharmacologic amphetamines, physiologic neuromodulators. *J Neurochem* 2004;**90**:257–71.

34. Premont RT, Gainetdinov RR, Caron MG. Following the trace of elusive amines. *Proc Natl Acad Sci USA* 2001;**98**:9474–5.

35. Nelson DA, Tolbert MD, Singh SJ, Bost KL. Expression of neuronal trace amine-associated receptor (Taar) mRNAs in leukocytes. *J Neuroimmunol* 2007;**192**:21–30.

36. Fiedorowicz JG, Swartz KL. The role of monoamine oxidase inhibitors in current psychiatric practice. *J Psychiatr Pract* 2004;**10**:239–48.

37. Brinsden MJ, Shaw IC. Do molecular structures of migraine drugs point to a common cause of this elusive disease and suggest future drug designs? *Pharmaceut Med* 2015;**29**:1–5.

38. Irsfeld M, Spadafore M, Prüß BM. β-phenylethylamine, a small molecule with a large impact. *Webmed Central* 2013;**4**:4409.

39. Stevenson PA, Rillich J. The decision to fight or flee—insights into underlying mechanism in crickets. *Front Neurosci* 2012;**6**:118.

40. Meinertzhagen IA, Lee CH. The genetic analysis of functional connectomics in Drosophila. *Adv Genet* 2012;**80**:99–151.

41. Pei Y, Lee J, Leo D, Gainetdinov RR, Hoener MC, Canales JJ. Activation of the trace amine-associated receptor 1 prevents relapse to cocaine seeking. *Neuropsychopharmacology* 2014;**39**:2299–308.

42. Panas MW, Xie Z, Panas HN, Hoener MC, Vallender EJ, Miller GM. Trace amine associated receptor 1 signaling in activated lymphocytes. *J Neuroimmun Pharmacol* 2012;**7**:866–76.

43. Barger G, Walpole GS. Isolation of the pressor principles of putrid meat. *J Physiol* 1909;**38**:343–52.
44. Wolinsky TD, Swanson CJ, Smith KE, Zhong H, Borowsky B, Seeman P, et al. The Trace Amine 1 receptor knockout mouse: an animal model with relevance to schizophrenia. *Genes Brain Behav* 2007;**6**:628–39.
45. Revel FG, Moreau JL, Gainetdinov RR, Bradaia A, Sotnikova TD, Mory R, et al. TAAR1 activation modulates monoaminergic neurotransmission, preventing hyperdopaminergic and hypogluatmatergic activity. *Proc Natl Acad Sci USA* 2011;**108**:8485–90.
46. Liberles SD, Buck LB. A second class of chemosensory receptors in the olfactory epithelium. *Nature* 2006;**442**: 645–50.
47. Cacabelos R, Fernández-Novoa L, Martínez-Bouza R, et al. Future trends in the pharmacogenomics of brain disorders and dementia: influence of APOE and CYP2DE variants. *Pharmaceuticals* 2010;**3**:3040–100.
48. Zupancic M, Mahajan A, Handa K. Dementia with Lewy bodies: diagnosis and management for primary care providers. *Prim Care Companion CNS Disord* 2011;**13** PCC.11r01190.
49. Lakhan SE, Kirchgessner A. Prescription stimulants in individuals with and without attention deficit hyperactivity disorder: misuse, cognitive impact, and adverse effects. *Brain Behav* 2012;**2**:661–77.
50. Shen H, Luo Y, Yu SJ, Wang Y. Enhanced neurodegeneration after a high dose of methamphetamine in Adenosine A3 receptor Null mutant mice. *Neuroscience* 2011;**194**:170–80.
51. Budney AJ, Roffman R, Stephens RS, Walker D. Marijuana dependence and its treatment. *Addict Sci Clin Pract* 2007;**4**:4–16.
52. Minnes S, Lang A, Singer L. Prenatal tobacco, marijuana, stimulant, and opiate exposure: outcomes and practice implications. *Addict Sci Clin Pract* 2011;**6**:57–70.
53. Costagliola C, Parmeggiani F, Semeraro F, Sebastiani A. Selective serotonin reuptake inhibitors: a review of its effects on intraocular pressure. *Curr Neuropharmacol* 2008;**6**:293–310.
54. Bradaia A, Trube G, Stalder H, Norcross RD, Ozmen L, Wettstein JG, et al. The selective antagonist EPPTB reveals TAAR1-mediated regulatory mechanisms in dopaminergic neurons of the mesolimbic system. *Proc Natl Acad Sci USA* 2009;**106**:20081–6.
55. Schmitt KC, Rothman RB, Reith MEA. Nonclassical pharmacology of the dopamine transporter: atypical inhibitors, allosteric modulators, and partial substrates. *J Pharmacol Exp Ther* 2013;**346**:2–10.
56. Beuming T, Kniazeff J, Bergmann ML, et al. The binding sites for cocaine and dopamine in the dopamine transporter overlap. *Nat Neurosci* 2008;**11**:780–9.
57. Haile CN, Kosten TR, Kosten TA. Pharmacogenetic treatments for drug addiction: cocaine, amphetamine and methamphetamine. *Am J Drug Alcohol Abuse* 2009;**35**:161–77.
58. Tritsch NX, Oh WJ, Gu C, Sabatini BL. Midbrain dopamine neurons sustain inhibitory transmission using plasma membrane uptake of GABA, not synthesis. *eLife* 2014;**3**:e01936.
59. Al-Hasani R, Bruchas MR. Molecular mechanisms of opioid receptor-dependent signaling and behavior. *Anesthesiology* 2011;**115**:1363–81.
60. Marambaud P, Dreses-Werringloer U, Vingtdeux V. Calcium signaling in neurodegeneration. *Mol Neurodegeneration* 2009;**4**:20.
61. Boulton AA. Amines and theories in pychiatry. *Lancet* 1971;**2**:7871.
62. Boulton AA. Identification, distribution, metabolism, and function of meta and para tyramine, phenylethylamine and tryptamine in brain. *Adv Biochem Psychopharmacol* 1976;**15**:57–67.
63. Boulton AA. Trace amines and the neurosciences. In: Boulton AA, Baker GB, Dewhurst WG, Sandler M, editors. *Neurobiology of the trace amines*. Clifton, NJ: Human Press; 1984. p. 13–24.
64. Huffman JC, Celano CM, Januzzi JL. The relationship between depression, anxiety, and cardiovascular outcomes in patients with acute coronary syndromes. *Neuropsychiatr Dis Treat* 2010;**6**:123–36.
65. Drevets WC, Price JL, Furey ML. Brain structural and functional abnormalities in mood disorders: implications for neurocircuitry models of depression. *Brain Struct Funct* 2008;**213**:93–118.
66. Javitt DC, Spencer KM, Thaker GK, Winterer G, Hajós M. Neurophysiological biomarkers for drug development in schizophrenia. *Nat Rev Drug Discov* 2008;**7**:68–83.
67. Dichter GS, Damiano CA, Allen JA. Reward circuitry dysfunction in psychiatric and neurodevelopmental disorders and genetic syndromes: animal models and clinical findings. *J Neurodevelop Disord* 2012;**4**:19.
68. Pandey UB, Nichols CD. Human disease models in *Drosophila melanogaster* and the role of the fly in therapeutic drug discovery. *Pharmacol Rev* 2011;**63**:411–36.
69. Cheng MH, Block E, Hu F, Cobanoglu MC, Sorkin A, Bahar I. Insights into the modulation of dopamine transporter function by amphetamine, orphenadrine, and cocaine binding. *Front Neurol* 2015;**6**:134.

70. Dwyer DS, Aamodt E, Cohen B, Buttner EA. Drug elucidation: invertebrate genetics sheds new light on the molecular targets of CNS drugs. *Front Pharmacol* 2014;**5**:177.

71. Bortolato M, Chen K, Shih JC. Monoamine oxidase inactivation: from pathophysiology to therapeutics. *Adv Drug Deliv Rev* 2008;**60**:1527–33.

72. Tallman KR, Grandy DK. A decade of pharma discovery delivers new tools targeting trace amine-associated receptor 1. *Neuropsychopharmacology* 2012;**37**:2553–4.

73. Fuchs S, Rende E, Crisanti A, Nolan T. Disruption of aminergic signalling reveals novel compounds with distinct inhibitory effects on mosquito reproduction, locomotor function and survival. *Sci Rep* 2014;**4**:5526.

74. Sturm VE, Levenson RW. Alexithymia in neurodegenerative disease. *Neurocase* 2011;**17**:242–50.

75. Tansey MG, Goldberg MS. Neuroinflammation in Parkinson's disease: its role in neuronal death and implications for therapeutic intervention. *Neurobiol Dis* 2010;**37**:510–8.

76. Caito SW, Milatovic D, Hill KE, Aschner M, Burk RF, Valentine WM. Progression of neurodegeneration and morphologic changes in the brains of juvenile mice with selenoprotein P deleted. *Brain Res* 2011;**1398**:1–12.

77. Nah J, Yuan J, Jung YK. Autophagy in neurodegenerative diseases: from mechanism to therapeutic approach. *Mol Cell* 2015;**38**:381–9.

78. Ledonne A, Berretta N, Davoli A, Rizzo GR, Bernardi G, Mercuri NB. Electrophysiological effects of trace amines on mesencephalic dopaminergic neurons. *Front Syst Neurosci* 2011;**5**:56.

79. Alvarsson A, Zhang X, Stan TL, Schintu N, Kadkhodaei B, Millan MJ, et al. Modulation by trace amine-associated receptor 1 of experimental Parkinsonism, l-DOPA responsivity, and glutamatergic neurotransmission. *J Neurosci* 2015;**35**:14057–69.

80. Espinoza S, Salahpour A, Masri B, et al. Functional interaction between trace amine-associated receptor 1 and dopamine D2 receptor. *Mol Pharmacol* 2011;**80**:416–25.

81. Nagatsu T. The catecholamine system in health and disease-relation to tyrosine 3-monooxygenase and other catecholamine-synthesizing enzymes. *Proc Jap Acad Ser B Phys Biol Sci* 2006;**82**:388–415.

82. Boulton AA, Baker GB, Dewhurst WG, Sandler M, editors. *Neurobiology of the trace amines: analytical, physiological, pharmacological, behavioral, and clinical aspects*. Clifton, NJ: Humana Press; 1984.

83. Baker GB. Chronic administration of monoamine oxidase inhibitors: implications for interactions between trace amines and catecholamines. In: Dahlstrom A, Belmaker RH, Sandler M, editors. *Progress in catecholamine research. Part A: basic aspects and peripheral mechanisms*. New York, NY: Alan R. Liss; 1988.

84. Berry MD. The potential of trace amines and their receptors for treating neurological and psychiatric diseases. *Rev Recent Clin Trial* 2007;**2**:3–19.

85. Dourish CT. An observational analysis of the behavioural effects of β—phenylethylamine in isolated and grouped mice. *Prog Neuropsychopharmacol Biol Psychiatry* 1982;**6**:143–58.

86. Greenshaw AJ, Juorio AV, Boulton AA. Behavioral and neurochemical effects of deprenyl and β-phenylethylamine in Wistar rats. *Brain Res Bull* 1985;**15**:183–9.

87. Janssen PA, Leysen JE, Megens AA, Awouters FH. Does phenylethylamine act as an endogenous amphetamine in some patients? *Int J Neuropsychopharmcol* 1999;**2**:229–40.

88. Borison RL, Mosnaim AD, Sabelli HC. Brain 2-phenylethylamine as a major mediator for the central actions of amphetamine and methylphenidate. *Life Sci* 1975;**17**:1331–43.

89. O'Reilly RL, Davis BA. Phenylethylamine and schizophrenia. *Prog Neuropsychopharmacol Biol Psychiatry* 1994;**18**:63–75.

90. Potkin SG, Karoum F, Chuang LW, et al. Phenylethylamine in paranoid chronic schizophrenia. *Science* 1979;**206**:470–1.

91. Shirkande S, O'Reilly R, Davis B, Durden D, Malcolm D. Plasma phenylethylamine levels of schizophrenic patients. *Can J Psychiatry* 1995;**40**:221.

92. O'Reilly R, Davis BA, Durden DA, et al. Plasma phenylethylamine in schizophrenic patients. *Biol Psychiatry* 1991;**30**:145–50.

93. Sandler M, Reynolds GP. Does phenylethylamine cause schizophrenia? *Lancet* 1976;**1**:70–1.

94. Gilka L. Schizophrenia, a disorder of tryptophan metabolism. *Acta Psychiatr Scand Suppl* 1975;**258**:1–83.

95. Revel FG, Moreau JL, Pouzet B, et al. A new perspective for schizophrenia: TAAR1 agonists reveal antipsychotic- and antidepressant-like activity, improve cognition and control body weight. *Mol Psychiatry* 2013;**18**:543–56.

96. Aridon P, D'Andrea G, Rigamonti A, Leone M, Casari G, Bussone G. Elusive amines and cluster headache: mutational analysis of trace amine receptor cluster on chromosome 6q23. *Neurol Sci* 2004;**25**:S279–80.

97. D'Andrea G, Terrazzino S, Leon A, Fortin D, Perini F, Granella F, et al. Elevated levels of circulating trace amines in primary headaches. *Neurology* 2004;**62**:1701–5.

98. D'Andrea G, Granella F, Leone M, Perini F, Farruggio A, Bussone G. Abnormal platelet trace amine profiles in migraine with and without aura. *Cephalalgia* 2006;**26**:968–72.

99. D'Andrea G, D'Amico D, Bussone G, Bolner A, Aguggia M, Saracco MG, et al. The role of tyrosine metabolism in the pathogenesis of chronic migraine. *Cephalalgia* 2013;**33**:932–7.

100. Sever PS. False transmitters and migraine. *Lancet* 1979;**1**:333.

101. Lapin IP. Beta-phenylethylamine (PEA): an endogenous anxiogen? Three series of experimental data. *Biol Psychiatry* 1990;**28**:997–1003.

102. Paulos MA, Tessel RE. Excretion of beta-phenethylamine is elevated in humans after profound stress. *Science* 1982;**215**:1127–9.

103. Paetsch PR, Baker GB, Greenshaw AJ. Induction of functional downregulation of beta-adrenoceptors in rats by 2-phenylethylamine. *J Pharm Sci* 1993;**82**:22–4.

104. Szabo A, Billett E. Turner J. Phenylethylamine, a possible link to the antidepressant effects of exercise? *Br J Sports Med* 2001;**35**:342–3.

105. Paterson IA, Juorio AV, Boulton AA. 2-Phenylethylamine: a modulator of catecholamine transmission in the mammalian central nervous system? *J Neurochem* 1990;**55**:1827–37.

106. Youdim MB, Riederer PF. A review of the mechanisms and role of monoamine oxidase inhibitors in Parkinson's disease. *Neurology* 2004;**63**:S32–5.

107. Juorio AV. Drug-induced changes in the central metabolism of tyramine an dother trace monoamines: their possible role in brain functions. In: Boulton AA, Baker GB, Dewhurst WG, Sandler AD, editors. *Nuerobiology of the trace amines*. Clifton, NJ: Humana Press; 1984.

108. Juorio AV, Greenshaw AJ, Zhu MY, Paterson IA. The effects of some neuroleptics and d-amphetamine on striatal 2-phenylethylamine in the mouse. *Gen Pharmacol* 1991;**22**:407–13.

109. Baker GB, Greenshaw AJ. Effects of long-term administration of antidepressants and neuroleptics on receptors in the central nervous system. *Cell Mol Neurobiol* 1989;**9**:1–44.

110. Baumeister AA, Hawkins MF, Uzelac SM. The myth of reserpineinduced depression: role in the historical development of the monoamine hypothesis. *J Hist Neurosci* 2003;**12**:207–20.

111. Boulton AA, Juorio AV, Philips SR, Wu PH. The effects of reserpine and 6-hydroxydopamine on the concentrations of some arylalkylamines in rat brain. *Br J Pharmacol* 1977;**59**:209–14.

112. Sabelli H, Fink P, Fawcett J, Tom C. Sustained antidepressant effect of PEA replacement. *J Neuropsychiatry Clin Neurosci* 1996;**8**:168–71.

113. Sandler M, Ruthven CR, Goodwin BL, et al. Deficient production of tyramine and octopamine in cases of depression. *Nature* 1979;**278**:357–8.

114. Anderson GM, Gerner RH, Cohen DJ, Fairbanks L. Central tryptamine turnover in depression, schizophrenia, and anorexia: measurement of indoleacetic acid in cerebrospinal fluid. *Biol Psychiatry* 1984;**19**:1427–35.

115. Revel FG, Moreau JL, Gainetdinov RR, et al. Trace amine-associated receptor 1 partial agonism reveals novel paradigm for neuropsychiatric therapeutics. *Biol Psychiatry* 2012;**72**:934–42.

116. Sabelli HC, Vazquez AJ, Mosnaim AD, Madrid-Pedemonte L. 2-Phenylethylamine as a possible mediator for Δ^9-tetrahydrocannabinol-induced stimulation. *Nature* 1974;**248**:144–5.

117. Turner D. A review of the use of modafinil for attention-deficit hyperactivity disorder. *Expert Rev Neurother* 2006;**6**:455–68.

118. Madras BK, Xie Z, Lin Z, et al. Modafinil occupies dopamine and norepinephrine transporters in vivo and modulates the transporters and trace amine activity in vitro. *J Pharmacol Exp Ther* 2006;**319**:561–9.

119. Zametkin AJ, Karoum F, Rapoport JL, Brown GL, Wyatt RJ. Phenylethylamine excretion in attention deficit disorder. *J Am Acad Child Psychiatry* 1984;**23**:310–4.

120. Kusaga A, Yamashita Y, Koeda T, et al. Increased urine phenylethylamine after methylphenidate treatment in children with ADHD. *Ann Neurol* 2002;**52**:372–4.

121. Baker GB, Bornstein RA, Rouget AC, et al. Phenylethylaminergic mechanisms in attention-deficit disorder. *Biol Psychiatry* 1991;**9**:15–22.

β-Phenylethylamine-Class Trace Amines in Neuropsychiatric Disorders: A Brief Historical Perspective

A.D. Mosnaim[1] and M.E. Wolf[2]

[1]Department of Cellular and Molecular Pharmacology,
Neuroimmunopharmacology Laboratory, The Chicago Medical School at
Rosalind Franklin University of Medicine and Science, Chicago, IL,
United States [2]International Neuropsychiatry Consultants,
Highland Park, IL, United States

INTRODUCTION

Early animal research demonstrating the peripheral sympathomimetic effects of phenylethylamine (PEA) led to steady interest in its behavioral and pharmacological properties.[1] However, suggestion of a pathophysiological role in humans for this amine came much later after Jepson et al.[2] reported marked changes in its 24-h urinary excretion by phenylketonuric patients. This finding was supported by Fisher et al.,[3,4] who extended this work to other neuropsychiatric disorders such as depression and schizophrenia, eventually leading to suggestions that changes in PEA could be used as a "biological marker" for these conditions. However, efforts to use variations of urinary PEA levels as a reflection of changes in brain PEA function received much skepticism reflecting, among other considerations, the difficulties in replicating reported results, failure to convincingly show this amine to be endogenous in humans and, most importantly, growing doubts of the existence of a valid relationship between levels of neuroamine in the brain (including PEA) and their urinary concentration.[5]

It took several years to unequivocally demonstrate PEA's natural occurrence in human fluids and tissues, including brain,[6] and then it was promptly recognized as the simplest known aromatic endogenous biogenic amine. This work led to renewed speculation on this amine function in health and disease, and to the formulation of the PEA hypothesis of affective disorders,[7] which has long served as a general framework when discussing the role of PEA-like trace amines (TA) in mental disorders.[8] PEA molecular structure was soon identified as the basic building block critical for an extensive number of endogenous and exogenous bioactive PEA-class compounds. Many of these substances were already used, or were in the process of review for approval, by the Federal Drug Administration for medicinal purposes, among others, as sympathomimetic, stimulant, bronchodilator, and attention deficit hyperactivity disorders.[9] Others, including an increasing number of PEA derivatives are consumed, largely illegally, for recreational purposes; most notably α-methyl PEA (amphetamine) and further substituted amphetamine substances, for example, methamphetamine, methylenedioxy PEAs, cathinone (β-keto-amphetamine), and synthetic cathinones (bath salts).[10] Most recently, some small amounts of potentially toxic PEA-analogues, for example, β-methylphenylethylamine, have been identified in over-the-counter products advertised as "weight-loss adjuvants," and as "memory and general performance enhancers."[11–13]

PEA was also identified as the parent compound of phenylethanolamine, octopamines, and the tyramines, a group of endogenous amines originally called noncatecholic PEAs, to chemically differentiate them from dopamine and epinephrine, and now commonly known as trace amines (TAs), to emphasize their low plasma and tissue levels relative to those observed for the classical neurotransmitter amines, for example, catecholamines and serotonin.[14] Currently, the general name of TA has been expanded to include a growing number of exogenous biologically active amines showing wide differences in chemical structure as well as behavioral and pharmacological profile, for example, lysergic acid diethylamide (LSD).[15]

The early research, based largely on a number of noncontrolled studies, suggested that PEA is involved in the etiology and/or pathophysiology of a number of neuropsychiatric diseases including depression, schizophrenia, migraine, and Parkinsonism.[3,8,16] However, testing these proposals proved to be difficult on account of a number of practical considerations, for example, PEA showed very low plasma and tissue (including brain) levels in comparison to catecholamines and serotonin, a very short degradation half-life, and required elaborated analytical procedures for its determination.[5,14] These early methodological issues slowed initial progress in this area. We shall now review research data supporting a role for these TAs in neuropsychiatric disorders.

TRACE AMINES AND SCHIZOPHRENIA

Early suggestions that some forms of schizophrenia might be associated with an abnormal response to PEA resulting, at least in part, from increased bioavailability and/or changes in receptor sensitivity to this amine, were put forward based on its closely related chemical structure and behavioral profile to amphetamine.[17] Intake of this α-methyl substituted PEA analogue by humans may result in a psychosis closely resembling an acute episode of paranoid schizophrenia, and its use by schizophrenic patients may lead to exacerbation of their illness. It has been noted that PEA meets several of the criteria required to be considered a "schizogen."[18] Both amphetamine psychosis and schizophrenia respond to treatment with antipsychotic agents shown in animal studies to block PEA's behavioral effects. Thus, amphetamine- and PEA-induced stereotypies are generally considered as animal models for schizophrenia useful to screen neuroleptics since the ability of these drugs to block stereotyped behavior correlates well with their antipsychotic potency.[8,19] A number of PEA-class TAs, including various recently studied monomethylated and monohalogenated substituted derivatives, also elicit amphetamine-like behaviors which are inhibited by reserpine, blocked by neuroleptics, and enhanced by the MAO inhibitor pargyline, indicating these actions to involve catecholaminergic mechanisms. In general, these studies provide supporting evidence to postulate PEA as the "endogenous amphetamine."[13]

Various attempts to use changes in PEA urinary levels as a biological marker for schizophrenia have not produced convincing results. These include open clinical studies reporting 24-h PEA excretion to be higher in schizophrenic patients, higher in paranoid chronic schizophrenics than in non-paranoid chronic schizophrenics and normal controls, and to remain unchanged in a group of depressed patients with psychotic behavior.[20] Reports of significantly decreased platelet MAO activity in schizophrenics, particularly paranoid schizophrenic patients, leading to increased PEA availability, need further corroboration.[19]

Of special note are the studies by two independent groups led by Borowsky[21] and Grandy and Bunzow[22] leading to the cloning of a novel G-protein-coupled receptor family (trace amine-associated receptor, TAAR) that specifically binds PEA and various other TAs stimulating a remarkable new interest in the field.[21–25] Ikemoto suggested that a decrease in TA synthesizing D-neurons in the striatum and nucleus accumbens of schizophrenics with its resultant reduced TAAR stimulation may lead to mesolimbic dopamine hyperactivity that characterizes schizophrenia. These studies spurted new interest in the neurobiology of D-cells as they may provide novel insights into the pathogenesis of this condition.[26] Of special interest is the recent TAAR mapping on chromosome 6q23 as this site falls in the previously identified SCZD schizophrenia locus that links with increased susceptibility to this illness.[27]

TRACE AMINES AND AFFECTIVE DISORDERS

The PEA hypothesis of affective disorders, based on early results derived from animal and human studies, essentially states that this amine and its metabolites modulate affective behavior and that certain forms of endogenous depression are associated with a decrease in its brain levels or turnover.[7] Thus, the administration of PEA and various PEA-class substances has been shown to produce behavioral stimulation, to alter electroencephalogram tracing, and to antagonize reserpine-induced depression. Monoamine oxidase inhibitors, tricyclics (desipramine), and other antidepressant agents increase PEA brain levels which, in turn, are decreased by depression-causing drugs such as reserpine and α-methyl DOPA.[5,13,14]

However, suggestions that changes in PEA urinary excretion could be used as a biological marker for depression, including significant variations between subgroups of psychotic depressives or neurotic depressives, met with little success.[19] Attempts to use PEA precursors, for example, phenylalanine or tyrosine, for the treatment of depression have failed to produce reproducible results.[28,29]

TRACE AMINES AND EXTRAPYRAMIDAL DISORDERS

Various lines of animal research suggest a role for PEA in the pathophysiology of extrapyramidal disorders. Besides being concentrated in the caudate, its brain levels are either reduced by drugs which induce Parkinsonism (reserpine, α-methyl DOPA) or increased after the administration of the dopamine precursor L-DOPA.[8] PEA administration elicits stereotyped and choreic-like behaviors in rodents, relieves reserpine-induced Parkinsonism, and markedly potentiates the choreic-like stereotypies produced by high dosages of L-DOPA in mice pretreated with MAOI.[30] The possible role of PEA in the etiology of tardive dyskinesia ought to consider the unusual clinical findings of the coexistence of both Parkinsonism and tardive dyskinesia in the same individual, and the reversion from tardive dyskinesia to Parkinsonism and vice versa with the use of drugs.[31]

There are a few case reports in the literature describing larger-dose amphetamine-induced chorea in both adults and children which was reversible after drug discontinuation.[32,33] Furthermore, baclofen, a drug usually viewed as a GABA derivative used in the treatment of spasticity and dystonias, could also be considered as a PEA-class agent, raising the possibility that some of its effects may involve phenylethylaminergic mechanisms.[34] Recent research has focused on the role of TAARs' interaction with a variety of TAs (PEA, amphetamines, and dopamine metabolites) in modulating movement control and the actions of antiparkinsonian drugs; this promising approach uses an experimental paradigm involving a novel model of acute dopamine deficiency (DDD mice) and TAAR1 knockout mice.[35]

TRACE AMINES AND MIGRAINE

A role for vasoactive TA in the etiology of migraine has been long assumed following reports of the precipitation of this condition following the ingestion of a long list of natural and prepared foods and beverages allegedly containing various amounts of PEA and/or tyramine.[36] Administration of these substances, oral intake or injection, has been claimed to induce migraine attacks. Furthermore, reports of a significant decreased in platelet MAO activity during a migraine episode, with its proposed associated increase in circulating TA, appeared to corroborate these claims, and various attempts have been made to use MAO activity as a biological marker for migraine.[37] However, rigorous studies failed to show meaningful levels of PEA and/or tyramine in a large number of foodstuffs claimed to precipitate migraine;[38,39] a few foods samples, for example, raw shrimp and oysters allowed to decompose over time at room temperature, had only minimal increases in amine content (upto µg/g).[39] Similarly, neither platelet MAO activity nor the concentration of circulating TA studied in migraineurs, either outside or during an acute migraine episode, was statistically different from population sex-, race- and age-matched controls.[40]

TRACE AMINES AND ATTENTION DEFICIT HYPERACTIVITY DISORDERS

A possible role of TAs in attention deficit hyperactivity disorders (ADHDs) was proposed after PEA and amphetamine were shown to have similar chemical structure and shared the ability to elicit comparable stimulant behaviors in humans and animals.[8,13,19] Like amphetamine(s), which are used as therapeutic agents for ADHD, PEAs are good ligands for TA receptors, and modafinil, which has been shown to have beneficial effects in adult ADHD, has been reported to enhance PEA activity at TAAR1;[41] it has been suggested that this drug potentiates TAAR1 catecholamine-releasing actions by enhancing PEA access to these intracellular receptors via dopamine and norepinephrine transporters. Under baseline conditions transgenic mice with an overactive TAAR1 gene show "normal" behavior, and were hyposensitive to the stimulant effects of either PEA or amphetamine, further emphasizing the similar mechanism of action of these amines.[42]

NOVEL MONOSUBSTITUTED PEA DERIVATIVES: COULD THESE TRACE AMINES LEAD TO NEW PHARMACOLOGICAL TREATMENTS?

Our recent work has focused on a group of monohalogenated (p-Br-, p-Cl-, p-F, p-I) and monomethylated PEA analogues, the latter including various structural amphetamine isomers, for example, o-, p-, β-, and N-methyl PEA. These TAs share, albeit with different efficacy, a similar behavioral and pharmacological profile with amphetamine and PEA. In animal studies they elicit stimulant behaviors,[13] produce measurable analgesic effects,[43] show comparable neurotoxicity,[44] and present similar LD50 values, brain-uptake index, and brain distribution;[45] moreover, with the exclusion of amphetamine, they have a short half-life (min) reflecting their rapid oxidative degradation, mostly by MAO-B.[46]

At the 2012 Summer Olympics, PEAs and other TAs were reported to be employed as sport performance enhancers stirring controversy and attracting renewed interest in the field leading to updated rules regarding their use.[47] Although there is little information, let alone scientifically credible data, on the effects of these monosubstituted PEAs in humans, one could speculate that they may produce symptoms similar to those observed after PEA itself and amphetamine, including autonomic effects and behavioral manifestations.[48] Their use poses other obvious dangers due to their potential interaction, not only with other PEA-class compounds, but also with other drugs of abuse known to modulate PEA-associated behavior, including cocaine, morphine, and (-)-trans-delta[9]-tetrahydrocannabinol.[49,50] In animal experiments (mice and rabbits) this latter substance induced a marked and selective increase in brain PEA levels, which may mediate the euphoriant effects reported by marijuana users. In turn PEA enhances the stereotyped excitement and aggression induced by marijuana.[49,50] Despite the risks involved, these compounds are increasingly web-advertised and sold for a wide variety of conditions, including their alleged effectiveness as weight-reducing agents, mood improvers, and sport enhancers.[51] In fact, a growing number of PEA-class substances, generally known to have substantial sympathomimetic, cardiovascular, pulmonary and central nervous system (including psychotic, convulsion, and addiction potential) properties,[9] are included "as supplements" in essentially unregulated so-called energy formulations, posing significant public health risks.[11,12,52,53] The current FDA investigation of the possible health consequences of recent findings detailing the hitherto unknown presence of different β-MePEA and N,N-dimethylPEA quantities in Health Store's "healthy supplements" is being extended to other PEAs. Definitively, there is a problem with quality control on the preparation of nutritional supplements which need to be addressed by the industry with stakeholder input (see recent State of Oregon vs General Nutrition Corp.[54]).

An initial evaluation of these substances would tend to discourage further research on their possible beneficial medical uses. However, of special note are the recent studies identifying the TAAR1 as a novel integrator of metabolic control which acts on key endocrine organs such as the pancreas and small intestine, contributing to the control of glucose metabolism and body weight.[55] These authors conclude that TAAR1 thus qualifies as a novel target for the treatment of type 2 diabetes and obesity, two conditions that constitute major worldwide public health concerns.

Learning from the development of an important number of PEA-derived medicines, it could be argued that introducing changes to the basic PEA skeleton of these substances could produce new molecular entities with unexpected, desirable pharmacological profiles. Among alterations to the basic amine structure one could introduce one or more of the following: (1) changes in the length of the side chain; (2) substitution on the primary nitrogen atom; (3) substitution of the α-, and β-carbon atoms; and (4) further substitution of the phenyl ring.[9] It should be noted, for example, that the stimulant behavior of N-MePEA is changed to a depressant effect by introducing a second methyl group (to form) N,N-di-MePEA on the nitrogen atom, or o-MePEA and p-IPEA have a markedly different pharmacological profile than the other monomethylated or halogenated PEA analogues.[13]

Testing these "novel" TAs for conditions known to respond to amphetamine, that is, attention deficit hyperactivity disorder, weight loss, and narcolepsy would be a logical start.[9] In fact, and similar to PEA and amphetamine, a number of mechanisms of action could be responsible, to various degrees, for the range of biological effects produced by these TAs. Some of these actions appear to involve as yet not fully characterized biological mechanisms, which may include binding to specific PEA receptors,[56–58] or acting as "nonspecific neuromodulators."[9] Others involve direct or indirect catecholamine and serotonin release and/or "redistribution" (false neurotransmitters),[14,59] peptidergic and/or alpha

2-adrenoceptors mediated processes,[60] and inhibition of the dopamine transporter.[61,62] Most recently, research has been focused on TA binding to trace amine-associated receptors (TAARs), a collection of intracellular receptors expressed in both rats and humans which, despite their similarities to adrenergic receptors, are associated with the membrane fraction of cells but not the cell surface membrane.[60] The TAARs family of G-protein-coupled receptors contains three major groupings encompassing nine receptor subtypes.[21,22,24,25] PEA-class TAs belong to the group of neurotransmitters/neuromodulators that appear to modify catecholamine's activity at low concentrations and evoke their release at higher levels; their effects as a selective endogenous agonist at the TAAR subtypes 1 and 2 are thought to explain, at least in part, their role in interacting with adrenergic and dopaminergic transmission; overexpressing or blocking cell expression of the dopamine transporter increases or decreases PEA actions on TAAR1.[24,63]

CONCLUSION

In the last two decades we have witnessed major advances in the understanding of the molecular biology and genetics of TA that might lead to the rational design of newer, safer, and effective medications for various medical disorders.

Acknowledgments

The authors are most grateful to their numerous colleagues and coauthors (faculty, students, and staff of The Chicago Medical School and elsewhere) who contributed to their work on the biology of trace amines, in particular phenylethylamine. They are most appreciative to the various federal agencies, the State of Illinois, private foundations and pharmaceutical industry that supported this research.

The authors have no conflict of interest to declare.

References

1. Barger A, Dale JJ. Chemical structure and sympathomimetic action of amines. *J Physiol (London)* 1910;**41**:19–59.

2. Jepson JB, Lovenberg W, Zaltsman A, Oates S, Sjoerdsma A, Udenfriend S. Amine metabolism studied in normal and phenylketonuric humans by monoamine oxidase inhibition. *Biochem J* 1960;**74**:5P.

3. Fisher E, Spatz H, Heller B, Reggiani H. Phenylethylamine content of human urine and rat brain. Its alteration in pathogenic conditions and after drug administration. *Experientia* 1972;**28**:307–8.

4. Karoum F, Nasrallah H, Potkin S, Chuang L, Moyer-Schwing J, Phillips I, et al. Mass fragmentography of phenylethylamine, m- and p-tyramine and related amines in plasma, cerebrospinal fluid, urine, and brain. *J Neurochem* 1979;**33**(1):201–12.

5. Mosnaim AD, Wolf ME. *Non-catecholic phenylethylamines. Part 1: phenylethylamine: biological mechanisms and clinical aspects.* New York: Marcel Dekker; 1978.

6. Inwang EE, Mosnaim AD, Sabelli HC. Isolation and characterization of phenylethylamine and phenylethanolamine from human brain. *J Neurochem* 1973;**20**:1469–73.

7. Sabelli HC, Mosnaim AD. Phenylethylamine hypothesis of affective behavior. *Am J Psychiatry* 1974;**131**:695–9.

8. Wolf ME, Mosnaim AD. Phenylethylamine in neuropsychiatric disorders. *Gen Pharmacol* 1983;**14**(4):385–90.

9. Westfall TC, Westfall DP. Adrenergic agonists and antagonists. In: Brunton LL, Chabner BA, Krollmann BC, editors. *Goodman and Gilman's the pharmacological basis of therapeutics*, 12th ed. Mcgraw Hill; p. 277–33.

10. Shulgin A, Shulgin A. *Phenethylamines I have known and loved: a chemical love story.* Berkeley (CA): Transform Press; 1979.

11. Pawar RS, Grundel E, Fardin-Kia AR, Rader JI. Determination of selected biogenic amines in *Acacia rigidula* plant materials and dietary supplements using LC–MS/MS methods. *J Pharm Biomed Analysis* 2014;**88**:457–66.

12. Cohen PA, Bloszies C, Yee C, Gerona R. An amphetamine isomer whose efficacy and safety in humans has never been studied, β-methylphenylethylamine (BMPEA), is found in multiple dietary supplements. *Drug Testing Analysis* 2015 [Epub ahead of print].

13. Mosnaim AD, Hudzik T, Wolf ME. Behavioral effects of β-phenylethylamine and various monomethylated and monohalogenated analogs in mice are mediated by catecholaminergic mechanisms. *Am J Ther* 2015;**22**(6):412–22.

14. Mosnaim AD, Wolf ME. *Non-catecholic phenylethylamines. Part 2: phenylethanolamine, tyramines and octopamine.* New York: Marcel Dekker; 1980.

15. Berry MD. The potential of trace amines and their receptors for treating neurological and psychiatric diseases. *Rev Recent Clin Trials* 2007;**2**:3–19.

16. Boulton AA. Trace amines and mental disorders. *Can J Neurol Sci* 1980;**7**:261–3.

17. Sandler M, Reynolds GP. Does phenylethylamine cause schizophrenia? *Lancet* 1976;**1**:70–1.

18. Wyatt RJ, Potkin SG, Gillin JC, Murphy DL. Enzymes involved in phenylethylamine and catecholamine metabolism in schizophrenics and controls. In: Lipton W, DiMascio A, editors. *Psychopharmacology: a review of progress 1967–1976.* New York: Raven Press; 1977.

19. Potkin SG, Karoum F, Chwang LW, Cannon-Spoor HE, Philips L, Wyatt RJ. Phenylethylamine in paranoid chronic schizophrenia. *Science* 1979;**206**:470–1.

20. Karoum F, Potkin SG, Murphy D, Wyatt RJ. Quantitation and metabolism of phenylethylamine and tyramine's three isomer's in man. In: Mosnaim AD, Wolf ME, editors. *Noncatecholic phenylethylamines. Part 2, Tyramines, phenylethanolamine and octopamines.* New York: Marcel Dekker; 1980. p. 177–82.

21. Borowsky B, Adham N, Jones KA, Raddatz R, Artymyshyn R, Ogozalek KL, et al. Trace amines; identification of a family of mammalian G-protein-coupled receptors. *Proc Natl Acad Sci USA* 2001;**98**:8966–71.

22. Bunzow JR, Sonders MS, Arttamangkul S, Harrison LM, Zhang G, Quigley DI, et al. Amphetamine, 3,4methylenedioxymethamphetamine, lysergic acid diethylamide, and metabolites of the catecholamine neurotransmitters are agonists of a rat trace amine receptor. *Mol Pharmacol* 2001;**60**:1181–8.

23. Maguire JJ, Barriocanal FP, Davenport AP. Are vasoconstrictor responses to tyramine in human blood vessels, in vitro, mediated by the orphan trace amine receptor TIR1. *Br J Pharmacol* 2002;**131**:71P.

24. Lindemann L, Hoener MC. A renaissance in trace amines inspired by a novel GPCR family. *Trends Pharmacol Sci* 2005;**26**(5):274–81. <http://www.ncbi.nlm.nih.gov/pubmed/15860375>.

25. Lewin AH. Receptors of mammalian trace amines. *AAPS J* 2006;**8**:e138–45. Available at: <www.aapsj.org. Article 16>.

26. Ikemoto K. D-Cell hypothesis: pathogenesis of mesolimbic dopamine hyperactivity of schizophrenia. *J Behavioral Brain Sci* 2012;**2**:411–4.

27. Cao Q, Martinez M, Zhang J, Sanders AR, Badner JA, Cravchik A, et al. Suggestive evidence for a schizophrenia susceptibility locus on chromosome 6q and a confirmation in an independent series of pedigrees. *Genomics* 1997;**43**(1):1–8.

28. Gelenberg AJ, Doller-Wojcik J, Growdon JH. Choline and lecithin in the treatment of tardive dyskinesia: preliminary results from a pilot study. *Am J Psychiatry* 1979;**136**(6):772–6.

29. Yaryura-Tobias JA, Heller B, Spatz H, Fisher E. Phenylalanine for endogenous depression. *J Orthomol Psychiatry* 1974;**3**:80–1.

30. Duty S, Jenner P. Animal models of Parkinson's disease: a source of novel treatments and clues to the cause of the disease. *Br J Pharmacol* 2011;**164**(4):1357–10391.

31. Wolf ME, Chevesich E, Lehrer E, Mosnaim AD. The clinical association of tardive dyskinesia and drug induced Parkinsonism. *Biol Psychiatry* 1983;**13**(10):1181–8.

32. Morgan JC, Winter WC, Wooten GF. Amphetamine-induced chorea in attention deficit-hyperactive disorder. *Mov Disord* 2004;**19**(7):840–2.

33. Ford JB, Albertson TE, Owen KP, Sutter ME. Acute sustained chorea in children after supratherapeutic dosing of amphetamine-derived medications. *Pediat Neurol* 2012;**47**(3):216–8.

34. Wolf ME, Keener S, Mathis P, Mosnaim AD. Phenylethylamine-like properties of Baclofen. *Neuropsychobiol* 1983;**9**:219–22.

35. Sotnikova TD, Zorina OI, Ghisi V, Caron MG, Gainetdinov RR. Trace amine associated receptor 1 and movement control. *Parkinsonism Relat Disord* 2008;**14**(Suppl. 2):S99–S102.

36. McCabe BJ. Dietary migraine and other pressor amines in MAOI regimens: a review. *J Am Diet Assoc* 1982;**86**:1059–64.

37. Hannington E. Preliminary report on tyramine headache. *Br Med J* 1967;**ii**:550–1.

38. Gardner D, Shulman KI, Walker SE, Tailor S. The making of a user friendly MAOI diet. *J Clin Psychiatry* 1996;**57**:99–104.

39. Mosnaim AD, Freitag F, Ignacio R, Salas MA, Karoum F, Wolf ME, et al. Apparent lack of correlation between tyramine and phenylethylamine content and the occurrence of food-precipitated migraine. Reexamination of a variety of food products frequently consumed in the United States and commonly restricted in tyramine-free diets. *Headache Q Curr Treat Res* 1996;**7**(3):239–49.

40. Mosnaim AD, Wolf ME, Curr M, Mosnaim JM, Freitag F, Diamond S. Myths in migraine research: 1. Migraine and cluster headache patients are characterized by significant alterations in platelet monoamine oxidase activity. *Headache Q Curr Treat Res* 1996;**7**(3):225–34.

41. Madras B, Xie X, Zhicheng L, Jassen A, Panas H, Lynch L, et al. Modafinil occupies dopamine and norepinephrine transporters in vivo and modulates the transporters and trace amine activity in vitro. *J Pharmacol Exp Ther* 2006;**319**(2):561–6.

42. Achat-Mendes C, Lynch LJ, Sullivan KA, Vallender EJ, Miller GM. Augmentation of methamphetamine-induced behaviors in transgenic mice lacking the trace amine-associated receptor 1. *Pharmacol Biochem Behav* 2012;**101**(2):2001–7.

43. Mosnaim AD, Hudzik T, Wolf ME. Analgesic effects of β-phenylethylamine and various methylated derivatives in mice. *Neurochem Res* 2014;**39**(9):1675–80.

44. Mosnaim AD, Vazquez AJ, Nair V. Studies on the neurotoxicology of phenylethylamine derivatives. 1st World Congress Toxicology and Environmental Health, Washington DC. *J Am Coll Toxicol* 1982;**77**(1):124. [Abstract 152].

45. Mosnaim AD, Callaghan OH, Hudzik T, Wolf ME. Rat brain-uptake index for phenylethylamine and various monomethylated derivatives. *Neurochem Res* 2013;**38**(4):842–6.

46. Mosnaim AD, Wolf ME, Zeller EA. Degradation kinetics by MAO of PEA derivatives: a model for the molecular basis of their analgesic and behavioral effects? In: Boulton A, Baker G, Dewhurst W, Sandler M, editors. *Neurobiology of trace amines*. New Jersey: Humana Press; 1984. p. 299–306.

47. www.wada-ama.org *World anti-doping agency* [accessed October 2015].

48. *"Amphetamines" Merck manual for health care professionals*. <www.merckmanuals.com/professional/special_subjects/drug_use_and_dependence/amphetamines> [accessed June 2015].

49. Sabelli HC, Vazquez AJ, Mosnaim AD, Madrid-Pedemonte L. 2-Phenylethylamine as a possible mediator for delta^9-tetrahydrocannabinol-induced stimulation. *Nature* 1974;**248**:144–5.

50. Sabelli HC, Mosnaim AD, Vazquez AJ, Madrid-Pedmonte L. Delta^9 tetrahydrocannabinol-induced increase in brain 2-phenylethylamine: its possible role in the psychological effect of marihuana. In: Singh JM, Lal H, editors. *Drug addiction/volume 3: neurobiology and influences on behavior*. New York: Stratton; 1974. p. 271–84.

51. www.healthremedies.com [accessed October 2015, Copyrights 2008–2015].

52. ElSohly MA, Gul W. LC–MS-MS Analysis of dietary supplements for *N*-ethyl-α-ethyl-phenethylamine (ETH), *N*, *N*-diethylphenethylamine and phenylethylamine. *J Anal Toxicol* 2014;**38**(2):63–72.

53. Venhuis B, Keizers P, Van Riel A, de Kaste D. A cocktail of synthetic stimulants found in a dietary supplement associated with serious adverse effects. *Drug Testing Analysis* 2014;**6**(6):578–81.

54. *State of Oregon vs General Nutrition Corporation 2015* <http://www.usatoday.com/story/news/2015/10/22/oregon-lawsuit-gnc-supplements/74344318/>.

55. Raab S, Wang H, Uhles S, Cole N, Alvarez-Sanchez R, Künnecke B, et al. Incretin-like effects of small molecule trace amine-associated receptor 1 agonist. *Mol Metabol* 2015;**5**(1):47–56. [Epub November 2015].

56. Hansen TR, Greenberg J, Mosnaim AD. Direct effect of phenylethylamine upon isolated vascular smooth muscle of the rat. *Eur J Pharmacol* 1980;**63**:95–101.

57. Hauger R, Skolnick P, Martin L. Specific 3Hbeta-phenylethylamine binding sites in rat brain. *Eur J Pharmacol* 1982;**83**:147–8.

58. Fehler M, Broadley KJ, Ford WR, Kidd EJ. Identification of trace-amine associated receptors (TAAR) in the rat aorta and their role in vasoconstriction by β-phenylethylamine. *Naunyn Schmiedebergs Arch Pharmacol* 2010;**82**(4):385–98.

59. Sloviter RS, Connor JD, Dimaano BP, Drust EG. Para-halogenated phenylethylamines: similar serotonergic effects in rats by different mechanisms. *Pharmacol Biochem Behav* 1980;**13**(2):283–6.

60. *Independent drug monitoring unit, UK*. <www.idmu.co.uk/amphetpain.htm> [accessed June 2015].

61. Parker EM, Cubeddu LX. Comparative effects of amphetamine, phenylethylamine and related drugs on dopamine efflux, dopamine uptake and mazindol binding. *J Pharmacol Exp Ther* 1988;**245**:199–210.
62. Xie Z, Miller GM. Beta-phenylethylamine alters monoamine transporter function via trace amine associated receptor 1: implication for modulatory roles of trace amines in brain. *J Pharmacol Exp Ther* 2008;**325**(2):617–28.
63. Berry MD. Mammalian central nervous system trace amines. Pharmacologic amphetamines, physiologic neuromodulators. *J Neurochem* 2004;**90**:257–71.

CHAPTER

20

Involvement of So-Called D-Neuron (Trace Amine Neuron) in the Pathogenesis of Schizophrenia: D-Cell Hypothesis

K. Ikemoto

Department of Psychiatry, Iwaki Kyoritsu General Hospital, Iwaki, Japan

OUTLINE

Trace Amines and Neurological Disorders
DOI: http://dx.doi.org/10.1016/B978-0-12-803603-7.00020-3

INTRODUCTION

Schizophrenia is a mental illness, which afflicts approximately 1% of the world's population, and manifests delusion, hallucination, disorganized thought, flattened affect, and impaired cognitive processes. Dopamine (DA) dysfunction,[1,2] glutamate dysfunction,[3,4] neurodevelopmental deficits,[5,6] or neural stem cell (NSC) dysfunction,[7,8] are well-known hypotheses for the etiology of schizophrenia. The DA dysfunction hypothesis suggested that mesolimbic DA hyperactivity caused positive symptoms such as the paranoid-hallucinatory state of schizophrenia.[1,2] It is also explained by the efficacy of DA D2 blockers for paranoid-hallucinatory state and also by hallucinogenic acts of DA stimulants including methamphetamine or amphetamine.[1,2] Glutamate dysfunction theory was induced by the fact that intake of phencyclidine (PCP), an antagonist of N-methyl-D-aspartate (NMDA) receptor, produces the equivalent negative symptoms of schizophrenia, such as withdrawal or flattened affect, as well as the positive symptoms.[3,4] The neurodevelopmental deficits hypothesis implicates that schizophrenia is the consequence of prenatal abnormalities resulting from the interaction of genetic and environmental factors.[5,6] NSC dysfunction has also been shown to be a cause of schizophrenia.[7,8] Although mesolimbic DA hyperactivity[1,2] has been well documented in the pathogenesis of schizophrenia, the molecular basis for this mechanism has not yet been detailed. In this chapter, the author shows the rationale for the reduction of putative so-called D-neurons (trace amine (TA) neurons), that is, ligand neurons of TA-associated receptor, type 1 (TAAR1), in the striatum in the pathogenesis of mesolimbic DA hyperactivity of schizophrenia.[9] The novel hypothesis, "D-cell hypothesis of schizophrenia," is a pivotal theory to link the NSC dysfunction hypothesis with the DA hypothesis in the etiology of schizophrenia.

SO-CALLED D-NEURON (TRACE AMINE (TA) NEURON)

Definition of "D-Neuron"

TA neurons in the rat central nervous system (CNS) were described by Jaeger et al. in 1983.[10] Initially, they defined "the non-monoaminergic aromatic L-amino acid decarboxylase (AADC)-containing cell," and called it the "D-cell."[10] AADC is an equivalent enzyme to dopa decarboxylase (DDC). The D-cell contains AADC but not dopaminergic nor serotonergic.[10] Therefore, it is natural that the D-cell is thought to produce TAs,[11,12] such as β-phenylethylamine (PEA), tyramine, tryptamine, and octopamine. AADC is the rate-limiting enzyme for TA synthesis. However, it is confusing that these TAs are also "monoamines," as each one has one amino residue. It would be better to use the nomenclature of

"TA cells" for D-cells, and "TA neurons" for D-neurons. In this chapter, the author uses the words, D-cell and D-neuron, signifying TA cell and TA neuron, respectively.

Anatomy and Species Differences

The localizations of D-neurons were specified into 14 groups, from D1 (the spinal cord) to D14 (the bed nucleus of stria terminalis) in caudo-rostral orders of the rat central nervous system using AADC immunohistochemistry (Fig. 20.1).[13] In this usage, the classification term "D" means decarboxylation. In rodents,[14,15] a small number of D-cells in the striatum were rostrally described and confirmed to be neurons by electron microscopic observation.[14,15] The author reported in 1997, "dopa-decarboxylating neurons specific to the human striatum,"[16–19] that is, "D-neurons" in the human striatum[18,20] (classified to be D15) and the nucleus accumbens (Acc, D16), though monkey striatum did not contain D-neurons in these areas (Figs. 20.2 and 20.3A).[16] By using human postmortem brain materials, D-neurons have also been described in the basal forebrain (D17)[17] and the cerebral cortex (D18).[22] In humans, the D-neuron system is far developed in the forebrain (Fig. 20.1).[19]

Corresponding to the anatomical nomenclature of amine neurons, that is, A group for catecholamine neurons (A1–A16), B group for serotonergic neurons (B1–B14), and C group for epinephrine (adrenergic) neurons (C1–C3), the D group is used as the classification term for TA neurons (D1–D18) (see Table 20.1).[13,16–20,22]

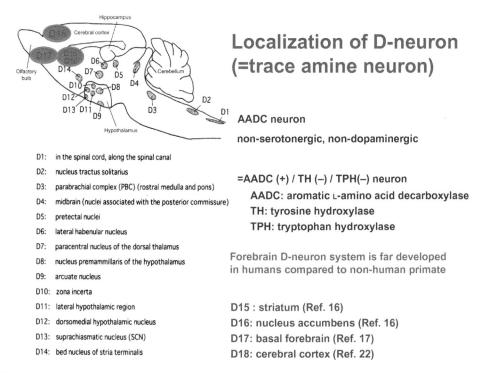

Localization of D-neuron (=trace amine neuron)

AADC neuron

non-serotonergic, non-dopaminergic

=AADC (+) / TH (−) / TPH(−) neuron
  AADC: aromatic L-amino acid decarboxylase
  TH: tyrosine hydroxylase
  TPH: tryptophan hydroxylase

Forebrain D-neuron system is far developed in humans compared to non-human primate

D15 : striatum (Ref. 16)
D16: nucleus accumbens (Ref. 16)
D17: basal forebrain (Ref. 17)
D18: cerebral cortex (Ref. 22)

D1:   in the spinal cord, along the spinal canal
D2:   nucleus tractus solitarius
D3:   parabrachial complex (PBC) (rostral medulla and pons)
D4:   midbrain (nuclei associated with the posterior commissure)
D5:   pretectal nuclei
D6:   lateral habenular nucleus
D7:   paracentral nucleus of the dorsal thalamus
D8:   nucleus premammillaris of the hypothalamus
D9:   arcuate nucleus
D10: zona incerta
D11: lateral hypothalamic region
D12: dorsomedial hypothalamic nucleus
D13: suprachiasmatic nucleus (SCN)
D14: bed nucleus of stria terminalis

FIGURE 20.1   Anatomical localization of D-neurons (trace amine (TA) neurons).

Importance of striatal D-neuron (D15, D16)

1) Species differences: Non-human primates (marmosets, macaque monkeys) lack striatal D-neurons.

2) Subventricular neural stem cell (NSC) origin

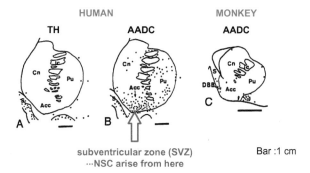

HUMAN   MONKEY
TH   AADC   AADC

subventricular zone (SVZ)
···NSC arise from here

Bar :1 cm

Each dot represents a neuronal cell body in a 50 μm-thick coronal section immunostained by TH or AADC antibody.

S, septum; Cn, caudate nucleus; Pu, putamen;
Acc, nucleus accumbens; DBB, nucleus of diagonal band of Broca

FIGURE 20.2 Importance of striatal D-neuron (D15, D16). Nonhuman primates (marmosets, macaque monkeys) do not contain striatal D-neurons (compare B and C). Anatomical localization of the nucleus accummbens (Acc, D16) is overlapped with the subventricular zone (SVZ) of lateral ventricle, the region where NSC arise.[21]

Lack of D-neuron in striatum of schizophrenia

(A) D-neurons in human Nucleus Accumbens (Acc) (D16)
D-neuron = AADC-only neuron = trace amine neuron

intrinsic neurons? or projection neurons?

(B)

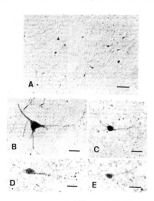

A: 100 μm B-E: 25 μm

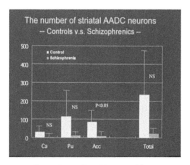

The number of striatal AADC neurons
-- Controls v.s. Schizophrenics --

Control: *n*=5 (27–64 years old)
Schizophrenia: *n*=6 (51–78 years old)

FIGURE 20.3 Lack of striatal D-neuron in striatum of schizophrenia. (A) D-neuron in the nucleus accumbens (Acc, D16) of postmortem brain without any detectable neuropsychiatric diseases.[16] The specimen was immunostained by using an antibody against aromatic L-amino acid decarboxylase (AADC).[16] (B) The number of AADC-immunostained neurons (=D-neurons) is reduced in the striatum (D15) and Acc (D16) of postmortem brains of patients with schizophrenia. As the average number of AADC-positive neurons per one section of 50 μm thick in the striatum reduced in the brains with longer postmortem period to death (PMI), analysis was performed by using fresh brain samples with PMI less than 8 h. Control: $n = 5$ (27–64 years old); schizophrenia: $n = 6$ (51–78 years old). In the Acc, the reduction of D-neurons was significant ($P < 0.05$).[9] The number of examined postmortem brains is still small, due to difficulty in obtaining postmortem brain specimens suitable enough for D-neuron visualization. An international patent, "Methodology for Visualization of Human D-cell," was required in 2014 by The Patent Cooperation Treaty (PCT).

TABLE 20.1   Nomenclature of Amine Neuron Groups—Localization of Neural Cell Bodies

A group	Norepinephrine (NE)	(A1–A7)
	Dopamine (DA)	(A8–A16)
B group	Serotonin (5-HT)	(B1–B14)
C group	Epinephrine (EN)	(C1–C3)
D group	AADC(+)	
	DA(−) 5-HT(−)	
	Trace amines (TA)	(D1–D18)

(=*d*ecarboxylation)

Lack of D-Neurons in Striatum and Nucleus Accumbens of Schizophrenia

In 2003, by using pathological and legal autopsy brains of patients with schizophrenia, reduction of D-neurons in the striatum (D15) and Acc (D16) of patients with schizophrenia [9,20] was also shown (Fig. 20.3B).

TRACE AMINE-ASSOCIATED RECEPTOR, TYPE 1 (TAAR1)

Functions

Cloning of TA receptors in 2001,[23,24] elicited enormous efforts for exploring signal transduction of these G-protein-coupled receptors whose genes are located on chromosome focus 6q23.1.[25] The receptors have been shown to colocalize with DA or epinephrine transporters in monoamine neurons and to modulate the functions of monoamines.[26–28] The TAAR1 having a large number of ligands, including PEA, tyramine, 3-iodothyronamine, 3-methoxytyramine, normetanephrine, and psychostimulants, for example, methamphetamine, 3,4-methylenedioxymethamphetamine (MDMA), and lysergic acid diethylamide (LSD),[23,25,28] has become a target receptor for exploring novel neuroleptics.[29,30]

TAAR1 knockout mice showed schizophrenia-like behaviors with a deficit in prepulse inhibition (Fig. 20.4).[31,32] TAAR1 knockout mice showed greater locomotor response to amphetamine and released more DA (and norepinephrine) in response to amphetamine than wildtype mice.[31] It has been shown that TAAR1 has a thermoregulatory function.[32] Interestingly, TAAR1 is the only human receptor that has been shown to bind endogenous TAs.[33]

Acts of TAAR1 on Ventral Tegmental Area (VTA) Dopamine (DA) Neuron

Importantly, it was clarified that increased stimulation of TAAR1 receptors on cell membranes of DA neurons in the midbrain ventral tegmental area (VTA) reduced the firing frequency of VTA DA neurons (Fig. 20.4).[29–33] This made the author suspect the existence of a critical role of TAAR1 stimulation decrease for mesolimbic DA hyperactivity in schizophrenia.

Mesolimbic dopamine (DA) system

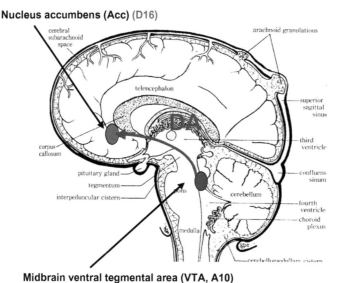

Nucleus accumbens (Acc) (D16)

Reduced TAAR1 stimulation increases firing frequency of VTA DA neurons.

TA receptor knockout mice show increase of spontaneous firing rate of VTA DA neurons, impaired prepulse inhibition, and are assumed to be a schizophrenia model animal.

Midbrain ventral tegmental area (VTA, A10)

FIGURE 20.4   Scheme of the mesolimbic dopamine system.

A NEW "D-CELL HYPOTHESIS" OF SCHIZOPHRENIA

Critical Theory for the Molecular Basis of Mesolimbic Dopamine Hyperactivity

A new theory, "D-cell hypothesis," to explain mesolimbic DA hyperactivity in the pathogenesis of schizophrenia is shown in Fig. 20.1. In the brains of patients with schizophrenia, dysfunction of NSC in the subventricular zone of lateral ventricle causes D-neuron decrease in the striatum and Acc.[8,21,33] This induces TA decrease in these nuclei, though direct evidence has not yet been demonstrated. Enlargement of the lateral ventricle,[35,36] a usual finding documented in brain imaging studies of schizophrenia, is probably due to NSC dysfunction in the subventricular zone.[7,8]

The reduction of TAAR1 stimulation on DA terminals of VTA DA neurons, caused by TA decrease, would increase the firing frequency of VTA DA neurons.[21,30,32] This increases DA release and DA turnover in the Acc,[2] resulting in mesolimbic DA hyperactivity. It has been shown that D2 stimulation of NSC inhibited forebrain NSC proliferation.[21,37] Striatal DA hyperactivity may accelerate D-neuron decrease, which accelerates hyperactivity of the mesolimbic DA system. Actions of D2-blocking agents in pharmacotherapy of schizophrenia might be explained by blocking the inhibition to forebrain NSC proliferations, and also by formation of TAAR1 ligands, such as 3-methoxytyramine and normetanephrine.[39] It is consistent with clinical evidence that initial pharmacotherapy using D2 antagonists has been proven to be critical for preventing progressive pathognomonic procedures of schizophrenia.[39]

"D-Cell hypothesis" of schizophrenia
Link of "NSC hypothesis" to "DA hypothesis"

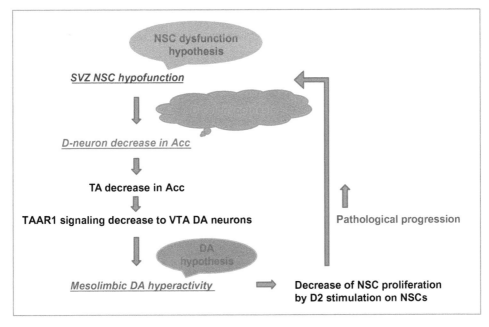

FIGURE 20.5 Scheme of D-cell hypothesis (trace amine (TA) hypothesis) of schizophrenia. In schizophrenia brain, dysfunction of neural stem cells (NSC) in the subventricular zone (SVZ) of lateral ventricle causes a D-neuron decrease in the striatum and nucleus accumbens (Acc).[8,21] This induces TA decrease in these nuclei and TAAR1 stimulation decrease onto DA terminals of VTA DA neurons, causing a firing frequency increase in VTA DA neurons.[21,30,34] This increases DA release and DA turnover in the Acc, being the molecular basis of mesolimbic DA hyperactivity of schizophrenia. Striatal DA hyperactivity causes excessive D2 stimulation of NSC and inhibits forebrain NSC proliferation,[21,37] which accelerates D-neuron decrease and accelerates mesolimbic DA hyperactivity. The rationale is that lack of striatal D-neuron, due to subventricular zone NSC dysfunction, is pivotal in explaining mesolimbic DA hyperactivity of schizophrenia.

D-Cell Hypothesis Explains Disease Progression of Schizophrenia, Leading to Establish Novel Therapeutic Strategies

The D-cell hypothesis not only links the DA hypothesis with the NSC dysfunction hypothesis, but also explains the mechanisms of disease progression of schizophrenia as shown in Figs. 20.5 and 20.6. To inhibit this cycle of pathological progression, the intervention shown in Fig. 20.6 is supposed to be effective.

1. TAAR1 agonists and partial agonists (Fig. 20.6)
   Early studies have shown formation of some TAAR1 ligands by administration of D2 antagonists, including haloperidol and chlorpromazine.[38] In recent animal studies, the effectiveness of TAAR1 ligands for schizophrenia-like symptoms of schizophrenia model animals has been shown.[30]

Strategies for novel pharmacotherapy

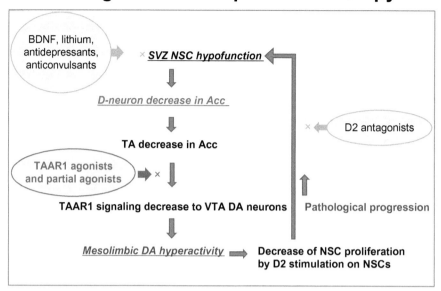

FIGURE 20.6   From a view point of pharmacotherapy, the intervention shown in the scheme is supposed to be effective in inhibiting the cycle of pathological progression.

- TAAR1 agonists or partial agonist
  TAAR1 is a prospective neuroleptic target receptor.[29,30]
- D2 antagonists
  - Early intervention for first-episode schizophrenia by D2 blockers inhibits this cycle.[39]
  - Chronic D2 blocker administration has a preventive effect for recurrence of psychoses.
  - D2 blockers increase TAAR1 agonist.[38]
- Neurotrophic substances
  Brain-derived neurotrophic factor (BDNF), lithium, anticonvulsants, antidepressants, having neurotrophic effects, activate NSC functions.[40]

2. D2 antagonists

Duration of untreated psychosis is a predictor of the long-term outcome of schizophrenia. The importance of early intervention for first-episode schizophrenia by using a D2 antagonist has been emphasized. Chronic D2 blocker administration has a preventive effect for recurrence of psychoses. D2 antagonists may block disease progression as shown in Fig. 20.6. D2 antagonists have dual actions for inhibiting this cycle of disease progression by also forming some TAAR1 ligands (3-methoxytyramine, normetanephrine) which may increase TAAR1 stimulation as shown in Fig. 20.6.[38]

3. Neurotrophic substances

Disease progression would be inhibited by neurotrophic substances (Fig. 20.6), for example, brain-derived neurotrophic factor (BDNF), lithium, anticonvulsants, or antidepressants. These substances, having neurotrophic effects, activate NSC functions,[40] and inhibit striatoaccumbal D-neuron decrease.

4. Intranasal administration of drugs, expecting retrograde transport of neuroactive substances or their precursors through the olfactory bulb, might be a novel therapeutic strategy. It is a possible preferable method of administration, as it is devoid of gastrointestinal side effects.[41–43] In this context, further investigation remains to be performed.

Some Evidences Supporting the D-Cell Hypothesis of Schizophrenia

Although it has not yet been detailed which type of TA in the human CNS is related to psychiatric symptoms, clinical and/or pharmacological observations may enable us to determine the critical type of TA. Furthermore, the type of TA that is synthesized in human striatal D-neurons has not yet been clarified.

Early in 1974, Sabelli and Mosnaim proposed the "phenylethylamine hypothesis of affective behavior,"[44] indicating the involvement of TA in animal behaviors. PEA, with a similar chemical structure to methamphetamine, is the most probable TA which affects psychiatric symptoms. One of the initial clinical symptoms frequently observed in first-episode schizophrenia is the disturbance of the sleep–wake rhythm, that is, insomnia and daytime hypersomnia. As PEA is the specific substrate for monoamine oxidase, type B (MAOB), MAOB knockout mice contained elevated levels of PEA in the striatum by 8–10 times that of controls.[45] Clinically, the MAOB inhibitor, selegiline, ameliorates daytime sleepiness of narcolepsy or other neuropsychiatric diseases. This is explained by PEA increase due to inhibition of PEA degradation.

The D-neuron decrease in the striatum of schizophrenia[9] due to NSC dysfunction, causes striatal TA decrease. The author's postmortem brain study has shown an increased DNA methylation rate of MAOB gene in Acc of schizophrenia.[46] This may be the compensation for a PEA decrease caused by lack of D-neurons in Acc.

From the aspect of food intake, PEA is included in chocolate. A high incidence of a chocolate habit of Nobel Prizewinners, that is, eating chocolate more than twice a week, has been reported.[48] PEA is supposed to be related to higher mental functions. However, too much chocolate intake in children is generally restricted, possibly aimed at preventing D-neuron downregulation.

Carlsson et al.[38] reported that administration of D2 antagonists such as chlorpromazine and haloperidol increased TAAR1 ligands, including 3-methotytyramine and normetanephrine. This indicates that the molecular basis of efficacy of D2 antagonists may be effects also via TAAR1 stimulation by 3-methotyramine and/or normetanephrine.

Ventricular enlargement in brain imaging of patients with schizophrenia[35,36] may be a similar phenomenon to the D-neuron decrease in the striatum of schizophrenia,[9] both of which support NSC dysfunction hypothesis of schizophrenia. A decreased level of plasma brain-derived neurotrophic factor (BDNF) in schizophrenia[43] is also related to NSC dysfunction.

Some evidence supporting D-cell hypothesis of schizophrenia is summarized in Table 20.2.

TABLE 20.2   Some Evidences Supporting the D-cell Hypothesis

NEURAL STEM CELL (NSC)

1. NSC dysfunction hypothesis of schizophrenia[5,6]
2. Ventricular enlargement in brain imaging of patients with schizophrenia[35–36]
3. D-neurons decrease in Acc of postmortem brains of patients with schizophrenia[9]
4. Plasma brain-derived neurotrophic factor (BDNF) decrease in patients with schizophrenia[40]

PEA AND MAOB, TRACE AMINE (TA) DEGRADING ENZYME

1. PEA hypothesis of affective bahavior[44]
2. MAOB KO mice showed striatal PEA elevation by 10 times that of controls[45]
3. MAOB inhibitor, selegiline, ameliorates daytime sleepiness of narcolepsy or other neuropsychiatric diseases. … by PEA increase?
4. In schizophrenia, insomnia and daytime sleepiness can be frequently observed as initial symptoms. … by PEA decrease?
5. Increased rate of DNA methylation of MAOB gene in Acc of some cases of schizophrenia of both sexes. … Possible compensation for TA reduction?[46]

TRACE AMINE (TA)

1. Sleep–wake rhythm disturbance of patients with schizophrenia. (insomnia and day time hypersomnia)
2. TA neuron (=D-neuron) decrease in postmortem brains of schizophrenics[9]
3. Decreased anxiety-related behaviors in PEA-elevated MAOB KO mice[47]
4. Chocolate habit of Nobel Prizewinners[48]
5. Children chocolate intake be said to restrict. … to prevent D-neuron downregulation?

D-Cell Hypothesis: Link of NSC Dysfunction Hypothesis and Dopamine Hypothesis of Schizophrenia

The "D-cell hypothesis," which is proposed by a postmortem brain study of schizophrenia, explains the molecular mechanism of mesolimbic DA hyperactivity of schizophrenia, linking the NSC dysfunction hypothesis with the DA hypothesis. Such D-cell-involved etiological dynamism in schizophrenia may exist in a wide spectrum of mental illnesses, and also in neurological illnesses.[49] As shown in Fig. 20.6, NSC functions affect not only D-neuron activity, but also the clinical courses and prognoses of neuropsychiatric illnesses. NSC regulation, such as NSC protection, as well as NSC activation, is supposed to be critical for prevention and improvement of prognoses of neuropsychiatric illnesses.

CONCLUSION

The D-neuron, that is, the TA neuron, is a clue to the pathogenesis of neuropsychiatric illnesses. The rationale is that the D-cell hypothesis of schizophrenia is a pivotal theory to link the NSC dysfunction hypothesis with the DA hypothesis, which explains the molecular basis of mesolimbic DA hyperactivity of schizophrenia. Exploration of NSC- and D-neuron-mediated signal transduction of normal and/or disease state(s) is critical for the future direction of neuropsychiatric research.

Acknowledgments

This present study was supported by Grants-in-Aids for Scientific Research from Japan Society for the Promotion of Science (C1-10680713, C1-12680740, C-22591265), Research Resource Network (RRN), and Sumitomo Pharmaceutical Corporation, and by INSERM U52, CNRS ERS 5645, Claude Bernard University (France), Shiga University of Medical Science (Japan), Fujita Health University (Japan), Clinical Research Institute of National Minami Hanamaki Hospital (Japan), Fukushima Medical University (Japan), and Iwaki City in Japan. The author is grateful to Emeritus Professor Michal Jouvet, Dr Kunio Kitahama (INSERM U52, CNRS ERS 5645, Department of Experimental Medicine, Claude Bernard University, France), and Professor Shin-Ichi Niwa (Fukushima Medical University).

References

1. Hokfelt T, Ljungdahl A, Fuxe K, Johansson O. Dopamine nerve terminals in the rat limbic cortex: aspects of the dopamine hypothesis of schizophrenia. *Science* 1974;**184**:177–9.
2. Toru M, Nishikawa T, Mataga N, Takashima M. Dopamine metabolism increases in post-mortem schizophrenic basal ganglia. *J Neural Transm* 1982;**54**:181–91.
3. Watis L, Chen SH, Chua HC, Chong SA, Sim K. Glutamatergic abnormalities of the thalamus in schizophrenia: a systematic review. *J Neural Transm* 2008;**115**:493–511.
4. Olbrich HM, Valerius G, Rüsch N, Buchert M, Thiel T, Hennig J, et al. Frontolimbic glutamate alterations in first episode schizophrenia: evidence from a magnetic resonance spectroscopy study. *World J Biol Psychiatry* 2008;**9**:59–63.
5. Christison GW, Casanova MF, Weinberger DR, Rawlings R, Kleinman JE. A quantitative investigation of hippocampal pyramidal cell size, shape, and variability of orientation in schizophrenia. *Arch Gen Psychiatry* 1989;**46**:1027–32.
6. McGlashan TH, Hoffman RE. Schizophrenia as a disorder of developmentally reduced synaptic connectivity. *Arch Gen Psychiatry* 2000;**57**:637–48.
7. Duan X, Chang JH, Ge S, et al. Disrupted-in-schizophrenia 1 regulates integration of newly generated neurons in the adult brain. *Cell* 2007;**130**:1146–58.
8. Reif A, Fritzen S, Finger M, et al. Neural stem cell proliferation is decreased in schizophrenia, but not in depression. *Mol Psychiatry* 2006;**11**:514–22.
9. Ikemoto K, Nishimura A, Oda T, Nagatsu I, Nishi K. Number of striatal D-neurons is reduced in autopsy brains of schizophrenics. *Leg Med (Tokyo)* 2003(Suppl. 1):S221–224.
10. Jaeger CB, Teitelman G, Joh TH, Albert VR, Park DH, Reis DJ. Some neurons of the rat central nervous system contain aromatic-L-amino-acid decarboxylase but not monoamines. *Science* 1983;**219**:1233–5.
11. Boulton AA. Amines and theories in psychiatry. *Lancet* 1974;**2**:52–3.
12. Boulton AA, Juorio AV. The tyramines: are they involved in the psychoses? *Biol Psychiatry* 1979;**14**:413–9.
13. Jaeger CB, Ruggiero DA, Albert VR, Park DH, Joh TH, Reis DJ. Aromatic L-amino acid decarboxylase in the rat brain: immunocytochemical localization in neurons of the rat brain stem. *Neuroscience* 1984;**11**:691–713.
14. Tashiro Y, Kaneko T, Sugimoto T, Nagatsu I, Kikuchi H, Mizuno N. Striatal neurons with aromatic L-amino acid decarboxylase-like immunoreactivity in the rat. *Neurosci Lett* 1989;**100**:29–34.
15. Mura A, Linder JC, Young SJ, Groves PM. Striatal cells containing aromatic L-amino acid decarboxylase: an immunohistochemical comparison with other classes of striatal neurons. *Neuroscience* 2000;**98**:501–11.
16. Ikemoto K, Kitahama K, Jouvet A, et al. Demonstration of L-dopa decarboxylating neurons specific to human striatum. *Neurosci Lett* 1997;**232**:111–4.
17. Ikemoto K, Nagatsu I, Kitahama K, et al. A dopamine-synthesizing cell group demonstrated in the human basal forebrain by dual labeling immunohistochemical technique of tyrosine hydroxylase and aromatic L-amino acid decarboxylase. *Neurosci Lett* 1998;**243**:129–32.
18. Kitahama K, Ikemoto K, Jouvet A, Nagatsu I, Sakamoto N, Pearson J. Aromatic L-amino acid decarboxylase and tyrosine hydroxylase immunohistochemistry in the adult human hypothalamus. *J Chem Neuroanat* 1998;**16**:43–55.
19. Kitahama K, Ikemoto K, Jouvet A, et al. Aromatic L-amino acid decarboxylase-immunoreactive structures in human midbrain, pons, and medulla. *J Chem Neuroanat* 2009;**38**:130–40.
20. Ikemoto K. Significance of human striatal D-neurons: implications in neuropsychiatric functions. *Prog Neuropsychopharmacol Biol Psychiatry* 2004;**28**:429–34.

21. Sanai N, Tramontin AD, Quiñones-Hinojosa A, et al. Unique astrocyte ribbon in adult human brain contains neural stem cells but lacks chain migration. *Nature* 2004;**427**:740–4.

22. Ikemoto K, Kitahama K, Jouvet M, et al. Tyrosine hydroxylase and aromatic L-amino acid decarboxylase do not coexist in neurons in the human anterior cingulate cortex. *Neurosci Lett* 1999;**269**:37–40.

23. Bunzow JR, Sonders MS, Arttamangkul S, et al. Amphetamine, 3,4-methylenedioxymethamphetamine, lysergic acid diethylamide, and metabolites of the catecholamine neurotransmitters are agonists of a rat trace amine receptor. *Mol Pharmacol* 2001;**60**:1181–8.

24. Borowsky B, Adham N, Jones KA, et al. Trace amines: identification of a family of mammalian G protein-coupled receptors. *Proc Natl Acad Sci USA* 2001;**98**:8966–71.

25. Miller GM. The emerging role of trace amine-associated receptor 1 in the functional regulation of monoamine transporters and dopaminergic activity. *J Neurochem* 2011;**116**:164–76.

26. Xie Z, Miller GM. Trace amine-associated receptor 1 as a monoaminergic modulator in brain. *Biochem Pharmacol* 2009;**78**:1095–104.

27. Lindemann L, Meyer CA, Jeanneau K, et al. Trace amine-associated receptor 1 modulates dopaminergic activity. *J Pharmacol Exp Ther* 2008;**324**:948–56.

28. Zucchi R, Chiellini G, Scanlan TS, Grandy DK. Trace amine-associated receptors and their ligands. *Br J Pharmacol* 2006;**149**:967–78.

29. Bradaia A, Trube G, Stalder H, et al. The selective antagonist EPPTB reveals TAAR1-mediated regulatory mechanisms in dopaminergic neurons of the mesolimbic system. *Proc Natl Acad Sci USA* 2009;**106**:20081–6.

30. Revel FG, Moreau JL, Hoener MC, et al. A new perspective for schizophrenia: TAAR1 agonists reveal antipsychotic- and antidepressant-like activity, improve cognition and control body weight. *Mol Psychiatry* 2013;**18**:543–56.

31. Panas HN, Lynch LJ, Vallender EJ, et al. Normal thermoregulatory responses to 3-iodothyronamine, trace amines and amphetamine-like psychostimulants in trace amine associated receptor 1 knockout mice. *J Neurosci Res* 2010;**88**:1962–9.

32. Wolinsky TD, Swanson CJ, Smith KE, et al. The trace amine 1 receptor knockout mouse: an animal model with relevance to schizophrenia. *Genes Brain Behav* 2007;**6**:628–39.

33. Ikemoto K. NSC-induced D-neurons are decreased in striatum of schizophrenia: Possible cause of mesolimbic dopamine hyperactivity. *Stem cell Discov* 2012;**2**:58–61.

34. Lam VM, Espinoza S, Gerasimov AS, et al. In-vivo pharmacology of trace-amine associated receptor 1. *Eur J Pharmacol* 2015;**763**:136–42. http://dx.doi.org/10.1016/j.ejphar.2015.06.026. Epub 2015 Jun 17.

35. Degreef G, Ashtari M, Bogerts B, et al. Volumes of ventricular system subdivisions measured from magnetic resonance images in first-episode schizophrenic patients. *Arch Gen Psychiatry* 1992;**49**:531–7.

36. Horga G, Bernacer J, Dusi N, et al. Correlations between ventricular enlargement and gray and white matter volumes of cortex, thalamus, striatum, and internal capsule in schizophrenia. *Eur Arch Psychiatry Clin Neurosci* 2011;**261**:467–76.

37. Kippin TE, Kapur S, van der Kooy D. Dopamine specifically inhibits forebrain neural stem cell proliferation, suggesting a novel effect of antipsychotic drugs. *J Neurosci* 2005;**25**:5815–23.

38. Carlsson A, Lindqvist M. Effect of chlorpromazine or haloperidol on formation of 3-methoxytyramine and normetanephrine in mouse brain. *Acra Pharmacol Toxicol (Copenh)* 1963;**20**:140–4.

39. Penttilä M, Jääskeläinen E, Hirvonen N, Isohanni M, Miettunen J. Duration of untreated psychosis as predictor of long-term outcome in schizophrenia: systematic review and meta-analysis. *Br J Psychiatry* 2014;**205**: 88–94. http://dx.doi.org/10.1192/bjp.bp.113.127753.

40. Fernandes BS, Steiner J, Berk M, et al. Peripheral brain-derived neurotrophic factor in schizophrenia and the role of antipsychotics: meta-analysis and implications. *Mol Psychiatry* 2015;**20**:1108–19. http://dx.doi.org/10.1038/mp.2014.117.

41. Piazza J, Hoare T, Molinaro L, et al. Haloperidol-loaded intranasally administere lectin functionalized poly(ethylene glycol)-block-poly(D,L)-lactic-co-glycolic acid (PEG-PLGA) nanoparticles for the treatment of schizophrenia. *Eur J Pharm Biopharm* 2014;**87**:30–9. http://dx.doi.org/10.1016/j.ejpb.2014.02.007.

42. Wen Z, Yan Z, Hu K, et al. Odorranalectin-conjugated nanoparticles: preparation, brain delivery and pharmacodynamic study on Parkinson's disease following intranasal administration. *J Control Release* 2011;**151**: 131–8. http://dx.doi.org/10.1016/j.jconrel.2011.02.022.

43. Ikemoto K, Nishi K, Nishimura A. Lectin-positive spherical deposit (SPD) in the molecular layer of hippocampal dentate gyrus of dementia, Down's syndrome, schizophrenia. *J Alzheimers Dis Parkinsonism* 2014;**4**:1000169. http://dx.doi.org/10.4172/2161-0460.1000169.

44. Sabelli HC, Mosnaim AD. Phenylethylamine hypothesis of affective behavior. *Am J Psychiatry* 1974;**131**:695–9.
45. Grimsby J, Toth M, Chen K, et al. Increased stress response and beta-phenylethylamine in MAOB-deficient mice. *Nat Genet* 1997;**17**:206–10.
46. Yang QH, Ikemoto K, Nishino S, et al. DNA methylation of the Monoamine Oxidases A and B genes in postmortem brains of subjects with schizophrenia. *OJPsych* 2012;**2**:374–83. http://dx.doi.org/10.4236/ojpsych.2012.224053.
47. Bortolato M, Godar SC, Davarian S, et al. Behavioral disinhibition and reduced anxiety-like behaviors in monoamine oxidase B-deficient mice. *Neuropsychopharmacology* 2009;**34**:2746–57. http://dx.doi.org/10.1038/npp.2009.118.
48. Golomb BA. Chocolate habits of Nobel prizewinners. *Nature* 2013;**499**:409.
49. Bachmann RF, Schloesser RJ, Gould TD, Manji HK. Mood stabilizers target cellular plasticity and resilience cascades: implications for the development of novel therapeutics. *Mol Neurobiol* 2005;**32**:173–202.

3-Iodothyronamine, a New Chapter in Thyroid Story: Implications in Learning Processes

A. Laurino[1] and L. Raimondi[2]

[1]Department of Neuroscience, NEUROFARBA, Section of Pharmacology, University of Florence, Florence, Italy [2]Pharmacology Unit, Department of Neuroscience, Drug Sciences and Child Health, University of Florence, Florence, Italy

INTRODUCTION

Thyroid hormone metabolism generates a family of derivatives which can be classified according to their chemical structure as thyronines (amino acidic), thyronamines (primary amines), and thyroacetic acids, each at different degrees of iodination. The discovery of the physiopathological significance of tissue levels of such compounds recently became an attractive area of research interest because of their possible implications as markers of thyroid disease or pathogenic events in their clinical manifestations.

Even if the synthetic pathways involved in the generation of thyroid hormone derivatives remain to be clarified, a plethora of iodinated compounds became potentially available for local or systemic effects in target tissues of thyroid hormone action. This possibility offers the rationale for investigating thyroid hormone metabolite functions in relation to those of thyroid hormone. Since thyroid hormone metabolites do not exert genomic functions the growing hypothesis is that they could represent the nongenomic harm of thyroid hormone.[1]

Pharmacological evidence collected around T1AM and 3-iothyroacetic acid (TA1) suggested they could behave as hormones as well as neuromodulators interfering with animal behavior and metabolism (Figure 21.1). Collectively such evidence points to the importance of measuring their tissue levels and identifying their molecular targets to open new perspectives in the diagnosis, the care of thyroid diseases, and their neurological manifestations.

Concerning the potential role of T1AM in neurological disorders we will discuss some of the main findings we collected from pharmacological studies focusing on the relationship between T1AM and histamine. The discussion will be arranged in the following sections:

1. In the first section the inclusion of T1AM in the category of the trace amine will be discussed.
2. In the second section the pharmacological features of T1AM and of its oxidative metabolite, with particular respect of their effect on memory, will be summarized.
3. In the third section, the relationship between T1AM and its oxidative metabolite (3-iodothyroacetic acid, TA1) as neuromodulators of the histaminergic system will be discussed.
4. In the final section, the potential diagnostic and/or therapeutic role of evaluating TA1/T1AM in neurodegenerative diseases will be discussed.

T1AM HAS SOME BUT NOT ALL THE FEATURES OF A TRACE AMINE

3-Iodothyronamine (T1AM) is a monoiodinated thyronamine found circulating in mammals and accumulating in several tissues including the brain. Considering its low levels recovered in tissues (nanomolar range), T1AM has been included within the category of the trace amines. Paradigmatically, trace amines are endogenous primary amines related to the classical monoaminergic mediators and, in some case, byproducts of aminergic synthesis. Evidence indicates that T1AM has some, but not all, of the characteristics to be considered as such.

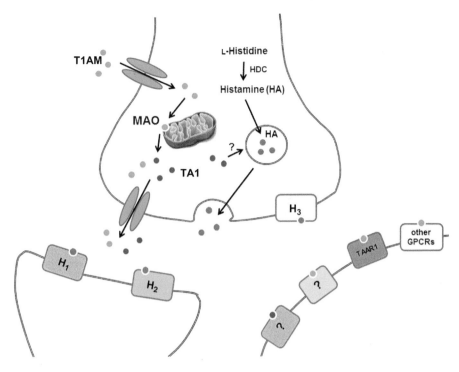

FIGURE 21.1 T1AM behaves as a neuromodulator of the histaminergic system. T1AM can enter cells and be oxidatively deaminated by MAO activity to TA1. TA1 promotes HA release or it can exit the cells and interact with unknown targets. HA released stimulates its own receptors and produces effects. T1AM, 3-iodothyronamine; TA1, 3-iodothyroacetic acid; HA, histamine; H_1, H_2 and H_3, histamine type 1, 2, and 3 receptors; MAO, mitochondrial monoamine oxidase; HDC, histidine decarboxylase; TAAR1, trace amine-associated receptor type 1; GPCRs, G-protein-coupled receptors.

The exact origin of T1AM remains elusive[2] but, irrespective of the initial substrate, decarboxylase and deiodinase activities are essential for producing T1AM and all the thyronamines. Differently from other trace amines, the decarboxylase activity involved in T1AM synthesis could be not the aromatic amino acid decarboxylase. Furthermore, T1AM enters inside cells with a mechanism different from that chosen by other monoamines or thyroid hormones.[3]

Once inside cells, T1AM undergoes enzyme degradation by the activity of ubiquitous enzymes, including monoamine oxidase (MAO) and deiodinases, producing 3-iodothyroacteic acid (TA1) and thyronamine (T0AM) respectively, or conjugated to organic acids. The measurement of endogenous brain levels of T1AM and TA1 indicated that TA1 represents 1.7% of amine levels. After pharmacological administration of T1AM, a concomitant increase in T1AM and TA1 brain levels is achieved but their ratio is reset to the same (physiological) value (1.7).[4] This result confirms that T1AM can be converted into TA1 in vivo and that, at condition of thyroid homeostasis, the two metabolites have to maintain a constant reciprocal relationship of concentration. This ratio might change at the condition of thyroid diseases and then assume diagnostic value. Moreover, differently from classical trace amines, TA1 and T0AM are still pharmacologically active and experimental evidence indicates that TA1 formation is an essential event in T1AM pharmacological effects.[4]

The discovery of a family of G-protein-coupled receptors, present in several isoforms, selectively activated by trace amines (TAARs) has prompted a resurgence of interest and forced a re-evaluation of the potential physiological relevance of these compounds.

As the canonical trace amines, T1AM does activate TAARs. In fact, T1AM was first described as a high-affinity ligand for TAAR1[5] and, more recently for type 8 (TAAR8). At the latter, T1AM would behave as an inverse agonist.[6]

However, the T1AM skeleton includes a β-phenylethylamine-like structure, a feature which potentially allows T1AM to recognize multiple cell targets including other G-coupled receptors and ion channels. In fact, T1AM binds in the micromolar range to the pancreatic alfa-2 adrenoreceptor, likely behaving as an inverse agonist.[7,8] Almost in parallel, Khajavi et al.[9] described the capacity of T1AM to modulate TRPM8 of conjunctival cells, thus introducing the possibility that T1AM might modulate store operated ion channels. Activation of such targets encourages the investigation of T1AM effects on functions governed by sensory neurons activation.

PHARMACOLOGICAL FEATURES OF T1AM AND OF ITS OXIDATIVE METABOLITE IN RODENTS

Activation of ion channels could explain the rapid timing of occurrence of the effects of pharmacologically administered T1AM and the nonlinear dose–effect relationship often observed. In addition, whether all the potential targets described are involved in the pharmacological effects of T1AM remains to be demonstrated.[10] To complicate the picture, the range of affinity values shown by the amine towards some such targets does not always correlate with the potency of the amine in vivo. In particular, it was reported that T1AM reduced body temperature when injected into mice at the dose of 25–100 mg/kg.[5] This effect was ascribed to the activation of TAAR1. Of note, for this target T1AM showed a potency within the nanomolar range of concentrations. Therefore, it is difficult to understand why, to obtain the hypothermic effect, it is necessary to inject mice with mg/kg T1AM. The pharmacokinetic features of T1AM can only explain in part this discrepancy.

Instead, T1AM is also pharmacologically active at much lower doses. In fact, we reported T1AM modified mice behavior also when administered at doses close to its physiological levels (µg/kg). At this latter condition, most of the behavioral and metabolic effects induced by T1AM were prevented or reduced in animals pretreated with clorgyline, an MAO inhibitor. Such a condition prevents the oxidative deamination of T1AM reducing the formation of TA1, which might, directly or indirectly, be part of the T1AM pharmacological effects.

In detail, the pharmacological injection of µg/kg T1AM modified mice behavior and metabolism including feeding (producing hypo- or hyperphagia, depending on the doses injected), stimulation of learning and memory, reduction of the threshold to painful stimuli, increased plasma glycemia raising a transient insulin resistance mediated by glucagon release.[11] Interestingly enough, all these effects were reduced or prevented by clorgyline (2.5 mg/kg), a monoamine oxidase inhibitor, pretreatment. In addition, T1AM effects did not always follow a linear dose–effect relationship.[11-13]

Getting further inside the relationship between T1AM and TA1, we found that TA1, injected into mice at doses equimolar to those used for T1AM, induced most of the effects observed injecting T1AM. In fact, TA1 exerted a prolearning but also an amnestic behavior without giving consolidation of memory, it increased plasma glycemia, and it reduced the threshold to hot stimuli.[13] In addition, at the doses simulating memory, TA1 reverted scopolamine- and diazepam-induced amnesia (unpublished data).

T1AM AND TA1 BEHAVE AS NEUROMODULATORS OF THE HISTAMINERGIC SYSTEM

Having ascertained that TA1 injection reproduced most of the effects described for T1AM, we then tried to investigate the mechanism underlying such effects. In line with this, we found that TA1 prolearning and hyperalgesic effects were modulated by histaminergic antagonists and that hyperalgesia was not observed in histidine decarboxylase knockout mice (HDC$^{-/-}$).[12,13] These findings indicated strongly the participation of histamine in TA1 hyperalgesia and in its prolearning effect. The next issue was to verify whether T1AM effects were also modulated by histaminergic antagonists or reduced in a mouse strain lacking histamine (HDC knockout mice). Evidence indicated it was so, corroborating the hypothesis that T1AM, throughout the biotransformation into TA1, was part of the same signaling network linking the thyroid with histamine. Of course these results also corroborated the hypothesis that TA1 was the active metabolite of T1AM responsible for hyperlgesia, hyperglycemia, and memory-enhancing.

Paradigmatically, a neuromodulator is a compound which does not produce any effects per se, but it enhances the effect of a specific neurotransmitter. In addition, a neuromodulator is a compound released from a neuron which causes no change in the excitability of postsynaptic cells in the absence of neurotransmitters. Furthermore, a neuromodulator acts to modify the action (increase or decrease) of a coexisting neurotransmitter.

Trace amines function just as neuromodulators of the monoaminergic systems fulfilling an important role in amplification/reinforcement mechanisms of the monoaminergic tone. Intuitively, such functions have important implications in brain function, including such that as basal monoaminergic tone fluctuates trace amine synthesis is altered to maintain the status quo. Accordingly, thyronamine tissue levels are expected to be altered in the case of thyroid diseases but, until now, this assumption remains to be demonstrated. However, at the light of our results, alteration of T1AM levels are also expected to produce modifications of the functions governed by histamine. Histamine is classically referred to as one among the most potent inflammatory, pruritogenic mediators. Furthermore, histamine impacts on the sleep/arousal cycle, on pain, on memory, and on feeding.[14,15] Consistently, TA1 elicits itch, reduction of the threshold to noxious and to painful heat stimuli, and it stimulates learning with a mechanism involving histamine release.[12] As far as we known, the system T1AM/TA1 can be considered as the first endogenous neuromodulator of the histaminergic system.

As has already been mentioned, T1AM stimulates learning and memory and TA1 stimulates learning but not memory consolidation. Furthermore, T1AM and TA1 can revert amnesia by scopolamine and diazepam (unpublished results). These latter findings offer the rationale for sustaining a role for these metabolites in memory impairment conditions typical of neurodegenerative disorders but also a hypothetic mechanism explaining memory disorders associated with thyroid diseases.[16]

T1AM and TA1 effects on memory were obtained when the drugs were given before the training session of the passive avoidance task and these effects are dependent on histamine release and on MAO activity. This finding sustains the role of homeostatic T1AM levels in the preservation of memory circuits by means of histamine release.

T1AM and TA1 also Revert Scopolamine- and Diazepam-Induced Amnesia

It is rather well established that the cholinergic system, other than the histaminergic system, is essential for learning and memory consolidation. In fact, pharmacological inhibition or pathological degeneration of cholinergic neurons or, alternatively, activation of inhibitory signals, including GABAergic neurons, reduces the efficacy of memory circuits mimicking aging and memory impairments typical of neurodegenerative disorders. Accordingly, muscarinic receptor antagonists, including scopolamine, and activators of GABA-A receptors, including diazepam, are often used to produce experimental models of amnesia.

T1AM, given to mice, reverted amnesia induced by low, not analgesic or sedative, doses of scopolamine or diazepam. Again, the antiamnesic effect of T1AM was prevented in mice pretreated with clorgyline, thus reinforcing the evidence that TA1 is the active principle mediating both the prolearning and the antiamnesic effects of T1AM. Interestingly enough, T1AM was more potent as antiamnesic than as prolearning, thus indicating that the removal of the cholinergic tone, by using scopolamine, amplified T1AM effects. The mechanism might relay in the production of TA1 which, in turn, activated the histaminergic system. Accordingly, histaminergic drugs proved antiamnesic efficacy at different tasks including diazepam-induced amnesia.[17]

Actually, TA1 produced amnesia at 0.4 μg/kg and stimulation of learning at 1.32 and 4 μg/kg, doses at which T1AM resulted only in prolearning. Locally, at conditions of limited distribution, T1AM has the potential to be completely transformed into TA1, with a small percentage of T1AM converted into T0AM.[4] The difference between T1AM and TA1 behavior on learning and memory might relay just on T0AM production. In any case, again, the system T1AM–TA1–histamine represented an endogenous regulator of neuronal circuits implicated in learning and memory. Currently, the upstream mechanism responsible for TA1-induced release of histamine is unknown. In this respect, recent evidence points to the role of histamine type 3 receptor (H_3R) antagonists as novel tools for treating cognitive impairments.[18] The H_3R is a presynaptic auto- and/or heteroreceptor regulating the synthesis and/or release of histamine[19] as well as of a variety of other neurotransmitters including acetylcholine, glutamate, and GABA.[20] The possibility that TA1 worked as an antagonist at H_3R cannot be ruled out even if the chemical structure of TA1 makes this hypothesis unlikely. Rather, the TA1 structure recalls the possibility of activating GABAergic or glutamatergic receptors, without excluding any interactions at channels activated by H^+. Up to now, we can disregard that TA1 interacts at GABA-A receptor and at its allosteric sites (unpublished data). In line with the role of histaminergic neurons which are the neurons of the attention, of the awaking if activated before the challenge with any negative, depressive, signals, TA1 given before the training session reverted diazepam-induced (or scopolamine) amnesia. These finding indicate that activation of the histaminergic neurons can circumvent inhibitory tones including the GABAergic and the blockade of the cholinergic systems.

This finding points to the importance of respecting the timing of T1AM/TA1 administration to evoke the release of histamine and to obtain an increase in attention and awaking behavior. Consequently one can speculate that thyroid hormone metabolites and not only thyroid hormone, could somehow be implicated in maintaining learning circuits in healthy conditions. This statement needs to be supported by further experimental and clinical data.

THE HISTAMINE–THYROID AXES AS A NOVEL POTENTIAL TARGET FOR DIAGNOSIS OR THERAPY OF NEURODEGENERATIVE DISORDERS

The existence of a causal relationship between thyroid as well as neuronal histamine signaling dysregulation and the onset of neurodegenerative diseases remains to be established in appropriate studies. Even less is known about the levels of thyroid hormone metabolites, in particular TA1/T1AM, in neurodegenerative diseases, either at experimental or clinical settings.

Thyroid hormone has an incontrovertible prenatal role on the harmonious development of the brain. However, thyroid hormone has important actions in the adult brain too. In fact, hypo- and hyperthyroidism associate with neuropsychiatric complaints and symptoms, including increased susceptibility to depression and reductions in health-related quality of life. Neuropsychiatric symptoms refer to a spectrum of emotional, melancholic depression, dementia, and cognitive problems that are directly related to changes in the brain secondary to multiple factors, including the direct effects of thyroid disease, as well as changes in hormone levels. These symptoms tend to improve with treatment and normalization to an euthyroid state, thus indirectly suggesting thyroid hormone homeostasis can play a role in the onset of such symptoms. Furthermore, the importance of homeostatic hormone levels in neuropsychiatry would be further sustained by the impact of subclinical hypothyroidism on mood and cognitive functions.[20] Patients with subclinical hypothyroidism often experience neurological complications including cognition and sleep–arousal cycle impairments, altered threshold to noxious and heat stimuli, itch, and effects related to altered cholinergic transmission. Despite these, the relationship between TSH and/or T4 plasma levels and neurological deficit scores remains to be demonstrated making inconsistent the clinical assessment of a causal relationship between thyroid hormone levels and cognitive decline, including aging.[21,22] This notwithstanding that in a population of old women, low T(4) levels, but still within the euthyroid range, were found associated with a greater risk of cognitive decline over a 3-year period[23] and with future dementia risk.[24,25] Other evidence indicates that hypothyroidism doubled the risk of Alzhemier disease.[26] Interestingly, in this disease, the activity of enzymes involved in thyroid hormone synthesis and metabolism, including deiodinases and monoamine oxidases, has been documented in several brain areas.[27,28] At these conditions, irrespective of the synthetic pathway of T1AM, the TA1/T1AM is expected to change locally, potentially contributing to affect behavior controlled by specific brain area, including cognition. This possibility warrants investigations constituting an important evidence of the fact that, independently of T4 levels, the thyroid–histamine axis might change secondary to pathology-induced alterations of thyroid hormone metabolism. Therefore, even if thyroid function is within physiological limits, the activity of histaminergic neurons might be altered.

Furthermore, it is thought that the role of the histaminergic system in neurodegeneration is likely secondary, with histaminergic neurons located at the hypothalamus from which they innervate most brain areas. Instead this localization is compatible with the concept that histamine is involved in general central regulatory mechanisms. However, the neuronal histaminergic system has been reported to be dysregulated in neurodegenerative disorders including Parkionson[29,30] and Alzheimer diseases.[31] Some reports indicate that histamine concentrations increased in various brain regions of Parkinson disease patients,[32] likely for pathology-induced inactivation of histamine-metabolizing enzymes including N-methyl transferase. On the contrary, histamine levels were found reduced in Alzheimer disease patients.[33] Taken together all this evidence makes it very likely that alterations of the histaminergic system participate in the pathogenesis of symptoms including disturbed sleep, memory and alimentary impairments. In particular, from the point of view of the thyroid–histamine axis, and independently on the fluctuation of histamine levels, local T1AM levels might be determinant in the regulation of histaminergic neuron firing. Thus the maintenance of a correct link between thyroid and the histaminergic neurons, through the mediation of T1AM, might constitute a warrant of preservation of behavioral circuits, including those of memory and learning.

This potential role of T1AM would be even more sustained by the fact that current treatment options for neurodegenerative diseases are highly limited and drugs used have several side-effects and/or are only partially effective in ameliorating the cognitive symptoms. Administration of T1AM could work as prospected for H_3 antagonists[34] and T1AM could represent an add-on therapy in counteracting cognitive and other behavioral disorders when administered at the very early phases of diseases. In addition, another advantage of T1AM would be represented by its pharmacokinetic features, including the fact it is a prodrug, which makes available TA1 in the central nervous system after its systemic administration. In this respect, the important peripheral metabolism of T1AM would ensure high bioavailability of TA1 at the central nervous system. Of course to make our hypothesis consistent, the evaluation of the TA1/T1AM in the brain of healthy and pathological individuals and the identification of active doses of T1AM is absolutely compelling.

CONCLUSION

The evaluation of TA1/T1AM, resuming thyroid hormone metabolism, could potentially represent a novel marker of neurodegenerative disease diagnosis, an index of its severity but also a target for drugs sustaining cognitive functions.

To this goal, more extensive data on T1AM/TA1 levels in neurodegenerative diseases should be obtained in order to validate the diagnostic significance of such a ratio and to assess conditions for supplementing T1AM to selected patients.

References

1. Davis PJ, Davis FB. Mechanisms of nongenomic actions of thyroid hormone. *Thyroid* 1996;**6**:497–504. Review.
2. Hackenmueller S, Marchini M, Saba A, Zucchi R, Scanlan TS. Biosynthesis of 3-iodothyronamine (T1AM) is dependent on the sodium-iodide symporter and thyroperoxidase but does not involve extrathyroidal metabolism of T4. *Endocrinology* 2012;**153**:5659–67.
3. Ianculescu AG, Giacomini KM, Scanlan TS. Identification and characterization of 3-Iodothyronamine intracellular transport. *Endocrinology* 2009;**150**:1991–9.
4. Laurino A, De Siena G, Saba A, Chiellini G, Landucci E, Zucchi R, et al. In the brain of mice, 3-iodothyronamine (T1AM) is converted into 3-iodothyroacetic acid (TA1) and it is included within the signaling network connecting thyroid hormone metabolites with histamine. *Eur J Pharmacol* 2015;**716**:130–4.
5. Scanlan TS, Suchland KL, Hart ME, Chiellini G, Huang Y, Kruzich PJ, et al. 3-Iodothyronamine is an endogenous and rapid-acting derivative of thyroid hormone. *Nat Med* 2004;**10**:638–42.
6. Dinter J, Mühlhaus J, Wienchol CL, Yi CX, Nürnberg D, Morin S, et al. Inverse agonistic action of 3-iodothyronamine at the human trace amine-associated receptor 5. *PLoS One* 2015;**10**:e0117774.
7. Regard JB, Kataoka H, David AC, Camerer E, Yin L, Zheng YW, et al. Probing cell type–specific functions of Gi in vivo identifies GPCR regulators of insulin secretion. *J Clin Inv* 2007;**117**:4034–43.
8. Dinter J, Mühlhaus J, Jacobi SF, Wienchol CL, Cöster M, Meister J, et al. 3-iodothyronamine differentially modulates alpha-2A adrenergic receptor-mediated signaling. *J Mol Endocrinol* 2015;**15**:0003.
9. Khajavi N, Reinach PS, Slavi N, Skrzypski M, Lucius A, Strauß O, et al. Thyronamine induces TRPM8 channel activation in human conjunctival epithelial cells. *Cell Signal* 2015;**27**:315–25.
10. Chiellini G, Erba P, Carnicelli V, Manfredi C, Frascarelli S, Ghelardoni S, et al. Distribution of exogenous [125I]-3-iodothyronamine in mouse in vivo: relationship with trace amine-associated receptors. *J Endocrinol* 2012;**213**:223–30.
11. Manni ME, DeSiena G, Saba A, Marchini M, Dicembrini I, Bigagli E, et al. 3-Iodothyronamine: a modulator of the hypothalamus–pancreas–thyroid axes in mice. *Br J Pharmacol* 2012;**166**:650–8.
12. Laurino A, De Siena G, Resta F, Masi A, Musilli C, Zucchi R, et al. 3-iodothyroacetic acid, a metabolite of thyroid hormone, induces itch and reduces threshold to noxious and to painful heat stimuli in mice. *Br J Pharmacol* 2015;**172**:1859–68.
13. Musilli C, De Siena G, Manni ME, Logli A, Landucci E, Zucchi R, et al. Histamine mediates behavioral and metabolic effects of 3-iodothyroacetic acid (TA1), an endogenous end product of thyroid hormone metabolism. *Br J Pharmacol* 2015;**172**:1859–68.
14. Kajihara Y, Murakami M, Imagawa T, Otsuguro K, Ito S, Ohta T. Histamine potentiates acid-induced responses mediating transient receptor potential V1 in mouse primary sensory neurons. *Neuroscience* 2010;**166**:292–304.
15. Kohler CA, da Silva WC, Benetti F, Sartori Bonini J. Histaminergic mechanisms for modulation of memory systems. *Neural Plast* 2011;**2011**:328602.
16. Rivas M, Naranjo JR. Thyroid hormones, learning and memory. *Genes Brain Behav* 2007;**6**(Suppl. 1):40–4.
17. Blandina P, Efoudebe M, Cenni G, Mannaioni P, Passani MB. Acetylcholine, histamine, and cognition: two sides of the same coin. *Learn Mem* 2004;**11**:1–8.
18. Schneider EH, Neumann D, Seifert R. Modulation of behavior by the histaminergic system: lessons from HDC-, H3R- and H4R-deficient mice. *Neurosci Biobehav Rev* 2001;**447**:101–21.
19. Singh M, Jadhav HR. Histamine H3 receptor function and ligands: recent developments. *Mini Rev Med Chem* 2013;**13**:47–57.
20. Bhowmik M, Khanam R, Vohora D. Histamine H3 receptor antagonists in relation to epilepsy and neurodegeneration: a systemic consideration of recent progress and perspectives. *Br J Pharmacol* 2012;**167**:1398–414.
21. Hogervorst E, Huppert F, Matthews FE, Brayne C. Thyroid function and cognitive decline in the MRC Cognitive Function and Ageing Study. *Psychoneuroendocrinology* 2008;**33**:1013–22.
22. Wijsman LW, de Craen AJ, Trompet S, Gussekloo J, Stott DJ, Rodondi N, et al. Subclinical thyroid dysfunction and cognitive decline in old age. *PLoS One* 2013;**8**:e59199.
23. Volpato S, Guralnik JM, Fried LP, Remaley AT, Cappola AR, Launer LJ. Serum thyroxine level and cognitive decline in euthyroid older women. *Neurology* 2002;**58**:1055–61.
24. Ganguli M, Burmeister LA, Seaberg EC, Belle S, DeKosky ST. Association between dementia and elevated TSH: a community based study. *Biol Psychiatry* 1996;**40**:714–25.

25. Tan ZS, Beiser A, Vasan RS, Au R, Auerbach S, Kiel DP, et al. Thyroid function and the risk of Alzheimer disease: the Framingham study. *Arch Intern Med* 2008;**168**:1514–20.
26. O'Barr SA, Oh JS, Ma C, Brent GA, Schultz JJ. Thyroid hormone regulates endogenous amyloid-beta precursor protein gene expression and processing in both in vitro and in vivo models. *Thyroid* 2006;**16**:1207–13.
27. Sparks DL, Van Woeltz M, Markesbery WR. Alterations in brain monoamine oxidase activity in aging, Alzheimer's disease, and Pick's disease. *Arch Neurol* 1991;**48**:718–21.
28. Courtin F, Zrouri H, Lamirand A, Li WW, Mercier G, Schumacher M, et al. Thyroid hormone deiodinases in the central and peripheral nervous system. *Thyroid* 2005;**15**:931–42.
29. Anichtchik OV, Rinne JO, Kalimo H, Panula P. An altered histaminergic innervation of the substantia nigra in Parkinson's disease. *Exp Neurol* 2000;**163**:20–30.
30. Anichtchik OV, Peitsaro N, Rinne JO, Kalimo H, Panula P. Distribution and modulation of histamine H(3) receptors in basal ganglia and frontal cortex of healthy controls and patients with Parkinson's disease. *Neurobiol Dis* 2001;**8**:707–16.
31. Trillo L, Dasb D, Hsieh W, Medina B, Moghadam S, Lin B, et al. Ascending monoaminergic systems alterations in Alzheimer's disease. Translating basic science into clinical care. *Neurosci Biobehav Rev* 2013;**37**:1363–79.
32. Rinne JO, Anichtchik OV, Eriksson KS, Kaslin J, Tuomisto L, Kalimo H, et al. Increased brain histamine levels in Parkinson's disease but not in multiple system atrophy. *J Neurochem* 2002;**81**:954–60.
33. Shan L, Bossers K, Unmehopa U, Bao AM, Swaab DF. Alterations in the histaminergic system in Alzheimer's disease: a postmortem study. *Neurobiol Aging* 2012;**33**:2585–98.
34. Ellembroek BA, Ghiabi B. The other side of the histamine H$_3$ receptor. *Trends Neurosci* 2014;**37**:191–9.

Trace Amine Receptors and Mood Disorders: Focusing on Depression

C.-U. Pae[1,2,3] and S.-M. Wang[1,4,5]

[1]Department of Psychiatry, The Catholic University of Korea College of Medicine, Seoul, Republic of Korea [2]Department of Psychiatry and Behavioral Sciences, Duke University Medical Center, Durham, NC, United States [3]Department of Psychiatry, Bucheon St. Mary's Hospital, Wonmi-Gu, Bucheon, Kyeonggi-Do, Republic of Korea [4]International Health Care Center, Seoul St. Mary's Hospital, The Catholic University of Korea College of Medicine, Seoul, Republic of Korea [5]Department of Psychiatry, Seoul St. Mary's Hospital, Seocho-Gu, Seoul, Republic of Korea

Trace Amines and Neurological Disorders
DOI: http://dx.doi.org/10.1016/B978-0-12-803603-7.00022-7

INTRODUCTION

Major depressive disorder (MDD) is a complex mental condition including physical, emotional, cognitive and functional impairments. It is characterized by one or more major depressive episodes (MDE) which are not due to a medical condition, medication, abused substance or psychotic diseases. The MDE must have either depressed mood or loss of interest and at least five additional symptoms are also required for the diagnosis: (1) depressed mood; (2) loss of interest or pleasure; (3) significant change in appetite/weight; (4) insomnia/hypersomnia; (5) psychomotor agitation/slowing; (6) fatigue/loss of energy; (7) feelings of worthlessness/inappropriate guilt; (8) inability to concentrate/indecisiveness; and (9) recurrent thoughts of death/suicide. The pathophysiology of MDD has not yet been fully elucidated. Biological, genetic, personality, and psychosocial factors may interact and lead to the development of diverse mood disorders including MDD.

The principal treatment of MDD is pharmacological approaches including antidepressants since the 1950s. However, the action mechanism of antidepressants is still also barely understood. Despite a lack of exact understanding on their mechanism of action, the traditional concept is that enhancement of the monoamine neurotransmitters in the central nervous system (CNS) is clearly linked to the antidepressant effects, supporting evidence for the "monoamine hypothesis" of depression.[1]

The "monoamine hypothesis" of MDD may assert that the underlying biological or neuroanatomical basis for MDD is a deficiency of central norepinephrine (NE), dopamine (DA), and/or serotonin (5-hydroxytryptamine, 5-HT) systems and that targeting this neuronal lesion with an antidepressant would tend to restore normal function in patients with MDD.[2] With advanced technologies, the monoamine hypothesis has also modified and evolved as the evidence has changed over the years.[2] When patients with MDD are given antidepressants, the treatment effects are not prompt but gradual, supporting a new concept of adaptive model of neurotransmitter receptors in the CNS. Most antidepressants presented their efficacy not only for MDD but also for the treatment of various mental disorders such as anxiety disorders, eating disorders, and obsessive-compulsive disorder, indicating that monoamine imbalance is not confined to mood disorder.[2,3] Furthermore, the antidepressant effect is not sufficient for treatment of all patients with MDD, only approximately 50–70% of such patients may respond to antidepressant treatment based on well-controlled clinical trial data.[4–6] Theoretically, the development of broad-spectrum antidepressants such as triple reuptake inhibitors simultaneously inhibiting the reuptake of 5-HT, NE, and DA from the synapse may be beneficial in treatment of patients with MDD since the additive effect of enhancing neurotransmission in all three monoamine systems may lead to improved efficacy and rapid onset of antidepressant response.[7] However, such efforts have not been successful. In addition, there is no proof of class effects among various antidepressants regardless of their main neurotransmitter involved in action mechanisms (according to basic action on DA, NE, and/or 5-HT).

Thus, the monoamine hypothesis is still waiting for a definite and comprehensive re-evaluation, and other neurotransmitters may be clearly involved. To address this topic, researchers' attention has been recently attracted by amines which can be found in traces in the brain, so-called "trace amines" (TAs), as they potentially play a regulatory role for traditional monoamines.

TAS AND DEPRESSION

The recent discovery and advancement in characterization of molecular properties of TAs brought attention to researchers investigating neurotransmitter disturbance and mood disorders. A number of research findings have been proposed to be potential evidences indicating that TAs may be involved in the development and treatment of mood disorders.

The TAs are a family of endogenous compounds that are structurally similar to traditional monoamine neurotransmitters. The major TAs include β-phenylethylamine (β-PEA), tyramine (TYR), tryptamine (TRYP), octopamine (OA), phenylacetic acid (PhAA), and indolacetic acid (IAA), which are presented in trace levels in the CNS in relation with NE, DA, and 5-HT).[8,9] Under physiological conditions, they are released in an activity-dependent manner and do not alter the electrical excitability of neurons in the absence of other neurotransmitters.[10] For this reason they have also been termed "false neurotransmitters."[11] Although endogenous levels of TAs are usually several hundred-fold lower than those of traditional amines (<10 nM), they have a similar biosynthesis rate resulting in an extremely rapid rate of turnover (endogenous pool half-life of approximately 30 s) due to their lack of a vesicle storage mechanism.[10]

They are synthesized from amino acid precursors by aromatic amino acid decarboxylase and are enzymatically degraded by the action of monoamine oxidase (MAO) A and B.[10] Their levels are also dramatically increased when MAOs (MAO-A and MAO-B) are inhibited or MAO genes are deleted in animal knockout models.[12] One of the older forms of antidepressants, MAO inhibitors (MAOI), are known to increase the functional availability of monoamine and improve neurotransmission at catecholamine- and/or indoleamine-mediated synapses in the CNS, indicating that TA metabolism may be involved in the action mechanism of MAOIs.[13]

Colocalization of TAs and Traditional Neurotransmitters

TAs are distributed throughout the CNS and also colocalized with the origins and terminal projections of DA, NE, and 5-HT neurotransmitters system.[13] This heterogenous and adjacent distribution of TAs around traditional monoamine neurotransmitters may allow us to assume that their interactive action mechanisms in the development of depressive disorders as well as conceive TAs as potential and novel therapeutic target for mood disorders. For instance, the globus pallidus shows high levels of β-PEA and TYR, whereas TRYP levels are relatively low.[13,14] TRYP and OA are highly expressed in the dorsal and medial raphe, as well as in the locus coeruleus, substantia nigra–ventral tegmental area (VTA), and in additional mesencephalic regions.[10,14,15] OA immunoreactivity is present in more limbic forebrain regions (ie, amygdala, hippocampus, globus pallidus, caudate-putamen, and nucleus accumbens).[8,13,15] However, TAs exert their functions by being also distributed in other CNS sites not containing traditional neurotransmitters, indicating a more complex action mechanism of TAs in the CNS. The clear understanding on the colocalization between TAs and traditional neurotransmitters will help to reveal the exact role of TAs as neuromodulators, neurotransmitters, or simple cotransmitters with adjacent neurotransmitters.[13] Similar to TAs, a heterogeneous and complex distribution of the TA1 receptor has also been found today.[10]

Alteration of TAs in Depressive Disorder

There has been some evidence suggesting alteration of TAs and/or their metabolites in patients with mood disorders, although some inconsistent findings are also currently available.[16,17]

The β-PEA hypothesis has been already postulated, proposing that it may regulate mood and affect. Fischer et al.[18] has firstly reported significantly lower β-PEA excretion in depressed patients than in controls. Subsequent researches[13,19–21] have also supported Fisher et al.'s previous finding, although others have failed to do so.[22,23] Such discrepancies between studies may reflect the complex diagnosis of depression and different clinical factors among study samples (ie, age, intraindividual variability in β-PEA excretion, duration of illness, first onset of diseases, and drug influence) as well as different laboratory techniques to detect β-PEA. Likewise, research findings supporting significant difference of plasma level of β-PEA between depressive patients and controls has had a dearth of data until now.[24] According to Nakagawara's study,[24] there were no differences between depressive patients and controls in terms of plasma level of β-PEA. In addition, the plasma MHPG levels in depressive patients were also similar to those in healthy controls. However, there was a significant negative correlation between plasma β-PEA levels and plasma MHPG levels in depressive patients, while it was not observed in the controls, indicating a clear role of β-PEA on noradrenergic functional activity in depressed patients.

It is unclear whether or not the β-PEA administration or its derivatives may directly affect mood symptoms in patients with MDD since there has been a paucity of such researches. According to a long-term follow-up study (20 and 50 weeks),[25] approximately 86% of patients who received β-PEA administration (10–60 mg/day) responded and maintained its clinical response, without serious adverse events, indicating that β-PEA may be a useful and potential alternative therapeutic agent for maintenance treatment of depressive disorder. β-PEA administration was also effective in the treatment of those who failed to show clinical response to the previous standard treatments. Additionally, it improves mood as rapidly as amphetamine but does not produce tolerance. Some anecdotal pilot studies have proposed its potential as an alternative treatment to current antidepressant treatment.[26–28] In Beckmann et al.'s study, 75–200 mg/day phenylalanine (PAA) was given to 20 depressive inpatients that showed unsuccessful antidepressant treatment and diagnosed by International Classification of Diseases (ICD) for 20 days, in which 20% and 40% of patients showed a good response and complete remission to PAA treatment, respectively. Interestingly, the core depressive symptoms including depressed mood, psychomotor retardation, and/or agitation presented better treatment response than anxiety, insomnia, somatic symptoms (ie, hypochondriasis) and compulsiveness. According to another study,[26] PAA of 50 or 100 mg/day was prescribed for patients who were nonresponsive to either Tricyclic antidepressants (TCAs) or MAOIs for 2 weeks, in which 74.0% of patients treated with PAA achieved complete remission without significant adverse events (AEs). This study suggested that PAA may be another option for patients with difficult-to-treat depression.

Another interesting point is that concentrations of β-PEA in CSF in Parkinson's disease (PD) were also found to be significantly lower than in controls.[29] The prevalence of depression in PD is estimated to be high (approximately 50.0%)[30] and the pathophysiologies of PD have been known to be shared with depressive disorders, in particular, loss of dopamine and NE innervation in the limbic system.[31] The plasma β-PEA concentrations in PD patients were also significantly lower than those in the control group. The concentrations of β-PEA showed differential levels by natural course of PD patients, the progressive group presented a downward trend over after 1 or 2 years, while the nonprogressive group retained same levels of β-PEA.

PhAA is the principal metabolite of β-PEA and thereby it should play a role as a surrogate marker of central β-PEA metabolism. Significantly lower urinary PhAA concentrations in depressive patients have been consistently proposed by some previous researches,[28,32,33] while others have failed to replicate such findings.[12,22] It should also be noted that severely depressed patients showed significantly lower plasma and urinary PhAA levels than did those with less severe depression, while the levels of β-PEA were not influenced by severity of depression, indicating that the levels of PhAA may be a more useful and sensitive indicator for accurate discrimination on the severity of depressive symptomatology.[34] However, similar to the findings of β-PEA studies, there have been inconsistent findings in quantitative and qualitative researches regarding the levels of PhAA in depressive disorders, indicating a need for more consistently supporting study results favorable to the potential role of PhAA for the development and treatment of depressive disorders. In fact, such anecdotal studies were mostly performed at the time that diagnostic criteria were not fully established, by which undetectable subpopulation bias may alter study results. We should also consider the fact that molecular and overall clinical study techniques at the time were less advanced compared to current methods (ie, interindividual and diurnal variation in the level of PhAA, gender effect, complex migration, and existence of PhAA in human body, sensitivity and specificity to detect PhAA level, etc.). When it comes to clinical factors in researches on PhAA for depressive disorders, the differential changes of levels of PhAA between depressive patients with acute, chronic and recurrent state in their clinical course were possibly not reflected in such previous studies.[17] Interestingly, despite DeLisi et al.[22] failing to replicate the differential levels between patients with depressive disorder and controls, significantly a lower level of PhAA was found in acute depressed and hospitalized patients along with suicidal ideation than in chronically ill patients, indicating PhAA excretion may be differential as the clinical status in depressive patients.

There has been a lack of study findings concerning the level changes of TYR in depressive disorder, compared with β-PEA and PhAA. It was proposed that TYR may have a role in the regulation of NE turnover since one study found that the excretion rate of TYR positively correlated with 24-h NE turnover in depressive patients as well as the TYR excretion being high in depressive patients.[35] Likewise, Sandler et al. successfully demonstrated a statistically significant decrease in urinary excretion of TYR in depressive patients compared with controls. Interestingly such difference was more evident in patients with more severe endogenous depression than those without.[23] However, Roy et al.[36] have reported that there was no significant difference in urinary levels of TYR between depressive patients and healthy controls; additionally no significant correlations among the depressive patients between the urinary excretion rates of TYR and the urinary excretion rates of NE and its metabolites were found either. This discrepancy may be due to different diagnosis of patients included in both studies (bipolar depression vs unipolar depression, respectively) and lack of controls of the former research. Other research groups[37,38] have also continuously failed to show increased urinary TYR excretion in depressive patients compared with controls.

Significantly less excretion of TRYP in patients with depression versus controls has been shown in some studies,[26] while others failed to do so.[18,39,40] TRYP excretion was also found to be significantly higher in the depressive phase compared with the treatment for the psychotic and total samples, while changes in TRYP excretion between the treatment and recovered phases were nonsignificant.[39]

Other TAs, including OA[23] and hydroxyphenylacetic acid (HpAA),[41–43] have also been known to be altered in patients with depressive disorders. There were no differences in the concentration of IAA between patients with depression and controls.[39,44]

Modulation Effects of TAs on Neurotransmitters

TAs have been known to influence the release or uptake of DA, NE, and 5-HT through diverse molecular events such as direct targeting on neurotransmitter vesicles and active transport regulation in plasma membrane.[9,45–47] Overall these classical modulatory effects of various TAs may occur by preactivation, simultaneous activation, or postactivation of specific biochemical molecules of the DA, NE, or 5-HT-related signal transduction system.[48]

Indeed, β-PEA was found to have prominent effects in release of DA, NE, and 5-HT in a number of previous researches.[45,47,49,50] According to one interesting study,[50] changes in functional activity onto traditional monoamine uptake and release was observed by modifications in the β-PEA structure: (1) uptake inhibition and release stimulation of DA and 5-HT by methylation in the α-position of β-PEA; (2) enhancement of 5-HT uptake and release by creation of p-Cl group in the β-PEA; (3) increased potency as an uptake inhibitor or a releaser of DA, 5-HT, and NE by production of phenolic–OH groups in the β-PEA; and (4) decreased potency in uptake inhibition or release of DA and 5-HT by introduction of β-OH groups in the β-PEA. This finding clearly indicates that β-PEA interferes in a complex way with the uptake and the release of principal traditional monoamines such as DA, NE, and 5-HT of the CNS under their interactions (ie, presence of absence of some neurotransmitters may be needed to potentiate the inhibition of uptake or release of others). Based on a previous research,[47] much lower concentrations of β-PEA than of phenelzine were required to stimulate the release of DA and 5-HT, indicating a more strong potency of PEA as a releaser of monoamines than phenelzine, a potent antidepressant. β-PEA has an additional effect on the regulation of different neurotransmitters (ie, promotes acetylcholine release via facilitation of the signal pathway of glutamatergic neurotransmission and blockade of GABAergic inhibitory signals in DA neurons).[51–53] Indeed, TYR and β-PEA effectively reduced gamma-aminobutyric acid (GABA)-B receptor-mediated presynaptic inhibition at the GABA-B-dependent presynaptic inhibition of GABAergic inputs to midbrain DA neurons, proposing further extending the role of TAs and a novel mechanism to control DA neurotransmission, potentially leading to better understanding of neurophysiology and pharmacodynamic aspects for development of newer antidepressants.[51] It has been well-known that GABAergic inhibitory interneurons play a major role in the regulation of traditional neurotransmitters in the treatment of depressive disorders in terms of antidepressant action mechanisms. For instance, the action mechanism of vortioxetine that is a newer antidepressant (approved in 2015 by the US FDA) with a multimodal action mechanism mainly through 5-HT transporter inhibition and regulation of multiple 5-HT receptors such as 5-HT1A, 5-HT1B, 5-HT1D, 5-HT3, and 5-HT7, via interacting with GABA interneurons leading to alteration of glutamate, DA, NE, and 5-HT in favor of improvement of mood symptoms.[54] It was also found that β-PEA may have an additional effect on the release of histamine that is involved in the antidepressant action mechanisms of certain antidepressants and atypical antipsychotics such as mirtazapine and quetiapine.[55]

TYR may not normally function in the brain as a releaser of the biogenic monoamines since turnover rate is too fast and its concentration in blood and tissues is kept low by rapid deamination by MAO-A/B.[13] However, when patients treated for depression with MAOIs consume certain foods rich in TYR, a severe hypertensive crisis can be provoked. The hypertensive reaction is caused by TYR-induced NE release, indicating a modulation effect of TYR on NE.[55] Likewise, actions of β-PEA leading to a neuromodulation effect by structural modification, α-methylation of TYR was also found to result in TYR analogs that are strong MAOIs.[13]

OA may also act as a functional modulator of NE activity in the CNS.[56] After DA-β-hydroxylase converts TYR into OA in the NE neuron terminal, OA partially substitutes NE and may act as a false neurotransmitter in the periphery, which is in line with weak activity of OA on postsynaptic α- and β-NE receptors.[13] Although synaptic release of neurotransmitters is usually dependent on provocation by activity-dependent Ca^{2+} entry into the nerve terminal, a recent study has shown that strong synaptic neuropeptide release could be evoked by OA alone without an influence of extracellular Ca^{2+}.[57] It should be noted that OA-evoked neuropeptide release also requires endoplasmic reticulum (ER) Ca^{2+} mobilization by some particular receptors during molecular events of signal cascades (ie, inositol trisphosphate receptor). Hence, this study proposes that a behaviorally important neuromodulator mainly uses synergistic cAMP-dependent protein kinase and ER Ca^{2+} signaling to induce synaptic neuropeptide not depending on activity-dependent Ca^{2+} entry into the nerve terminal.

TRYP has an effect on release of 5-HT through modulation of concentration-related inhibition of potassium-evoked release of tritium.[58] Serotonin antagonists were found to block such effect of TRYP, indicating that it may act on a postsynaptic 5-HT receptor, and while the response to TRYP was also inhibited by tetrodotoxin, suggesting that TRYP may be acting indirectly via the release of a second neurotransmitter.[58] As observed in the findings from the structural modification of PEA and TYR, creation of a chloride or fluoride atom into the phenyl ring of TRYP also changes it from MAO substrates to MAOIs.[55,59,60]

The following study by Locock et al.[61] has also suggested the role of TAs as neuromodulators of traditional neurotransmitters by the modulation of the neurotransmitter binding site at the CNS. The differential effects of β-PEA, TYR, and TRYP on the serotonin binding sites (5-HT1 and 5-HT2) of rat cerebral cortex were evident as the measurement results of IC50 in a rat model, indicating that such TAs may have different functional roles with their different degrees of displacement of serotonin at 5-HT1 and 5-HT2 binding sites in the CNS. With the advancement of molecular biotechnologies, the availability of more detailed biological information on the neurotransmitter receptor binding profile of the TAs and their functional interaction mechanisms with DA, NE, and 5-HT binding sites may lead to better understanding of the pivotal roles of TAs in the modulation of neurotransmitters at the CNS as well as more information about the psychopharmacological and/or neurophysiological functions of the TAs as novel therapeutic targets for new-generation antidepressants.[61]

According to currently available data, the exact action mechanisms of various TAs about whether or not they may act as vesicular/nonvesicular releaser or indirect releasing agent is still elusive. It is also possible that at certain synapses, the TAs could functionally act as mixed-action agonists, not only directly stimulating postsynaptic receptors while simultaneously increasing the availability of neurotransmitters by the blockade of transporters but also increasing the release of neurotransmitters from presynaptic terminal neurons.[9,13] Intensive and extensive researches on the interactive action mechanisms between TAs and traditional neurotransmitters including neuronal cascades process and specific neurobiological molecules recruited by the subsequent signal transduction may enhance molecular understanding of the crucial functions of TAs in the CNS.

TAs and Antidepressants

It is also noteworthy that the antidepressant-associated changes of the CNS levels of TAs have been consistently reported. Administration of MAOIs,[62] such as phenelzine[63,64] and tranylcypromine,[65] results in a significant increase in the levels of some TAs in the CNS than of classical neurotransmitter amines.[9,17] In addition, β-PEA is an active metabolite of phenelzine that is known to influence noradrenergic neurotransmission.[47] In an animal study, the levels of four TAs in diencephalon and hippocampus were measured after tranylcypromine treatment.[66] Intriguingly the study found differential increases of TAs; β-PEA and TRYP showed greater increases than TYR.[66] In line with these findings, β-adrenoceptor downregulation, which is associated with the antidepressant effect and observed in most of other antidepressants, was found after administration of β-PEA, indicating that β-PEA may mediate the antidepressant effects of some MAOIs such as phenelzine.[67] However, there has been a great deal of controversy about whether the excretion of β-PEA is correlated with the clinical improvement of depression after antidepressant treatment; some researchers[63,68,69] have reported not only significantly high excretion of β-PEA but also significant correlation between β-PEA excretion and clinical response in patients with depression after antidepressant treatment, while others have failed to support those findings.[22,37,38,70,71] One interesting study[72] has investigated a putative link between β-PEA and the antidepressant effects of exercise. According to the results, the 24-h mean urinary concentration of PhAA, the metabolite of β-PEA, was increased by 14–572% compared with the values before exercise in 90% of patients. As PhAA implies β-PEA levels and β-PEA may have antidepressant effects and the antidepressant effects of exercise appear to be linked to increased β-PEA concentrations. In addition, an increased level of β-PEA was found to cause an enhanced modulation of the psychostimulant-like discriminative effects of (R)-(−)-deprenyl metabolites under certain conditions.[62]

The influence of antidepressant effects on TYR excretion or levels has been inconsistent but mainly negative to support its alteration after antidepressant treatments, although it has been thought to be a state marker of depression.[17] According to some previous researches investigating the free excretion of TYR after antidepressant treatment,[38,55,73] there was no significant correlation between excretion of TYR and clinical response, while the free excretion of TYR was significantly increased, indicating that the efficacy of MAOIs is dependent on MAO-A but not on direct moderating effects of TYR. It should also be noted that anecdotal data propose a positive correlation between low TYR excretion and clinical benefit after treatments with imipramine and phenelzine, Stewart et al.[74] have investigated whether or not abnormal TYR excretion may predict the clinical response of antidepressant therapy. In the study, responders were defined if they benefited from either imipramine or phenelzine, while nonresponders were defined if they received a trial on at least one of these medications and did not respond to either drug. According to the results, 59% of patients with normal TYR excretion responded to one of the medications, while 93% of patients with abnormal TYR excretion responded. However, the mean baseline MAO levels did not differ between responders and nonresponders to any individual treatment even after adjustment of gender effect. In a subsequent study using three TCAs,[75] a significant correlation between low excretion of TYR and clinical response to TCAs was also found; the excretion of TYR reached only 57.4% of that from nonresponders and there was no difference in clinical response between individual TCA.

Previous studies have found that the urinary PhAA excretion may be a potential state marker for predicting a clinical response to antidepressant treatment. A significant correlation between urinary excretion of PhAA and clinical improvement in depressive symptoms were consistently reported in a number of studies regardless of the type of antidepressants (ie, phenelzine, isocarboxazid, maprotiline, etc.). For instance, Sabelli et al.[28] reported no increased excretion of PhAA when patients did not responded to antidepressant treatment, while responders clearly showed a significant correlation between excretion of PhAA and clinical recovery. Davis et al.[76] have reported that treatment of the depressed patients with amitriptyline or fluoxetine over a 6-week period resulted in clinical improvement and in a significant increase in plasma PhAA concentrations. A significant negative correlation between rating scales of depressive disorders such as Beck and Hamilton depression rating scores and the concentrations of unconjugated, conjugated and total PhAA was also observed by that study. Other studies are also in line with those findings.[17,77,78]

The increased urinary excretion of TRYP after antidepressant treatment has also been found in some researches,[79] however, others failed to replicate such results.[39] The relationship between the excretion of IAA and antidepressant treatment is still obscure.[17,40,80] The relationship between HpAA excretion and improvement of depression has a dearth of data, only a few studies have found that low excretion of HpAA was correlated with severity of endogenous depression, where it was profound in cases with treatment resistant to TCAs.[81]

Trace Amine-Associated Receptors

Beyond a number of preclinical data suggesting the presynaptic and postsynaptic activity of TAs on the regulation of traditional neurotransmitters, their electrophysiological effects against such neurotransmitters may call into existence specific binding sites for the TAs.[13,82]

Although delayed discovery of cloned receptor proteins has limited advanced understanding on the accurate mechanistic roles in the regulation of neurotransmitters, recently, Borowsky et al.[83] have found 15 G protein-coupled receptors (GPCR) from human and rodent tissues with the use of a degenerate PCR approach. According to the study, four human receptors have been identified [TA1, TA3, TA4, and TA-5, the same as the orphan receptor PNR (putative neurotransmitter receptor)], whereas 14 rat receptors have been identified (TA1–TA4 and TA6–TA15). Another interesting point is the existence of remarkable interspecies differences, particularly the complete absence of any functional counterpart of the rodent TAAR7 orthologues in human.[84] They reported that two of those GPCRs potentially bind and/or are activated by TAs. Based on the study results,[83] TA1 was activated by TA and β-PEA possessing low affinity for TRYP, OA, and DA, while TA2 (currently TAAR4) was activated by TRYP and β-PEA. Other GPCRs, including TA3–TA15, had structural similarity, sharing a high degree of amino acid identity (62–96%), while the human 5-HT receptors share 28–63% of amino acid identities.[83] The fact that member homology between TA receptors and clustering of TA receptor genes around chromosome 6q23.2 may putatively indicate a relatively recent evolving of these GPCRs and residence on a potential hotspot for gene duplication events.[83] Human TA1 mRNA is expressed in low to trace levels in the nervous system, in detail: (1) low levels in amygdala and (2) trace levels in cerebellum, dorsal root ganglia, hippocampus, hypothalamus, medulla, pituitary, and pontine reticular formation.[83] The expression of TA1 mRNA in human amygdala is intriguing in light of evidence suggesting a role of TAs in the etiology and the treatment of mood disorders.[83] The expression of mouse TA1 mRNA in the dorsal raphe, locus ceruleus, and VTA also indicates that TAs may modulate the activity of 5-HT, NE, and DA systems and further supports a role for TA receptors in the regulation of mood.[83] All those CNS regions are involved in the development and treatment of depressive disorders under interactive communication with traditional neurotransmitters.

In an immediately successive study by Borowsky et al.,[83] the rat TA1 was activated by various TAs, such as TYR and β-PEA, and also had a clear effect on enhancement of cAMP production as did other TAs, indicating that TAs may also serve as the endogenous ligands of a novel intercellular signaling system.[85]

Lindemann et al.[84] replicated the results of previous studies and located the gene loci of the TA receptors[83]: chromosome 6q23.2 for human, chromosome 1p12 for rat, and chromosome 10 for mouse. Intriguingly, chromosome 6q regions are close to vulnerability genes of depression spectrum disorders[86] as well as bipolar disorder[87] and schizophrenia.[88] Some researches have proposed possible involvement of trace amine-associated receptors (TAARs) gene polymorphisms in the development and treatment of mood disorders.[89–91] The rs6903874 T/T carriers had a statistically significant better improvement, and rs6937506 C/C genotype was found to be more frequent in patients without a history of suicide attempt (incomplete or unsuccessful suicide). Haplotype analyses confirmed the association with suicide attempt behavior being haplotype G-T at SNPs rs7452939 and rs6937506 at risk of suicide. These results suggest a possible role of TAAR6 in antidepressant response and suicide behavior in patients with depressive disorder.[91] In addition, they completed the identification of all members of trace amine GPCRs in human, chimpanzee, rat, and mouse and observed remarkable interspecies differences, even between human and chimpanzee as stated before. The new nomenclature system including subcategorization by subfamily members has also been set up by Lindemann et al., which improved the ambiguities and contradictions of the previous naming system.[83] Table 22.1 represents the current nomenclature system of the TAARs. The proposed distinction of three subgroups is based on the phylogenetic relationships, the pharmacophore similarity analysis, and other pharmacological profiles among the TAAR family.[92]

Monoamine autoreceptors colocalize with monoamine transporters on presynaptic membranes of monoaminergic neurons and give negative feedback regulation of monoamine neurotransmitter release by the availability of monoamines in the synapse.[93] For instance,

TABLE 22.1 Classification of Trace Amine-Associated Receptors (TAARs) in Humans

Group	Old Name	New Name	Synthetic Selective (Partial) Agonist	Swiss-Prot/RefSeq
Group 1	TA1, TAR1, TRAR1	TAAR1	RO5166017; RO5256390; RO5263397; RO4992479; RO5073012	Q96RJ0 NP_612200
	GPR58	TAAR2		Q9P1P5 NP_001028252 or NP_055441
	GPR57P	TAAR3		Q9P1P4
	TA2 P, 5-HT-4P	TAAR4		–
Group 2	PNR	TAAR5		O14804 NP_003958
Group 3	TA4, TRAR4	TAAR6		Q96R18 NP_778237
		TAAR7		–
	TA-5, TAR5, TRAR5, GPR102	TAAR8		Q969N4 NP_444508
	TA3, TAR3, TRAR3	TAAR9		Q96R19 NP_778227

Information has been taken from Borowsky B, Adham N, Jones KA, et al. Trace amines: identification of a family of mammalian G protein-coupled receptors. Proc Natl Acad Sci USA 2001;98(16):8966–71; Lindemann L, Ebeling M, Kratochwil NA, Bunzow JR, Grandy DK, Hoener MC. Trace amine-associated receptors form structurally and functionally distinct subfamilies of novel G protein-coupled receptors. Genomics 2005;85(3):372–85; and Lindemann L, Hoener MC. A renaissance in trace amines inspired by a novel GPCR family. Trends Pharmacol Sci 2005;26(5):274–81.

D2/5-HT1A autoreceptor stimulation negatively regulates dopamine transporter/5-HT transporter function and expression.[93,94] For instance, the action mechanism of selective serotonin reuptake inhibitors is also influenced by the interaction of serotonin availability in the synapse and negative feedback of the 5-HT-2A autoreceptor on the presynaptic serotonergic neuron (ie, delayed antidepressant effect and refiring of 5-HT neuron).

Recent pharmacological studies have produced the scientific relevance and potential of TAARs, mainly focusing on TAAR1, a subtype of TAARs, as a prospective target receptor for novel antidepressants. Previously, modulation effects of TAAR1 focused on DA transmission.[95–97] However, numerous studies have also produced positive results regarding simultaneous modulatory effects of TAAR1 on both NE and 5-HT neurotransmission today.[98–100] Interestingly, Xie and Miller[93] have evaluated whether TAs interact with monoamine autoreceptors, including D2, adrenergic α-2A and α-2B, and 5-HT1A and 5-HT1B receptors for regulation of the TAAR1 signaling process. In that study, DA, NE, and 5-HT autoreceptors were unresponsive to TAs, while β-PEA regulates the function of DA, NE, and 5-HT transporters through interaction with TAAR1 but not with DA, NE, and 5-HT autoreceptors. β-PEA did not change DA, NE, and 5-HT uptake without the existence of TAAR1 (ie, TAAR1 knockout mouse synaptosomes) but did significantly prohibit DA, NE, and 5-HT uptake in the presence of TAAR1 (ie, TAAR1 wild-type mouse synaptosomes). These findings clearly indicate that TAAR1 is a mediator of effects of TAs (ie, β-PEA) on the DA, NE, and 5-HT transporters, indicating a different mode of action as an antidepressant.

Traditional neurotransmitters were also found to activate TAAR1.[97] The DA, NE, and 5-HT were capable of enhancing corticotropin-releasing factor (CRF)-luciferase expression in a TAAR1-dependent manner with cAMP-associated signal transduction.[96,100] Hence, we may assume that traditional neurotransmitters may also act as TAAR1 agonists. However, coexpression of TAAR1 with such traditional neurotransmitter autoreceptors reduced the CRF-luciferase expression induced by trace level of each traditional neurotransmitter. In addition, such attenuation was completely blocked by a specific monoamine autoreceptor antagonist. Traditional neurotransmitters could change the functional activities of DA, NE, and 5-HT transporters through interaction with either TAAR1 or monoamine autoreceptors. Accordingly, TAAR1, and DA, NE, and 5-HT autoreceptor-associated signal transduction processes may also be putatively provoked by counteracting each other.[97]

Despite currently available data suggesting that TAARs may act as neuromodulators, the action mechanism at the molecular level has been still obscured since the exclusive TA action on neurotransmission has not yet been completely consistent and supported by the available findings. Previous researches proposed a high chance of potential existence of TA-sensitive receptors other than TAARs and that they may convey the pharmacological effects of TAs.[92,101,102] Competition studies using various neurotransmitter receptor agonists and antagonists indicate that [3H]β-PEA does not bind to adrenergic, muscarinic-cholinergic or dopaminergic receptors and contemporary antidepressants were not able to displace [3H]β-PEA binding from its specific binding sites.[102] It was found that amphetamines were able to activate TAAR1 but not capable of displacing [3H]β-PEA.[102] The [3H]β-PEA, [3H]TA, [3H]OA, and [3H]TRYP are potently displaced by several molecules which do not or only weakly activate TAAR1.[92,101] These findings may propose a possible presence of numerous binding sites and distinct characteristics of TAs substantially different from traditional neurotransmitters.[102]

Despite still being in their infancy, novel molecular agents that can activate or inhibit TAAR1 have been actively investigated and thereby some agonists have been available to test the action mechanism of TAAR1 today.[103,104] The full-agonist RO5256390 and the partial-agonist RO5263397 were found to inhibit psychostimulant-induced hyperactivity and resemble a brain activation pattern of atypical antipsychotics.[98] TAAR1 agonists RO5263397 were capable of decreasing and preventing antipsychotic-related catalepsy, weight gain, and fat accumulation.[105] Recently, the molecule 1-(7-methoxy-2-methyl-1,2,3,4-tetrahydroisoquinolin-4-YL)-cyclohexanol, which is a novel β-PEA substituted molecule, was found to be involved in the L-arginine-nitric oxide-cyclic guanosine monophosphate pathway resulting in alteration of nitric oxide production[106]; nitric oxide exerts a crucial role in the neurobiology of depressive disorder. After discovery of the first selective TAAR1 agonist and antagonist, the RO5166017 and EPPTB, respectively, few molecules have been available. Thus identification of more new molecular entities able to act as ligands for targeting TAAR1 would be mandatory for providing a novel and exact pharmacological approach to reveal how to interact with traditional neurotransmitter systems in mood disorders.[103,104]

Procognitive and antidepressant-like effects were also proven after stimulation of TAAR1.[105] A number of studies have also replicated these positive and promising effects of TAAR1 (partial) agonist as antidepressants, antipsychotics, and anxiolytics.[98,99] According to Harmeier's study,[107] interaction of TAAR1 with D2R reduced βArr2 recruitment to D2R. In addition, cAMP signaling of TAAR1 was reduced while its βArr2 signaling was enhanced in the presence of D2R resulting in reduced GSK3β activation. These results demonstrate that βArr2 signaling may be an important pathway for TAAR1 function and that the activation of the TAAR1–D2R complex negatively modulates GSK3β signaling. Given that patients with mood disorders show an alteration of intracellular second messenger/signal transduction cascades such as GSK3β signaling, such a reduction of GSK3β signaling triggered by the interaction of D2R with activated TAAR1 further supports TAAR1 as a target for the treatment of mood disorders.[108,109] It was questionable whether TAAR1 can affect prefrontal cortex (PFC)-related processes and functions. Addressing this issue should be useful and may provide crucial information on the role of TAAR1 in depressive disorders since the aberration of PFC dysfunction of depressed patients has been continuously reported today[110]; furthermore, the PFC dysfunction is also critically related to cognitive function that is highly associated with residual symptoms in depressed patients leading to incomplete treatment response to contemporary antidepressants and frequent relapse. A recent study[109] has found strong evidence suggesting a potential role of TAAR1 in PFC, where distinct alteration of subunit and functional defect of glutamate N-methyl-D-aspartate (NMDA) receptors under a lack of TAAR1 in animal model. Such a study clearly points out the close association between TAAR1 expression and NMDA receptor-mediated glutamate transmission in the PFC, possibly linked to aberrant dysregulation of PFC observed in depressed patients as considerable neuroimaging and clinical evidence indicate poor functioning of the medial and orbital frontal cortex as well as amygdala, in depressive disorder.[111] The regulation effects of TAAR1 on the immunomodulation,[112] neurotrophic factors,[99] and stress axis[98] have also been demonstrated, which are also implicated in the development and treatment of depressive disorders.

Some medications against depressive disorders were found to utilize TAAR1 as a media for their therapeutic action. Modafinil and methylphenidate, known DA transporter inhibitors with the property of antidepressant augmentation effects proven through placebo-controlled clinical trials, augmented the enhancement of TAAR1 activity by β-PEA in DA and NE transporters expressing cells, indicating that modafinil could affect, albeit indirectly, β-PEA activation of the TAAR1 in dopamine and NE neurons.[113] Likewise contemporary antidepressants, both desipramine and citalopram, also showed similar effects observed in studies with modafinil and methylphenidate.[85,97] Although these antidepressants and augmentation agents are likely to present modulating effects via putative several action mechanisms, such previous studies may support the view that their delicate and complex action mechanisms in harmony with TAAR1 and traditional neurotransmitter transporters strongly warrant further scrutiny as important and novel molecular target for treating depressive disorders.

However, there has so far been a clear dearth of research data concerning other TAARs except for TAAR1 in the development and treatment of depressive disorders.[92] Fig. 22.1 illustrates a hypothetical action mechanism of TAAR1 for treatment of depressive disorders.

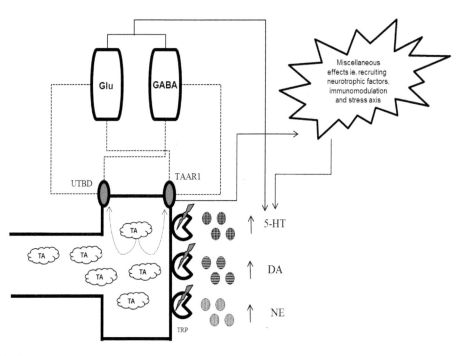

FIGURE 22.1   The speculative action mechanism of TAAR1 in the regulation of neurotransmitters. Glu, glutamine; GABA, gamma-aminobutyric acid; TAAR, trace amine-associated receptor; 5-HT, serotonin; DA, dopamine; NE, norepinephrine; TA, trace amine; UTBD, unidentified binding sites.

CONCLUSION

The TA (ie, β-PEA) hypothesis for depressive disorder was based on a number of preclinical and clinical studies showing alteration of its levels in various locations of the human body as well as changes in their excretion rates. Furthermore, such changes of TAs were also found to be associated with clinical improvement of depression and antidepressants (ie, MAOIs) had modulatory effects on TAs in depressed patients in relation to clinical outcomes. However, these observations remain equivocal. That is, a lot of questions to be explained exist as to whether TAs are state or trait biological markers for depressive disorders based on such inconsistent findings. Additionally, interactive actions with traditional neurotransmitters via some putative binding sites or TAARs have been proposed to fully express their effects as potential novel therapeutic targets for treatment of depression have also been proposed. Therefore this research field should be thoroughly further investigated and validated in future as well as obtaining specific molecular compounds to stimulate or inhibit TAARs in order to reveal or understand the clear mode of action of TAs and related receptors as therapeutic agents or their pathophysiological involvement in the development of depressive disorder.[103] Indeed the contribution of TAs to the molecular biological aspects in the pathophysiology and treatment of depressive disorders (ie, neuromodulation on monoamine reuptake inhibition) was skeptical until the discovery of unique binding sites for these putative neuromodulators (ie, RO5166017, the first selective TAAR1 agonist), however, modern progress in detection of such molecular targets and wider interest has been dramatically increasing, leading to a novel avenue of approach for rapid development to treat depressive disorder.[97] Currently available research findings may suggest that TAAR-mediated mechanisms of TAs in the functional regulation of traditional neurotransmitters and transporters may be a stepping-stone for the development of novel therapeutic agents to improve the medication of depressive disorders.[92]

Given well-known physiological functions of TAs to depressive disorders, a clearer understanding of the physiological relevance of TAs, more identification (ie, type and distribution) of human TAARs and their TAAR family, hidden binding sites of TAs other than TAARs, related signal transduction mechanisms, discovery of relevant physiological high-affinity ligands specific to individual TAARs, intricate and complex interaction of TAARs with other molecules not understood yet but involving pathophysiology of depressive disorders, and well-powered clinical trials using such novel therapeutic agents may truly shed light on the role of TAs and TAARs in the perspective of an etiological and therapeutic approach for depressive disorders.

Acknowledgments

This work was supported by a grant from the Korean Health Technology R&D Project, Ministry of Health & Welfare, Republic of Korea (HI12C0003).

References

1. Owens MJ. Selectivity of antidepressants: from the monoamine hypothesis of depression to the SSRI revolution and beyond. *J Clin Psychiatry* 2004;**65**(Suppl 4):5–10.

2. Hirschfeld RM. History and evolution of the monoamine hypothesis of depression. *J Clin Psychiatry* 2000;**61**(Suppl 6):4–6.

3. Stahl SM. Regulation of neurotransmitter receptors by desipramine and other antidepressant drugs: the neurotransmitter receptor hypothesis of antidepressant action. *J Clin Psychiatry* 1984;**45**(10 Pt 2):37–45.

4. Rush AJ. STAR*D: what have we learned? *Am J Psychiatry* 2007;**164**(2):201–4.

5. Wang SM, Han C, Pae CU. Criticisms of drugs in early development for the treatment of depression: what can be improved? *Expert Opin Investig Drugs* 2015;**24**(4):445–53.

6. Pae CU, Wang SM, Han C, et al. Vortioxetine: a meta-analysis of 12 short-term, randomised, placebo-controlled clinical trials for the treatment of major depressive disorder. *J Psychiatry Neurosci.* 2015;**40**:174–86.

7. Marks DM, Shah MJ, Patkar AA, Masand PS, Park GY, Pae CU. Serotonin-norepinephrine reuptake inhibitors for pain control: premise and promise. *Curr Neuropharmacol* 2009;**7**(4):331–6.

8. Boulton AA, Baker GB, Dewhurst WG, Sandler M, editors. *Neurobiology of the trace amines: analytical, physiological, pharmacological, behavioral, and clinical aspects.* Clifton, NJ: Humana Press; 1984.

9. Narang D, Tomlinson S, Holt A, Mousseau DD, Baker GB. Trace amines and their relevance to psychiatry and neurology: a brief overview. *Bull Clin Psychopharmacol* 2011;**21**:73–9.

10. Berry MD. Mammalian central nervous system trace amines. Pharmacologic amphetamines, physiologic neuromodulators. *J Neurochem* 2004;**90**(2):257–71.

11. Premont RT, Gainetdinov RR, Caron MG. Following the trace of elusive amines. *Proc Natl Acad Sci USA* 2001;**98**(17):9474–5.

12. Holschneider DP, Chen K, Seif I, Shih JC. Biochemical, behavioral, physiologic, and neurodevelopmental changes in mice deficient in monoamine oxidase A or B. *Brain Res Bull* 2001;**56**(5):453–62.

13. Burchett SA, Hicks TP. The mysterious trace amines: protean neuromodulators of synaptic transmission in mammalian brain. *Progress Neurobiol.* 2006;**79**(5–6):223–46.

14. Dabadie H, Mons N, Geffard M. Simultaneous detection of tryptamine and dopamine in rat substantia nigra and raphe nuclei using specific antibodies. *Brain Res* 1990;**512**(1):138–42.

15. Burchett S.A. *Immuno-histochemical localization of trace amines in rat brain* [Thesis]. Greensboro, NC: University of North Carolina at Greensboro; 1994.

16. Davis BA. Biogenic amines and their metabolites in body fluids of normal, psychiatric and neurological subjects. *J Chromatogr* 1989;**466**:89–218.

17. Davis BA, Boulton AA. The trace amines and their acidic metabolites in depression—an overview. *Prog Neuropsychopharmacol Biol Psychiatry* 1994;**18**(1):17–45.

18. Fischer E, Heller B, Miro AH. Beta-phenylethylamine in human urine. *Arzneimittel-Forschung* 1968;**18**(11):1486.

19. Boulton AA, Milward L. Separation, detection and quantitative analysis of urinary beta-phenylethylamine. *J Chromatogr* 1971;**57**(2):287–96.

20. Fischer E, Spatz H, Fernandez Labriola RS, Rodriguez Casanova EM, Spatz N. Quantitative gas-chromatographic determination and infrared spectrographic identification of urinary phenethylamine. *Biol Psychiatry* 1973;**7**(2):161–5.

21. Fischer E, Heller B. Phenethylamine as a neurohumoral agent in brain. *Behav Neuropsychiatry* 1972;**4**(3–4):8–11. passim.

22. DeLisi LE, Murphy DL, Karoum F, Mueller E, Targum S, Wyatt RJ. Phenylethylamine excretion in depression. *Psychiatry Res* 1984;**13**(3):193–201.

23. Sandler M, Ruthven CR, Goodwin BL, Reynolds GP, Rao VA, Coppen A. Deficient production of tyramine and octopamine in cases of depression. *Nature* 1979;**278**(5702):357–8.

24. Nakagawara M. Beta-phenylethylamine and noradrenergic function in depression. *Prog Neuropsychopharmacol Biol Psychiatry* 1992;**16**(1):45–53.

25. Sabelli H, Fink P, Fawcett J, Tom C. Sustained antidepressant effect of PEA replacement. *J Neuropsychiatry Clin Neurosci* 1996;**8**(2):168–71.

26. Fischer E, Heller B, Nachon M, Spatz H. Therapy of depression by phenylalanine. Preliminary note. *Arzneimittel-Forschung* 1975;**25**(1):132.

27. Beckmann H, Strauss MA, Ludolph E. Dl-phenylalanine in depressed patients: an open study. *J Neural Transm* 1977;**41**(2-3):123–34.

28. Sabelli HC, Fawcett J, Gusovsky F, et al. Clinical studies on the phenylethylamine hypothesis of affective disorder: urine and blood phenylacetic acid and phenylalanine dietary supplements. *J Clin Psychiatry* 1986;**47**(2):66–70.

29. Zhou G, Shoji H, Yamada S, Matsuishi T. Decreased beta-phenylethylamine in CSF in Parkinson's disease. *J Neurol Neurosurg Psychiatry* 1997;**63**(6):754–8.

30. Reijnders JS, Ehrt U, Weber WE, Aarsland D, Leentjens AF. A systematic review of prevalence studies of depression in Parkinson's disease. *Mov Disord* 2008;**23**(2):183–9. quiz 313.

31. Remy P, Doder M, Lees A, Turjanski N, Brooks D. Depression in Parkinson's disease: loss of dopamine and noradrenaline innervation in the limbic system. *Brain J Neurol* 2005;**128**(Pt 6):1314–22.

32. Sabelli HC, Fawcett J, Gusovsky F, Javaid J, Edwards J, Jeffriess H. Urinary phenyl acetate: a diagnostic test for depression? *Science* 1983;**220**(4602):1187–8.

33. Sabelli H, Fawcett J, Gusovsky F, Edwards J, Jeffriess H, Javaid J. Phenylacetic acid as an indicator in bipolar affective disorders. *J Clin Psychopharmacol* 1983;**3**(4):268–70.

34. Davis BA, Kennedy SH, D'Souza J, Durden DA, Goldbloom DS, Boulton AA. Correlations of plasma and urinary phenylacetic acid and phenylethylamine concentrations with eating behavior and mood rating scores in brofaromine-treated women with bulimia nervosa. *J Psychiatry Neurosci* 1994;**19**(4):282–8.

35. Linnoila M, Karoum F, Potter WZ. High positive correlation between urinary free tyramine excretion rate and "whole body" norepinephrine turnover in depressed patients. *Biol Psychiatry* 1982;**17**(9):1031–6.

36. Roy A, Linnoila M, Karoum F, Pickar D. Urinary excretion of free tyramine and of norepinephrine and its metabolites in unipolar depressed patients. *Biol Psychiatry* 1986;**21**(2):221–4.

37. Karoum F, Chuang LW, Eisler T, et al. Metabolism of (-) deprenyl to amphetamine and methamphetamine may be responsible for deprenyl's therapeutic benefit: a biochemical assessment. *Neurology* 1982;**32**(5):503–9.

38. Karoum F, Potkin SG, Murphy DL, Wyatt RJ. Quantitation and metabolism of phenylethylamine and tyramine's three isomers in humans *Noncatecholic phenylethylamines, Part 2: Phenylethanolamine, tyramines and octopamine*. New York: Marcel Dekker, Inc.; 1980.

39. McNamee HB, Moody JP, Naylor GJ. Indoleamine metabolism in affective disorders: excretion of tryptamine, indoleacetic acid and 5-hydroxy-indoleacetic acid in depressive states. *J Psychosomatic Res* 1972;**16**(1):63–70.

40. Coppen A, Shaw DM, Malleson A, Eccleston E, Gundy G. Tryptamine metabolism in depression. *Br J Psychiatry* 1965;**111**(479):993–8.

41. Zheng P, Wang Y, Chen L, et al. Identification and validation of urinary metabolite biomarkers for major depressive disorder. *Mol Cell Proteomics* 2013;**12**(1):207–14.

42. Yu PH, Bowen RC, Davis BA, Boulton AA. Platelet monoamine oxidase activity and trace acid levels in plasma of agoraphobic patients. *Acta Psychiatr Scand* 1983;**67**(3):188–94.

43. Kawabata M, Kobayashi K, Shohmori T. Determination of phenylacetic acid in cerebrospinal fluid by gas chromatography-mass spectrometry. *Acta Med Okayama* 1986;**40**(5):271–6.

44. Sarrias MJ, Artigas F, Martinez E, et al. Decreased plasma serotonin in melancholic patients: a study with clomipramine. *Biol Psychiatry* 1987;**22**(12):1429–38.

45. Baker GB, Hiob LE, Dewhurst WG. Effects of monoamine oxidase inhibitors on release of dopamine and 5-hydroxytryptamine from rat striatum in vitro. *Cell Molec Biol Includ Cyto-Enzymol* 1980;**26**(2):183–6.

46. Dyck LE. Release of radiolabeled dopamine, p-tyramine, and m-tyramine from rat striatal slices by some aminotetralins. *Neurochem Res* 1981;**6**(4):365–75.

47. Dyck LE. Release of monoamines from striatal slices by phenelzine and beta-phenylethylamine. *Prog Neuropsychopharmacol Biol Psychiatry* 1983;**7**(4–6):797–800.

48. Brinton RE. Neuromodulation: associative and nonlinear adaptation. *Brain Res Bull* 1990;**24**(5):651–8.

49. Baker GB, Raiteri M, Bertolini A, del Carmine R. Interaction of beta-phenethylamine with dopamine and noradrenaline in the central nervous system of the rat. *J Pharm Pharmacol* 1976;**28**(5):456–7.

50. Raiteri M, Del Carmine R, Bertollini A, Levi G. Effect of sympathomimetic amines on the synaptosomal transport of noradrenaline, dopamine and 5-hydroxytryptamine. *Eur J Pharmacol* 1977;**41**(2):133–43.

51. Berretta N, Giustizieri M, Bernardi G, Mercuri NB. Trace amines reduce GABA(B) receptor-mediated presynaptic inhibition at GABAergic synapses of the rat substantia nigra pars compacta. *Brain Res* 2005;**1062**(1–2):175–8.

52. Kato M, Ishida K, Chuma T, et al. Beta-phenylethylamine modulates acetylcholine release in the rat striatum: involvement of a dopamine D(2) receptor mechanism. *Eur J Pharmacol* 2001;**418**(1–2):65–71.

53. Ishida K, Murata M, Kato M, Utsunomiya I, Hoshi K, Taguchi K. Beta-phenylethylamine stimulates striatal acetylcholine release through activation of the AMPA glutamatergic pathway. *Biol Pharmaceut Bull* 2005;**28**(9):1626–9.

54. Tritschler L, Felice D, Colle R, et al. Vortioxetine for the treatment of major depressive disorder. *Expert Rev Clin Pharmacol* 2014;**7**(6):731–45.

55. Irsfeld M, Spadafore M, Pruss BM. Beta-phenylethylamine, a small molecule with a large impact. *WebmedCentral* 2013;**4**:9. pii: 4409.

56. Fletcher PJ, Paterson IA. M-octopamine injected into the paraventricular nucleus induces eating in rats: a comparison with noradrenaline-induced eating. *Br J Pharmacol* 1989;**97**(2):483–9.

57. Shakiryanova D, Zettel GM, Gu T, Hewes RS, Levitan ES. Synaptic neuropeptide release induced by octopamine without Ca2+ entry into the nerve terminal. *Proc Natl Acad Sci USA* 2011;**108**(11):4477–81.

58. Ennis C, Cox B. The effect of tryptamine on serotonin release from hypothalamic slices is mediated by a cholinergic interneurone. *Psychopharmacology (Berl)* 1982;**78**(1):85–8.

59. Kinemuchi H, Arai Y. Selective inhibition of monoamine oxidase A and B by two substrate-analogues, 5-fluoro-alpha-methyltryptamine and p-chloro-beta-methylphenethylamine. *Res Commun Chem Pathol Pharmacol* 1986;**54**(1):125–8.

60. Kinemuchi H, Arai Y, Toyoshima Y, Tadano T, Kisara K. Studies on 5-fluoro-alpha-methyltryptamine and p-chloro-beta-methylphenethylamine: determination of the MAO-A or MAO-B selective inhibition in vitro. *Jap J Pharmacol* 1988;**46**(2):197–9.

61. Locock RA, Baker GB, Coutts RT, Dewhurst WG. Displacement of serotonin from binding sites in rat cortex: the effects of biogenic "trace" amines. *Prog Neuropsychopharmacol Biol Psychiatry* 1984;**8**(4–6):701–4.

62. Yasar S, Justinova Z, Lee SH, Stefanski R, Goldberg SR, Tanda G. Metabolic transformation plays a primary role in the psychostimulant-like discriminative-stimulus effects of selegiline [(R)-(-)-deprenyl]. *J Pharmacol Exp Ther* 2006;**317**(1):387–94.

63. McGrath PJ, Cooper TB, Quitkin FM, Klein DF. Effects of imipramine and phenelzine on plasma PEA levels. *Psychiatry Res* 1988;**26**(2):239.

64. McManus DJ, Mousseau DD, Paetsch PR, Wishart TB, Greenshaw AJ. Beta-adrenoceptors and antidepressants: possible 2-phenylethylamine mediation of chronic phenelzine effects. *Biol Psychiatry* 1991;**30**(11):1122–30.

65. Baker GB, Nazarali AJ, Coutts RT, Micetich RG, Hall TW. Brain levels of 5-hydroxytryptamine, tryptamine and 2-phenylethylamine in the rat after administration of N-cyanoethyltranylcypromine. *Prog Neuropsychopharmacol Biol Psychiatry* 1984;**8**(4–6):657–60.

66. Philips SR, Baker GB, McKim HR. Effects of tranylcypromine on the concentration of some trace amines in the diencephalon and hippocampus of the rat. *Experientia* 1980;**36**(2):241–2.

67. Paetsch PR, Baker GB, Greenshaw AJ. Induction of functional down-regulation of beta-adrenoceptors in rats by 2-phenylethylamine. *J Pharmac Sci* 1993;**82**(1):22–4.

68. Nazarali AJ, Baker GB, Coutts RT, Yeung JM, Rao TS. Rapid analysis of beta-phenylethylamine in tissues and body fluids utilizing pentafluorobenzoylation followed by electron-capture gas chromatography. *Prog Neuropsychopharmacol Biol Psychiatry* 1987;**11**(2–3):251–8.

69. Fischer E, Spatz H, Saavedra JM, Reggiani H, Miro AH, Heller B. Urinary elimination of phenethylamine. *Biol Psychiatry* 1972;**5**(2):139–47.

70. Liebowitz MR, Karoum F, Quitkin FM, et al. Biochemical effects of L-deprenyl in atypical depressives. *Biol Psychiatry* 1985;**20**(5):558–65.

71. Murphy DL, Cohen RM, Siever LJ, et al. Clinical and laboratory studies with selective monoamine-oxidase-inhibiting drugs. Implications for hypothesized neurotransmitter changes associated with depression and antidepressant drug effects. *Modern Probl Pharmacopsychiatry* 1983;**19**:287–303.

72. Szabo A, Billett E, Turner J. Phenylethylamine, a possible link to the antidepressant effects of exercise? *Br J Sports Med* 2001;**35**(5):342–3.

73. Murphy DL, Karoum F, Pickar D, et al. Differential trace amine alterations in individuals receiving acetylenic inhibitors of MAO-A (clorgyline) or MAO-B (selegiline and pargyline). *J Neu Transmission Suppl* 1998;**52**:39–48.

74. Stewart JW, Harrison W, Cooper TB, Quitkin FM. Tyramine sulfate excretion may be a better predictor of antidepressant response than monoamine oxidase activity. *Psychiatry Res* 1988;**25**(2):195–201.

75. Hale AS, Sandler M, Hannah P, Bridges PK. Tyramine conjugation test for prediction of treatment response in depressed patients. *Lancet* 1989;**1**(8632):234–6.

76. Davis BA, Boulton AA, Yu PH, et al. Longitudinal effect of amitriptyline and fluoxetine treatment on plasma phenylacetic acid concentrations in depression. *Biol Psychiatry* 1991;**30**(6):600–8.

77. Fawcett J, Sabelli H, Gusvosky F, Epstein P, Javaid J, Jeffries H. Phenylethylaminic mechanisms in maprotiline antidepressant effect. *Fed Proc* 1983;**42**:1164.

78. Gusovsky F, Fawcett J, Javaid JI, Jeffriess H, Sabelli H. A high-pressure liquid chromatographic method for plasma phenylacetic acid, a putative marker for depressive disorders. *Analyt Biochem* 1985;**145**(1):101–5.

79. Prange Jr. AJ, Wilson IC, Knox AE, et al. Thyroid-imipramine clinical and chemical interaction: evidence for a receptor deficit in depression. *J Psychiatric Res* 1972;**9**(3):187–205.

80. Siwers B, Ringberger VA, Tuck JR, Sjoqvist F. Initial clinical trial based on biochemical methodology of zimelidine (a serotonin uptake inhibitor) in depressed patients. *Clin Pharmacol Ther* 1977;**21**(2):194–200.

81. Sandler M, Ruthven CR, Goodwin BL, Reynolds GP, Rao VA, Coppen A. Trace amine deficit in depressive illness: the phenylalanine connexion. *Acta Psychiatr Scand Suppl* 1980;**280**:29–39.

82. Nguyen TV, Juorio AV. Binding sites for brain trace amines. *Cell Mol Neurobiol* 1989;**9**(3):297–311.

83. Borowsky B, Adham N, Jones KA, et al. Trace amines: identification of a family of mammalian G protein-coupled receptors. *Proc Natl Acad Sci USA* 2001;**98**(16):8966–71.

84. Lindemann L, Ebeling M, Kratochwil NA, Bunzow JR, Grandy DK, Hoener MC. Trace amine-associated receptors form structurally and functionally distinct subfamilies of novel G protein-coupled receptors. *Genomics* 2005;**85**(3):372–85.

85. Bunzow JR, Sonders MS, Arttamangkul S, et al. Amphetamine, 3,4-methylenedioxymethamphetamine, lysergic acid diethylamide, and metabolites of the catecholamine neurotransmitters are agonists of a rat trace amine receptor. *Molec Pharmacol* 2001;**60**(6):1181–8.

86. Zubenko GS, Maher BS, Hughes III HB, Zubenko WN, Scott Stiffler J, Marazita ML. Genome-wide linkage survey for genetic loci that affect the risk of suicide attempts in families with recurrent, early-onset, major depression. *Am J Med Genet Part B Neuropsychiatr Genet Official Public Int Soc Psychiatric Genet* 2004;**129B**(1):47–54.

87. Freudenberg-Hua Y, Freudenberg J, Kluck N, Cichon S, Propping P, Nothen MM. Single nucleotide variation analysis in 65 candidate genes for CNS disorders in a representative sample of the European population. *Genome Res* 2003;**13**(10):2271–6.

88. Schwab SG, Hallmayer J, Albus M, et al. A genome-wide autosomal screen for schizophrenia susceptibility loci in 71 families with affected siblings: support for loci on chromosome 10p and 6. *Mol Psychiatry* 2000;**5**(6):638–49.

89. Pae CU, Yu HS, Amann D, et al. Association of the trace amine associated receptor 6 (TAAR6) gene with schizophrenia and bipolar disorder in a Korean case control sample. *J Psychiatr Res* 2008;**42**(1):35–40.

90. Pae CU, Drago A, Forlani M, Patkar AA, Serretti A. Investigation of an epistastic effect between a set of TAAR6 and HSP-70 genes variations and major mood disorders. *Am J Med Genet Part B Neuropsychiatr Genet Official Public Int Soc Psychiatric Genet* 2010;**153B**(2):680–3.

91. Pae CU, Drago A, Kim JJ, et al. TAAR6 variations possibly associated with antidepressant response and suicidal behavior. *Psychiatry Res* 2010;**180**(1):20–4.

92. Lindemann L, Hoener MC. A renaissance in trace amines inspired by a novel GPCR family. *Trends Pharmacol Sci* 2005;**26**(5):274–81.

93. Xie Z, Miller GM. Beta-phenylethylamine alters monoamine transporter function via trace amine-associated receptor 1: implication for modulatory roles of trace amines in brain. *J Pharmacol Exp Ther* 2008;**325**(2):617–28.

94. Hjorth S, Bengtsson HJ, Kullberg A, Carlzon D, Peilot H, Auerbach SB. Serotonin autoreceptor function and antidepressant drug action. *J Psychopharmacol* 2000;**14**(2):177–85.

95. Lindemann L, Meyer CA, Jeanneau K, et al. Trace amine-associated receptor 1 modulates dopaminergic activity. *J Pharmacol Exp Ther* 2008;**324**(3):948–56.

96. Xie Z, Miller GM. Trace amine-associated receptor 1 is a modulator of the dopamine transporter. *J Pharmacol Exp Ther* 2007;**321**(1):128–36.

97. Miller GM. The emerging role of trace amine-associated receptor 1 in the functional regulation of monoamine transporters and dopaminergic activity. *J Neurochem* 2011;**116**(2):164–76.

98. Revel FG, Moreau JL, Gainetdinov RR, et al. TAAR1 activation modulates monoaminergic neurotransmission, preventing hyperdopaminergic and hypoglutamatergic activity. *Proc Natl Acad Sci USA* 2011;**108**(20):8485–90.

99. Revel FG, Moreau JL, Gainetdinov RR, et al. Trace amine-associated receptor 1 partial agonism reveals novel paradigm for neuropsychiatric therapeutics. *Biol Psychiatry* 2012;**72**(11):934–42.

100. Xie Z, Westmoreland SV, Miller GM. Modulation of monoamine transporters by common biogenic amines via trace amine-associated receptor 1 and monoamine autoreceptors in human embryonic kidney 293 cells and brain synaptosomes. *J Pharmacol Exp Ther* 2008;**325**(2):629–40.

101. Vaccari A. High affinity binding of [3H]-tyramine in the central nervous system. *Br J Pharmacol* 1986;**89**(1):15–25.

102. Hauger RL, Skolnick P, Paul SM. Specific [3H] beta-phenylethylamine binding sites in rat brain. *Eur J Pharmacol* 1982;**83**(1–2):147–8.

103. Cichero E, Espinoza S, Franchini S, et al. Further insights into the pharmacology of the human trace amine-associated receptors: discovery of novel ligands for TAAR1 by a virtual screening approach. *Chem Biol Drug Design* 2014;**84**(6):712–20.

104. Cichero E, Espinoza S, Gainetdinov RR, Brasili L, Fossa P. Insights into the structure and pharmacology of the human trace amine-associated receptor 1 (hTAAR1): homology modelling and docking studies. *Chem Biol Drug Design* 2013;**81**(4):509–16.

105. Revel FG, Moreau JL, Pouzet B, et al. A new perspective for schizophrenia: TAAR1 agonists reveal antipsychotic- and antidepressant-like activity, improve cognition and control body weight. *Mol Psychiatry* 2013;**18**(5):543–56.

106. Dhir A, Kulkarni SK. Antidepressant-like effect of 1-(7-methoxy-2-methyl-1,2,3,4-tetrahydro-isoquinolin-4-YL)-cyclohexanol, a putative trace amine receptor ligand involves l-arginine-nitric oxide-cyclic guanosine monophosphate pathway. *Neurosci Lett* 2011;**503**(2):120–4.

107. Harmeier A, Obermueller S, Meyer CA, et al. Trace amine-associated receptor 1 activation silences GSK3beta signaling of TAAR1 and D2R heteromers. *Eur Neuropsychopharmacol* 2015;**25**:2049–61.

108. Pandey GN, Ren X, Rizavi HS, Dwivedi Y. Glycogen synthase kinase-3beta in the platelets of patients with mood disorders: effect of treatment. *J Psychiatric Res* 2010;**44**(3):143–8.

109. Karege F, Perroud N, Burkhardt S, et al. Protein levels of beta-catenin and activation state of glycogen synthase kinase-3beta in major depression. A study with postmortem prefrontal cortex. *J Affect Disord* 2012;**136**(1–2):185–8.

110. George MS, Ketter TA, Post RM. Prefrontal cortex dysfunction in clinical depression. *Depression* 1994;**2**:59–72.

111. Murray EA, Wise SP, Drevets WC. Localization of dysfunction in major depressive disorder: prefrontal cortex and amygdala. *Biol Psychiatry* 2011;**69**(12):e43–54.

112. Sriram U, Cenna JM, Haldar B, et al. Methamphetamine induces trace amine-associated receptor 1 (TAAR1) expression in human T lymphocytes: role in immunomodulation. *J Leukoc Biol* 2016;**99**:213–23.

113. Madras BK, Xie Z, Lin Z, et al. Modafinil occupies dopamine and norepinephrine transporters in vivo and modulates the transporters and trace amine activity in vitro. *J Pharmacol Exp Ther* 2006;**319**(2):561–9.

Trace Amine-Associated Receptor 1: Implications for Treating Stimulant Drug Addiction

L.J. Wallace

Division of Pharmacology, College of Pharmacy, The Ohio State University,
Columbus, OH, United States

OUTLINE

Trace Amines and Neurological Disorders
DOI: http://dx.doi.org/10.1016/B978-0-12-803603-7.00023-9

INTRODUCTION

All drugs with addiction potential increase levels of extracellular dopamine in nucleus accumbens and striatum.[1] Populations of trace amine-related receptor 1 (TAAR1) are associated with dopamine cell body and terminal areas and have a role in modulation of dopamine signaling (see chapter: Effects of Trace Amines on the Dopaminergic Mesencephalic System). The primary target of the stimulant drugs, cocaine and amphetamine, is the dopamine transporter,[2–6] and the increase in extracellular dopamine resulting from interference with recycling of dopamine is thought to be a necessary step in the development of addiction to these drugs.[7–12] This raises the possibility that manipulation of TARR1 in dopamine systems might impact addiction liability of stimulant drugs.

EFFECTS OF STIMULANT DRUGS

Stimulant drugs derived this moniker because of their overall stimulant effect in humans. Two important members of this drug family are amphetamine and cocaine. People using lower doses of these drugs experience an increase in confidence, a mild behavioral activation, an increase in talkativeness, appetite suppression, and an increase in heart rate and blood pressure. Lower doses of these drugs are effective treatments for attention deficit hyperactivity disorder. Higher doses produce a tendency towards aggression, participating in stereotyped behaviors, and occasionally paranoid thoughts. A relatively small percentage of users will develop a pattern of use characterized by loss of control. Such users may utilize the drug in a binge fashion, in which doses are successively repeated each time the effects begin to wane until the user runs out of drug or energy. When a person who has been using stimulant drugs regularly and frequently for a period of time stops, they experience withdrawal. Withdrawal symptoms are broadly characterized as depression, with anhedonia being a prominent feature. Withdrawal symptoms last a few days to a week. One characteristic of former users of stimulant drugs is that cues relating to prior drug use have acquired motivational value. Thus, a former user might see a picture of a person using cocaine and experience intense desire to want to resume drug use. This can be a powerful trigger for relapse. Human experience suggests that the magnitude of the motivational power of cues increases over a few months' time after last use of drug.

The phenomena of tolerance and sensitization show interesting patterns with continuing use of stimulant drugs. Tolerance develops to appetite suppressant, cardiovascular, and the confidence and feel good effects. In contrast, sensitization, a phenomenon where the intensity of effect gets bigger as the drug is continually used, develops for stereotypic behaviors and paranoia/psychotic-like effects. Because of this, sensitization studies in rodents are sometimes used as a model for stimulant-associated addictive behaviors.

A number of animal studies have examined neural plasticity developed in response to stimulant drug use.[13,14] These studies identify a constellation of alterations during chronic drug use, a different constellation during withdrawal, and even some alterations in brain structure occurring more than a week after last use of drug.[15–18] This latter phenomenon corresponds with the fact that the motivational value of cues associated with former drug use increases with time after stopping drug use. One important fact relating to stimulant drug-induced plasticity is that alterations in synaptic strength are not limited to dopaminergic neurons and/or receptors, the site of initial primary action of stimulant drugs, but rather occur in many regions of the brain involving many different types of neuron.[13]

TREATMENT OF ADDICTION AND ANIMAL MODELS

In treating addiction, one can select from several options. Treatment assumes that a person is currently or has previously used drug in a manner that leads to lack of control over drug use. One treatment strategy for a current user is to allow the person to continue to use drug but to better manage use. A pharmacological intervention in this case would result in either decreased use (but not elimination) of the addicting drug or in attenuated level of undesirable effects. A second strategy is to substitute one addicting drug for another characterized by fewer adverse effects (such as substituting methadone or buprenorphine for heroin). A third strategy for a current user is to add a drug that decreases the desirability of continuing use of addicting drug, thus helping wean off the addicting drug. For former users who are currently drug-free, the goal is to prevent relapse. Three major factors are triggers for relapse. One is stress. A second is the taking of any drug with potential addiction liability. The third is exposure to cues associated with previous drug use. Such cues have motivational power in that they can cause a craving to resume use of the drug. Pharmacological strategies are mostly aimed at lessening the motivational power of cues, although some are aimed at lessening the impact of ingestion of a drug with addiction liability.

Several animal models are available for evaluating effects of drugs on addiction-associated behaviors. These will be itemized following the order of strategies for drug addiction provided in the previous paragraph. Determining whether a potential treatment drug attenuates stimulant-induced locomotor activity is a common paradigm to estimate ability to lessen a stimulant drug's effect. Two paradigms are commonly used as surrogate markers of stimulant-induced neuroplasticity. Locomotor sensitization involves daily doses of stimulant drug. With appropriate dose and environmental conditions, the amount of stimulant-induced locomotor activity increases with time. The augmented effect persists for weeks after the end of the training period. Potential treatment drugs can be paired with the stimulant drug during training to determine whether they interfere with the development of sensitization. Alternatively, potential treatment drugs are paired with the stimulant drug during the test session to determine whether they interfere with the expression of sensitization. Conditioned place preference is a test of the ability of a drug to induce a learned association with the location in which the drug is experienced. In this paradigm, animals are administered drug in one of two adjacent chambers with different sensory stimuli (different color on the walls, different flooring/bedding, etc.) and saline in the other chamber. After an appropriate number of training sessions, the animal is given the choice to explore both chambers via an opening between them. Training using drugs with addiction potential develops an association with the drug-paired chamber such that the animal will spend more time there on test days in the absence of drug. One can evaluate a test agent for impact on development of place preference by pairing its administration with that of the addicting drug on test days and evaluating place preference in absence of drugs. One can evaluate a test agent for impact on expression of place preference by not using the agent during the training phase and using it alone on test days. Attenuation of expression of place preference suggests the possibility that the test agent might be efficacious when taken during the time when one is giving up drugs. To test whether a potential treatment drug can decrease use of an addicting drug, drug self-administration paradigms are used. One can train an animal to self-administer drug and then determine whether the experimental agent decreases the amount of drug self-administration. In a variant to test for ability to prevent relapse, animals are subjected to an extinction period after learning to self-administer drug. Exposure to a drug-associated cue, drug, or stress reliability reinstates drug-seeking behaviors. The test drug can be given during the extinction period and/or just prior to exposure to a cue, stress, or drug to determine ability to lessen relapse. Finally, to determine if one drug will substitute for another, one can carry out drug discrimination studies. In this paradigm, an animal is trained to press a lever after receiving a drug. After the animal is well-trained, one can administer a different drug to the animal to determine whether its interoceptive cues cause the animal to press the lever.

ADDICTION-RELATED STUDIES USING GENETICALLY ALTERED MICE

The first hints for a potential involvement of TAAR1 in stimulant addiction came from studies using mice lacking TAAR1. Such studies showed somewhat enhanced amphetamine- and methamphetamine-stimulated locomotor activity[19–21] and elevation of extracellular dopamine[20,21] in TAAR1 knockout mice. In addition, the extent of methamphetamine-induced place preference was increased and rate of extinction slowed in TAAR1 knockout mice.[19] Mice with genetic alterations resulting in loss of TAAR1 function show higher voluntary intake of methamphetamine.[22] Such data showing that amphetamine effects are potentiated in knockout mice suggest that receptor agonists might attenuate amphetamine effects.

ADDICTION-RELATED STUDIES USING TAAR1 AGONISTS

Agonist and antagonist drugs are required to probe the hypothesis that manipulations of a particular receptor activity have efficacy in some aspect of addiction treatment. Transgenic receptor knockout animals have some validity in demonstrating that a receptor is necessary for expression of an addiction-associated behavior; however, these models are limited by compensatory adaptations occurring during development and by inability to evaluate low-dose (partial receptor blockade in the case of antagonist drugs) effects. Thus, the ability to rigorously evaluate the hypothesis that manipulation of TAAR1 activity can be used in the treatment of addiction has only been recently available with the advent of drugs with selectivity for TAAR1 receptors. Four chemicals with a high affinity (1–31 nM depending on whether the receptor is human, mouse, rat, or monkey) and high selectivity (greater than 100-fold selectivity for more than 100 receptors, with the exception of as low as

10-fold selectivity for the imidazoline receptor (again depending on whether the receptor is human, mouse, rat, or monkey)), have been described.[23–26] RO5166017 and RO5256390 are full agonists (measured by extent of accumulation of cAMP as compared to that stimulated by β-phenethylamine), while RO5203648 and RO5263397 are partial agonists.

The four TAAR1 agonists have been evaluated in several rodent addiction models. The results are summarized as follows.

Acute locomotor effects

- All four TAAR1 agonists attenuate acute cocaine-induced locomotor activity in most[24–26] but not all[27,28] studies.
- RO5203648 (partial agonist) at very high dose attenuates amphetamine-induced locomotor activity in rats,[26] although this result depends on the time period over which locomotor activity is measured.[29]

Development of locomotor sensitization

- RO5263397 (partial agonist) and RO5203648 (partial agonist) lessen development of locomotor sensitization to amphetamine.[29,30]

Expression of locomotor sensitization

- RO523397 (partial agonist) lessens expression of locomotor sensitization to cocaine.[27]
- RO5203648 (partial agonist) lessens expression of locomotor sensitization to methamphetamine.[29]

Conditioned place preference

- RO523397 (partial agonist) lessens expression of cocaine-induced place preference but does not attenuate development of place preference.[28]

Self-administration

- RO523397 (partial agonist) lessens self-administration of cocaine in high but not low fixed-ratio paradigms.[28]
- RO5263397 (partial agonist) and RO5203648 (partial agonist) lessen methamphetamine self-administration.[29,30]

Reinstatement of drug-seeking behavior: cocaine

- RO523397 (partial agonist) and RO5203648 (partial agonist) lessen cocaine-induced reinstatement of drug-seeking behavior.[28,31]
- RO5203648 (partial agonist) and RO5256390 (full agonist) lessen cue-induced reinstatement of drug-seeking behavior.[31]
- RO5203648 (partial agonist) alters patterns in breakpoint experiments.[31]

Reinstatement of drug-seeking behavior: methamphetamine

- RO5263397 (partial agonist) lessens methamphetamine-induced reinstatement of drug-seeking behavior.[30]
- RO5263397 (partial agonist) lessens cue-induced reinstatement of drug-seeking behavior.[30]

CONCLUSIONS FROM THE DRUG STUDIES

The studies show that TAAR1 receptor agonists lessen cocaine, amphetamine, and methamphetamine acute effects as well as development of locomotor sensitization to the extent that these behaviors have been examined. One caveat is that time course of effect has not been adequately controlled. In the one study in which locomotor activity was monitored over a 3-h time block, methamphetamine-induced locomotor activity was initially high but decreased across the observation period, a pattern customarily seen after this drug. However, in the presence of RO5203648, the initial response to methamphetamine was blunted but no decrease in locomotor activity was observed across the 3-h observation period.[29] Thus, the TAAR1 agonist would decrease methamphetamine-induced locomotor activity if the observation period is the first 30 min after administration of drug, have no effect if the observation period is for 3 h after administration of drug, and increase locomotor activity if the observation period is 2½–3 h after administration of drug.

The studies consistently show an attenuation of learned behaviors evaluated after use of chronic drugs. Thus, TAAR1 agonists lessen expression of sensitized locomotor activity, lessen expression of place preference, and lessen reinstatement of drug-seeking behaviors elicited by cues or by re-exposure to stimulant drug.

IMPLICATIONS AND ISSUES

One interesting observation is that both the full- and the partial-agonist drugs have the same spectrum of effects in these models of addiction-related behaviors. This contrasts with their effects on rate of dopamine cell firing in brain slices in vitro where the full agonists decrease the rate of firing[24,25] and the partial agonists increase the rate of firing.[24,26] Assuming this finding extends to in vivo, it suggests that the mechanism for attenuation of addiction-associated behaviors is likely independent of the rate of dopamine neuron firing. This suggests the TAAR1 agonists are affecting behaviors at locations downstream from the dopamine system.

One study evaluated the effects of RO5203648 on methamphetamine-elicited increases in dopamine levels. The time to peak effect was somewhat delayed, and the peak effect was somewhat decreased.[29] However, the magnitude of increase in extracellular dopamine in the presence of the TAAR1 agonist was still substantially greater than that elicited by cocaine, morphine, and other drugs with addiction liability. The fact that dopamine levels in the presence of the TAAR1 agonist are still high is another evidence suggesting that TAAR1 agonist attenuation of acute effects of amphetamines occurs downstream from the dopamine system.

Amphetamine and methamphetamine are potent agonists for TAAR1 receptors,[21,32] with ED50 value for interaction with the mouse receptor in in vitro studies comparable to those of the four available TAAR1 agonists[21] and approximately 10-fold less potent at the rat receptor.[32] Thus, behaviorally relevant doses of amphetamines should occupy and activate all TAAR1 receptors. This suggests that TAAR1 agonists should have no impact on the acute effects of amphetamine, meaning the TAAR1 agonists would not be effective as a treatment to lessen the impact of amphetamine while someone is still using the stimulant drug. However, one of the TAAR1 receptor partial agonists is reported to decrease amphetamine-stimulated hyperactivity in rats.[26] However, the dose required for this effect was high, and lower doses that did attenuate cocaine-elicited locomotor activity did not affect amphetamine-elicited locomotor activity. This suggests that activation of the TAAR1 receptor is required for amphetamine-induced hyperactivity and that the TAAR1 receptor partial agonist is out-competing amphetamine for occupancy at the receptor and thus diminishing the extent of receptor activation.

The extrapolation from rodent to human subjects always has some uncertainty. In the case of the TAAR1 receptor, there is substantial variability in the amino acid sequence of the human, monkey, rat, and mouse receptor.[32,33] Thus, although all four agonists have high potency and selectivity for both human and mouse TAAR1 receptors, one might need to be careful extrapolating from mouse to human.

There are a number of studies that should be done to provide additional information pertinent to how TAAR1 receptor agonists might be used in treating stimulant drug-induced addictions. For example, data on timing of drug administration and duration of drug effect data are needed. If the action of the TAAR1 agonists reduces the motivational power of cues, stress, or exposure to drug without reversing the alteration of circuits controlling this phenomenon, then TAAR1 agonists would have to be taken continually in order to prevent relapse. On the other hand, if TAAR1 agonists taken during the incubation period for development of craving after termination of stimulant drug use can attenuate the augmentation of craving, then the TAAR1 agonist drug therapy could be administered for a relatively short but critical period of time. As the nature of the stimulant drug-induced neuroplasticity varies as a function of the drug regimen to which animals are exposed, experiments should be done using a variety of stimulant drug self-administration schedules. Neurobiological data on effects of TAAR1 drugs on stimulant-induced neuroplasticity would be very interesting.

There are a number of compounds that have similar effects to TAAR1 and have been more extensively evaluated in rodent models. One example is the mGluR5 agonist, MPEP, which decreases cocaine self-administration[34–36] and attenuates reinstatement of drug-induced behaviors elicited by a priming injection of cocaine[35] or cocaine-associated cues.[37] Another example is N-acetylcysteine, which has efficacy in preventing cue and drug-priming reinstatement of drug-seeking behaviors as well as reversing cocaine-induced neuroplasticities.[38–43] However, addiction is so complex and multifaceted that a variety of potential therapeutic agents should be concurrently pursued. It will be interesting to see which compounds are most efficacious as additional studies are done with TAAR1 agonists and other compounds relative to efficacy in various addiction models.

One other factor argues in favor of continued evaluation of TAAR1 agonists in treating stimulant addiction. At this point, theoretical considerations suggest the TAAR1 agonists could have the advantage of fewer adverse effects, as the compounds have very high potency and selectivity and the number and distribution of receptors is smaller. Assuming all compounds have equal efficacy, the treatment of choice would then be the one with the fewest adverse effects.

CONCLUSION

In summary, a limited number of studies show that agonists at TAAR1 receptors show efficacy in animal models of relevance to treatment of stimulant addiction. These studies should be continued to validate that the drugs have efficacy in a variety of paradigms, to determine timing and duration of treatment needed to lessen relapse, and to determine whether stimulant-induced neuroplasticities are reversed by the TAAR1 agonist treatment.

References

1. Di Chiara G, Imperato A. Drugs abused by humans preferentially increase synaptic dopamine concentrations in the mesolimbic system of freely moving rats. *Proc Natl Acad Sci USA* 1988;**85**:5274–78.
2. Amara SG, Kuhar MJ. Neurotransmitter transporters—recent progress. *Ann Rev Neurosci* 1993;**16**:73–93.
3. Kuczenski R. Biochemical actions of amphetamine and other stimulants. In: Creese I, editor. *Stimulants: neurochemical, behavioral, and clinical perspectives*. New York, NY: Raven Press; 1983. p. 31–61.
4. Ritz MC, Lamb RJ, Goldberg SR, Kuhar MJ. Cocaine receptors on dopamine transporters are related to self-administration of cocaine. *Science* 1987;**237**(4819):1219–23.
5. Sulzer D, Sonders MS, Poulsen NW, Galli A. Mechanisms of neurotransmitter release by amphetamines: a review. *Prog Neurobiol* 2005;**75**(6):406–33.
6. Wise RA, Bozarth MA. A psychomotor stimulant theory of addiction. *Psychol Rev* 1987;**94**(4):469–92.
7. Brown MTC, Bellone C, Mameli M, et al. Drug-driven AMPA receptor redistribution mimicked by selective dopamine neuron stimulation. *PLoS One* 2010;**5**(12):e15870.
8. Jones SR, Gainetdinov RR, Wightman RM, Caron MG. Mechanisms of amphetamine action revealed in mice lacking the dopamine transporter. *J Neurosci* 1998;**18**(6):1979–86.
9. Koob GF, Bloom FE. Cellular and molecular mechanisms of drug dependence. *Science* 1988;**242**:715–23.
10. Self DW, Nestler EJ. Molecular mechanisms of drug reinforcement and addiction. *Annu Rev Neurosci* 1995;**18**:463–95.
11. Wise RA. Addictive drugs and brain stimulation reward. *Annu Rev Neurosci* 1996;**19**:319–40.
12. Chen R, Tilley MR, Wei H, Zhou FW, Zhou FM, Ching S, Quan N, Stephens RL, Hill ER, Nottoli T, Han DD, and Gu HH. Abolished cocaine reward in mice with a cocaine-insensitive dopamine transporter. *Proc Natl Acad Sci USA* 2006;**103**(24):9333–8.
13. Kalivas PW, Volkow ND. The neural basis of addiction: a pathology of motivation and choice. *Am J Psychiatry* 2005;**162**(8):1403–13.
14. Luescher C, Malenka RC. Drug-evoked synaptic plasticity in addiction: from molecular changes to circuit remodeling. *Neuron* 2011;**69**(4):650–63.
15. Conrad KL, Tseng KY, Uejima JL, et al. Formation of accumbens GluR2-lacking AMPA receptors mediates incubation of cocaine craving. *Nature* 2008;**454**(7200):118–24.
16. Grimm JW, Hope BT, Wise RA, Shaham Y. Neuroadaptation—incubation of cocaine craving after withdrawal. *Nature* 2001;**412**(6843):141–2.
17. Neisewander JL, Baker DA, Fuchs RA, Tran-Nguyen LTL, Palmer A, Marshall JF. Fos protein expression and cocaine-seeking behavior in rats after exposure to a cocaine self-administration environment. *J Neurosci* 2000;**20**(2):798–805.
18. Pickens CL, Airavaara M, Theberge F, Fanous S, Hope BT, Shaham Y. Neurobiology of the incubation of drug craving. *Trends Neurosci* 2011;**34**(8):411–20.
19. Achat-Mendes C, Lynch LJ, Sullivan KA, Vallender EJ, Miller GM. Augmentation of methamphetamine-induced behaviors in transgenic mice lacking the trace amine-associated receptor 1. *Pharmacol Biochem Behav* 2012;**101**(2):201–7.
20. Lindemann L, Meyer CA, Jeanneau K, et al. Trace amine-associated receptor 1 modulates dopaminergic activity. *J Pharmacol Exp Ther* 2008;**324**(3):948–56.
21. Wolinsky TD, Swanson CJ, Smith KE, et al. The trace amine 1 receptor knockout mouse: an animal model with relevance to schizophrenia. *Genes Brain Behav* 2007;**6**(7):628–39.
22. Harkness JH, Shi X, Janowsky A, Phillips TJ. Trace amine-associated receptor 1 regulation of methamphetamine intake and related traits. *Neuropsychopharmacology* 2015;**40**(9):2175–84.
23. Bradaia A, Trube G, Stalder H, et al. The selective antagonist EPPTb reveals TAAR1-mediated regulatory mechanisms in dopaminergic neurons of the mesolimbic system. *Proc Natl Acad Sci USA* 2009;**106**(47):20081–6.

24. Revel FG, Moreau JL, Pouzet B, et al. A new perspective for schizophrenia: TAAR1 agonists reveal antipsychotic- and antidepressant-like activity, improve cognition and control body weight. *Mol Psychiatry* 2013;**18**(5):543–56.
25. Revel FG, Moreau J-L, Gainetdinov RR, et al. TAAR1 activation modulates monoaminergic neurotransmission, preventing hyperdopaminergic and hypoglutamatergic activity. *Proc Natl Acad Sci USA* 2011;**108**(20):8485–90.
26. Revel FG, Moreau J-L, Gainetdinov RR, et al. Trace amine-associated receptor 1 partial agonism reveals novel paradigm for neuropsychiatric therapeutics. *Biol Psychiatry* 2012;**72**(11):934–42.
27. Thorn DA, Zhang C, Zhang Y, Li J-X. The trace amine associated receptor 1 agonist RO5263397 attenuates the induction of cocaine behavioral sensitization in rats. *Neurosci Lett* 2014;**566**:67–71.
28. Thorn DA, Jing L, Qiu Y, et al. Effects of the trace amine-associated receptor 1 agonist RO5263397 on abuse-related effects of cocaine in rats. *Neuropsychopharmacology* 2014;**39**(10):2309–16.
29. Cotter R, Pei Y, Mus L, et al. The trace amine-associated receptor 1 modulates methamphetamine's neurochemical and behavioral effects. *Front Neurosci* 2015;**9**:39.
30. Jing L, Zhang Y, Li JX. Effects of the trace amine associated receptor 1 agonist RO263397 on abuse-related behavioral indices of methamphetamine in rats. *Int J Neuropsychopharmacol* 2015;**18**(4):1–7.
31. Pei Y, Lee J, Leo D, Gainetdinov RR, Hoener MC, Canales JJ. Activation of the trace amine-associated receptor 1 prevents relapse to cocaine seeking. *Neuropsychopharmacology* 2014;**39**(10):2299–308.
32. Bunzow JR, Sonders MS, Arttamangkul S, et al. Amphetamine, 3,4-methylenedioxymethamphetamine, lysergic acid diethylamide, and metabolites of the catecholamine neurotransmitters are agonists of a rat trace amine receptor. *Mol Pharmacol* 2001;**60**(6):1181–8.
33. Lindemann L, Ebeling M, Kratochwil NA, Bunzow JR, Grandy DK, Hoener MC. Trace amine-associated receptors form structurally and functionally distinct subfamilies of novel G protein-coupled receptors. *Genomics* 2005;**85**:372–85.
34. Kenny PJ, Boutrel B, Gasparini F, Koob GF, Markou A. Metabotropic glutamate 5 receptor blockade may attenuate cocaine self-administration by decreasing brain reward function in rats. *Psychopharmacology* 2005;**179**(1):247–54.
35. Lee B, Platt DM, Rowlett JK, Adewale AS, Spealman RD. Attenuation of behavioral effects of cocaine by the metabotropic glutamate receptor 5 antagonist 2-methyl-6-(phenylethynyl)-pyridine in squirrel monkeys: comparison with dizocilpine. *J Pharmacol Exp Ther* 2005;**312**(3):1232–40.
36. Platt DM, Rowlett JK, Spealman RD. Attenuation of cocaine self-administration in squirrel monkeys following repeated administration of the mGluR5 antagonist MPEP: comparison with dizocilpine. *Psychopharmacology* 2008;**200**(2):167–76.
37. Backstrom P, Hyytia P. Ionotropic and metabotropic glutamate receptor antagonism attenuates cue-induced cocaine seeking. *Neuropsychopharmacology* 2006;**31**(4):778–86.
38. Amen SL, Piacentine LB, Ahmad ME, et al. Repeated N-acetyl cysteine reduces cocaine seeking in rodents and craving in cocaine-dependent humans. *Neuropsychopharmacology* 2011;**36**(4):871–8.
39. Corbit LH, Chieng BC, Balleine BW. Effects of repeated cocaine exposure on habit learning and reversal by N-acetylcysteine. *Neuropsychopharmacology* 2014;**39**(8):1893–901.
40. Madayag A, Lobner D, Kau KS, et al. Repeated N-acetylcysteine administration alters plasticity-dependent effects of cocaine. *J Neurosci* 2007;**27**(51):13968–76.
41. Moussawi K, Pacchioni A, Moran M, et al. N-acetylcysteine reverses cocaine-induced metaplasticity. *Nat Neurosci* 2009;**12**(2):182–9.
42. Reichel CM, Moussawi K, Do PH, Kalivas PW, See RE. Chronic N-acetylcysteine during abstinence or extinction after cocaine self-administration produces enduring reductions in drug seeking. *J Pharmacol Exp Ther* 2011;**337**(2):487–93.
43. Reissner KJ, Kalivas PW. Using glutamate homeostasis as a target for treating addictive disorders. *Behav Pharmacol* 2010;**21**(5–6):514–22.

Trace Amines and Their Potential Role in Primary Headaches: An Overview

T. Farooqui

Department of Entomology, The Ohio State University, Columbus, OH,
United States

INTRODUCTION

Headaches are defined as a pain located in the head being over the eyes, ears, neck, and in the back of skull. Headaches are generally divided into two types: primary and secondary. The primary headaches are far more common than secondary headaches. The primary headaches are benign, recurrent, not dangerous, and not caused by underlying disease or structural problems, but may be caused by a combination of multiple factors including genetic, developmental, and environmental risk factors. Secondary headaches are relatively

Trace Amines and Neurological Disorders
DOI: http://dx.doi.org/10.1016/B978-0-12-803603-7.00024-0

rare, may be dangerous, and result due to abnormal conditions or underlying diseases such as brain tumors, traumatic brain injury, vascular disorders, aneurysms, meningitis, and encephalitis. The most common primary headaches include cluster, migraine, and tension-type headaches.

Cluster headache is considered as a neurological disorder, in which attacks are rapid in the onset with very severe and recurrent pain, confined to one side of the head in the trigeminal nerve area.[1] The accompanying ipsilateral cranial autonomic features (eye watering, nasal congestion, and redness or drooping eyelids swelling around the eye) associated with cluster headache are typically confined to unilateral head pain. According to the International Headache Society, the criteria for cluster headache include attacks of severe or very severe, strictly unilateral pain, which is orbital, supraorbital, or temporal, lasting 15–180 min, and occurring from once every other day to eight times daily.[2] Cluster headache has been reported in all age groups; however, the onset of symptoms most commonly occurs between the second and fourth decades of life.[3–5] Cluster headache attacks often occur periodically; spontaneous remissions may interrupt active periods of pain. However, ~10–15% of chronic cluster headache never remits. The prevalence of cluster headache is likely to be at least one person per 500 and occurs more in men.[6] Risk factors for cluster headache include: (1) sex—men are more likely to have cluster headaches; (2) age—mostly between 20–50 years; (3) genetics factors—genetic factors play a role in the disease; (4) ethnic ancestry—mainly people of African ancestry; (5) smoking—many people who get cluster headache are smokers; and (6) alcohol consumption—if a person is at risk, alcohol can trigger an attack.

Migraine is another neurological condition characterized by unilateral or bilateral severe headache. This painful recurrence of headache frequently occurs, affecting ~20% of the population. Migraine is significantly higher in women than men, and is not associated with socioeconomic status.[7] Migraine is subclassified into two forms: (1) classical migraine or migraine with aura (MWA), and (2) common migraine or migraine without aura (MWoA). In MWA the headache itself is preceded by an aura consisting of visual problems, generally blind spots in one or both eyes, seeing zigzag patterns or flashing lights, hallucinations, numbness or tingling in the face or hands (10–30 min before an attack); whereas in MWoA there is no premonitory aura. Migraine lasts 4–70 h, commonly accompanied by nausea, vomiting, an unpleasant perception (osmophobia), and also extreme sensitivity to light (photophobia) and sound (phonophobia). Osmophobia, a peculiar symptom of migraine, discriminates adequately between cluster and migraine patients.[8] The etiology of cluster and migraine headaches is not completely understood; however, they can be differentiated by their length and frequency of recurrence (Table 24.1).[7,9,10] Risk factors for migraine headache include: (1) sex—more prevalent in females than males with a ratio of 3:1; (2) age—between 15–45, most people experience their first migraine during adolescence; (3) genetic factors—about 70–80% of patients with migraine have a family history; and (4) hormonal changes—due to changes in female hormone levels, headaches begin just before or shortly after onset of menstruation.

The third most common type of primary headache is tension-type headache, which is bilateral, mild-to-moderate head pain, and often not severe enough to warrant neurological referral. It can be treated with nonsteroidal anti-inflammatory drugs (NSAIDs) or acetaminophen, but will not be discussed here. Despite numerous hypotheses postulated for cluster

TABLE 24.1   Comparison Between Cluster and Migraine Headaches

Types of Headaches	Cranial Autonomic Symptoms	Severity/Quality of Pain	Laterality	Attack Frequency	Attack Duration	Response to Indometacin
Cluster headache	Present	Very severe/sharp stabbing, throbbing	Strictly unilateral	1–3/day	15–180 min	No effect
Migraine headache	Unusual	Moderate to severe/throbbing (but may be steady)	Often unilateral (but may be bilateral)	Varies	4–70 h	No

Information adapted from Refs. 7,9.

and migraine headaches, the pathophysiological mechanisms for these headaches remain obscure. Therefore, their management is challenging because only a few effective treatments are available and high doses may be required to control the headache, compromising patients' adherence to treatments.

The cluster and migraine headaches may affect individuals from childhood to old age and are most troublesome in the productive years of life, thus generating an economic burden for both society and healthcare systems. According to Global Burden of Disease studies, primary headaches are among the top 10 disorders causing significant disability. The purpose of this overview is to discuss the pathophysiology, and genetic factors associated with cluster and migraine headaches. Information on the possible contribution and role of neurotransmitters, neuromodulators, neuropeptides, and trace amines in the pathogenesis of cluster and migraine headaches has also been presented.

PATHOPHYSIOLOGY OF CLUSTER AND MIGRAINE HEADACHES

Cluster headache occurs in bouts (daily for 6–12 weeks, once a year or 2 years, or disappears entirely for months or years then reoccurs). Cluster headache is not caused by any underlying structural pathology and belongs to the group of trigeminal-autonomic cephalalgias (TACs), whose pathophysiologies have not been adequately defined. There are two types of cluster headaches: (1) episodic cluster headache (ECH) and (2) chronic cluster headache (CCH) that are distinguished by the frequency of headaches.[11] ECH attacks occur in clusters and last for weeks to months (with remissions of 1–4 years) in ~80% of patients; whereas, CCH attacks occur for more than 1 year (without remissions) in ~15–20% of patients (Table 24.2). Cluster headache is accompanied by characteristic associated unilateral symptoms, such as tearing, nasal congestion, and/or rhinorrhea (excessive discharge of mucus from the nose), eyelid edema, miosis (excessive constriction of the pupil of the eye), and/or ptosis (drooping of an upper eyelid).[12] It also develops a sense of restlessness and agitation. There are many theories for cluster headache, but three are well accepted with regard to its mechanism (Fig. 24.1A): (1) pain is always located periorbitally frontally, implicating nociceptive mechanisms involving the trigeminal nerve; (2) activation of the trigeminovascular system and the cranial parasympathetic nervous system, resulting in ipsilateral autonomic manifestations due to involvement of both parasympathetic discharge

TABLE 24.2   Ipsilateral Cranial Autonomic Features, Forms, and Brain Regions Involved in Cluster and Migraine Headaches

Headaches	Types	Ipsilateral Cranial Autonomic Features	Brain Regions
Cluster	ECH: (~80%) two attacks for at least 7 days that last for several weeks or months, and are separated by a remission period of several months to many years CCH: (~20%) 10–15 continuous attacks without remission that occur between one and eight times per day	Conjunctival injection, lacrimation, rhinorrhea, nasal congestion, eyelid edema, forehead and facial sweating, mitosis, and ptosis	Posterior hypothalamus
Migraine	EM: (57%) duration 0–14 headache days/month CM: (43%) duration 15 or more headache days/month MWoA: in ~75% cases MWA: in ~25% of cases	Nausea, vomiting, fever, chills, aching, sweating and sensitivity to light, sound and/or exacerbation of pain by head movement	Hypothalamus and subcortical structures (diencephalic and brainstem nuclei)

Information adapted from Refs. 10–13.

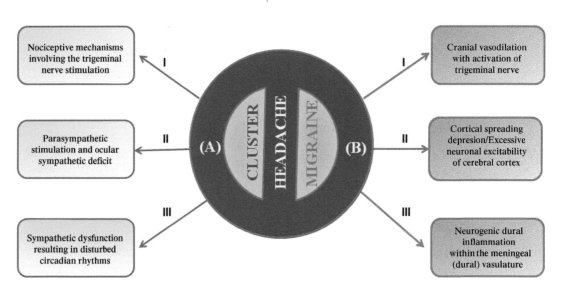

FIGURE 24.1   Theories underlying cluster and migraine headaches. (A) Cluster headache: (I) Trigeminal nerve stimulation results in the release of inflammatory peptides causing vascular dilation. (II) Parasympathetic stimulation and ocular sympathetic deficit giving rise to autonomic disturbance. (III) Sympathetic dysfunction that brings changes in the sleep–wake schedule leaving a significant effect on cluster headaches. (B) Migraine: (I) Vascular theory due to cerebral vasodilation. (II) Neurological theory in which abnormal neurological firing causes cortical spreading depression. (III) Neurogenic dural inflammation that activates trigeminal sensory nerve fibers, resulting in the inflammation of dura mater and extracranial tissues, but not of the brain.

(lacrimation and rhinorrhoea) and sympathetic deficit (ptosis and miosis); and (3) the periodicity of these attacks and seasonal recurrence of the cluster periods, suggesting involvement of a biological clock within hypothalamus (sympathetic dysfunction) associated with the hypercortisolism, hypotestosteronism, and reduced melatonin levels in the active cluster patient.[13–16] Moreover, histamine release, an increase in mast cells, and genetic factors may contribute to cluster headache. The calcium-channel-blocking agent (verapamil) and inflammation-suppressing corticosteroids (prednisone) are used for the prevention of cluster headache.[12,17] Other most effective pharmacological treatment options to abort the attack of acute cluster attack include subcutaneous serotonin receptor ($5\text{-}HT_{1B}/5\text{-}HT_{1D}$) agonists such as sumatriptan, intranasal zolmitriptan, and high-flow oxygen.[18] In addition, neuromodulation with either occipital nerve stimulation, or deep brain stimulation of the hypothalamus is an alternative treatment strategy for treating drug-resistant chronic cluster headache.[12]

Migraine headache is a disabling, recurring headache, which frequently occurs only on one side of the head. Over 56% of respondents have episodic migraine (EM), while the remaining reported having chronic migraine (CM) (Table 24.2). CM is characterized by 15 or more headache days per month with symptoms lasting greater than 4h; whereas EM is characterized as less than 15 headache days per month. The patterns of treatment response for EM and CM differ, which raises the possibility of both overlapping and distinct biological mechanisms.[19] Migraine often begins with warning signs and aura. Once the headache develops, it typically throbs, intensifies with an increase in intracranial pressure, and presents itself in association with nausea, vomiting, and abnormal sensitivity to light, noise, and/or exacerbation of pain by head movement.[20]

Once the migraine develops, it typically throbs, intensifies with an increase in intracranial pressure, and presents itself in association with nausea, vomiting, and abnormal sensitivity to light, noise, and smell. It can also be accompanied by abnormal skin sensitivity (allodynia) and muscle tenderness. The pathophysiology of migraine is not clearly understood. However, three theories have been suggested with regard to migraine mechanism (Fig. 24.1B): (1) according to traditional vascular theory, the vasoconstriction of cerebral blood vessels occurs during an aura, whereas dilation of cerebral vessels causes the headache; (2) the neuronal theory suggests that the excessive excitement of nerve cells in the cerebral cortex "cortical spreading depression" activates trigeminal vascular sensitization, which causes reduction in brain electrical activity and decreases blood flow, leading to migraine; and (3) the neurogenic dural inflammation theory states that activation of trigeminal sensory fibers leads to a painful neurogenic inflammation within the meningeal (dural) vasculature mediated by inflammatory neuropeptide (calcitonin gene-related peptide (CGRP)), substance P (SP) and neurokinin-A (NK-A) release from trigeminal sensory fibers.[21–24] Among these theories the neurovascular hypothesis is the generally accepted theory, which states that migraine is mediated by prolonged activation of meningeal nociceptors, which are located in the dura mater and vessels.[25] The neurovascular theory centers on activation of the trigeminovascular system. It consists of pseudounipolar neurons in the trigeminal ganglion that has first-order afferent neurons innervating the pial and dural meningeal vessels, and efferent projections synapsing with second-order neurons in the spinal cord trigeminal nucleus caudalis, which provides projections to several higher brain centers, including the posterior thalamus, hypothalamus, and cortex.[25] Activation of perivascular trigeminal nerves within meninges causes the release of CGRP, phosphorylation of extracellular signal-regulated kinase (ERK), SP and NK-A, leading to a series of peripheral and central events, including inflammation and peripheral/central sensitization.[26,27] In addition to CGRP, other neuropeptides, and cyclooxygenase-2 (COX-2) are important peripheral and central mediators of inflammation and pain. COX-2 not only increases levels of inflammatory mediators (prostaglandin E_2) in the brain, but also contributes to the severity of pain responses in inflammatory pain.[27,28] Regardless of the hypothesis by which a migraine is initiated, most theories support the idea that the blood vessels in the brain undergo a temporary cerebral vasoconstriction, followed by a period of prolonged cerebral vasodilation. This is also supported by cerebral blood flow studies, showing initial hypoperfusion followed by hyperperfusion.[29] Thus, migraine appears to be a neurovascular disorder involving local vasodilation of intracranial, extracerebral blood vessels and simultaneous stimulation of surrounding trigeminal sensory nervous pain pathway that leads to headache.[28] Low serotonin levels have been observed in migraine patients; however, cluster headaches have no change in their serotonin levels but show an increase in blood histamine concentration coincident with their headache. A migraine-active region has been pointed out in the brainstem, whereas for the cluster headache a specific active locus is implicated in the posterior hypothalamus.[30] The triptans (tryptamine-based drugs) are selective serotonin receptor agonists, which are preferred over ergot alkaloids (nonspecific serotonin receptor agonists) for acute treatment of migraine. The prophylactic (preventative) treatments for migraine include calcium channel blockers, 5-HT2 receptor antagonists, β-adrenoceptor blockers, and γ-amino butyric acid (GABA) agonist.[31] Despite such progress, many of these treatments are nonspecific and not always effective.

GENETICS OF PRIMARY HEADACHES

The aim of genetic studies is to identify corresponding key proteins in order to unravel molecular pathways underlying primary headache disorders, in order to discover new therapeutic targets. Gene discoveries in migraine headaches (genetic component higher in MWA than in MWoA) have provided most insight for understanding the underling mechanism; however, gene discoveries in cluster headache types are still in their infancy (Table 24.3).

Genetics of Cluster Headache

Cluster headache is rare compared to other types of headaches, therefore it may be an inherited disorder.[51] Its familial occurrence was recognized in 1995, implicating that its inheritance is likely to be autosomal dominant with low penetrance in some families, and it may also be autosomal recessive or multifactorial in other families.[52] Genetic epidemiological surveys have shown that first-degree relatives of cluster headache patients are at 5–18-fold increased risk and second-degree relatives are at one- to threefold increased risk of having cluster headache than the general population, indicating a genetic component.[6]

At present, the type and the number of genes involved in cluster headache are still unclear. The hypocretins (Hcrts) are neuropeptides, which are synthesized by neurons located in the hypocampus. Hcrts regulate arousal, wakefulness, and appetite. A number of recent studies in experimental animals have shown that Hcrts are involved in pain

TABLE 24.3   Genes Involved in Cluster and Migraine Headaches

Headaches	Protein	Gene Mutation	References
Cluster headache	Hypocretin receptor 2 involved in pain modulation within CNS	HCRTR2	32,33
Cluster headache	Neuronal nitric oxide synthase and calcium channel genes – No mutation found in these genes	NOS, CACNA1A	34–38
Migraine	Hypocretin receptor 1 involved in pain modulation within the CNS	HCRTR1	39
(FHM type 1) (FHM type 2) (FHM type 3)	– Neuronal calcium channel – Na$^+$/K$^+$-ATPase α-2 – Sodium channel gene	CACNA1A ATP1A2 SCN1A	40
Common migraine	– Modulates glutamate homeostasis	rs1835740	41,42
Common migraine	– Lipoprotein receptor (LRP1) modulates synaptic transmission in neuronal glutamate signaling	rs11172113 (LRP1)	41,43
Migraine without aura	– Hypocretin receptor 1 involved in pain modulation within CNS	HCRTR1	39,44
Migraine with and without aura	– Methylene tetrahydrofolate reductase	MTHFR	45,46
Common migraine	– Located in close proximity to TRPM8, which codes for a cold and pain sensor, neuropathic pain – Specific association with migraine pathophysiology	rs10166942 (TRPM8)	47
Common migraine	– Specific association with migraine	rs2651899 (PRDM16)	47
Migraine with aura	– Nominal associations found for single nucleotide polymorphisms within the gene	rs2651899 (PRDM16)	48
Migraine with & without aura	– Potassium channel, subfamily K member 18	KCNK18	44,49
Common migraine	– TRPV1 receptor activation releases CGRP, and causes vasodilation	TRPV1	44,50
Common migraine	– Role as a thermosensor	TRPV3	44,50

FHM, Familial hemiplegic migraine, rare autosomal dominant form of migraine with aura; TRPV1, transient related potential vanilloid type 1 receptor; TRPV3, transient related potential vanilloid type 3 receptor; CGRP, calcitonin gene-related peptide.

modulation within the CNS, and suggest the presence of a link between these peptides and nociceptive phenomena observed in primary headaches. Hcrt-1 and Hcrt-2[32,33] are peptides derived by proteolytic cleavage from the same 130 amino acid precursor peptide. The hypocretins bind to G-protein-coupled receptors (GPCRs) including HCRTR1 and HCRTR2. A significant association between a 1246G-A polymorphism (rs2653349) in the HCRTR2 gene (602393), and cluster headache has been independently reported by two groups.[32,33] This gene is located on chromosome 6p11 and involved in the regulation of chronobiological rhythms. No association was observed between *HCRTR2* gene polymorphism and migraine. Nitric oxide (NO) plays a critical role in the regulation of cerebral and extracerebral cranial blood flow and arterial diameters, and is also involved with nociceptive processing. The prevention of NO production by competitively inhibiting nitric oxide synthases (NOS) with L-NG-monomethylarginine (L-NMMA), has been shown to effectively treat attacks of MWoA, as well as chronic tension-type headache and cluster headache.[34] However, evaluation of several polymorphisms in three subtypes of *NOS* genes resulted in no difference between cluster headache patients and controls.[35] Similarly, no mutation in the *CACNA1A* gene, which encodes for a P/Q-type neuronal calcium channel, has been reported in cluster headache patients, suggesting that cluster headache is not caused by mutations in the *CACNA1A* gene.[36,37] In contrast, mutations in the *CACNA1A* gene are shown to facilitate the initiation of cortical spreading depression, a mechanism underlying MWA.[1] Another study of a Japanese man with cluster headache has shown mutation in mitochondrial transfer *RNAleu (UUR)* gene at nucleotide pair 3243.[53] However, such a mutation was not detected in Italian and German patients with cluster headache.[54,55]

The identification of genes for cluster headache seems to be a difficult task (Table 24.3) because most families reported have few affected members and genetic heterogeneity is likely. Additional studies are needed to search for the susceptible genes involved with cluster headache.

Genetics of Migraine Headache

The three first genes, identified from 1996 to 2005, all encode ion-channel transporters: a neuronal calcium channel (*CACNA1A*, FHM-type 1), a glial sodium/potassium pump (*ATP1A2*, FHM-type 2), and a neuronal sodium channel (*SCN1A*, FHM-type 3) were identified by studying familial hemiplegic migraine (FHM), a variety of migraine with motor aura.[40] Functional studies performed in cellular and/or transgenic animal models suggest that mutations in *CACNA1A* (α-1 subunit of a calcium channel called CaV2.1) and *ATP1A2* (α-2 subunit of a protein known as Na^+/K^+ ATPase) facilitated the initiation of cortical spreading depression, the mechanism underlying MWA, and most likely increased neuronal excitability via enhanced glutamatergic neurotransmission in FHM. However, there is ongoing debate about whether the genes involved for ion channels are specific for migraine, the accompanying hemiplegic aura, or possibly for both.[41] Genome-wide association studies (GWAS) have already identified over a dozen migraine-associated variants pointing at genes that cluster in pathways for glutamatergic neurotransmission, synaptic function, pain sensing, metalloproteinases, and the vasculature involved in neuronal and vascular mechanisms, suggesting that enhanced glutamatergic neurotransmission and abnormal vascular function as key migraine mechanisms.[56,57] A GWAS in a huge population of migraineurs has identified a minor allele of common migraine susceptibility variant (rs1835740) on chromosome 8q22.1, which is located between two genes, the astrocyte elevated gene1 (*MTDH/ AEG-1*) and plasma glutamate carboxypeptidase (*PGCP*), involved in glutamate homeostasis, implicating the involvement of glutamate pathways in migraine pathogenesis.[42] Hypocretin receptor 1 gene (*HCRTR1*) has been represented as a genetic susceptibility factor for migraine without aura, suggesting that it may have a role in the pathophysiology of migraine.[39,44] Migraine with aura has previously been shown to have a significant comorbidity with stroke, making the vascular class of genes a priority for migraine studies. One study identified methylenetetrahydrofolate reductase gene (*MTHFR*) and its role in MWA, showing three single nucleotide polymorphisms and its association with migraine in the Norfolk Island population, reinforcing the potential role of *MTHFR* in migraine susceptibility.[45] This is confirmed by another study demonstrating that *MTHFR* polymorphism is responsible for reduction in methylenetetrahydrofolate reductase enzyme activity in folate metabolism, which may be a genetic susceptibility factor for migraine.[46] The LDL receptor-related protein (LRP1) is involved in modulating synaptic transmission. Tissue-specific gene deletion studies suggest LRP1protein's role in the vasculature, central nervous system, macrophages, and adipocytes. *LRP1* gene is expressed in many tissues including brain and vasculature.[43] TRPM8 encodes a sensor for cold and cold-induced burning pain, primarily expressed in sensory neurons and dorsal root ganglion neurons, and therefore a target for neuropathic pain. Migraine has some characteristics of neuropathic pain disorders, therefore TRPM8 may be a pathophysiological link between both pain syndromes.[47] A potential role of PRDM16 protein in migraine is unclear.[43,47] In a population-based genome-wide analysis three SNPs specifically rs2651899 (*PRDM16*), rs10166942 (*TRMP8*), and rs11172113 (*LRP1*) were among the top associations with migraine.[47] In addition, a potential role for other genes encoding the potassium channel, subfamily K member 18 (*KCNK18*), and cation channels such as transient-related potential vanilloid type 1 (*TRPV1*) and type 3 (*TRPV3*), has been suggested in migraine headaches.[47,49] Briefly, transient receptor potential (TRP) is a superfamily of nonselective cationic channels that contribute to membrane depolarization and activation of second messenger signaling cascades due to the influx of Na^+ and Ca^{2+} by responding to a variety of endogenous conditions (such as temperature, changes in extracellular osmolarity, chemical mediators, and low pH). Activation of TRP channels promotes excitation of nociceptive afferent fibers that may potentially lead to migraine pain by promoting excitation of nociceptive afferent fibers.[50] Collective studies support the notion that migraine headaches are complex, multifactorial disorders (Table 24.3).

Additional research is needed to better elucidate the involvement of genes in different types and/or subtypes of primary headaches and to evaluate their therapeutic strategies.

PHARMACOLOGY OF PRIMARY HEADACHES

Neuropeptides in Primary Headaches

Both cluster and migraine headaches involve activation of the trigeminovascular system. CGRP is expressed throughout the trigeminovascular system and acts as a vasodilator. CGRP is a key neuropeptide, which is involved in the activation of the trigeminovascular nociceptive system. In addition to the activation of the trigeminovascular system, there is a release of the parasympathetic vasoactive intestinal peptide (VIP) involved in the activation of the cranial parasympathetic nervous system that is associated with facial vasomotor symptoms (nasal congestion and rhinorrhea) associated with cluster headache.[58] Upon treatment with triptans (5-HT$_{1B/1D}$ receptor agonists), head pain subsides and CGRP release becomes normal, implicating the involvement of sensory and parasympathetic mechanisms in cluster headache.[58]

Pathogenesis of migraine involves different brain structures: the trigeminal nerve with nuclei located in the brain stem, vascular system, the cerebral cortex, and diverse mechanisms and pathological processes involving uncontrolled activation of the trigeminal nerve, vasoconstriction followed by CGRP-mediated vasodilatation, and cortical spreading depression, a process which causes disruptions of transmembrane ionic gradients, increasing the intracellular concentrations of Ca^{2+} and the extracellular concentrations of H^+, K^+, glutamate, arachidonic acid (AA), and NO in the synaptic cleft.[59] These molecules are able to activate perivascular and meningeal nociceptors of trigeminovascular system and also central trigeminovascular neurons in the spinal trigeminal nucleus. These processes are mediated by pathologically altered concentrations of extracellular signaling molecules and neurotransmitters, resulting in vasodilation and an increase in cerebral blood flow, with plasma protein extravasation and mast cell degranulation on the dura with the release of proinflammatory factors that finally result in an inflammatory reaction known as neurogenic inflammation and pain. There is some evidence confirming the involvement of purines in the above-mentioned processes throughout the central nervous system.[60]

Migraine is a neurovascular disorder associated with dysfunction of the cerebral nerves and blood vessels. CGRP, the most potent peptidergic dilator of peripheral and cerebral blood vessels, is released during severe migraine attacks, and has long been postulated to play an integral role in the pathophysiology of migraine. Administration of triptans (sumatriptan) causes the headache to subside and the levels of CGRP to normalize.[47] Moreover, CGRP mechanism blockade, either by CGRP receptor antagonists or by monoclonal antibodies, can have a preventive effect in migraine, suggesting it to be a promising therapeutic target.[61,62]

Catecholamines in Primary Headaches

Catecholamines are characterized by a catechol group (a benzene ring with two hydroxyl groups) attached with an amine (nitrogen-containing) group. Catecholamines originate from an amino acid tyrosine in the brain, adrenal medulla, and by some sympathetic nerve fibers in the peripheral nervous system by four enzymatic reactions: (1) initial reaction in catecholamine synthesis is catalyzed by tyrosine hydroxylase (TH), an enzyme that uses tetrahydrobiopterin and molecular oxygen, and catalyzes the hydroxylation of tyrosine to L-3,4-dihydroxyphenylalanine (L-DOPA), a rate-limiting step in the synthesis of catecholamines; (2) DOPA decarboylase (DDC or aromatic L-amino acid decarboxylase) catalyzes the decarboxylation of L-DOPA to dopamine; (3) dopamine β-hydroxylase (DBH), a major ascorbate and cu^{2+} dependent enzyme, which converts dopamine to norepinephrine; and (4) the last reaction is catalyzed by phenylethanolamine N-methyltransferase (PNMT) that utilizes S-adenosyl homocysteine as methyl donor and catalyzes the conversion of norepinephrine to epinephrine (Fig. 24.2).[63]

Significantly lower levels of norepinephrine have been reported in the cerebrospinal fluid (CSF) of cluster headache patients, and these values in CSF correlate with plasma norepinephrine levels.[64] Higher platelet levels of dopamine have been observed in cluster headache and both types of migraine patients than in control subjects; however, cluster headache patients showed the highest levels of platelet dopamine levels, suggesting impaired

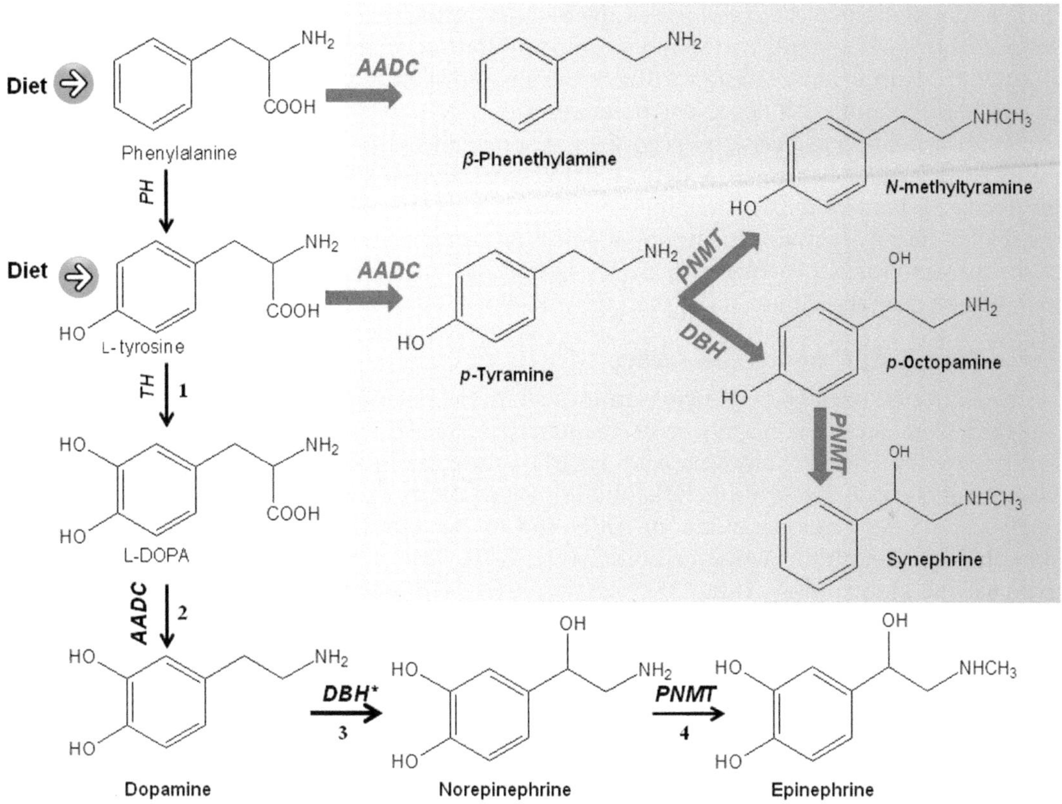

FIGURE 24.2 Biosynthesis of catecholamines and trace amines. Black arrows represent reactions in catecholamines synthesis. Red arrows indicate reactions in trace amines synthesis. *PH*, Phenylalanine hydroylase; *TH*, tyrosine hydroxylase; *AADC*, aromatic amino acid decarboxylase; *DBH*, dopamine β-hydroxylase; *PNMT*, phenylethanolamine N-methyl transferase; human hepatic microsomes and human cytochrome P450 (CYP) isoform (CYP2D).

dopaminergic system in these primary headaches.[65] DBH enzyme activity is reduced in migraine.[66] Moreover, increased plasma dopamine levels and decreased plasma norepinephrine levels were found at menses in migraineurs relative to normal subjects.[67] A significantly altered allelic association of the DBH gene with typical migraine susceptibility, and insertion/deletion polymorphism at DBH showing association with MWA, clearly suggesting that abnormalities in dopamine and norepinephrine levels play an important role in the pathogenesis of migraine.[68,69] Nothing is known about DBH activity in cluster headache patients. However, low levels of norepinephrine have been found in platelets (both in active and remission phases), in plasma, and in CSF (in active period) of cluster headache patients.[70,71] A statistically significant correlation occurs between norepinephrine levels and clinical features of the pain attacks, including duration, intensity, and frequency, implicating a pathophysiological involvement of the sympathetic nervous system in cluster headache patients.

The hypothalamus is considered as the key site in the pathophysiology of primary headaches.[72,73] It is well established that dopamine inhibits prolactin release.[74] A reduction in prolactin levels during acute migraine due to increased dopaminergic activity suggests potential use of antidopaminergic drugs for relieving migraine headache.[75] An altered regulation of prolactin secretion has been observed during active cluster periods as well as symptom-free intervals in cluster headache patients.[75] However, increased dopaminergic activity in cluster headache may lie within the broader frame of a hypothalamic derangement, involving a complex interplay between the hypothalamus, neuroendocrinological parameters, and activity of the autonomic nervous system. These findings suggest that a defect in brain tyrosine metabolism may contribute to the pathophysiological mechanism of primary headaches.

Trace Amines in Primary Headaches

Trace amines, such as β-phenethylamine, tyramine, octopamine, and synephrine are synthesized from aromatic amino acids in mammals.[76] β-Phenethylamine and tyramine are synthesized by aromatic L-amino acid decarboxylase catalyzing the conversion of L-phenylalanine and L-tyrosine to β-phenylethylamine and tyramine, respectively. Tyramine can be further converted to octopamine via DBH and to N-methylated derivative of octopamine, synephrine, through the action of PNMT (Fig. 24.2). Trace amines then can be metabolized primarily by monoamine oxidase.[76]

Elevated levels of plasma trace amine levels have been found in both cluster headache and migraine patients, suggesting the disruption in biogenic amine metabolism is one of the characteristics of primary headache sufferers.[77] However, these alterations are more prominent in patients with cluster headache than migraine patients, implicating sympathetic or hypothalamic dysfunction. Abnormal levels of dopamine and trace amines may contribute to the metabolic events that predispose to the occurrence of these headaches.[78] Tryptamine is derived from tryptophan, and is involved in modulating the function of the pain matrix serotonergic system. Tryptamine levels in plasma of chronic migraine and chronic tension-type headache are very low.[79] Decreased levels of serotonin have been observed in migraine patients.[30] Therefore, low tryptamine plasma levels found in chronic migraine and chronic tension-type headache patients suggest that these two primary headaches should be characterized by a common insufficient serotoninergic control.[80]

MOLECULAR MECHANISMS UNDERLYING PRIMARY HEADACHES

Both cluster and migraine headaches result due to activation of the trigeminovascular system, primarily releasing CGRP from sensory nerves. CGRP receptor (CGRP-R) is most commonly coupled to Gsα to increase cAMP levels, and NO release from trigeminal ganglion, involving adenylate cyclase (AC), cAMP-dependent protein kinase A (PKA), and nitric oxide synthase (NOS), resulting in vasodilation, causing primary headache (Fig. 24.3). Neurogenic inflammation is caused by activation of sensory nerve fibers and mediated by the release of neuropeptides, such as CGRP, from the primary sensory nociceptive fibers. The CGRP synthesis and release can be mediated by activation of mitogen-activated protein kinase (MAPK) pathways (Fig. 24.3), which can also be positively modulated by endogenous inflammatory substances such as TNF-α and inhibited by 5-HT1B/1D receptor agonists.[81]

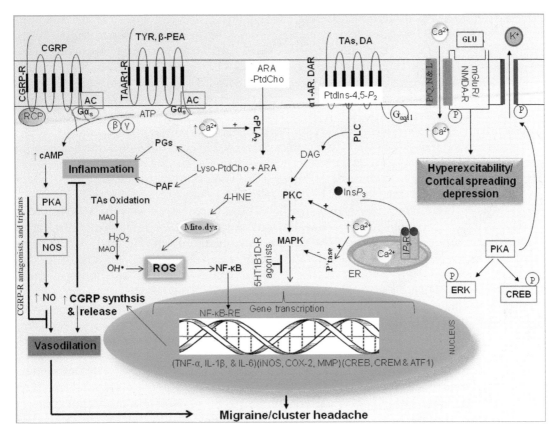

FIGURE 24.3 Hypothetical molecular mechanisms underlying cluster and/or migraine headaches, both sharing a common cause (calcitonin gene-related peptide, CGRP) release that begins after activation of trigeminal nerve. *AC*, Adenylate cyclase; *PKA*, cAMP-dependent protein kinase A; *NO*, nitric oxide; *NOS*, nitric oxide synthase; *cPLA₂*, phospholipase A₂; *PKC*, protein kinase C; *MAPK*, Map kinase; *P'tase*, phosphatase; *ERK*, extracellular signal-regulated kinase; *NF-κB*, nuclear factor kappa-light-chain-enhancer of activated B cells; *PAF*, platelet-activating factor; *PGs*, prostaglandins; *ROS*, reactive oxygen species; *CGRP*, calcitonin gene-related peptide; *RCP*, CGRP-receptor component protein; *ARA*, arachidonic acid; *ARA-PtdCho*, arachidonic acid-phosphatidylcholine; *PtdIns-4,5-P2*, phosphatidylinositol 4,5-bisphosphate; *InsP₃*, inositol (1,4,5)-trisphosphate; *5HT1-R*, serotonin 5-HT1 receptors; *TAs*, trace amines; *Tyr*, tyramine; *β-PEA*, β-phenylethylamine; *DA*, dopamine; *Glu*, glutamate; *DHA*, docosahexaenoic acid; *Mito.dys*, mitochondrial dysfunction; *TNFα*, tumor necrosis factor α; *IL-1β*, interleukin-1β; *IL-6*, interleukin-6; *COX2*, cyclooxygenase-2; *MMP*, matrix metalloproteinase; *CREB*, cAMP-dependent protein kinase; *CREM*, cAMP-responsive element modulator; *ATF1*, activating transcription factor 1; *ER*, endoplasmic reticulum.

Abnormalities in the synthesis of neurotransmitters (high dopamine concentration, low norepinepherine) and very high levels of neuromodulators (octopamine and synephrine) in plasma of episodic migraine without aura and cluster headache sufferers have been hypothesized to be due to an imbalance between the neurotransmitter and neuromodulator metabolic abnormalities.[80] Thus, the metabolic shift in tyrosine metabolism is favored by the neuronal hyperexcitability (high glutamate) and by mitochondrial dysfunction in the brain of migraine and possibly cluster headache sufferers.[82] Trace amines exert their role in several physiological processes including modulation of aminergic neurotransmission. In human brain, TAAR1 is the only receptor that binds to endogenous trace amines and negatively regulates dopamine transmission, implicating TAAR1, a novel target for neuropsychiatric disorders.[83] TAAR1, and to a lesser extent, TAAR4 receptors respond to tyramine and β-PEA; whereas, other trace amines bind with TAAR1 with much less potency.[84] Activation of TAAR1 increases AC activity, catalyzing a conversion of ATP to cAMP (Fig. 24.3).

Dopamine and adrenergic receptors are coupled with enzymes of phospholipid catabolism via G-proteins. These enzymes include phospholipase A_2 (PLA$_2$) and phospholipase C (PLC). Activation of these receptors leads to the stimulation of PLA$_2$ and PLC, resulting in the production of AA (Fig. 24.3). DA-ergic nigrostriatal neurons are also rich in glutamate receptors, which may contribute to excitotoxicity. Hyperstimulation of glutamate receptors results in extensive influx of Ca^{2+} that activates PLA$_2$, resulting in production of AA from neural membrane glycerophospholipids. Glutamate-mediated release of AA also involves participation of the PLC/diacylglycerol lipase pathway. Enzymic oxidation of AA through cyclooxygenase results in the production of eicosanoids that promote inflammation (Fig. 24.3). Nonenzymic oxidation of AA produces ROS that activates NF-κB, a key regulator of neuronal death (Fig. 24.3). ROS interacts with subunits of NF-κB (p50 and p65) and promotes the translocation of NF-κB from the cytoplasm to the nucleus. In the nucleus, NF-κB promotes the transcription of genes that encode number of proteins including many enzymes cyclooxygenase-2 (COX-2), inducible nitric oxide synthase (iNOS), superoxide dismutase (SOD), soluble phospholipase A_2 (sPLA$_2$), matrix metalloprotease (MMP), vascular cell adhesion molecule-1 (VCAM-1), and cytokines (TNF-α, IL-6, and IL-10)). These mediators not only contribute to oxidative stress (a disturbance in the cellular pro-oxidant/antioxidant ratio), but also induce neuroinflammation, thus it possibly results in elevation in CGRP synthesis and release, leading to primary headaches (Fig. 24.3). Recent studies have reported some connection between different types of migraine and mitochondrial dysfunction.[85] This includes (1) mitochondrial dysfunction resulting in high intracellular Ca^{2+} and production of excessive free radicals, and deficient oxidative phosphorylation, ultimately causing energy failure in neurons and astrocytes, triggering migraine mechanisms including spreading depression, (2) morphologic mitochondria found in migraine sufferers, (3) genetic evidence showing two common mitochondrial DNA polymorphisms (16519C→T and 3010G→A) associated with pediatric cyclic vomiting syndrome and migraine, and (4) several therapeutic agents (such as riboflavin B2, coenzyme Q10, magnesium, niacin, carnitine, and lipoic acid), which exert a positive effect on mitochondrial metabolism are also effective in treating migraines.[85] Thus more research is needed to investigate whether high circulating trace amine levels produce abnormal biochemical phenotypes, and whether this abnormality, either alone or in association with other factors (crosslinking among different signaling pathways), leads to increased susceptibility for cluster and/or migraine headaches. This may facilitate the development of more specific therapeutic drugs to reduce the frequency of cluster and or migraine attacks.

CONCLUSION

Both cluster and migraine headaches are neurovascular disorders. The key differentiating factors between these two types of primary headaches include lateralization, attack duration, periodicity of the syndrome, and the nature of associated symptoms. The active region in the cluster headache by functional studies has been suggested to be the "posterior hypothalamus," a structure that modulates nociceptive and autonomic pathways, and contains circadian pacemaker cells. However, structural imaging data have provided conflicting results, thus the ultimate role of the hypothalamus in the pathophysiology of cluster headache remains enigmatic. The active regions in migraine involve subcortical structures (including diencephalic and brainstem nuclei) that modulate the perception of activation of the trigeminovascular system, which carries sensory information from the cranial vasculature to the hypothalamus. Both headaches are clinically distinct, and have biochemical differences between them. In migraine sufferers there is an increase in blood serotonin level at the onset, thus they have increased neuropeptide release, which results in vasodilation, and later deficiency in serotonin. Therefore serotonin and its receptor agonists subside the vasodilatory headache by lowering neuropeptide release in part through presynaptic inhibition of the cranial sensory nerves. In contrast, plasma histamine levels rise significantly during cluster headache; whereas, plasma serotonin levels show no change or slight, nonsignificant elevation. Increased number of mast cells in cluster headache sufferers during attacks supports histamine-mediated vasodilatory action.

At present, there is neither a unifying hypothesis to explain all the clinical particularities of primary headaches nor is there any clear understanding about the molecular mechanisms underlying these headaches. The complexity of interactions due to genetic and environmental factors taking place in the sensory neuronal network with the mediation of all different neurotransmitters/neuromodulators/neuropeptides involved gives the measure of the extreme difficulty connected with the knowledge about the pathogenesis of these headaches. Therefore, this unfinished puzzle is still far from being fully solved. Since cluster and migraine headaches are multifactorial diseases; therefore, the neuromodulatory role of trace amines by binding to TAAR1 as well as dopaminergic and/or noradrenergic receptors may be just one of the risk factors, which could play a possible role in the pathogenesis of these headaches. Thus, developing novel molecules selectively targeting the corresponding receptors could be used as potential therapeutic drugs for the treatment of cluster and migraine headaches.

References

1. Nesbitt AD, Goadsby PJ. Cluster headache. *Br Med J* 2012;**344**:e2407.
2. Matharu M. Cluster headache. *BMJ Clin Evid* 2010;**2010**:1212.
3. Klapper JA, Klapper A, Voss T. The misdiagnosis of cluster headache: a nonclinic, population-based, Internet survey. *Headache* 2000;**40**:730–3.
4. Bahra A, May A, Goadsby PJ. Cluster headache: a prospective clinical study with diagnostic implications. *Neurology* 2002;**58**:354–61.
5. Manzoni GC, Terzano MG, Bono G, et al. Cluster headache-clinical findings in 180 patients. *Cephalalgia* 1983;**3**:21–30.
6. Russell MB. Epidemiology and genetics of cluster headache. *Lancet Neurol* 2004;**3**(5):279–83.
7. Launer LJ, Terwindt GM, Ferrari MD. The prevalence and characteristics of migraine in a population based cohort: the GEM study. *Neurology* 1999;**53**:537–42.
8. Zanchin G, Dainese F, Mainardi F, Mampreso E, Perin C, et al. Osmophobia in primary headaches. *J Headache Pain* 2005;**6**(4):213–5.
9. Lipton RB, Bigal ME, Steiner TJ, Silberstein SD, Olesen J. Classification of primary headaches. *Neurology* 2004;**63**:427–35.
10. Goadsby PJ. Recent advances in understanding migraine mechanisms, molecules and therapeutics. *Trends Mol Med* 2007;**13**:39–44.

11. ICHD–II the internal classification of headache and disorders: 2nd edition. *Cephalalgia* 2004; **24**:9–160.

12. Tfelt-Hansen PC, Jensen RH. Management of cluster headache. *CNS Drugs* 2012;**26**(7):571–80.

13. Bussone G, Usai S. Trigeminal autonomic cephalalgias: from pathophysiology to clinical aspects. *Neurol Sci* 2004;**25**(Suppl. 3):S74–6.

14. Goadsby PJ, Edvinsson L. Human in vivo evidence for trigeminovascular activation in cluster headache. Neuropeptide changes and effects of acute attacks therapies. *Brain* 1994;**117**(Pt 3):427–34.

15. Stillman M. Steroid hormones in cluster headaches. *Curr Pain Headache Rep* 2006;**10**(2):147–52.

16. Edvinsson L. Both neurogenic and vascular causes of primary headache. *Lakartidningen* 2001;**98**(39) 441766-4183.

17. Shimzu T. New treatments for cluster headache. *Rinsho Shinkeigaku* 2013;**53**(11):1131–3.

18. Becker WJ. Cluster headache: conventional pharmacological management. *Headache* 2013;**53**(7):1191–6.

19. Katsarava Z, Buse DC, Manack AN, Lipton RB. Defining the differences between episodic migraine and chronic migraine. *Curr Pain Headache Rep* 2012;**16**:86–92.

20. Vuković V, Strineka M, Lovrenčić-Huzjan A, Demarin V. Migraine – pathophysiology of pain. Rad 504. *Med Sci* 2009;**33**:33–41.

21. Araki N. Migraine. *Jpn Med Assoc J* 2004;**47**(3):124–9.

22. Williamson DJ, Hargreaves RJ. Neurogenic inflammation in the context of migraine. *Microsc Res Tech* 2001;**53**(3):167–78.

23. Peroutka SJ. Neurogenic inflammation and migraine: implications for the therapeutics. *Mol Interv* 2005;**5**(5):304–11.

24. Lauritzen M. Pathophysiology of the migraine aura. The spreading depression theory. *Brain* 1994;**11** (Pt 1):199–210.

25. Kaiser EA, Russo AF. CGRP and migraine: could PACAP play a role too? *Neuropeptides* 2013;**47**:451–61.

26. Ramachandran R, Bhatt DK, Ploug KB, Hay-Schmidt A, Jansen-Olesen I, Gupta S, et al. Nitric oxide synthase, calcitonin gene-related peptide and NK-1 receptor mechanisms are involved in GTN-induced neuronal activation. *Cephalgia* 2014;**34**(2):136–47.

27. Dong X, Hu Y, Chen J. Role of phosphorylated extracellular signal-regulated kinase, calcitonin gene-related peptide and cyclooxygenase-2 in experimental rat models of migraine. *Mol Med Rep* 2015;**12**:1803–9.

28. Kawabata A. Prostaglandin E2 and pain – an update. *Biol Pharm Bull* 2011;**34**:1170–3.

29. Sakai F, Meyer JS. Regional cerebral hemodynamics during migraine and cluster headache measured by the 133Xe inhalation method. *Headache* 1978;**18**:122–32.

30. Aggarwal M, Puri V, Puri S. Serotonin and CGRP in migraine. *Ann Neurosci* 2012;**19**(2):88–94.

31. Arulmozhi DK, Veeranjaneyulu A, Bodhankar SL. Migraine: current therapeutic targets and future avenues. *Curr Vasc Pharmacol* 2006;**4**(2):117–28.

32. Schurks M, Kurth T, Geissler I, Tessmann G, Diener H-C, Rosskopf D. Cluster headache is associated with the G1246A polymorphism in the hypocretin receptor 2 gene. *Neurology* 2006;**66**:1917–9.

33. Rainero I, Gallone S, Valfre W, Ferrero M, Angilella G, Rivoiro C, et al. A polymorphism of the hypocretin receptor 2 gene is associated with cluster headache. *Neurology* 2004;**63**:1286–8.

34. Olsen J. The role the role of nitric oxide (NO) in migraine, tension-type headache and cluster headache. *Pharmacol Ther* 2008;**120**(2):157–71.

35. Sjöstrand C, Modin H, Masterman T, Ekbom K, Waldenlind E, Hillert J. Analysis of the nitric oxide synthase genes in cluster headache. *Cephalalgia* 2002;**22**:758–64.

36. Sjöstrand C, Giedratis V, Ekbom K, Waldenlind E, Hillert J. *CACNA1A* gene polymorphisms in cluster headache. *Cephalalgia* 2001;**21**:953–8.

37. Haan J, van Vliet JA, Kors EE, Terwindt GM, Vermeulen FL, van den Maagdenberg AM, et al. No involvement of the calcium channel gene (*CACNA1A*) in a family with cluster headache. *Cephalalgia* 2001;**21**:959–62.

38. Ophoff RA, Terwindt GM, Vergouwe MN, van Eijk R, Oefner PJ, et al. Familial hemiplegic migraine and episodic ataxia type-2 are caused by mutations in the Ca^{2+}channel gene CACNL1A4. *Cell* 1996;**87**:543–52.

39. Rainero I, Rubino E, Gallone S, Fenoglio P, Picci LR, Giobbe L, et al. Evidence for an association between migraine and the hypocretin receptor 1 gene. *J Headache Pain* 2011;**12**(2):193–9.

40. Ducros A. Genetics of migraine. *Rev Neurol* 2013;**169**(5):360–71.

41. Schürks M. Genetics of migraine in the age of genome-wide association studies. *J Headache Pain* 2012;**13**(1):1–9.

42. Anttila V, Stefansson H, Kallela M, Todt U, Terwindt GM, et al. Genome-wide association study of migraine implicates a common susceptibility variant on 8q22.1. *Nat Genet* 2010;**42**(10):869–73.

43. Lillis AP, Van Duyn LB, Murphy-Ullrich JE, Strickland DK. LDL receptor-related protein 1: unique tissue-specific functions revealed by selective gene knockout studies. *Physiol Rev* 2008;**88**:887–918.

44. Rainero I, Rubino E, Paemeleire K, Gai A, Vacca A, De Martino P, et al. Genes and primary headaches: discovering new potential therapeutic targets. *J Headache Pain* 2013;**14**:61.

45. Stuart S, Cox HC, Lea RA, Griffiths LR. The role of the MTHFR gene in migraine. *Headache* 2012;**52**(3):515–20.

46. Liu R, Geng P, Ma M, Yu S, Yang M, He M, et al. MTHFR C677T polymorphism and migraine risk: a meta-analysis. *J Neurol Sci* 2014;**336**(1–2):68–73.

47. Chasman D, Schürks M, Anttila V, de Vries B, Schminke U, Launer LJ, et al. Genome-wide association study reveals three susceptibility loci for common migraine in the general population. *Nat Genet* 2011;**43**(7):695–8.

48. Sintas C, Fernández-Morales J, Vila-Pueyo M, Narberhaus B, Arenas C, et al. Replication study of previous migraine genome-wide association study findings in a Spanish sample of migrainewith aura. *Cephalalgia* 2015;**35**(9):76–782.

49. Lafrenière RG, Cader MZ, Poulin JF, Andres-Enguix I, Simoneau M, et al. A dominant-negative mutation in the TRESK potassium channel is linked to familial migraine with aura. *Nat Med* 2010;**16**(10):1157–60.

50. Dussor G, Yan J, Xie JY, Ossipov MH, Dodick DW, et al. Targeting TRP channels for novel migraine therapeutics. *ACS Chem Neurosci* 2014;**5**(11):1085–96.

51. D'Alessandro R, Gamberini G, Benassi G, Morganti G, Cortelli P, Lugaresi E. Cluster headache in the Republic of San Marino. *Cephalalgia* 1986;**6**:159–62.

52. Russell MB, Andersson PG, Thomsen LL, Iselius L. Cluster headache is an autosomal dominantly inherited disorder in some families: a complex segregation analysis. *J Med Genet* 1995;**32**:954–6.

53. Shimomura T, Kitano A, Marukawa H, Mishima K, Isoe K, Adachi Y, et al. Point mutation in platelet mitochondrial tRNA (Leu(UUR)) in patient with cluster headache. *Lancet* 1994;**344**(8922):625.

54. Cortelli P, Zacchini A, Barboni P, Malpassi P, Carelli V, Montagna P. Lack of association between mitochondrial tRNA (Leu(UUR)) point mutation and cluster headache. *Lancet* 1995;**345**:1120–1.

55. Seibel P, Grünewald T, Gundolla A, Diener HC, Reichmann H. Investigation on the mitochondrial transfer RNA(Leu)(UUR) in blood cells from patients with cluster headache. *J Neurol* 1996;**243**:305–7.

56. Ferrari MD, Klever RR, Terwindt GM, Ayata C, van den Maagdenberg AM. Migraine pathophysiology: lessons from mouse models and human genetics. *Lancet Neurol* 2015;**14**(1):65–80.

57. Tolner EA, Houben T, Terwindt GM, de Vries B, Ferrari MD, van den Maagdenberg AM. From migraine genes to mechanisms. *Pain* 2015;**156**(Suppl. 1):S64–S174.

58. Edvinsson L, Uddman R. Neurobiology in primary headaches. *Brain Res Brain Res Rev* 2005;**48**(3):438–56.

59. Villalón CM, Centurión D, Valdivia LF, de Vries P, Saxena PR. Migraine: pathophysiology, pharmacology, treatment and future trends. *Curr Vasc Pharmacol* 2003;**1**(1):1–44.

60. Burnstock G, Ralevic V. Purinergic signaling and blood vessels in health and disease. *Pharmacol Rev* 2013;**66**(1):102–92.

61. Olesen J, Diener H, Husstedt I, et al. Calcitonin gene-related peptide receptor antagonist BIBN 4096 BS for the acute treatment of migraine. *N Eng J Med* 2004;**350**:1104–10.

62. Wrobel Goldberg S, Silberstein SD. Targeting CGRP: a new era for migraine treatment. *CNS Drugs* 2015;**29**(6):443–62.

63. Molinoff PB, Axelrod J. Biochemistry of catecholamines. *Annu Rev Biochem* 1971; **40**:465–500.

64. Strittmatter M, Hamann GF, Blaes F, Grauer M, Fischer C, Hoffmann KH. Reduced sympathetic nervous system activity during the cluster period of cluster-headache. *Schweiz Med Wochenschr* 1996;**126**(24):1054–61.

65. D'Andrea G, Granella F, Perini F, Farruggio A, Leone M, Bussone G. Platelet levels of dopamine are increased in migraine and cluster headache. *Headache* 2006;**46**(4):585–91.

66. Gallai V, Gaiti A, Sarchielli P, Coata G, Trequattrini A, Paciaroni M. Evidence for an altered dopamine beta-hydroxylase activity in migraine and tensiontype headache. *Acta Neurol Scand* 1992;**86**:403–6.

67. Nagel-Leiby S, Welch KM, D'Andrea G, Grunfeld S, Brown E. Event-related slow potentials and associated catecholamine function in migraine. *Cephalalgia* 1990;**10**:317–29.

68. Lea RA, Dohy A, Jordan K, Quinlan S, Brimage PJ, Griffiths LR. Evidence for allelic association of the dopamine beta-hydroxylase gene (DBH) with susceptibility to typical migraine. *Neurogenetics* 2000;**3**:35–40.

69. Fernandez F, Lea RA, Colson NJ, Bellis C, Quinlan S, et al. Association between a 19 bp deletion polymorphism at the dopamine beta-hydroxylase (DBH) locus and migraine with aura. *J Neurol Sci* 2006;**251**(1–2):118–23.

70. Martignoni E, Blandini F, Sances G, et al. Platelet and plasma catecholamines levels in migraine patients: evidence for a menstrually related variability of the noradrenergic tone. *Biogenic Amines* 1994;**10**:227–37.

71. Strittmatter M, Hamann GF, Grauer M, Fischer C, Bales F, et al. Altered activity of the sympathetic nervous system and changes in the balance of hypophyseal, pituitary and adrenal hormones in patients with cluster headache. *Neuroreport* 1996;**7**:1229–34.

72. Leone M, Bussone G. A review of hormonal findings in cluster headache. Evidence for hypothalamic involvement. *Cephalalgia* 1993;**13**:309–17.

73. Peres M, del Rio MS, Seabra M, Tufik S, Abucham J, et al. Hypothalamic involvement in chronic migraine. *J Neurol Neurosurg Psychiatry* 2001;**71**(6):747–51.

74. Masoud SA, Fakharian E. Serum prolactin and migraine. *Ann Saudi Med* 2005;**25**(6):489–4491.

75. Waldenlind E, Gustafsson SA. Prolactin in cluster headache: diurnal secretion, response to thyrotropin-releasing hormone, and relation to sex steroids and gonadotropins. *Cephalalgia* 1987;**7**(1):43–54.

76. Berry MD. Mammalian central nervous system trace amines. Pharmacologic amphetamines, physiologic neuromodulators. *J Neurochem* 2004;**90**(2):257–71.

77. D'Andrea G, Perini F, Terrazzino S, Nordera GP. Contributions of biochemistry to the pathogenesis of primary headaches. *Neurol Sci* 2004(Suppl. 3):S89–92.

78. D'Andrea G, Nordera GP, Perini F, Allais G, Granella F. Biochemistry of neuromodulation in primary headaches: focus on anomalies of tyrosine metabolism. *Neurol Sci* 2007;**28**(Suppl. 2):S94–6.

79. D'Andrea G, D'Amico D, Bussone G, Bolner A, Aguggia M, et al. Tryptamine levels are low in plasma of chronic migraine and chronic tension-type headache. *Neurol Sci* 2014;**35**(12):1941–5.

80. D'Andrea G, Cevoli S, Colavito D, Leon A. Biochemistry of primary headaches: role of tyrosine and tryptophan metabolism. *Neurol Sci* 2015;**36**(Suppl. 1):17–22.

81. Durham PL. Calcitonin gene-related peptide (CGRP) and migraine. *Headache* 2006;**46**(Suppl. 1):S3–S8.

82. D'Andrea G, D'Arrigo A, Dalle Carbonare M, Leon A. Pathogenesis of migraine: role of neuromodulators. *Headache* 2012;**52**(7):1155–63.

83. Lam VM, Espinoza S, Gerasimov AS, Gainetdinov RR, Salahpour A, et al. In-vivo pharmacology of trace-amine associated receptor 1. *Eur J Pharmacol* 2015;**763**(Pt B):136–42.

84. Lindemann L, Hoener MC. A renaissance in trace amines inspired by a novel GPCR family. *Trends Pharmacol Sci* 2005;**26**:274–81.

85. Yorns Jr WR, Hardison HH. Mitochondrial dysfunction in migraine. *Sem Pediatr Neurol* 2013;**20**(3):181–93.

PERSPECTIVE

25

Perspective and Directions for Future Research on Trace Amines and Neurological Disorders

T. Farooqui[1] and A.A. Farooqui[2]

[1]Department of Entomology, The Ohio State University, Columbus, OH,
United States [2]Department of Molecular and Cellular Biochemistry,
The Ohio State University, Columbus, OH, United States

INTRODUCTION

Trace amines were initially referred as "microamines" by Bouton in 1971 due to their very low concentrations (0.1–100 ng/g) as compared with classical biogenic amine neurotransmitters in the mammalian brain.[1–3] Trace amines are biologically active amines, which include β-phenylethylamine (β-PEA), tryptamine (TRY), p- and m-tyramine (TYR), p and m-octopamine (OCT), and the synephrines (SYN). Trace amines occur in all species so far examined, ranging from lower invertebrates (such as star fish) to higher invertebrates (such as mollusks and arthropods) up to higher mammals, including humans. In the mammalian nervous system, they are synthesized at rates comparable to classical monoamines, but due to their short half-life they are detectable only at trace levels, lower than classical biogenic amines, such as norepinephrine (NE), dopamine (DA), and serotonin (5-HT).[4] Trace amines are structurally, metabolically, physiologically, and pharmacologically similar to the classical monoamine transmitters, and are synthesized from the same precursor amino acids.[5] Trace amines are primary amines, which are produced directly by an enzymatic decarboxylation of precursor amino acids (L-tyrosine, L-phenylalanine, and L-tryptophan) catalyzed by an enzyme called aromatic L-amino acid decarboxylase (AADC; EC 4.1.1.28), giving rise to p-TYR, β-PEA, and TRY, respectively.[6–11] Since trace amines are substrates for monoamine oxidase enzymes (two isoforms: MAO-A and MAO-B) they can therefore be metabolized quickly, with a short half-life of ~30 s. Each isoform displays a characteristic preference for the specific trace amine. For example, β-PEA is a selective substrate for MAO-B. It is rapidly degraded by deamination reaction and oxidation of the amino group.[12–14] Due to less substrate specificity, other trace amines can be metabolized by both MAO isomers.[15,16] In addition to trace amine synthesis within parent monoamine neurotransmitter systems (L-tyrosine, L-phenylalanine and L-tryptophan), another member of this family, such as 3-iodothyronamine (3-T_1AM) is produced from thyroid hormone through decarboxylation and deiodination (Fig. 25.1). 3-T_1AM has been detected in traceable (pico- to nanomolar) concentrations in human blood.[17,18] Its functional effects in vivo are opposite to those produced on a longer timescale by thyroid hormones, including rapid and profound reduction in body temperature, heart rate, and metabolism. Due to structural similarities between 3-T_1AM and thyroxine, it was earlier suggested that 3-T_1AM may act as both a neuromodulator and a hormone-like molecule.[19]

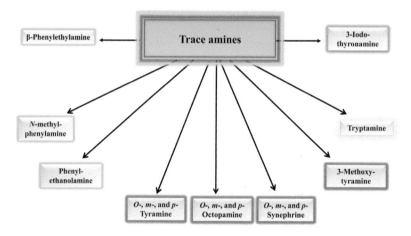

FIGURE 25.1   Trace amines in human nervous system. Trace amines constitute a group of phenyl or indolalkylamines that includes β-phenylethylamine, *ortho*-, *meta*-, and *para*-isomers of tyramine, octopamine, and synepherine, and tryptamine (a tryptophan metabolite) in humans. These are related to phenylalanine, tyrosine, and tryptophan. An additional member is 3-iodothyronamine, which belongs to a family of decarboxylated and deiodinated metabolites of the thyroid hormone (thyroxine).

Contrary to the situation in vertebrates, some trace amines (such as OCT and TYR) are the chief amines found in many invertebrate species. In insects, these trace amines play a major role in regulating insect physiology and a plethora of insect behaviors.[20-25] Moreover, these endogenous neuroactive amines can be obtained from the diet derived from herbal plants (such as ephedrine and cathinone), cocoa-based foods (such as chocolate), and from many fermented foods (such as wine, sausages, goat, and Dutch cheeses), and probiotic foods in which trace amines are generated by the high levels of lactic-acid-producing bacteria, such as *Lactobacillus*, *Lactococcus*, and *Enterococcus* species.[26] Collective evidence suggests that trace amines are found in all species so far examined.

Trace amines are not inactive byproducts of amino acid metabolism, but they are important neuromodulators associated with key physiological brain functions. Among monoamines, several investigators have demonstrated a close interplay between dopamine and trace amines, suggesting that trace amines may have relevant physiopathological implications due to their modulatory effect on dopaminergic signaling. Moreover, trace amines have attracted a lot of attention due to alterations in their levels reported in various neurological disorders (Fig. 25.2),[27-33] implicating the involvement of trace amines in the pathophysiology of monoaminergic systems. Although several trace amine-associated receptors (TAARs), belonging to a novel G-protein-coupled receptor (GPCR) family, have been identified,[34-37] the primary target of trace amines is TAAR1 because it is the best described receptor, and to a lesser extent TAAR4, which responds to typical trace amines.[36] TAAR1 is found in the brain, and it is activated by a wide spectrum of compounds (trace amines, common biogenic amines, and amphetamine-like psychostimulant drugs), and thus may play an important role in modulating brain monoamine systems.[38] Nevertheless, TAAR1 also binds with T_1AM, but normal thermoregulatory responses to T_1AM, trace amines (β-PEA), and amphetamine-like

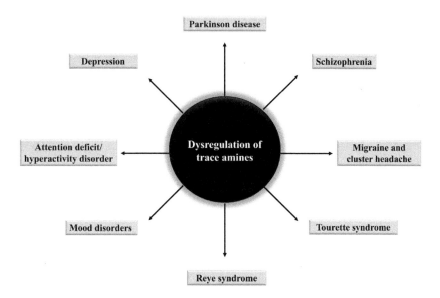

FIGURE 25.2   Alterations in the brain trace amine levels reported in various neurological disorders.

psychostimulants have been observed in TAAR1 knockout mice, suggesting that TAAR1 is not involved in mediation of thermoregulation.[39] Further investigations are required regarding its endogenous functions as well as its usefulness as a therapeutic agent.

In the invertebrate olfactory system, TAARs play an important role in modulation of neural plasticity in the olfactory pathway. The process of modulation of neural plasticity is similar to the neural components for odor perception with the vertebrate brain.[40–42] The similarities in the reward-processing systems of mammals and insects reinforce the utility of insects as a comparative neuroscience model.[43] Olfactory dysfunction is among the earliest nonmotor features and early "preclinical" sign of various neurological disorders, including Parkinson disease (PD), Alzheimer disease (AD), Huntington disease (HD), amyotrophic lateral sclerosis (ALS), Down syndrome, and the Parkinson-dementia complex of Guam, schizophrenia, and other related disorders;[44–54] which may be due to involvement of a common pathological substrate in these diseases. The mechanisms underlying olfactory dysfunction are currently unknown. Thus future research is needed to better define a potential link between olfactory dysfunction and trace amine-mediated pathological processes associated with these disorders.

NEUROLOGICAL DISORDERS

Neurological disorders are diseases of the central and peripheral nervous systems that occur due to structural, biochemical, or electrical abnormalities in the brain, spinal cord, or other nerves, resulting in a range of symptoms, including loss of sensation, confusion, poor coordination, altered levels of consciousness, and seizures. Neurological disorders are found among all age groups and in all geographical regions. The World Health Organization estimated that neurological disorders currently affect as many as a billion people worldwide, and this number is going to increase considerably in coming years. Therefore, unless immediate action is taken globally, the neurological burden is expected to become an even more serious and unmanageable threat to public health.

Major neurological disorders are classified into three groups including neurotraumatic diseases, neurodegenerative diseases, and neuropsychiatric diseases.[55] Neurotraumatic diseases include stroke, traumatic brain injury (TBI), and spinal cord injury (SCI) (Table 25.1). Stroke is a metabolic trauma caused by blockage of blood flow to the brain, TBI and SCI are caused by accidental falls, and motor cycle and car accidents. Neurodegenerative diseases are characterized by neuronal loss of specific neuronal population associated with cognitive and motor functions. Examples of neurodegenerative diseases are AD, PD, HD, and ALS.[55,56] Neuropsychiatric diseases are characterized by behavioral or psychological difficulties caused by alterations in neurotransmitters and their neuronal circuits related to abnormality in mitochondrial function, and redox imbalance.[57,58] Examples of neuropsychiatric disorders are depression, schizophrenia, bipolar affective disorders, autism, mood disorders, attention-deficit disorder, and tardive dyskinesia.

Neurological diseases are caused by structural, neurochemical, and electrophysiological changes in the brain, spinal cord, and nerves leading to neuronal injury and death. The symptoms of neurological disorders include paralysis, muscle weakness, poor coordination, seizures, loss of sensation, confusion, pain, and loss of memory. Neurodegeneration

TABLE 25.1   Classification of Neurological Disorders

Neurotraumatic Diseases	Neurodegenerative Diseases	Neuropsychiatric Diseases
Stroke	Alzheimer disease	Depression
Spinal cord injury	Parkinson disease	Schizophrenia
Traumatic brain injury	Amyotrophic lateral sclerosis	Bipolar disorders
Epilepsy	Prion diseases	Autism
	Multiple sclerosis	Mood disorders
		Attention-deficit disorders
		Dementia

Summarized from Farooqui AA. Neurochemical aspects of neurotraumatic and neurodegenerative diseases. New York: Springer; 2010.

TABLE 25.2   Differences Among Intensity of Oxidative Stress, Neuroinflammation, and Neurodegeneration Among Neurological Disorders

Diseases	Oxidative Stress	Neuroinflammation	Neurodegeneration Time
Neurotraumatic diseases	Acute	Acute	Rapid (hours to days)
Neurodegenerative diseases	Chronic	Chronic	Progressive (years)
Neuropsychiatric diseases	Mild	Mild	Slow (years)

Summarized from Farooqui AA. Neurochemical aspects of neurotraumatic and neurodegenerative diseases. New York: Springer; 2010.

in neurological disorders is a multifactorial process involving (1) abnormal protein dynamics with defective protein degradation and aggregation, (2) increased levels of reactive oxygen species, (3) impaired bioenergetics and mitochondrial dysfunction, and (4) onset of excitotoxicity, and neuroinflammation. These processes are supported by changes in blood flow and generation of proinflammatory prostaglandins and release of proinflammatory cytokines (tumor necrosis factor α, interleukin-1β, and interleukin-6) and chemokines (MCP1 and CXCL3).[55,59] Neuronal death in most neurological diseases is accompanied by the upregulation of interplay (crosstalk) among excitotoxicity, oxidative stress, and neuroinflammation.[55,60] In neurotraumatic diseases the onset of excitotoxicity, oxidative stress, and neuroinflammation occurs rapidly (in a matter of hours to days) because of blockage of blood flow, sudden lack of oxygen, rapid decrease in ATP, disturbance in transmembrane potential, and sudden collapse of ion gradients at a very early stage (Table 25.2).[55] In contrast, in neurodegenerative diseases, blood flow is decreased to the brain tissue leading to the partial availability of oxygen and nutrients to neural cells. These processes contribute to some reduction in levels of ATP. The presence of ATP not only provides the maintenance of cellular ionic homeostasis, but also leads to a slow rate of interactions among excitotoxicity, oxidative stress, and neuroinflammation.[55] The pathogenesis of neuropsychiatric diseases involves the disturbance in neurotransmitters (dopamine, norepinephrine, and serotonin), increase in chronic inflammatory processes, defects in neurogenesis, decrease in synaptic plasticity, mitochondrial dysfunction, and mild increase in oxidative stress and neuroinflammation, and redox imbalance. Moreover, changes in neuropeptides (vasopressin), cytokines, and gene–environmental interactions may also contribute to the pathogenesis of neuropsychiatric diseases.[57,58,61]

Considerable interest has developed in trace amines due to dysregulation in trace amines in patients with depression and psychotic episodes, posttreatment with monoamine oxidase inhibitors, or postconsumption of high trace amines diet, suggesting that the concentration of trace amines may be considerably higher at neuronal synapses than predicted by steady-state measures, implicating some pathophysiological role. Therefore, understanding of molecular mechanisms, and developing selective agonists and antagonists for TAARs, may serve as prime targets in drug development in the context of several neurological diseases.

TAARs AND THEIR RELEVANCE TO NEUROLOGICAL DISORDERS

As mentioned in previous chapters, trace amines are found in all mammals examined so far, including humans. They are structurally related to classic monoaminergic neurotransmitters, and may play significant roles in coordinating biogenic monoamine-based synaptic physiology. At high concentrations, trace amines have well-characterized presynaptic "amphetamine-like" effects on monoamine release, reuptake, and biosynthesis; whereas, at lower concentrations, they have been reported to possess postsynaptic modulatory effects, potentiating the activity of other neurotransmitters such as dopamine and serotonin.[62] A novel family of GPCRs is known as "TAARs" because the majority of these GPCRs are not activated by trace amines.[34,35] These receptors are located on both presynaptic and postsynaptic cells. TAARs represent putative targets not only for trace amines, and other amines, but also for a variety of monoaminergic compounds such as amphetamines and monoamine metabolites. Mammalian TAARS are comprised of three subgroups (TAAR1–4, TAAR5, and TAAR6–9), which are phylogenetically and functionally distinct from other GPCR families, as well as from OCT and TYR receptors in invertebrates.[36] With the exception of TAAR1 and TAAR4, none of the other TAARs is responsive to classical trace amines, including p-tyramine, β-phenylethylamine, octopamine, and tryptamine. Out of nine TAARs, humans possess six putative functional TAAR genes and three TAAR pseudogenes (Table 25.3).[36,63–77] The best-described member of the TAARs family is TAAR1, which is relatively conserved and found in all studied species. This receptor is activated by trace amines, and by drugs of abuse such as lysergic acid diethylamide (LSD, one of the most potent mood-changing drugs), and psychostimulants, such as methamphetamine (chemically similar to methamphetamine), and MDMA (3,4-methylenedioxy-methamphetamine), which is commonly known as "ecstasy" or "molly." TAAR1 plays a functional role in the regulation of brain monoamines and in mediating the action of amphetamine-like psychostimulants.[63] TAAR1 is a focus of extensive research, particularly because it can be activated by a variety of monoaminergic compounds including trace amines, amphetamines, and dopamine metabolites.

TAAR1 knockout mice display a deficit in prepulse inhibition (an ability to suppress the magnitude of startle induced by an incoming acoustic signal), which is typically impaired in schizophrenic patients, thereby *TAAR1* knockout mice have been proposed as an animal model for schizophrenia. Moreover, TAAR1 inhibits locomotor activity via downmodulation of dopaminergic neurotransmission,[32] exploring the possibility for using it as a target for studying PD. Moreover, TAAR1 is potently activated by 3-T_1AM, and is thereby considered as a specific 3-T_1AM receptor, and it can interact with the adrenergic system.[78] However, TAAR1 does not mediate the thermoregulatory responses of 3-T_1AM nor those of β-PEA or amphetamine-like psychostimulants.[78] It has been speculated that iodothyronines, including 3-T_1AM, may act physiologically as an endogenous adrenergic-blocking neuromodulator in the central noradrenergic system.[78] In addition to TAAR1, other 3-T_1AM targets, such as dopamine and norepinephrine transporters, and vesicular mono-amine transporter 2, might also be involved in neuromodulation.[79] However, this hypothesis still needs experimental validation.

All TAARs (except TAAR1) are localized in the OE with little to no expression in the brain or other tissues of the body.[68] Multiple laboratories have now reported its expression in various brain areas as well as in the peripheral organs.[34,35,80–82] TAAR4 is now considered as a GPCR pseudogene in humans (Table 25.3). However, *TAAR4* gene is found on

TABLE 25.3   Mammalian TAAR Ligands and Expression

Mammalian TAARs	Ligands	Expression
TAAR1[a,63–69]	• Trace amines (β-PEA, NMPEA, *p*-TYR) • Classical monoamines (DA, dopamine, serotonin, histamine) • Psychoactive substances (MDMA, amphetamine, LSD) • Thyronamines (3-T1AM) • The mTAAR1 and hTAAR1 respond to several volatile amines	• Expressed in the brain (amygdala) and several peripheral organs, as well as in circulating lymphocytes • Not expressed in the human or mouse OE
TAAR2[a,65,70]	• 2-PEA, TYR, and T1AM	• Coexpressed with TAAR1 in a subpopulation of blood PMN cells
TAAR3[b,36,69,70]	• Murine TAAR3 recognizes volatile amines	• Occurs as a pseudogene in some individuals but not others
TAAR4[b,66,69–71,83]	• Highly sensitive to amines • Exquisitely sensitive to 2-PEA • Responds to structurally diverse amines at high concentrations • Murine TAAR4 recognizes volatile amines	• Occurs as a pseudogene in humans, but occurs with TAAR3 as a functional gene in rodents • RT-PCR studies in humans confirm that TRAR4 is preferentially expressed in the brain regions implicated in the pathophysiology of schizophrenia • Gene on chromosome 6q23.2 associated with susceptibility to schizophrenia
TAAR5[a,66,69–73]	• Responds to tertiary amine *N,N*-DMEA, and TMA • 3-Iodothyronamine (inverse agonist) • Murine TAAR5 recognizes volatile amines • Human TAAR5 is specifically activated by TMA with less efficacy by DMEA	• Expressed in the human OE • Human basal ganglia and peripheral organs

(Continued)

TABLE 25.3   (Continued)

Mammalian TAARs	Ligands	Expression
TAAR6[a,62,74,75]	• Chemoreceptor for volatile odorants	• Expressed in the human OE • Highest levels in the amygdala and hippocampus, and peripheral organs
TAAR7[b,76]	–	• Occurs as a pseudogene in humans, but as a functional gene in rodents
TAAR8[a,66,70,77]	• Chemoreceptor for volatile odorants • 3-T1AM is not a ligand for mouse and human trace amine-associated receptor 8 (Taar8b, TAAR8) • Basal G(i/o) signaling activity	• Detected in trace levels in the human OE, but particularly located in amygdala • Taar8b transcripts appear to be scarce in brain tissue
TAAR9[a,36,69,70]	• Chemoreceptor for volatile odorants	• Human CNS and peripheral organs

Information is summarized from Lindemann L, Hoener MC. A renaissance in trace amines inspired by a novel GPCR family. Trends Pharmacol Sci 2005;26(5):274–81. Miller GM. The emerging role of trace amine associated receptor 1 in the functional regulation of monoamine transporters and dopaminergic activity. J Neurochem 2011;11(2)164–76. Scanlan TS, Suchland KL, Hart ME, Chiellini G, Huang Y, et al. 3-Iodothyronamine is an endogenous and rapid-acting derivative of thyroid hormone. Nat Med 2004;10:638–42. Babusyte A, Kotthoff M, Fiedler J, Krautwurst D. Biogenic amines activate blood leukocytes via trace amine-associated receptors TAAR1 and TAAR2. J Leukoc Biol 2013;93:387–94. Carnicelli V, Santoro A, Sellari-Franceschini S, Berrettini S, Zucchi R. Expression of trace amine-associated receptors in human nasal mucosa. Chemosensory Percept 2010;3:99–107. Zucchi R, Chiellini G, Scanlan TS, Grandy DK. Trace amine-associated receptors and their ligands. Br J Pharmacol 2006;149:967–78. Liberles SD, Buck LB. A second class of chemosensory receptors in the olfactory epithelium. Nature 2006;442:645–50. Liberles SD. Trace amine-associated receptors are olfactory receptors in vertebrates. Ann N Y Acad Sci 2009;1170:168–72. Maguire JJ, Parker WA, Foord SM, Bonner TI, Neubig RR, et al. International Union of Pharmacology. LXXII. Recommendations for trace amine receptor nomenclature. Pharmacol Rev 2009;61:1–8. Wallrabenstein I, Kuklan J, Weber L, Zborala S, Werner M, et al. Human trace amine-associated receptor TAAR5 can be activated by trimethylamine. PLoS One 2013;8:e54950. Stäubert C, Boselt I, Bohnekamp J, Rompler H, Enard W, et al. Structural and functional evolution of the trace amine-associated receptors TAAR3, TAAR4, and TAAR5 in primates. PLoS One 2010;5:e11133. Dinter J, Mühlhaus J, Wienchol CL, Yi CX, Nürnberg D, et al. Inverse agonistic action of 3-iodothyronamine at the human trace amine-associated receptor 5. PLoS One 2015;10:e0117774. Pae CU, Yu HS, Amann D, Kim JJ, Lee CU, et al. Association of the trace amine associated receptor 6 (TAAR6) gene with schizophrenia and bipolar disorder in a Korean case control sample. J Psychiatr Res 2008;42:35–40. Pae CU, Drago A, Mandellllo L, De Ronchelli L, Serretti A. TAAR 6 and HSP-70 variations associated with bipolar disorder. Neurosci Lett 2009;465(3):257–61. Ferrero DM, Wacker D, Roque MA, Baldwin MW, Stevens RC, et al. Agonists for 13 Trace Amine-Associated Receptors provide insight into the molecular basis of odor selectivity. ACS Chem Biol 2012;7:1184–9. Mühlhaus J, Dinter J, Nürnberg D, Rehders M, Depke M, et al. Analysis of human TAAR8 and murine Taar8b mediated signaling pathways and expression profile. Int J Mol Sci 2014;15:20638–55. Espinoza S, Salahpour A, Masri B, Sotnikova TD, Messa M, et al. Functional interaction between trace amine-associated receptor 1 and dopamine D2 receptor. Mol Pharmacol 2011;80:416–25.[85]

OE, Olfactory epithelium; TAAR, Trace amine-associated receptor; TYR, Tyramine; β-PEA, β- phenylethylamine; NMPEA, N-methyl phenethylamine; TMA, Trimethylamine; DMEA, Dimethylethylamine; DA, Dopamine; 5-HT, Serotonin; His, Histamine; 3-T1AM, 3-iodothyronamine; PMN, Polymorphonuclear leukocytes; mTAAR1, Mouse TAAR1; hTAAR1, Human TAAR1; RT-PCR, Reverse transcription polymerase chain reaction.

[a]Functional TAAR in humans.

[b]Pseudogene in humans.

chromosome 6q23.2, which is associated with susceptibility to schizophrenia. These findings are consistent and also valid for human and animal models of toxic psychosis and are in agreement with the expression pattern of *TRAR4* in frontal cortex, amygdala, and hippocampus.[83] It is possible that an actual GPCR *TAAR4* gene has missed by gene-prediction programs. The exact mechanisms of disease mediated by TRAR4 remain to be elucidated. Other functional TAARs (such as TAAR5, TAAR6, and TAAR9), have been detected in human central nervous system (CNS) and peripheral organs, and TAAR5 is in human basal

TABLE 25.4 Opposing Roles of Octopamine and Tyramine in Modulating Insect Behaviors

Trace Amines	Role	Insect
Octopamine	Increases flying	Honeybees[84]
Tyramine	Decreases flying	Honeybees[84]
Octopamine	Decreased octopamine results in abnormal locomotion	Drosophila larva[86]
Tyramine	Elevated tyramine results in abnormal locomotion	Drosophila larva[86]
Activation of AmTYR1 with tyramine	Decrease in forskolin-induced cAMP production	Apis mellifera[87,88]
Activation of AmOA1 with octopamine	Increase in intracellular cAMP levels	Apis mellifera[89]
Octopamine	Mediates the associative odor learning	Apis mellifera[90,91]
Tyramine	Increases sensitivity for sucrose and rate of habituation in foragers	Apis mellifera[92,93]
Octopamine	Influences division of labor in honey bee colonies	Apis mellifera[94]
Tyramine	Does not increase the number of new foragers	Apis mellifera[94]
Octopamine	Does not enhance ovarian development or ester production, but induces production of 10HDA, inducing foraging behavior of workers	Queenless worker honeybees Apis mellifera[95]
Tyramine	Enhances ovary development and the production of esters in the Dufour's gland and 9HDA in the mandibular glands, facilitating worker reproductive dominance	Queenless worker honeybees Apis mellifera[95]
Octopamine (activates OARα signaling)	Attractive behavioral plasticity in gregarious locusts	Locusta migratoria[96]
Tyramine (activates TAR signaling)	Repulsive behavioral plasticity in solitary locusts	Locusta migratoria[96]

AmTYR1, Apis mellifera tyramine receptor; AmOA1, Apis mellifera octopamine receptor; OARα, Octopamine receptor α; TAR, tyramine receptor; 9-HDA, queen component, 9-hydroxy-(E)-2-decenoic acid; 10-HAD, 10-hydroxy-2-decenoic acid; worker major component.

ganglia.[34,84] Collectively, based on the gene type (normal or pseudogene), mRNA expression, area specificity (human central nervous system or peripheral organs), and ligand specificity of specific receptor subtype in TAAR family (Table 25.3) may help with determining the functional role of each TAAR subtype in the spectrum of neurological diseases.

In insects, octopamine and tyramine play various important physiological roles not only in peripheral organs, learning, and memory, but also in circadian rhythm, by binding with octopamine and tyramine receptors. Octopamine and tyramine have antagonistic effects in invertebrates (Table 25.4),[84,86–96] similar to epinephrine and norepinephrine in vertebrates. However, invertebrate trace amine-activated receptors (octopamine receptor and tyramine receptor) are not closely related to vertebrate TAARs but rather more similar to receptors activated by serotonin, in particular 5HT1 receptors.[67] Thus it appears that vertebrate TAARs have not evolved from the invertebrate receptors for trace amines.

GPCRs SIGNALING

GPCRs are seven transmembrane domain receptors. GPCRs perceive extracellular signals and transduce them to heterotrimeric G proteins (Gα, Gβ/Gγ subunits), which further transduce these signals intracellularly to appropriate downstream effectors, and thus are key players in various signaling pathways.[97] The signal or primary stimulus could be light, hormone, odorant, antigen, neurotransmitter or the surface of another cell, which transport into the cell via membrane receptor, through a signal transduction triad (receptor/transducer/effector). The second messengers involved in these processes include Ca^{2+}, cyclic adenosine monophosphate (cAMP), and cyclic guanosine monophosphate (cGMP), inositol-1,4,5-triphosphate (IP3), diacylglycerol (DAG), and arachidonic acid (AA). The triad is responsible for converting the signal from first to second messenger, which may contribute to the regulation of protein kinases or phosphatases in the cytoplasm. The target of the signal may be enzymes, receptors, ion channels, and finally transcription factors, which ultimately control the gene expression.[97] There are two principal signal transduction pathways involving the GPCRs: (1) cAMP-dependent signaling pathway (also known as adenylyl cyclase (AC) pathway) in which activated $G_{s\alpha}$ subunit binds to and activates an enzyme called AC that in turn catalyzes the conversion of ATP into cAMP, and (2) phosphatidylinositol-mediated signaling pathway (also known as phospholipase C (PLC) pathway), in which activated $G_{q\alpha}$ subunit binds to and activates an enzyme called PLC, which hydrolyzes phosphatidylinositol 4,5-bisphosphate (PIP_2) to DAG and IP3,[98] releasing Ca^{2+} from intracellular stores. Despite the number and diversity of GPCRs, their ligands, and the processes they govern, GPCRs couple to intracellular signaling pathways via four families of G proteins: ($G_{\alpha s}$, $G_{\alpha i}/G_{\alpha 0}$, $G_{\alpha q}/G_{\alpha 11}$, and $G_{\alpha 12}/G_{\alpha 13}$).[99] This classification is based on their α-subunits that define the basic properties of a heterotrimeric G protein.[99] GPCRs (such as three biogenic amine receptors (5-hydroxytryptamine (5-HT)1A, 5-HT4, and dopamine D2), a peptide receptor (NK1), and a chemokine receptor (CCR3) have been a rich source of targets for all modern pharmaceuticals due to their involvement in regulation of many physiological and disease processes.[99–102] By understanding GPCR signaling, one can validate agonist response and directly determine antagonists and inverse agonists.

TAAR-MEDIATED SIGNALING

TAARs belong to a family of GPCRs, that originated prior to the emergence of jawed vertebrates.[103] They are evolutionarily conserved throughout vertebrates, including humans.[37,67,80,103,104] TAARs are distantly related to biogenic amine receptors, but these receptors function as vertebrate olfactory receptors.[105]

TAAR1 is an amine-activated stimulatory $G\alpha_s$-coupled GPCR, which is located in several peripheral organs, circulating lymphocytes, and in the mammalian brain.[34,35,106] Trace amines (β-PEA and tyramine), metabolites of catecholamines and iodothyronamines have been shown to interact with TAAR1 via coupling to $G_{\alpha s}$ and activate AC, resulting in cAMP production in a fibroblast-like cell line derived from monkey kidney tissue (COS-7),[34] and human embryonic kidney 293 (HEK293) cells,[37] and via the promiscuous G_q protein ($G\alpha_{16}$) in Chinese hamster ovary (CHO) cells,[35] and in RD-HGA16 cells to mobilization of intracellular Ca^{2+} via coupling to G_q coupled by GPCRs, activates PLC, increases IP3, resulting in the release of Ca^{2+} from intracellular stores.[107,108] It is established that TAAR1 couple with Gs proteins resulting in the stimulation of AC, or via the promiscuous G_q protein resulting in mobilization of intracellular Ca^{2+} in different cell types, suggesting that TAAR1 couples with different G proteins in different cells. It is also possible that different TAAR subtypes might couple with different G proteins. The cardiac effects of thyronamines are not consistent with increased cAMP, thus in the heart it may involve different signaling pathway such as changes in tyrosine kinase/phosphatase activity.

Moreover, homo- and heterodimerization is a common feature of many GPCRs, which may result in the modulation of their functions.[109] The biochemical mechanism of interaction between TAAR1 and dopamine D2 receptor (D2R), and the role of such interaction in D2R-related signaling have been successfully demonstrated, suggesting that D2R antagonists enhance selectively a TAAR1-mediated β-PEA increase in cAMP.[85] Moreover, TAAR1 and D2R form heterodimers when coexpressed in HEK293 cells, which was disrupted in the presence of haloperidol, an antipsychotic, that acts as D2R antagonist.[85] In addition, in mice lacking TAAR1, haloperidol-induced striatal c-Fos expression and catalepsy were significantly reduced. Taken together, suggesting that TAAR1 and D2R have functional and physical interactions that could be critical for the modulation of the dopaminergic system by TAAR1 in vivo.[85] This further suggests that both dopaminergic drugs and TAAR1 ligands offer therapeutical potential for treating dopamine-related disorders.[110] Although the TAAR4 cerebral role is still largely elusive, the observation that TAAR4 is a susceptibility gene for schizophrenia suggests its potential role in the modulation of the dopaminergic system in this disorder. However in primates, like TAAR1, TAAR4 is coupled to $G_{\alpha s}$, and enhances cAMP signaling.[70] TAAR6-mediated signaling in humans remains unclear. TAAR6 is unresponsive to methamphetamine in its ability to stimulate AC, or trigger mitogen-activated protein kinase/extracellular signal-regulated kinase phosphorylation, suggesting no direct association with brain monoaminergic neuronal function in rhesus monkey.[111] In comparison to catecholamines, histamine, and serotonin (Table 25.5),[112–118] the signaling information about TAARs, excluding TAAR1, is still in its infancy (Table 25.5).[34,35,63–71,106,108,111]

Collectively, the predominant signaling pathway for TAARs may be: (1) upon stimulation by trace amines and psychoactive substances, TAAR1 receptor activates $G_{\alpha s}$/AC pathway, which resulting in cAMP production, or (2) upon stimulation by odorants binding to other TAAR subtypes (except TAAR1), activation of G_{olf}/AC pathway results in cAMP production, which regulates opening of the cyclic nucleotide-gated ion channel, allowing an influx of both Na^+ and Ca^{2+} ions into the cell and efflux of Cl^- ions, leading to the depolarization of the cell.[112–114]

CONCLUSION

Trace amines are produced by many, if not all, genera of prokaryotes and eukaryotes and are not inactive byproducts of amino acid metabolism. They are structurally related to the classical biogenic amine neurotransmitters (DA, NE, EPI, and 5-HT). Moreover, some trace amines (phenethylamine and N-methylphenthylamine) are chemically related to the synthetic phenylethylamine psychostimulant (amphetamine), and its analogues, which by

TABLE 25.5   Physiological Ligands of Several G-Protein-Coupled Receptors

Endogenous Ligand	Receptor	G Protein (Subclass)	References
Epinephrine, norepinephrine	$\alpha_{1A}, \alpha_{1B}, \alpha_{1D}$	$G_{q/11}$	112
	$\alpha_{2A}, \alpha_{2B}, \alpha_{2C}$	$G_{i/0}$	112
	$\beta_1, \beta_2, \beta_3$	G_s	113
Dopamine	D_1, D_5	G_s	114
	D_2, D_3, D_4	$G_{i/0}$	114
Histamine	H_1	$G_{q/11}$	115
	H_2	G_s	115
	H_3, H_4	$G_{i/0}$	115
Serotonin	$5HT_{1A/B/D/E/F}$	$G_{i/0}$	116
	$5HT_{2A/B/C}$	$G_{q/11}$	116
	$5\text{-}HT_4, 5HT_6, 5HT_7$	G_s	116–118
	$5\text{-}HT_{5A/B}$	$G_{i/0}, G_s$	116
Trace amines (such as β-PEA, TYR, and 3-T₁AM)	TAAR1	$G_{\alpha s}$ G_q ($G_{\alpha 16}$)	34,35,37, 106–108
Trace amines (such as 2-PEA, TYR, and T1AM)	TAAR2 (coexpressed with TAAR1)	–	65,70
Volatile amines	TARR3[a]	• $G_{\alpha s}$ in primates • Expressed but not functional in humans	36,68–70
Volatile amines	TAAR4[a]	• $G_{\alpha s}$ in primates • Expressed but not functional in humans	36,66, 68–71,106,108
Volatile amines, tertiary amines (N,N-DMEA and TMA)	TAAR5	• $G_{\alpha s}$ in primates • Not functional in humans	34,66–73
Volatile odorants (trace amines could be endogenous agonists?)	TAAR6	• Receptor signaling in humans remain unclear • Lack of either G_s or G_i-linked signaling in response to monoamines or methamphetamine in rhesus monkey	62,74,75,111
–	TAAR7[a]	• Not found in humans	76
Volatile odorants (T1AM, not a ligand)	TAAR8 (in humans and murine)	• Basal $G_{i/0}$ signaling	77
Volatile odorants (trace amines could be endogenous ligands?)	TAAR9 (in humans)	–	36,69,70

TMA, Trimethylamine; N,N-DMEA, dimethylethylamine; β-PEA, β-phenylethylamine; TYR, tyramine; 3-T1AM, thyronamines; TAARs, trace amine-associated receptors; AC, adenylate cyclase; pseudogenes.

[a]Out of nine human TAAR subtypes, the best-characterized subtype is TAAR1 that activates the G(s) protein/AC pathway upon stimulation by TAAR1 agonists (trace amines and psychoactive substances). However, relatively little information is currently available on the ligand recognition properties, biochemistry, and function of other TAAR subtypes. TAAR3 and TAAR4 exhibit a disrupted open reading frame in humans; whereas in primates these receptors recognize volatile amines and signal via G_s.

binding to TAAR1 result in increasing biogenic amine and excitatory neurotransmitter activity in the brain. The dopaminergic system is critically involved in the control of key physiological functions, including voluntary movement and posture, motivated behaviors, and different cognitive functions; the nigrostriatal pathway involvement in the control of motor functions; and the mesoaccumbal and mesocortical pathways involvement in reward, will, and cognitive functions. Several investigators have demonstrated a close interplay among monoamines, particularly between dopamine and trace amines, implicating that trace amines may have relevant physiopathological implications due to their modulatory effect on dopaminergic signaling.

The best-characterized member of the TAAR family is TAAR1. Both TAAR1 and D2R can potently modulate each other's activity. TAAR4 is a pseudogene in humans and many primates. The polymorphism in TAAR4 suggests the gene for TAAR4 may be responsible for susceptibility to schizophrenia, but further replication of these findings is needed. Several investigators have also reported an association between TAAR6 and susceptibility to schizophrenia and bipolar affective disorder in humans. However, endogenous TAAR6 agonists and the receptor signaling profile and brain distribution remain unclear.

Moreover, a better understanding of the biochemical pathway(s) responsible for 3-T_1AM synthesis is required. Understanding 3-T_1AM-mediated signaling may attribute some effects to its precursor (thyroid hormone) itself, and thereby it may explore a new perspective providing new targets for potential therapeutic interventions in metabolic, endocrine, and neurological disorders. Thus the critical research issues for the near future may include clarifying the role of TAAR1 versus other receptors in the response to 3-T_1AM and dissecting the underlying transduction pathways. From the metabolic side, it stimulates lipid catabolism and induces, in general, anti-insulin responses; from the neurological side, it has been reported to favor learning and memory, modulate sleep and feeding, and decrease the pain threshold. Available evidence suggests that 3-T_1AM can be regarded as a neuromodulator. However, the molecular details of its actions, and the underlying transduction pathways remain to be determined.

TAAR1 is an important drug target for treating the psychiatric and neurodegenerative disorders. However, at present no selective ligands to identify TAAR1-specific signaling mechanisms are available; therefore, identification of selective ligands of the TAAR1 can be critical for investigating the functional role of this receptor in mammalian physiology in vivo and/or managing human disorders in which abnormalities in trace amine physiology may occur. Once complete signaling about all TAARs is unraveled and adequate pharmacological tools become available, important new therapeutic opportunities may definitely result for using them as prime candidates for targets in drug development in the context of several neurological diseases. Olfactory dysfunction is one of the earliest symptoms, which is effected in most neurological disorders. Since the functional organization of the olfactory system is remarkably similar in organisms ranging from insects to mammals, therefore insect model systems will also represent an extremely good solution to resolve some difficult signaling problems.

Overall, TAAR1 research is very interesting in modulating the dopaminergic transmission point of view, in which antagonizing TAAR1 with specific antagonists may potentially enhance L-dopa effect in PD, and activating TAAR1 with specific agonists may offer therapeutic potential for treating other neurological disorders, such as schizophrenia, addiction, or ADHD. However, before organizing clinical trials it is necessary to fully understand trace-amine-mediated signaling. Neurological disorders are multifactorial pathological conditions involving multiple pathways. It is possible that trace amines aid pathogenesis of neurological disorders not only by regulating "monoaminergic signaling" but also by combination with other neurotransmitters (glutamate, GABA, and dopamine), which intensify signal transduction processes closely associated with neural cell survival as well as neurodegeneration.

References

1. Boulton AA. Amines and theories in Pychiatry. *Lancet* 1971;**2**:7871.
2. Usdin E, Sandler M, editors. *Trace amines and the brain*. New York: Dekker; 1976.
3. Schmitt R, Nasse O. Beitrag zur Keuntnifs des tyrosins. *Liebigs Ann Chem* 1865;**133**:211–6.
4. Usdin E, Sandler M, editors. *Trace amines and the brain*. New York: Dekker; 1984.
5. Saavedra JM. á-Phenylethylamine, phenylethanoamine, tyramine and octopamine. In: Trendelenburg U, Weiner N, editors. *Catecholamines II*. Berlin: Springer-Verlag; 1989. p. 181–210.
6. Boulton AA, Wu PH. Biosynthesis of cerebral phenolic amines. I. In vivo formation of *p*-tyramine, octopamine, and synephrine. *Can J Biochem* 1972;**50**(3):261–7.
7. Boulton AA, Dyck LE. Biosynthesis and excretion of *meta* and *para* tyramine in the rat. *Life Sci* 1974;**14**:2497–506.
8. Boulton AA, Wu PH. Biosynthesis of cerebral phenolic amines. II. In vivo regional formation of *p*-tyramine and octopamine from tyrosine and dopamine. *Can J Biochem* 1973;**51**(4):428–35.
9. Dyck LE, Yang CR, Boulton AA. The biosynthesis of *p*-tyramine, *m*-tyramine, and *b*-phenylethylamine by rat striatal slices. *J Neurosci Res* 1983;**10**(2):211–20.
10. Silkaitis RP, Mosnaim AD. Pathways linking l-phenylalanine and 2-phenylethylamine with *p*-tyramine in rabbit brain. *Brain Res* 1976;**114**(1):105–15.
11. Tallman JF, Saavedra JM, Axelrod J. Biosynthesis and metabolism of endogenous tyramine and its normal presence in sympathetic nerves. *J Pharmacol Exp Ther* 1976;**199**:216–21.
12. Yang HY, Neff NH. β-phenylethylamine: a specific substrate for type B monoamine oxidase of brain. *J Pharmacol Exp Ther* 1973;**187**(2):3365–71.
13. Paterson IA, Juorio AV, Boulton AA. 2-Phenylethylamine: a modulator of catecholamine transmission in the mammalian central nervous system? *J Neurochem* 1990;**55**:1827–37.
14. Berry MD. Mammalian central nervous system trace amines pharmacologic amphetamines, physiologic neuro-modulators. *J Neurochem* 2004;**90**:257–71.
15. Philips SR, Boulton AA. The effect of monoamine oxidase inhibitors on some arylalkylamines in rate striatum. *J Neurochem* 1979;**33**(1):159–67.
16. Durden DA. Kinetic measurements of the turnover rates of phenylethylamine and tryptamine in vivo in the rat brain. *J Neurochem* 1980;**34**(6):1725–32.
17. Hoefig CS, Kohrle J, Brabant G, Dixit K, Yap B, et al. Evidence for extrathyroidal formation of 3-iodothyronamine in humans as provided by a novel monoclonal antibody-based chemiluminescent serum immunoassay. *J Clin Endocrinol Metab* 2011;**96**:1864–72.
18. Galli E, Marchini M, Saba A, Berti S, Tonacchera M, et al. Detection of 3-iodothyronamine in human patients: a preliminary study. *J Clin Endocrinol Metab* 2012;**97**:E69–74.
19. Iancullescu AG, Scanlan TS. 3-Iodothyronamine (T(1)AM): a new chapter of thyroid hormone endocrinology? *Mol Biosyst* 2010;**6**(8):1338–13344.
20. Axelrod J, Saavedra JM. Octopamine. *Nature* 1977;**265**:501–4.
21. David JC, Coulon JF. Octopamine in invertebrates and vertebrates. A review. *Prog Neurobiol* 1985;**24**:141–85.
22. Evans PD. OctopamineKerkut G.A.Gilbert LI, editors. *Comprehensive insect physiology, biochemistry and pharmacology*, **11**. Oxford: Pergamon Press; 1985. p. 499–530.
23. Roeder T. Octopamine in invertebrates. *Prog Neurobiol* 1999;**59**:533–41.
24. Roeder T. Tyramine and octopamine: ruling behavior and metabolism. *Annu Rev Entomol* 2005;**50**:447–77.

25. Bleanau W, Baumann A. Aminergic signal transduction in invertebrates: focus on tyramine and octopamine receptors. *Recent Res Dev Neurochem* 2003;**6**(2003):225–40.

26. Broadley KJ. The vascular effects of trace amines and amphetamines. *Pharmacol Ther* 2010;**125**(3):363–75.

27. Boulton AA, Juorio AV, Downer RGH, editors. *Trace amines: comparative and clinical neurobiology (experimental and clinical neuroscience)*. Totowa, NJ: Humana; 1988.

28. Mosnaim AD, Wolf ME, editors. *Noncatecholic phenylethylamines*. New York: M. Dekker; 1978.

29. Baker GB, Bornstein RA, Yeragani VK. Trace amines and Tourette's syndrome. *Neurochem Res* 1993;**18**:951–6.

30. Premont RT, Gainetdino RR, Caron MG. Following the trace of elusive amines. *Proc Natl Acad Sci USA* 2001;**9**(17):9474–5.

31. Berry MD. The potential of trace amines and their receptors for treating neurological and psychiatric diseases. *Rev Recent Clin Trials* 2007;**2**(1):3–19.

32. Sotnikova TD, Zorina OI, Ghisi V, Caron MG, Gainetdinov RR. Trace amine associated receptor 1 and movement control. *Parkinsonism Relat Disord* 2008;**14**(Suppl. 2):S99–102.

33. Narang D, Tomlinson S, Holt A, Darre D, Mousseau DD, Baker GB. Trace amines and their relevance to psychiatry and neurology: a brief overview. *Bull Clin Psychopharmacol* 2011;**21**:73–9.

34. Borowsky B, Adham N, Jones KA, Raddatz R, Artymyshyn R, et al. Trace amines: identification of a family of mammalian G protein coupled receptors. *Proc Natl Acad Sci USA* 2001;**98**(16):8966–71.

35. Bunzow JR, Sonders MS, Arttamangkul S, Harrison LM, Zhang G, et al. Amphetamine, 3,4-methylenedioxymethamphetamine, lysergic acid diethylamide, and metabolites of the catecholamine neurotransmitters are agonists of a rat trace amine receptor. *Mol Pharmacol* 2001;**60**:1181–8.

36. Lindemann L, Hoener MC. A renaissance in trace amines inspired by a novel GPCR family. *Trends Pharmacol Sci* 2005;**26**(5):274–81.

37. Lindemann L, Ebeling M, Kratochwill NA, Bunzow JR, Grandy DK, et al. Trace amine-associated receptors form structurally and functionally distinct subfamilies of novel G protein-coupled receptors. *Genomics* 2005;**85**:372–85.

38. Xie Z, Miler GM. Trace amine-associated receptor 1 as a monoaminergic modulator in brain. *Biochem Pharmacol* 2009;**78**(9):1095–104.

39. Panas HN, Lynch LJ, Valender EJ, Xie Z, Chen G, et al. Normal thermoregulatory responses to 3-iodothyronamine, trace amines and amphetamine-like psychostimulants in trace amine associated receptor 1 knockout mice. *J Neurosci* 2010;**88**(9):1962–9.

40. Menzel R. Memory dynamics in the honeybee. *J Comp Physiol A* 1999;**185**:323–40.

41. Menzel R, Giurfa M. Cognitive architecture of amini-brain: the honeybee. *Trends Cogn Sci* 2001;**5**:62–71.

42. Frasnelli E, Haase A, Rigosi E, Anfora G, Rogers LJ. The bee as a model to investigate brain and behavioural asymmetries. *Insects* 2014;**5**:120–38.

43. Perry CJ, Barron AB. Neural mechanisms of reward in insects. *Annu Rev Entomol* 2013;**58**:543–62.

44. Doty RL. Olfactory dysfunction in Parkinson disease. *Nat Rev Neurol* 2012;**8**(6):329–39.

45. Hawkes C. Olfaction in neurodegenerative disorder. *Adv Otorhinolaryngol* 2006;**63**:133–51.

46. Stamps JJ, Bartoshuk LM, Heilman KM. A brief olfactory test for Alzheimer's disease. *J Neurol Sci* 2013;**333**:19–24.

47. Moberg PJ, Pearlson GD, Speedie LJ, Lipsey JR, Strauss ME, et al. Olfactory recognition: differential impairments in early and late Huntington's and Alzheimer's diseases. *J Clin Exp Neuropsychol* 1987;**9**:650–64.

48. Barrios FA, Gonzalez L, Favila R, Alonso ME, Salgado PM, et al. Olfaction and neurodegeneration in HD. *Neuroreport* 2007;**18**:73–6.

49. Rolet A, Magnin E, Millot JL, Berger E, Vidal C, et al. Olfactory dysfunction in multiple sclerosis: evidence of a decrease in different aspects of olfactory function. *Eur Neurol* 2013;**69**:166–70.

50. Fleiner F, Dahlslett SB, Schmidt Harms L, Goektas O. Olfactory and gustatory function in patients with multiple sclerosis. *Am J Rhinol Allergy* 2010;**24**:e93–7.

51. Ahlskog JE, Waring SC, Petersen RC, Esteban-Santillan C, Craig UK, O'Brien PC, et al. Olfactory dysfunction in Guamanian ALS, parkinsonism, and dementia. *Neurology* 1998;**51**:1672–7.

52. Moberg PJ, Agrin R, Gur RE, Gur RC, Turetsky BI, et al. Olfactory dysfunction in schizophrenia: a qualitative and quantitative review. *Neuropsychopharmacology* 1999;**21**:325–40.

53. Moberg PJ, Kamath V, Marchetto DM, Calkins ME, Doty RL, et al. Meta-analysis of olfactory function in schizophrenia, first-degree family members, and youths at-risk for psychosis. *Schizophr Bull* 2014;**40**:50–9.

54. Moberg PJ, Arnold SE, Doty RL, Gur RE, Balderston CC, et al. Olfactory functioning in schizophrenia: relationship to clinical, neuropsychological, and volumetric MRI measures. *J Clin Exp Neuropsychol* 2006;**28**(8): 1444–61.

55. Farooqui AA. *Neurochemical aspects of neurotraumatic and neurodegenerative diseases.* New York: Springer; 2010.

56. Sheikh S, Safia Haque E, Mir SS. Neurodegenerative diseases: multifactorial conformational diseases and their therapeutic interventions. *J Neurodegener Dis* 2013;**2013**:563481.

57. Drevets WC, Price JL, Furey ML. Brain structural and functional abnormalities in mood disorders: implications for neurocircuitry models of depression. *Brain Struct Funct* 2008;**213**:93–118.

58. Hurley LL, Tizabi Y. Neuroinflammation, neurodegeneration, and depression. *Neurotox Res* 2013;**23**:131–44.

59. Petrou M, Bohnen NI, Müller ML, Koeppe RA, Albin RL, Frey KA. Aβ-amyloid deposition in patients with Parkinson disease at risk for development of dementia. *Neurology* 2012;**79**:1161–7.

60. Golde TE. The therapeutic importance of understanding mechanisms of neuronal cell death in neurodegenerative disease. *Mol Neurodegener* 2009;**4**:8.

61. Dowlati Y, Herrmann N, Swardfager W, Liu H, Sham L, et al. A meta-analysis of cytokines in major depression. *Biol Psychiatry* 2010;**67**:446–57.

62. Burchett SA, Hicks TP. The mysterious trace amines: protean neuromodulators of synaptic transmission in mammalian brain. *Prog Neurobiol* 2006;**79**:223–46.

63. Miller GM. The emerging role of trace amine associated receptor 1 in the functional regulation of monoamine transporters and dopaminergic activity. *J Neurochem* 2011;**11**(2):164–76.

64. Scanlan TS, Suchland KL, Hart ME, Chiellini G, Huang Y, et al. 3-Iodothyronamine is an endogenous and rapid-acting derivative of thyroid hormone. *Nat Med* 2004;**10**:638–42.

65. Babusyte A, Kotthoff M, Fiedler J, Krautwurst D. Biogenic amines activate blood leukocytes via trace amine-associated receptors TAAR1 and TAAR2. *J Leukoc Biol* 2013;**93**:387–94.

66. Carnicelli V, Santoro A, Sellari-Franceschini S, Berrettini S, Zucchi R. Expression of trace amine-associated receptors in human nasal mucosa. *Chemosensory Percept* 2010;**3**:99–107.

67. Zucchi R, Chiellini G, Scanlan TS, Grandy DK. Trace amine-associated receptors and their ligands. *Br J Pharmacol* 2006;**149**:967–78.

68. Liberles SD, Buck LB. A second class of chemosensory receptors in the olfactory epithelium. *Nature* 2006;**442**:645–50.

69. Liberles SD. Trace amine-associated receptors are olfactory receptors in vertebrates. *Ann N Y Acad Sci* 2009;**1170**:168–72.

70. Maguire JJ, Parker WA, Foord SM, Bonner TI, Neubig RR, et al. International Union of Pharmacology. LXXII. Recommendations for trace amine receptor nomenclature. *Pharmacol Rev* 2009;**61**:1–8.

71. Wallrabenstein I, Kuklan J, Weber L, Zborala S, Werner M, et al. Human trace amine-associated receptor TAAR5 can be activated by trimethylamine. *PLoS One* 2013;**8**:e54950.

72. Stäubert C, Boselt I, Bohnekamp J, Rompler H, Enard W, et al. Structural and functional evolution of the trace amine-associated receptors TAAR3, TAAR4 and TAAR5 in primates. *PLoS One* 2010;**5**:e11133.

73. Dinter J, Mühlhaus J, Wienchol CL, Yi CX, Nürnberg D, et al. Inverse agonistic action of 3-iodothyronamine at the human trace amine-associated receptor 5. *PLoS One* 2015;**10**:e0117774.

74. Pae CU, Yu HS, Amann D, Kim JJ, Lee CU, et al. Association of the trace amine associated receptor 6 (TAAR6) gene with schizophrenia and bipolar disorder in a Korean case control sample. *J Psychiatr Res* 2008; **42**:35–40.

75. Pae CU, Drago A, Mandelllo L, De Ronchelli L, Serretti A. TAAR 6 and HSP-70 variations associated with bipolar disorder. *Neurosci Lett* 2009;**465**(3):257–61.

76. Ferrero DM, Wacker D, Roque MA, Baldwin MW, Stevens RC, et al. Agonists for 13 Trace Amine-Associated Receptors provide insight into the molecular basis of odor selectivity. *ACS Chem Biol* 2012;**7**:1184–9.

77. Mühlhaus J, Dinter J, Nürnberg D, Rehders M, Depke M, et al. Analysis of human TAAR8 and murine Taar8b mediated signaling pathways and expression profile. *Int J Mol Sci* 2014;**15**:20638–55.

78. Gompf HS, Greenberg JH, Aston-Jones G, Ianculescu AG, Scanlan TS, et al. 3-Monoiodothyronamine: the rationale for its action as an endogenous adrenergic-blocking neuromodulator. *Brain Res* 2010;**1351**:130–40.

79. Mariotti V, Melissari E, Iofrida C, Righi M, Di Russo M, et al. Modulation of gene expression by 3-iodothyronamine: genetic evidence for a lipolytic pattern. *PLoS One* 2014;**9**(11):e106923.

80. Grandy DK. Trace amine-associated receptor 1-Family archetype or iconoclast? *Pharmacol Ther* 2007;**116**: 355–90.

81. Xie Z, Westmoreland SV, Bahn ME, Chen GL, Yang H, et al. Rhesus monkey trace amine-associated receptor 1 signaling: enhancement by monoamine transporters and attenuation by the D2 autoreceptor in vitro. *J Pharmacol Exp Ther* 2007;**321**(1):116–27.

82. Lindemann L, Meyer CA, Jeanneau K, Bradaia A, Ozmen L, et al. Trace amine-associated receptor 1 modulates dopaminergic activity. *J Pharmacol Exp Ther* 2008;**324**:948–56.

83. Duan J, Martinez M, Sanders AR, et al. Polymorphisms in the trace amine receptor 4 (TRAR4) gene on chromosome 6q23.2 are associated with susceptibility to schizophrenia. *Am J Hum Genet* 2004;**75**(4):624–38.

84. Fussnecker BL, Smith BH, Mustard JA. Octopamine and tyramine influence the behavioral profile of locomotor activity in the honey bee (*Apis mellifera*). *J Insect Physiol* 2006;**52**(10):1083–92.

85. Espinoza S, Salahpour A, Masri B, Sotnikova TD, Messa M, et al. Functional interaction between trace amine-associated receptor 1 and dopamine D2 receptor. *Mol Pharmacol* 2011;**80**:416–25.

86. Saraswati S, Fox LE, Soll DR, Wu CF. Tyramine and octopamine have opposite effects on the locomotion of *Drosophila* larvae. *J Neurobiol* 2004;**58**:425–41.

87. Blenau W, Balfanz S, Baumann A. Amtyr1: characterization of a gene from honeybee (*Apis mellifera*) brain encoding a functional tyramine receptor. *J Neurochem* 2000;**74**(3):900–8.

88. Mustard JA, Kurshan PT, Hamilton IS, Blenau W, Mercer AR. Developmental expression of a tyramine receptor gene in the brain of the honey bee, Apis mellifera. *J Comp Neurol* 2005;**483**(1):66–75.

89. Grohmann L, Blenau W, Erber J, Ebert PR, Strünker T, et al. Molecular and functional characterization of an octopamine receptor from honeybee (Apis mellifera) brain. *J Neurochem* 2003;**86**(3):725–35.

90. Hammer M, Menzel R. Multiple sites of associative odor learning as revealed by local brain microinjections of octopamine in honeybees. *Learn Mem* 1998;**5**(1–2):146–56.

91. Farooqui T, Robinson K, Vaessin H, Smith BH. Modulation of early olfactory processing by an octopaminergic reinforcement pathway in the honeybee. *J Neurosci* 2003;**23**:5370–80.

92. Scheiner R, Plückhahn S, Öney B, Blenau W, Erber J. Behavioural pharmacology of octopamine, tyramine and dopamine in honey bees. *Behav Brain Res* 2002;**136**:545–53.

93. Braun G, Bicker G. Habituation of an appetitive reflex in the honeybee. *J Neurophysiol* 1992;**67**:588–98.

94. Schulz DJ, Robinson GE. Octopamine influences division of labor in honey bee colonies. *J Comp Physiol A* 2001;**187**(1):53–61.

95. Salomon M, Malka O, Meer RK, Heftz A. The role of tyramine and octopamine in the regulation of reproduction in queenless worker honeybees. *Naturwissenschaften* 2012;**99**(2):123–31.

96. Ma Z, Guo X, Lei H, Li T, Hao S, Kang L. Octopamine and tyramine respectively regulate attractive and repulsive behavior in locust phase changes. *Sci Rep* 2015;**5**:8036.

97. Filmore D. It's a GPCR world. *Modern Drug Discov (American Chemical Society)* 2004:24–8. <http://pubs.acs.org/subscribe/journals/mdd/v07/i11/html/1104feature_filmore.html>.

98. Gilman AG. G proteins: transducers of receptor-generated signals. *Annu Rev Biochem* 1987;**56**(1):615–49.

99. Wettschureck N, Offermanns S. Mammalian G proteins and their cell type specific functions. *Physiol Rev* 2005;**85**:1159–204.

100. Overington JP, Al-Lazikani B, Hopkins AL. How many drug targets are there? *Nat Rev Drug Discov* 2006;**5**(12):993–6.

101. Becker OM, Marantz Y, Shacham S, Inbal B, Heifetz A, et al. G protein-coupled receptors: *in silico* drug discovery in 3D. *Proc Natl Acad Sci USA* 2004;**101**(31):11304–9.

102. Hopkins AL, Groom CR. The druggable genome. *Nat Rev Drug Discov* 2002;**1**(9):727–30.

103. Hussain A, Saraiva LR, Korsching SI. Positive Darwinian selection and the birth of an olfactory receptor clade in teleosts. *Proc Natl Acad Sci USA* 2009;**106**:4313–8.

104. Tessarolo JA, Tabesh MJ, Nesbitt M, Davidson WS. Genomic organization and evolution of the trace amine-associated receptor (TAAR) repertoire in Atlantic salmon (*Salmo salar*). *G3* 2014;**4**:1135–41.

105. Liberles SD. Trace amine-associated receptors: ligands, neural circuits, and behaviors. *Curr Opin Neurobiol* 2015;**34C**:1–7.

106. Hart ME, Suchland KL, Miyakawa M, Bunzow JR, Grandy DK, et al. Trace amine-associated receptor agonists: synthesis and evaluation of thyronamines and related analogues. *J Med Chem* 2006;**49**(3):1101–12.

107. Lewin AH, Navarro HA, Mascarella SW. Structure-activity correlations for beta-phenethylamines at human trace amine receptor 1. *Bioorg Med Chem* 2008;**16**(15):7415–23.

108. Lewin AH, Miller GM, Gilmour B. Trace amine-associated receptor 1 is a stereoselective binding site for compounds in the amphetamine class. *Bioorg Med Chem* 2011;**19**(23):7044–8.

109. Milligan G. G protein-coupled receptor hetero-dimerization: contribution to pharmacology and function. *Br J Pharmacol* 2009;**158**:5–14.

110. Revel FG, Moreau JL, Gainetdinov RR, Bradaia A, Sotnikova TD, et al. TAAR1 activation modulates monoaminergic neurotransmission, preventing hyperdopaminergic and hypoglutamatergic activity. *Proc Natl Acad Sci USA* 2011;**108**(20):8485–90.

111. Xie Z, Vallender EJ, Yu N, Kierstein S, Yang Y, et al. Cloning, expression and functional analysis of rhesus monkey trace amine-associated receptor 6: evidence for lack of monoaminergic association. *Neurosci Res* 2008;**86**(15):3435–46.

112. Hein L. Alpha-adrenergic system. In: Offermanns S, Rosenthal W, editors. *Encyclopedic reference of molecular pharmacology*. Berlin: Springer; 2004. p. 27–30.

113. Lohse MJ. Beta-adrenergic system. In: Offermanns S, Rosenthal W, editors. *Encyclopedic reference of molecular pharmacology*. Berlin: Springer; 2004. p. 169–72.

114. Oak JN, Van Tol Dopamine HHM. Dopamine system. In: Offermanns S, Rosenthal W, editors. *Encyclopedic reference of molecular pharmacology*. Berlin: Springer; 2004. p. 310–5.

115. Hill SJ, Baker JG. Histaminergic system. In: Offermanns S, Rosenthal W, editors. *Encyclopedic reference of molecular pharmacology*. Berlin: Springer; 2004. p. 456–60.

116. Hoyer D, Clarke DE, Fozard JR, Hartig PR, Martin GR, et al. International Union of Pharmacology classification of receptors for 5-hydroxytryptamine (serotonin). *Pharmacol Rev* 1994;**46**:157–203.

117. Murphy PM. International Union of Pharmacology. XXX. Update on chemokine receptor nomenclature. *Pharmacol Rev* 2002;**54**:227–9.

118. Murphy PM, Baggiolini M, Charo IF, Hebert CA, Horuk R, et al. International union of pharmacology. XXII. Nomenclature for chemokine receptors. *Pharmacol Rev* 2000;**52**:145–76.

Index

Note: Page numbers followed by "*f*" and "*t*" refer to figures and tables, respectively.